Deformed Spacetime

## Fundamental Theories of Physics

*An International Book Series on The Fundamental Theories of Physics: Their Clarification, Development and Application*



Volume 157

# Deformed Spacetime

## Geometrizing Interactions in Four and Five Dimensions

*by*

**Fabio Cardone**
*Consiglio Nazionale delle Ricerche*
*Roma, Italy*

**Roberto Mignani**
*Università Roma Tre*
*Roma, Italy*

Springer

A C.I.P. Catalogue record for this book is available from the Library of Congress.

ISBN 978-1-4020-6282-7 (HB)
ISBN 978-1-4020-6283-4 (e-book)

Published by Springer,
P.O. Box 17, 3300 AA Dordrecht, The Netherlands.

*www.springer.com*

*Printed on acid-free paper*

# Contents

# PREFACE

*"............................human kind*
*Cannot bear very much reality."*
(T.S. Eliot: "Four Quartets – Burnt Norton")

*"Exegi monumentum aere perennius*
*........Non omnis moriar."*
(Quintus Horatius Flaccus: Carmina III, 30)

The possible solution to the fundamental issue concerning the structure of the physical world – an old-debated problem, starting from presocratic philosophers – greatly benefited in the twentieth century from the final merging of the separate streams of geometry and physics in a broad river of synthesis and progress. Actually, it can be traced back to Pythagoras himself the awareness of the intimate link (and the reciprocal feedback) between the physical measurements of times, spaces and distances and their mathematical representation in terms of relations among abstract geometrical entities. This leads for instance to attach a physical meaning, within Euclidean geometry, to the mere spatial relations between objects. However, only the basic work by Lorentz, Poincaré, Minkowski, and Einstein permitted to state that space–time is something more than the simple union of space and time. Indeed, the Special and General theories of Relativity allowed physicists and mathematicians to recognize the arena where physical phenomena take place as a 4D manifold, endowed with a global, Riemannian geometrical structure of Lorentzian signature, which links together space and time in an indissoluble bond. What's more, it was possible for the first time not only to use the mathematical language in order to express physical laws (in accordance with the fundamental teaching by Galilei), but even to identify a physical interaction – gravity – with a geometric property of space–time itself – Cauchy–Riemann curvature – namely *to geometrize physics*.

As is well known, many attempts at generalizing the 4D relativistic picture have been made later on. Such efforts are roughly of two types, according to the mathematical tools exploited and the physical purposes pursued. A possibility is assuming the existence of further dimensions, in order to build up unified schemes of the fundamental interactions. The prototype theory of this kind, due to Kaluza and Klein (KK), was based on a 5D space–time and aimed at unifying gravitation and electromagnetism in a single geometrical structure. In the KK scheme, the fifth dimension is compactified, i.e., "rolled up" and assumed so small as to be unobservable. The Kaluza–Klein formalism has been subsequently extended to more dimensions, in the hope of achieving unification of all interactions, including weak and strong forces.

In the second group of generalizations, the 4D space–time manifold is preserved, but it is equipped with a global and/or local geometry different from the Minkowskian or the Riemannian one. Such a kind of approach points essentially at account for possible violations of standard relativity and Lorentz invariance. In this connection, we have discussed in the past some physical phenomena, ruled by different fundamental interactions, whose experimental data seem to provide some intriguing evidence of a (local) breakdown of Lorentz invariance. All the phenomena considered show indeed an inadequacy of the Minkowski metric in describing them, at different energy scales and for the four fundamental interactions involved (electromagnetic, weak, gravitational and strong). On the contrary, they apparently admit a consistent interpretation in terms of a *deformed Minkowski space–time, with metric coefficients depending on the energy exchanged in the process considered.*

The analysis and the discussion of such experiments led us therefore to envisage a (4D) generalization of the (local) space–time structure which is based on an energy-dependent deformation of the usual Minkowski geometry. In such a framework, the corresponding deformed metrics obtained from the fit to experimental data provide an effective dynamic description of the interactions which rule the phenomena examined (at least at the energy scale and in the energy range considered).

This formalism (Deformed Special Relativity, DSR) permits therefore to implement for all fundamental interactions the Principle of Solidarity (formulated by the Italian mathematician Bruno Finzi), according to which each interaction produces its own space–time geometry. The properties of DSR, consequence of Finzi's principle, establish a connection of this scheme with Lorentzian Relativity (LR) (rather than with the Einsteinian one, ER). Let us recall that LR is the version of Special Relativity due to Lorentz and Poincaré, whose main points of departures from the Einstein view are the existence of a preferred reference frame and the nonuniqueness of the light speed and of the coordinate transformations connecting inertial observers. These features, rather than constituting drawbacks of LR with respect to the ER unifying principles of uniqueness and invariance, actually testify the more flexible mathematical structure of Lorentzian relativity, thus able to fit the diversified nature of the different physical forces. As a matter of fact, all the present experimental tests in favor of ER do support LR, too, whereas there are sound clues for the existence of a preferred frame (like the frame of isotropy of the 2.7 K cosmological background radiation).

Due to the deformation of the space–time metric it describes, DSR belongs therefore to the second kind of generalizations of Einstein's relativities. However, it must be noted that, within DSR, the energy exchanged in the process considered (i.e., the energy measured by the electromagnetic interaction with the detectors in Minkowski space) plays the role of a true dynamic variable. It represents therefore a characteristic parameter of the phenomenon considered (so that, for a given process, it cannot be changed at will). In other words, when describing a given process, *the deformed geometry of space–time* (in the interaction region where the process is occurring) *is "frozen" by those values of the metric coefficients corresponding to the energy value of the process itself.* Otherwise speaking, from a geometrical point of view, all goes on as if we were actually working on "slices" (sections) of a 5D space, in which the fifth dimension is just represented by the energy.

The leading idea is that the 4D, deformed, energy-dependent space–time is only a manifestation (a "shadow," to use the famous word of Minkowski) of a larger, 5D space, in which energy plays the role of extra dimension. By imposing Einstein (vacuum) equations to the 5D metric, we have been able to show that the deformed, energy-dependent, phenomenological metrics, derived by the analysis of the earlier quoted experiments, can be obtained as the relevant 4D slices at constant energy of particular solutions of Einstein equations in the 5D space.

The DSR approach shares therefore the features of either kind of theories generalizing Einstein's relativities we outlined earlier. Indeed, on one side, it is endowed with a 4D space–time metric of the Finsler type; on the other hand, it can be naturally embedded in a 5D space with energy as extra dimension. So, this formalism is a Kaluza–Klein-like one, whereby now the fifth coordinate is a physically sensible dimension. Thus, on this respect, it belongs to the class of noncompactified KK theories.

Aim of the present book is to introduce and discuss the physical foundations and the mathematical formalism of such a generalized KK scheme – we called *Deformed Relativity in Five Dimensions (DR5)* – namely to guide the reader in the exploration of the new space–time–energy land. The book is organized as follows. Part I provides an introduction to the physical grounds of DSR and gives the explicit expressions of the deformed metrics obtained, for the fundamental interactions, by the phenomenological analysis of the experimental data. In Part II, we discuss in detail the mathematical structure of the 4D deformed space–time (including a thorough analysis of the related Killing symmetries). In Part III, we describe the experiments, related to all four fundamental interactions, which test and confirm some of the predictions of the DSR formalism, thus providing a possible evidence for the deformation of space–time. Among them, let us quote the anomalous behavior of some photon systems (at variance with respect to standard electrodynamics and quantum mechanics); the measurement of the speed of propagation of gravitational effects; the effectiveness of ultrasounds in speeding up the decay of radioactive elements and in triggering nuclear reactions in liquid solutions. Moreover, such experiments point out the need for considering the energy as a fifth coordinate, thus casting a bridge toward the 5D formalism of DR5. In Part IV, we introduce the 5D scheme of DR5, based on a Riemannian space in which the 4D space–time is deformed (according to the DSR scheme) and the energy plays the role of an extra dimension. We write down the related 5D Einstein equations in vacuum, with all five metric coefficients depending on energy, and solve them in some cases of physical relevance. The solutions obtained and their physical meaning are discussed. We discuss the isometries of the 5D space by solving the related Killing equations for all four fundamental interactions. The dynamics of DR5 is dealt with in Part V, by considering the 5D geodesics equations in the main cases of physical interest. We also incidentally show how a particular solution leads to an intriguing relation which reminds the uncertainty relation between energy and time in quantum mechanics. Concluding remarks and possible further developments of the formalism are put forward.

**Acknowledgment** We are indebted to many colleagues and friends. First of all, it is both a duty and a pleasure to express our special gratitude to Prof. Alvyn van der Merwe, who was so kind to invite us to write this book for the Springer series *"Fundamental Theories of Physics"* of which he is the Editor in Chief. A small part of the book was used as a one-month lecture course delivered by F.C. in spring 2002 at the Mathematical Department of Messina University, on invitation by Enzo Ciancio and Liliana Restuccia, and we are pleased to thank them warmly. The development of the mathematical formalism of Deformed Relativity, in particular as far as its 5D generalization is concerned, greatly benefited from the authors' stimulating collaboration with Mauro Francaviglia and Alessio Marrani, we gratefully acknowledge. We have of course a great debt to the other co-authors of all our papers, Umberto Bartocci, Mario Gaspero, Vladislav S. Olkhovsky, Walter Perconti, Eliano Pessa, Andrea Petrucci, Renato Scrimaglio and Guido Spera. Thanks are also due to Gaetano Caricato, Sidney Coleman, Sheldon L. Glashow, Basil J. Hiley, Roman Jackiw, Maxim Yu. Khlopov, Rostislav V. Konoplich, Daniela Mugnai and Anedio Ranfagni for kind interest and useful discussions. One of us (F.C.) is at the same time deeply indebted and sincerely grateful to all people who has been involved in carrying out the experiments described in Part III of the book, namely: W. Perconti and A. Petrucci of C.N.R., for the experiments of Chaps. 13 and 15; the military technicians Antonio Aracu, Antonio Bellitto, Felice Contalbo, Pierluigi Muraglia and the civil researchers Giovanni Cherubini, A. Petrucci, Francesca Rosetto, G. Spera, for the experiments reported in Chaps. 16, 17. Last but not least, a special thank is due to Fabio Pistella, President of the Italian National Research Council (C.N.R.), for advise and encouragement. Moreover, we cannot forget our wives, Silvia Aquilani Cardone and Rossella Amato Mignani, who contributed to the book by their heroic displays of patience. Needless to say, the responsibility for any errors or omissions, as well as for infelicities of language and content, is entirely with the authors.

*Fabio Cardone* — *Roma*
*Roberto Mignani* — *December 2006*

# Part I

# PHYSICS OF DEFORMED SPACE-TIME

# 1

# The Principle of Solidarity: Geometrical Descriptions of Interactions

## 1.1 The Finzi Principle of Solidarity

In 1955 the Italian mathematician Bruno Finzi, in his contribution to the book "Fifty Years of Relativity" [1], stated his "Principle of Solidarity" (PS),[1] that sounds "It's (indeed) necessary to consider space–time TO BE SOLIDLY CONNECTED with the physical phenomena occurring in it, so that its features and its very nature do change with the features and the nature of those. In this way not only (as in classical and special-relativistic physics) space–time properties affect phenomena, but reciprocally phenomena do affect space–time properties. One thus recognizes in such an appealing "Principle of Solidarity" between phenomena and space–time that characteristic of mutual dependence between entities, which is peculiar to modern science." Moreover, referring to a generic $N$-dimensional space: "It can, a priori, *be pseudoeuclidean, Riemannian, non-Riemannian. But* – he wonders – *how is indeed the space–time where physical phenomena take place? Pseudoeuclidean, Riemannian, non-Riemannian, according to their nature, as requested by the principle of solidarity between space–time and phenomena occurring in it.*"

---

[1]It's quite difficult to express in English in a simple way the Italian words "solidarietà" and "solidale," used by Finzi to mean the feedback between space–time and interactions. A possible way to render them is to use "solidarity" and "solidly connected," respectively, – at the price of partially loosing the common root of the Italian words –, with the warning that what Finzi really means is that the very structure of space–time is determined by the physical phenomena which do take place in it.

Of course, Finzi's main purpose was to apply such a principle to Einstein's Theory of General Relativity, namely to the class of gravitational phenomena. However, its formulation is as general as possible, so to apply in principle to all the known physical interactions. Therefore, Finzi's PS is at the very ground of any attempt at geometrizing physics, i.e., describing physical forces in terms of the geometrical structure of space–time.

Such a project (pioneered by Einstein himself) revealed itself unsuccessful even when only two interactions were known, the electromagnetic and the gravitational one. It was fully abandoned starting from the middle of the twentieth century, due to the discovery of the two nuclear interactions, the weak and the strong one (apart from recent attempts based on string theory).

Let us notice that the Principle of Solidarity can in general apply also to interactions nonlocal and not derivable from a potential. Let us therefore clarify what it is meant by such terms in this book.

Essentially two definitions of nonlocality exist in literature. The first amounts to contradict the so-called *Einstein–Bell locality*, which can be stated as follows:

> *The elements of physical reality of a system cannot be affected instantaneously at a distance* (Einstein)

or

> *The probability of two measurements performed on events separated by a space-like interval is simply the product of the probabilities of the two measurements separately* (Bell).

It is easy to see that such a nonlocality (of quantum nature) is basically connected to the possibility of superluminal signals.

The second definition is related to the space–time functional dependence of the force. A force is local when it depends on a space–time *point* (or, better, on an *infinitesimal* neighborhood of the point); it is nonlocal when it depends on a whole (*finite*) space–time *region*. In the following, we shall just mean this latter definition whenever using the term nonlocal.

Let us stress, however, that "local" interaction (in the sense specified earlier) and "potential" interaction are *not* synonymous, in general. Once one fixes a space–time point, a local interaction is uniquely determined by an infinitesimal neighborhood of the point, whereas a potential interaction is just determined by the value the potential function takes at the point considered. Notice that the derivability from a potential requires the uniqueness of the potential function on the whole space–time region where the force field is defined.

An example of a nonlocal but potential interaction is provided by an interaction described by the potential

$$V(x_i) = \int \prod_{x_j \in \{x\}, j \neq i} \mathrm{d}x_j V(\{x\}), \tag{1.1}$$

where $\{x\}$ is the set of metric coordinates, the integration can be definite or indefinite (in the latter case, the potential will depend also on the geometry of the integration regions) and $V(\{x\})$ is regular enough to ensure its integrability (for instance, in the Riemann sense).

On the other side, the electromagnetic (e.m.) interaction associated to a magnetic monopole is an example of a local but nonpotential interaction. In this case, due to the presence of the singular Dirac string, the force field of the monopole is irrotational locally but not globally.[2] This implies that the monopole field is described by *many* (in general different) *local potentials*. By the nonuniqueness of the potential, the e.m. interaction of the magnetic monopole is nonpotential, but it is local indeed.

Apart from the earlier questions, the basic problem is how to implement Finzi's Principle of Solidarity for all interactions on a mere geometrical basis. Since, from an historical point of view, General Relativity (GR) is the only successful theoretical realization of geometrizing an interaction (the gravitational one), it is usually believed that the goal of geometrization of interactions can only be achieved by the tools of Riemannian spaces or of their suitable generalizations.

We want instead to show that implementing the Finzi principle can be obtained in the mere framework of Special Relativity, provided its very foundations are taken into proper account and suitably exploited. To this aim, let us analyze Special Relativity from an axiomatic standpoint.

## 1.2 An Axiomatic View to Special Relativity

Special Relativity (SR) is essentially grounded on the properties of space–time, i.e., isotropy of space and homogeneity of space and time (as a consequence of the equivalence of inertial frames) and on the principle of relativity.

The two basic postulates of SR in its axiomatic formulation are [2]:

1. *Space-time properties.* Space and time are homogeneous and space is isotropic.

2. *Principle of Relativity (PR).* All physical laws must be covariant when passing from an inertial reference frame $K$ to another frame $K'$, moving with constant velocity relative to $K$.

[2] In the language of differential geometry, the field of a Dirac magnetic monopole is associated to a differential form which is closed but not globally exact.

The second postulate can be traced back to Galilei himself, who of course enunciated and applied it with reference to the laws of classical mechanics (the only ones known at his times). In fact, the Relativity Principle contains implicitly (somewhat hidden, but actually easily understood after a moment's thought) the basic point that, for a correct formulation of SR, *it is necessary to specify the total class, $C_T$, of the physical phenomena to which the PR applies.* The importance of such a specification is easily seen if one thinks that, from an axiomatic viewpoint, the only difference between Galilean and Einsteinian relativities just consists in the choice of $C_T$ (i.e., the class of mechanical phenomena in the former case, and of mechanical and electromagnetic phenomena in the latter).

It is possible to show that, from the earlier two postulates, there follow – without any additional hypothesis – all the usual "principles" of SR, i.e., the "principle of reciprocity," the linearity of transformations between inertial frames, and the invariance of light speed in vacuum.

Concerning this last point, it can be shown in general that postulates 1 and 2 earlier imply the existence of an invariant, real quantity, having the dimensions of the square of a speed, whose value must be experimentally determined in the framework of the total class $C_T$ of the physical phenomena.[3] Such an invariant speed depends on the interaction (fundamental, or at least phenomenological) ruling the physical phenomenon considered. Therefore *there is, a priori, an invariant speed for every interaction*, namely, a maximal causal speed for every interaction.

All the formal machinery of SR in the Einsteinian sense (including Lorentz transformations and their implications, and the metric structure of space–time) is simply a consequence of the earlier two postulates and of the choice, for the total class of physical phenomena $C_T$, of the class of mechanical and electromagnetic phenomena.

If different explicit choices of $C_T$ are made, one gets a priori different realizations of the theory of relativity (in its abstract sense), each one embedded in the previous. Of course, the principle of relativity, together with the specification of the total class of phenomena considered, necessarily entails in all cases, for consistency, the uniqueness of the transformation equations connecting inertial reference frames.[4]

The attempt at including the class of nuclear and subnuclear phenomena in the total class of phenomena for which Special Relativity holds true is

---

[3]The invariant speed is obviously $\infty$ for Galilei's relativity, and $c$ (light speed in vacuum) for Einstein's relativity.

[4]The hypothesis of the existence a priori of different relativities for different interactions – formulated by Recami and one of the present authors (R.M.) on the basis of the above critical analysis of the foundations of Special Relativity – can be considered a generalization of the point of view advocated by Lorentz, according to which different interactions require different coordinate transformations between inertial reference frames.

therefore expected to imply a generalization of Minkowski metric, analogously to the generalization from the Euclidean to the Minkowski metric in going from mechanics to electrodynamics.

However, in order to avoid misunderstandings, it must be stressed that such an analogy with the extension of the Euclidean metric has to be understood not in the purely geometric meaning, but rather in the sense (as already stressed by Penrose [3]) of Euclidean geometry as a physical theory.

Indeed, the generalized metric must be equipped with a dynamic character and be not only a consequence, but also an effective description of (the interaction involved in) the class of phenomena considered. This allows one in this way to get a feedback between interactions and space–time structure, already accomplished for gravitation in General Relativity.

This complies with the "Principle of Solidarity" stated by Finzi in the form already quoted earlier, which can be embodied in the following third principle of Relativity:

3. *Principle of Solidarity (PS).* Each class of phenomena (namely, each interaction) determines its own space–time.

The fundamental problem is now: *How to endow the metric of the Minkowski space–time with a geometrical structure able to describe the interaction involved in a given process?* This is just the aim of this book.

## 1.3 Energy and the Finzi Principle

At present, General Relativity (GR) is the only successful theoretical realization of geometrizing an interaction (the gravitational one). As is well known, energy plays a fundamental role in GR, since the energy–momentum tensor of a given system is the very source of the gravitational field.

A moment's thought shows that this occurs actually also for other interactions. Let us remind, for instance, the case of Euclidean geometry in its intrinsic meaning of a theory of physical reality at its basic classical (macroscopic) level. In fact, it describes in a quantitative way, in mathematical language, the relations among measured physical entities – distances, in this case –, and therefore the physical space in which phenomena occur.

However, the measurement of distances depends on the motion of the body which actually performs the measurement. Such a dependence is indeed not on the *kind* of motion, but rather on the *energy* needed to let the body move, and on the *interaction* providing such energy. The measurement of time needs as well a periodic motion with constant frequency, and therefore it too depends on the energy and on the interaction.

This simple example shows how *energy does play a fundamental role in determining the very geometrical structure of space–time* (in analogy with the General-Relativistic case, where – as already noted – the energy-momentum tensor is the source of the gravitational field). Let us stress that such a viewpoint is very similar, on many respects, to the Ehlers–Pirani–Schild scheme [4] (based on the earlier work of Weyl), in which the geometry of space–time is operationally determined by using the trajectories of free-falling objects (geodesics). In this framework, the points of space–time become physically real in virtue of the geometrical relations between them, and the classical particle motion is exploited to obtain the geometry of space–time (the argument can be extended to quantum motion as well [5]).

Generalizing such an argument, we can state that exchanging energy between particles amounts to measure operationally their space–time separation.[5] Of course such a process depends on the interaction involved in the energy exchange; moreover, each exchange occurs at the maximal causal speed characteristic of the given interaction. It is therefore natural to assume that the measurement of distances, performed by the energy exchange according to a given interaction, realizes the "solidarity principle" between space–time and interactions at the microscopic scale.

By starting from such considerations, a possible way to implement Finzi's principle for *all* fundamental interactions is provided by the formalism of deformed special relativity (DSR) developed in the last decade of the twentieth century. It is based on a *deformation* of the Minkowski space, namely a space–time endowed with a metric whose coefficients just depend on energy (in the sense specified later on). Such an energy-dependent metric does assume a *dynamic role*, thus providing a geometrical description of the fundamental interaction considered and implementing the feedback between space–time structure and physical interactions which is just the content and the heritage of Finzi's principle.

The generalization of the Minkowski space implies, among the others, new, generalized transformation laws, which admit, as a suitable limit, the Lorentz transformations (just like Lorentz transformations represent a covering of the Galilei–Newton transformations) [6].

Then, the solidarity principle allows one to recover the basic features of the relativity theory in the Lorentz (not Einstein) view (Lorentzian relativity), namely different interactions entail different coordinate transformations and different invariant speeds (an in-depth discussion of this issue will be given in Sect. 3.3.7).

---

[5]Notice that, in this framework, a space–time point has only a mathematical (geometrical) meaning, since it physically corresponds to an energy insufficient to the motion (for the interaction considered).

# 2
# Description of Interactions by Energy-Dependent Metrics

We will now show how the dynamic role of the energy, in describing the structure of space–time, can be exploited in order to geometrize all four fundamental interactions, so to comply with the Finzi principle. As already stressed earlier, this can be achieved by suitably *deforming* space–time, according to what dictated by the energy involved in the process, ruled by the interaction considered. Speaking in a figurative language, we can say that in such a view space–time is not a rigid (and passive) background, but a sort of elastic carpet, able to change its shape according to the (energy of) the interaction involved, and to react in turn on the process, thus affecting its dynamics in an active way.

## 2.1 Deformed Minkowski Space–Time

In the attempt at a geometrical implementation of the Finzi principle, we have therefore – on the basis of the discussion of Chap. 1 – to take into account the role of energy in determining an interaction, and the different "relativities" obtained in correspondence to different classes of physical phenomena.

As is well known, the Minkowski metric[1]

[1] Throughout this book, unless otherwise specified, lower Latin indices take the values $\{1, 2, 3\}$ and label spatial dimensions, whereas lower Greek indices vary in the range $\{0, 1, 2, 3\}$, with 0 referring to the time dimension. Ordinary three-vectors are denoted in

$$g = \text{diag}(1, -1, -1, -1) \tag{2.1}$$

is a generalization of the Euclidean metric $\epsilon = \text{diag}(1, 1, 1)$. By the considerations of Chap. 1, we can assume that the metric describes, in an effective way, the interaction, and that there exist interactions more general than the electromagnetic ones (which, as well known, are long-range and derivable from a potential).

The simplest generalization of the space–time metric which accounts for such more general properties of interactions is provided by a *deformation*, $\eta$, of the Minkowski metric (2.1), defined as [6]

$$\eta = \text{diag}(b_0^2, -b_1^2, -b_2^2, -b_3^2). \tag{2.2}$$

Of course, from a formal point of view metric (2.2) is not new at all. Deformed Minkowski metrics of the same type have already been proposed in the past in various physical frameworks, starting from Finsler's generalization of Riemannian geometry [7] to Bogoslowski's anisotropic space–time [8] to isotopic Minkowski space [9]. A phenomenological deformation of the type (2.2) was also obtained by Nielsen and Picek [10] in the context of the electroweak theory. Moreover, although for quite different purposes, "quantum" deformed Minkowski spaces have been also considered in the context of quantum groups [11]. Leaving to later considerations the true specification of the exact meaning of the deformed metric (2.2) in our framework, let us right now stress two basic points.

1. Firstly, metric (2.2) is supposed to hold at a *local* (and not global) scale, i.e., to be valid not everywhere, but only in a suitable (local) space–time region (characteristic of both the system and the interaction considered). We shall therefore refer often to it as a "topical" deformed metric[2].

---

boldface. Upper Latin indices are used in the 5D framework, and assumed to take the values $\{0, 1, 2, 3, 5\}$. For brevity's sake, we shall denote simply by $x$ the (contravariant) four-vector $(x^0, x^1, x^2, x^3)$ whenever this notation cannot ingenerate confusion. Accordingly, the volume element $dx^0dx^1dx^2dx^3$ in Minkowski space will be denoted shortly by $d^4x$. Moreover, we adopt the signature $(+, -, -, -)$ for the 4D space–time, and employ the notation "ESC on" ("ESC off") to mean that the Einstein sum convention on repeated indices is (is not) used.

[2]Notice that the assumed local validity of (2.2) differentiates this approach from those based on Finsler's geometry or from the Bogoslowski's one (which, at least in their standard meaning, do consider deformed metrics at a *global* scale), and makes it similar, on some aspects, to the philosophy and methods of the isotopic generalizations of Minkowski spaces [9]. However, it is well known that Lie-isotopic theories rely in an essential way, from the mathematical standpoint, on (and are strictly characterized by) the very existence of the so-called isotopic unit. In the following, such a formal device will not be exploited (because unessential on all respects), so that the present formalism is not an isotopic one. Moreover, from a physical point of view, the isotopic formalism is expected to apply only to strong interactions. On the contrary, it will be assumed that the (effective) representation of interactions through the deformed metric (2.2) does hold for *all* kinds of interactions (at least for their nonlocal component). In spite of such basic

In the present case, the term "local" must be understood in the sense that a deformed metric of the kind (2.2) describes the geometry of a 4D variety attached at a point $x$ of the standard Minkowski space–time, in the same way as a local Lorentz frame is associated (as a tangent space) to each point of the (globally Riemannian) space of Einstein's GR. Another example, on some respects more similar to the present formalism, is provided by a space–time endowed with a vector fiber-bundle structure, where a Riemann space with constant curvature is attached at each point $x$.

2. Secondly, metric (2.2) is regarded to play a *dynamic role*. So, in order to comply with the solidarity principle, we assume that the parameters $b_\mu (\mu = 0, 1, 2, 3)$ are, in general, real and positive functions of a given set of observables $\{\mathcal{O}\}$ characterizing the system (in particular, of its total energy exchange, as specified later):

$$\{b_\mu\} = \{b_\mu(\{\mathcal{O}\})\} \in R_0^+, \ \forall \ \{\mathcal{O}\} \ . \tag{2.3}$$

The set $\{\mathcal{O}\}$ represents therefore, in general, a set of nonmetric variables $(\{x_{\text{n.m.}}\})$.

Equation (2.2) therefore becomes:

$$\begin{aligned} \eta_{\mu\nu} \quad &= \eta_{\mu\nu}(\{\mathcal{O}\}) \\ &= \text{diag}(b_0^2(\{\mathcal{O}\}), -b_1^2(\{\mathcal{O}\}), -b_2^2(\{\mathcal{O}\}), -b_3^2(\{\mathcal{O}\})) \\ &\stackrel{\text{ESC off}}{=} \delta_{\mu\nu}(\delta_{\mu 0} b_0^2(\{\mathcal{O}\}) - \delta_{\mu 1} b_1^2(\{\mathcal{O}\}) - \delta_{\mu 2} b_2^2(\{\mathcal{O}\}) - \delta_{\mu 3} b_3^2(\{\mathcal{O}\})). \end{aligned} \tag{2.4}$$

However, for the moment the deformation of the Minkowski space will be discussed only from a formal point of view, by disregarding the problem of the observables on which the coefficients $b_\mu$ actually depend (it will be faced later on).

It is now possible to define a generalized ("deformed") Minkowski space $\widetilde{M}(x, \eta(\{\mathcal{O}\}))$with the same local coordinates $x$ of $M$ (the four-vectors of the usual Minkowski space), but with metric given by the metric tensor $\eta$ (2.4). The generalized interval in $\widetilde{M}$ is therefore given by $(x^\mu = (x^0, x^1, x^2, x^3) = (ct, x, y, z)$, with $c$ being the usual light speed in vacuum) (ESC on) [6]:

$$\begin{aligned} \mathrm{d}\tilde{s^2}(\{\mathcal{O}\}) \quad &\equiv \\ &\equiv b_0^2(\{\mathcal{O}\})c^2\mathrm{d}t^2 - b_1^2(\{\mathcal{O}\})\left(\mathrm{d}x^1\right)^2 \\ &\quad -b_2^2(\{\mathcal{O}\})\left(\mathrm{d}x^2\right)^2 - b_3^2(\{\mathcal{O}\})\left(\mathrm{d}x^3\right)^2 \\ &= \eta_{\mu\nu}(\{\mathcal{O}\})\mathrm{d}x^\mu \mathrm{d}x^\nu = \mathrm{d}x * \mathrm{d}x. \end{aligned} \tag{2.5}$$

---

differences this formalism shares some common formal results – as we shall see in the following – with isotopic relativity (like the mathematical expression of the generalized Lorentz transformations).

The last step in (2.5) defines the scalar product $*$ in the deformed Minkowski space $\widetilde{M}$. Moreover, according to (2.5), we shall use the following notation for the deformed square norm of a four-vector:

$$|x|_*^2 \equiv x * x = \eta_{\mu\nu}(\{\mathcal{O}\}) x^\mu x^\nu = x^{\tilde{2}}. \tag{2.6}$$

In the following, in order to emphasize the dependence of the deformed Minkowski space on the set of observables $\{\mathcal{O}\}$, the notation $\widetilde{M}(\{\mathcal{O}\})$ will be also used.

In $\widetilde{M}$, it is possible a priori to consider two scalar products between three-vectors $\mathbf{v}_1$, $\mathbf{v}_2$: the standard, Euclidean product$\cdot$, defined by means of the metric tensor $g_{ik} = \delta_{ik}$, and the deformed one, induced by the deformed scalar product $*$ in $\widetilde{M}$, and defined by means of the metric tensor

$$-\eta_{ik}(\{\mathcal{O}\}) \overset{\text{ESC off}}{=} b_i^2(\{\mathcal{O}\})\delta_{ik},$$

(where the sign $-$ is obviously introduced in order to get a positive three-vector norm) as follows (cf. (2.5)):

$$\begin{aligned} \mathbf{v}_1 * \mathbf{v}_2 \quad &\equiv -\sum_{i=1}^{3} \eta_{ij}(\{\mathcal{O}\})\,(v_1)^i\,(v_2)^j \\ &= \sum_{i=1}^{3} b_i^2(\{\mathcal{O}\})\delta_{ij}\,(v_1)^i\,(v_2)^j \\ &= b_1^2(\{\mathcal{O}\})\,(v_1)^1\,(v_2)^1 + b_2^2(\{\mathcal{O}\})\,(v_1)^2\,(v_2)^2 + b_3^2(\{\mathcal{O}\})\,(v_1)^3\,(v_2)^3 . \end{aligned} \tag{2.7}$$

The 3D space embedded in $\widetilde{M}(\{\mathcal{O}\})$ and endowed with the (3D restriction of the) deformed scalar product $*$ will be denoted by $\widetilde{E}_3(\{\mathcal{O}\})$. Accordingly, parallelism and orthogonality of three-vectors can be defined either by the usual scalar product (that actually means one is working in $E_3 \subset M$) or by the deformed one. Moreover, in the following, $|\mathbf{v}|_*$ will denote the absolute value of a three-vector in $\widetilde{E}_3(\{\mathcal{O}\})$, whereas the notation $|\mathbf{v}| = v$ will be used for the norm of $\mathbf{v}$ with respect to the standard product $\cdot$ in the usual 3D Euclidean space $E_3$.

The existence of these two possible Euclidean structures may lead to ambiguities, if not carefully taken into account. For instance, the spatial unit vectors

$$\left\{\widehat{x^i_{\text{DSR}}(\{\mathcal{O}\})}\right\} \in \widetilde{E}_3(\{\mathcal{O}\}) \subset \widetilde{M}(\{\mathcal{O}\})$$

in the deformed Minkowski space *are a priori different* from the corresponding unit vectors

$$\left\{\widehat{x^i_{\text{SR}}}\right\} \in E_3 \subset M.$$

This is easily seen by noting that, by definition, a unit vector has unit norm, where the norm has to be evaluated according to the scalar product naturally induced by the metric considered. Then, in $\widetilde{M}(\{\mathcal{O}\})$

$$\widehat{x^0_{\mathrm{DSR}}(\{\mathcal{O}\})} \quad \Leftrightarrow g_{\mu\nu,\mathrm{DSR}}(\{\mathcal{O}\})\left(\widehat{x^0_{\mathrm{DSR}}(\{\mathcal{O}\})}\right)^{\mu}\left(\widehat{x^0_{\mathrm{DSR}}(\{\mathcal{O}\})}\right)^{\nu} = b_0^2(\{\mathcal{O}\})\left(\widehat{x^0_{\mathrm{DSR}}(\{\mathcal{O}\})}\right)^2 = 1; \tag{2.8}$$

$$\widehat{x^i_{\mathrm{DSR}}(\{\mathcal{O}\})} \quad \Leftrightarrow g_{\mu\nu,\mathrm{DSR}}(\{\mathcal{O}\})\left(\widehat{x^i_{\mathrm{DSR}}(\{\mathcal{O}\})}\right)^{\mu}\left(\widehat{x^i_{\mathrm{DSR}}(\{\mathcal{O}\})}\right)^{\nu} = -b_i^2(\{\mathcal{O}\})\left(\widehat{x^i_{\mathrm{DSR}}(\{\mathcal{O}\})}\right)^2 = 1, \tag{2.9}$$

where the purely imaginary nature of the nonzero components of $\widehat{x^i_{\mathrm{DSR}}(\{\mathcal{O}\})}$ is merely due to the 4D dimensional signature $(+,-,-,-)$ being used. Therefore, one gets the following relation between the two sets of coordinate unit vectors in $M$ and $\widetilde{M}(\{\mathcal{O}\})$:

$$b_0^2(\{\mathcal{O}\})\left(\widehat{x^0_{\mathrm{DSR}}(\{\mathcal{O}\})}\right)^2 = 1 = \left(\widehat{x^0_{\mathrm{SR}}}\right)^2 \Leftrightarrow \widehat{x^0_{\mathrm{DSR}}(\{\mathcal{O}\})} = b_0^{-1}(\{\mathcal{O}\})\widehat{x^0_{\mathrm{SR}}} \tag{2.10}$$

and (ESC off):

$$-b_i^2(\{\mathcal{O}\})\left(\widehat{x^i_{\mathrm{DSR}}(\{\mathcal{O}\})}\right)^2 = 1 = -\left(\widehat{x^i_{\mathrm{SR}}}\right)^2 \Leftrightarrow \widehat{x^i_{\mathrm{DSR}}(\{\mathcal{O}\})} = b_i^{-1}(\{\mathcal{O}\})\widehat{x^i_{\mathrm{SR}}}. \tag{2.11}$$

Summarizing (ESC off)

$$\widehat{x^{\mu}_{\mathrm{DSR}}(\{\mathcal{O}\})} = b_{\mu}^{-1}(\{\mathcal{O}\})\widehat{x^{\mu}_{\mathrm{SR}}} \ , \forall \mu = 0,1,2,3. \tag{2.12}$$

The orthogonality of the coordinate axes is of course expressed by

$$\widehat{x^{\mu}_{\mathrm{SR}}} \cdot \widehat{x^{\nu}_{\mathrm{SR}}} = \delta^{\mu\nu} \text{ and } \widehat{x^{\mu}_{\mathrm{DSR}}(\{\mathcal{O}\})} * \widehat{x^{\nu}_{\mathrm{DSR}}(\{\mathcal{O}\})} = \delta^{\mu\nu}.$$

It follows by the earlier relation that the unit vectors in the standard Minkowski space $M$ and in the deformed one $\widetilde{M}(\{\mathcal{O}\})$ are proportional, and therefore they specify the same directions in either space (this is due to very nature of the deformation $g \to \eta(\{\mathcal{O}\})$, preserving the diagonality of the metric tensor, while destroying its isochrony and spatial isotropy). However, since any coordinate unit vector is rescaled by a different coefficient in the metric deformation, vectors with the same components in the spaces $M$ and $\widetilde{M}(\{\mathcal{O}\})$ are *not* parallel to each other.

In the following, in order to evidence some implications of metric (2.4) not strictly related to its space anisotropy, we shall sometimes consider (for simplicity' sake and without loss of generality) an isotropic 3D space, i.e.,

$$b_1^2(\{\mathcal{O}\}) = b_2^2(\{\mathcal{O}\}) = b_3^2(\{\mathcal{O}\}) \equiv b^2(\{\mathcal{O}\}), \tag{2.13}$$

so that the corresponding deformed metric reads:

$$\begin{aligned} & \eta_{\mu\nu_{\rm ISO.}}(\{\mathcal{O}\}) \\ = \; & \mathrm{diag}(b_0^2(\{\mathcal{O}\}), -b^2(\{\mathcal{O}\}), -b^2(\{\mathcal{O}\}), -b^2(\{\mathcal{O}\})) \\ \overset{\text{ESC off}}{=} \; & \delta_{\mu\nu}\left[\delta_{\mu 0} b^2(\{\mathcal{O}\}) - (\delta_{\mu 1} + \delta_{\mu 1} + \delta_{\mu 1})\, b^2(\{\mathcal{O}\})\right]. \end{aligned} \quad (2.14)$$

It is worth to recall that the deformation of the metric, resulting in the interval (2.5), represents a geometrization of a suitable space–time region (corresponding to the physical system considered) that describes, in the average, the effect of nonlocal interactions on a test particle. It is clear that there exist infinitely many deformations of the Minkowski space (precisely, $\infty^4$), corresponding to the different possible choices of the parameters $b_\mu$, a priori different for each physical system.

Moreover, since the usual, "flat" Minkowski metric $g$ (2.1) is related in an essential way to the electromagnetic interaction, we shall always mean in the following – unless otherwise specified – that electromagnetic interactions imply the presence of a fully Minkowskian metric. Actually, as it will be seen, a deformed metric of the type (2.4) is required if one wants to account for possible nonlocal electromagnetic effects.

Once the mathematical body of our formalism is specified, one has now to give a physical soul to it, in order to comply with the Finzi principle. On the basis of the discussion of Sect. 1.3, *we have to take, as observable $\mathcal{O}$ on which the metric coefficients $b_\mu(\{\mathcal{O}\})$ depend, the total energy $E$ exchanged by the physical system considered during the interaction process:*

$$\{\mathcal{O}\} \equiv E \Leftrightarrow \{b_\mu(\{\mathcal{O}\})\} \equiv \{b_\mu(E)\}, \forall \mu = 0,1,2,3. \quad (2.15)$$

Actually, since all the functions $\{b_\mu\}$ are dimensionless, they must depend on a dimensionless variable. Then, one has to divide the energy $E$ by a constant $E_0$ (in general characteristic of each fundamental interaction), with dimensions of energy, so that:

$$\{b_\mu(\{\mathcal{O}\})\} \equiv \left\{ b_\mu\left(\frac{E}{E_0}\right)\right\}, \forall \mu = 0,1,2,3. \quad (2.16)$$

As it will be seen, $E_0$ has the meaning of a "threshold energy."

Thus, the distance measurement is accomplished by means of the deformed metric tensor function of the energy, given explicitly by

$$\begin{aligned} \eta_{\mu\nu}(E) \quad & = \mathrm{diag}(b_0^2(E), -b_1^2(E), -b_2^2(E), -b_3^2(E)) \overset{\text{ESC off}}{=} \\ & = \delta_{\mu\nu}(\delta_{\mu 0} b_0^2(E) - \delta_{\mu 1} b_1^2(E) - \delta_{\mu 2} b_2^2(E) - \delta_{\mu 3} b_3^2(E)). \end{aligned} \quad (2.17)$$

Any interaction can be therefore phenomenologically described by metric (2.17) in an *effective* way.This is true in general, but necessary in the case of nonlocal and nonpotential interactions. For force fields which admit a potential, such a description is complementary to the actual one.[3]

One is therefore led to put forward a revision of the concept of "geometrization of an interaction": each interaction produces its own metric, formally expressed by the metric tensor (2.17), but realized via different choices of the set of parameters $b_\mu(E)$. Otherwise said, the $b_\mu(E)$'s are peculiar to every given interaction. The statement that (2.17) provides us with a metric description of an interaction must be just understood in such a sense.

Therefore, the energy-dependent deformation of the Minkowski metric implements a generalization of the concept of geometrization of an interaction (in accordance with Finzi's principle). The GR theory implements a geometrization (at a *global* scale) of the gravitational interaction, based on its derivability from a potential and on the equivalence between the inertial mass of a body and its "gravitational charge." The formalism of energy-dependent metrics allows one instead to implement a geometrization (at a *local* scale) of any kind of interaction, at least on a phenomenological basis. As already stressed before, such a formalism applies, in principle, to both fundamental and phenomenological interactions, either potential (gravitational, electromagnetic) or nonpotential (strong, weak), *local* and *nonlocal* (in the sense already specified), for which either an Equivalence Principle holds (as it is the case of gravitation) or (in the more general case) the inertial mass of the body *is not* in general proportional to its charge in the force field considered (e.m., strong and weak interaction).

Let us explicitly stress that the theory of SR based on metric (2.4) has nothing to do with General Relativity . Indeed, in spite of the formal similarity between the interval (2.5), with the $b_\mu$ functions of the coordinates, and the metric structure of a Riemann space, in this framework no mention at all is made of the equivalence principle between mass and inertia, and among noninertial, accelerated frames. Moreover, General Relativity describes geometrization on a large-scale basis, whereas the special relativity with topical deformed metric describes local (small-scale) deformations of the metric structure (although the term "small scale" must be referred to the real dimensions of the physical system considered). But the basic difference is provided by the fact that actually the deformed Minkowski space $\widetilde{M}$ has zero curvature, as it is easily seen by remembering that, in a Riemann space, the scalar curvature is constructed from the derivatives, with respect to space–time coordinates, of the metric tensor. In others words, the space $\widetilde{M}$ is *intrinsically flat* – at least in a mathematical sense.

Namely, it would be possible, in principle, to find a change of coordinates, or a rescaling of the lengths, so as to recover the usual Minkowski space.

---

[3] As we shall see, an example is just provided by the gravitational interaction in the Newtonian limit.

However, such a possibility is only a mathematical, and not a physical one. This is related to the fact that the energy of the process is fixed, and cannot be changed at will. For that value of the energy, the metric coefficients do possess values different from unity, so that the corresponding space $\widetilde{M}$, for the given energy value, is actually different from the Minkowski one. The usual space–time $M$ is recovered for a special value $E_0$ of the energy (characteristic of any interaction), such that indeed

$$\eta(E_0) = g = \text{diag}(1, -1, -1, -1). \tag{2.18}$$

Such a value $E_0$ (which must be derived from the phenomenology) will be referred to as *the threshold energy of the interaction considered.* As we shall see, it is just the constant introduced in (2.16) by dimensional arguments.

As we shall see in Part II, the choice of the Minkowskian energy $E$ as the observable $\mathcal{O}$ on which the metric parameters depend gives the deformed Minkowski space $\widetilde{M}$ the structure of a Generalized Lagrange Space [12].

Special cases of the metric (2.17) correspond to

1. A space $\widetilde{M}$ spatially isotropic:

$$b_1(E) = b_2(E) = b_3(E) = b(E) \tag{2.19}$$

   (cf. (2.13), (2.14));

2. A space $\widetilde{M}$ locally conform with the Minkowski space $M$:

$$b_0(E) = b_1(E) = b_2(E) = b_3(E) = b(E). \tag{2.20}$$

   This is a particular case of the Miron–Tavakol metric [13] which satisfies the Ehlers–Pirani–Schild axiomatics.

## 2.2 Energy as Dynamic Variable

The basic point of the present way of geometrizing an interaction (thus implementing the Finzi legacy) consists in a "upsetting" of the space–time–energy parametrization. Whereas for potential interactions there exists a potential energy depending on the space–time metric coordinates, one has here to deal with a deformed metric tensor $\eta$, whose coefficients depend on the energy, that thus assumes a *dynamic* role. However, the identification of energy as the physical observable on which the metric must depend leaves open the question, what energy? Let us answer this question.

From the physical point of view, $E$ is the measured energy of the system, and thus a merely phenomenological variable. As is well known, all the present physically realizable detectors work via their electromagnetic interaction in the usual space–time $M$. This is why, in this formalism, the

Minkowski space and the e.m. interaction do play a fundamental role. The former is – as already stressed – the cornerstone on which to build up the generalization of Special Relativity based on the deformed metric (2.17). The latter is the comparison term for all fundamental interactions. Let us recall that they are strictly interrelated, since it is just electromagnetism which determines the Minkowski geometry. Then, stating that the measurement of $E$ occurs via the e.m. interaction amounts to say that it is measured in $M$. This ensures that the total energy is conserved, due the validity of the Hamilton theorem in Minkowski space. In summary, *$E$ has to be understood as the energy measured by the detectors through the e.m. interaction in Minkowskian conditions and under validity of total energy conservation.*

From the mathematical standpoint, $E$ has to be considered as a dynamic variable, because it specifies the dynamic behavior of the process under consideration, and, through the metric coefficients, provides us with a dynamic map – in the energy range of interest – of the interaction ruling the given process.

Let us notice that metric (2.17) plays, for nonpotential interactions, a role analogous to that of the Hamiltonian $H$ for a potential interaction. In particular, the metric tensor $\eta$ as well is not an input of the theory, but must be built up from the experimental knowledge of the physical data of the system concerned (in analogy with the specification of the Hamiltonian of a potential system). However, there are some differences between $\eta$ and $H$ worth to be stressed. Indeed, as is well known, $H$ represents the total energy $E_{\text{tot}}$ of the system irrespective of the value of $E_{\text{tot}}$ and the choice of the variables. On the contrary, $\eta(E)$ describes the variation in the measurements of space and time, in the physical system considered, as $E_{\text{tot}}$ changes; therefore, $\eta$ does depend on the numerical value of $H$, but not on its functional form. The explicit expression of $\eta$ depends only on the interaction involved.

It is moreover worth recalling that the use of an energy-dependent space–time metric can be traced back to Einstein himself, who generalized the Minkowski interval as follows

$$\mathrm{d}s^2 = \left(1 + \frac{2\phi}{c^2}\right) c^2 \mathrm{d}t^2 - (\mathrm{d}x^2 + \mathrm{d}y^2 + \mathrm{d}z^2), \tag{2.21}$$

(where $\phi$ is the Newtonian gravitational potential), in order to account for the modified rate of a clock in presence of a (weak) gravitational field.

One may be puzzled about the dependence of the metric on the energy, which is not an invariant under usual Lorentz transformations, but transforms like the time-component of a four vector.

Actually, energy has to be regarded, in this formalism, from two different points of view. One has, on one side, the energy as measured in full Minkowskian conditions, which, as such, behaves as a genuine four-vector under usual Lorentz transformations (in the sense that it changes in the usual way if we go, say, from the laboratory frame to another frame in uniform motion with respect to it). Once fixed the frame, one gets a measured value of the energy for a given process. This is the value which enters, as a parameter, in the expression (2.17) of the deformed metric. Such an energy, therefore, is no longer to be considered as a four vector in the deformed Minkowski space, but it is just a quantity whose value determines the deformed geometry of the process considered (or, otherwise speaking, which selects the deformed space–time we have to use to describe the phenomenon).[4]

The problem of a metric description of a given interaction is thus formally reduced to the determination of the coefficients $b_\mu(E)$ from the data on some physical system, whose dynamic behavior is ruled by the interaction considered.

[4]This different view to energy constitutes the basic point to building up a 5D space–time, in which $E$ does just represent the extra dimension (see Parts IV and V).

# 3
# Deformed Special Relativity

## 3.1 Postulates of Deformed Special Relativity

In order to develop the relativity theory in a deformed Minkowski space–time, one has to suitably generalize and clarify the basic concepts which are at the very foundation of SR.

Let us first of all define a "topical inertial frame":

1. A *topical "inertial" frame* (TIF) is a reference frame in which space–time is homogeneous, but spacc is not ncccssarily isotropic.

   Then, a *generalized principle of relativity*, or "principle of metric invariance," can be stated as follows:

2. All physical measurements within every topical "inertial" frame must be carried out via the *same* metric.

We named DSR [6] the generalization of SR based on the earlier two postulates, and whose space–time structure is given by the deformed Minkowski space $\widetilde{M}$ introduced in Sect. 2.2. The notation DSR4 has been also used in literature, in order to stress that the deformed Minkowski space we are concerned with is a 4D one (and to distinguish the theory from its 5D counterpart we shall discuss in Parts IV and V). Let us also warn the reader against confusing this formalism with a different generalization of SR, i.e., Doubly Special Relativity [14], that uses the same acronym. This latter theory is essentially based on the quantum deformation of the Poincar é algebra, precisely, its $\varkappa$-deformation. In such a kind of deformation, one essentially modifies the commutation relations of the Poincaré generators, whereas in the DSR framework the deformation concerns primarily the metrical structure of the space–time (although the Poincaré algebra is affected, too: see Part II). However, it is not clear at present if the two theories may have some points in common (for instance, the energy dependence of the metric in position space).

Moreover, henceforth we shall use the notation $g_{\mathrm{DSR}}$ for the metric tensor of DSR (in order to distinguish it from – but also to stress its affinities with – the standard Minkowskian metric tensor $g \equiv g_{\mathrm{SR}}$ ), so that (with reference to (2.17))

$$g_{\mu\nu,\mathrm{DSR}}(E) = \mathrm{diag}(b_0^2\,(E)\,, -b_1^2(E), -b_2^2(E), -b_3^2(E)) \tag{2.17a}$$

is the *covariant* deformed metric tensor of $\widetilde{M}$ , whereas

$$g^{\mu\nu}_{\mathrm{DSR}}(E) = \mathrm{diag}(b_0^{-2}\,(E)\,, -b_1^{-2}(E), -b_2^{-2}(E), -b_3^{-2}(E)) \tag{2.17b}$$

is its *contravariant* counterpart.

The corresponding deformed interval is of course

$$\mathrm{d}\tilde{s^2}(E) = g_{\mu\nu,\mathrm{DSR}}(E)\mathrm{d}x^\mu \mathrm{d}x^\nu = b_0^2(E)c^2\mathrm{d}t^2 - b_1^2(E)\mathrm{d}x^2 - b_2^2(E)\mathrm{d}y^2 - b_3^2(E)\mathrm{d}z^2. \tag{2.17c}$$

In matrix notation, one can write

$$\mathrm{d}X = \begin{pmatrix} \mathrm{d}x^0 \\ \mathrm{d}x^1 \\ \mathrm{d}x^2 \\ \mathrm{d}x^3 \end{pmatrix}. \tag{3.1}$$

Then, the deformed interval (2.17c) reads

$$\mathrm{d}\tilde{s^2}(E) = (\mathrm{d}X)^{\mathrm{T}}\, g_{\mathrm{DSR}}(E)\mathrm{d}X, \tag{3.2}$$

where the upper "T" denotes matrix transposition, and $g_{\mathrm{DSR}}(E)$ is the $4\times4$ matrix (2.17a).

Let us remark the mathematically self-evident, but physically basic, point that the generalized metric (2.17a) (and the corresponding interval) is clearly *not preserved by the usual Lorentz transformations*. If $\Lambda_{\mathrm{SR}}$ is the $4 \times 4$ matrix representing a standard Lorentz transformation, this amounts to say that the similarity transformation generated by $\Lambda_{\mathrm{SR}}$ does not preserve the deformed metric tensor $g_{\mathrm{DSR}}$:

$$(\Lambda_{\mathrm{SR}})^{\mathrm{T}}\, g_{\mathrm{DSR}} \Lambda_{\mathrm{SR}} \neq g_{\mathrm{DSR}}. \tag{3.3}$$

This is by no means an unexpected result, at the light of the axiomatic formulation of Special Relativity (see Sect. 1.2). However, as a consequence, the deformed metric structure of $\widetilde{M}$ *violates* the standard Lorentz invariance, characteristic of the usual Minkowski space–time $M$. In this sense, therefore, we can state that DSR is strictly related to (and able to describe) the possible breakdown of Lorentz invariance, since the deformed metrics are no longer kept invariant by the standard Lorentz transformations.

## 3.2 Maximal Causal Speed

As is well known, the maximal causal speed in $M$ is obtained by putting $\mathrm{d}s^2 = 0$, whence

$$\mathrm{d}s^2 = 0 \Leftrightarrow c^2\mathrm{d}t^2 - \mathrm{d}x^2 - \mathrm{d}y^2 - \mathrm{d}z^2 = 0 \Leftrightarrow \frac{\mathrm{d}x^2 + \mathrm{d}y^2 + \mathrm{d}z^2}{\mathrm{d}t^2} = c^2. \quad (3.4)$$

Then one interprets $c$ as the maximal causal speed along any direction of the (Euclidean) space $R^3$ (embedded in the pseudoeuclidean Minkowski space–time $M$). Such an interpretation is obviously based on the physical fact that $c$ coincides with the light speed in vacuum, and on the isotropy of $R^3$. Therefore $c$ represents the value of any of the three components of the maximal causal velocity vector (m.c.v.) of SR, $\mathbf{u}_{\mathrm{SR}}$, namely:

$$\mathbf{u}_{\mathrm{SR}} = (c, c, c). \quad (3.5)$$

Then, $c^2$ is not, in general, a square modulus, but the square of any component of $\mathbf{u}_{\mathrm{SR}}$, whose square modulus (with respect to the Euclidean scalar product $\cdot$), is instead:

$$|\mathbf{u}_{\mathrm{SR}}|^2 \equiv \sum_{i=1}^{3} \left(u^i_{\mathrm{SR}}\right)^2 = 3c^2, \quad (3.6)$$

so that

$$u^i_{\mathrm{SR}} = \frac{1}{\sqrt{3}} |\mathbf{u}_{\mathrm{SR}}|, \quad \forall i = 1, 2, 3. \quad (3.7)$$

The earlier procedure must be suitably modified in the DSR case, due to the space anisotropy of $\widetilde{M}$.

Actually, in order to sort out a single component of the three-vector m.c.v., in a general 4D special-relativistic theory (characterized by a diagonal metric tensor $\eta_{\mu\nu}(\{\mathcal{O}\})$, where $\{\mathcal{O}\}$ is a set of observables corresponding to nonmetrical variables), one has to exploit a "directional separation" (or "dimensional separation") method, which consists of the following three-step recipe (ESC off throughout):

1. Set $\mathrm{d}s^{\tilde{2}}$ equal to zero:

$$\mathrm{d}s^{\tilde{2}} = 0 \Leftrightarrow \eta_{00}(\{\mathcal{O}\})c^2\mathrm{d}t^2 + \sum_{i=1}^{3} \eta_{ii}(\{\mathcal{O}\})(\mathrm{d}x^i)^2 = 0; \quad (3.8)$$

2. In order to find the $i$th component $u^i(\{\mathcal{O}\})$ of the m.c.v., put $\mathrm{d}x^j = 0$ $(j \neq i)$, thus getting

$$\eta_{00}(\{\mathcal{O}\})c^2\mathrm{d}t^2 + \eta_{ii}(\{\mathcal{O}\})(\mathrm{d}x^i)^2 = 0; \tag{3.9}$$

3. Evidence on the l.h.s. of (3.9) a quantity with physical dimensions [space]/[time] = [velocity]; at this point, we have two different subcases:

   (I) One carries to the l.h.s. of (3.9) $\mathrm{d}x^i/\mathrm{d}t$, which amounts to consider the 3D Euclidean product $\cdot$, thus getting an *anisotropic* m.c.v.:

$$u^i(\{\mathcal{O}\}) \equiv \frac{\mathrm{d}x^i}{\mathrm{d}t} = \frac{(\eta_{00}(\{\mathcal{O}\}))^{1/2}}{(-\eta_{ii}(\{\mathcal{O}\}))^{1/2}}c\,, \qquad \forall i = 1,2,3; \tag{3.10}$$

   (II) One carries to the l.h.s. of (3.9) $(-\eta_{ii}(\{\mathcal{O}\}))^{1/2}\,\mathrm{d}x^i/\mathrm{d}t$, which amounts to consider the 3D deformed product $*$ defined by $-\eta_{ij}(\{\mathcal{O}\}) = \delta_{ij}\,|\eta_{ii}(\{\mathcal{O}\})|$, thus getting an *isotropic* m.c.v.:

$$u^i(\{\mathcal{O}\}) \equiv (-\eta_{ii}(\{\mathcal{O}\}))^{1/2}\frac{\mathrm{d}x^i}{\mathrm{d}t} = (\eta_{00}(\{\mathcal{O}\}))^{1/2}c\,, \quad \forall i = 1,2,3. \tag{3.11}$$

   The two subcases I and II differ essentially by the different way of implementing the space anisotropy. In the former case, the anisotropy is embedded in the definition of m.c.v.; in the latter one, in the scalar product.[1]

Specializing the earlier equations to the DSR framework, we get therefore, in the two subcases:

(I)

$$u^i_{\mathrm{DSR,I}}(E) \equiv u^i(E) = c\frac{b_0(E)}{b_i(E)}; \tag{3.12}$$

---

[1] Let us notice that the directionally separating procedure can be consistently applied only to (special or general relativistic) metrics which are fully diagonal. This is obviously due to the mixings between different space directions which arise in the case of nondiagonal metrics.

Of course, such a procedure gives (in either subcase) the same standard result when applied to SR. In fact:

$$\begin{aligned} u^i_{\mathrm{SR}} &= (-\eta_{ii})^{1/2}\frac{\mathrm{d}x^i}{\mathrm{d}t} = (\eta_{00})^{1/2}c = \frac{\mathrm{d}x^i}{\mathrm{d}t} \\ &= \frac{(\eta_{00})^{1/2}}{(-\eta_{ii})^{1/2}}c = c \quad \forall i = 1,2,3. \end{aligned}$$

$$|\mathbf{u}_{\text{DSR,I}}(E)| = \left( \sum_{i=1}^{3} \left( u^i_{\text{DSR,I}}(E) \right)^2 \right)^{1/2}$$
$$= cb_0(E) \left( \frac{1}{b_1^2(E)} + \frac{1}{b_2^2(E)} + \frac{1}{b_3^2(E)} \right)^{1/2}; \quad (3.13)$$

(*anisotropic m.c.v.* $\mathbf{u}$);[2]
(II)

$$u^i_{\text{DSR,II}}(E) \equiv w^i(E) = cb_0(E); \quad (3.14)$$

$$|\mathbf{u}_{\text{DSR,II}}(E)|_* = \left( \sum_{i=1}^{3} b_i^2(E) \left( u^i_{\text{DSR,II}}(E) \right)^2 \right)^{1/2}$$
$$= cb_0(E) \left( b_1^2(E) + b_2^2(E) + b_3^2(E) \right)^{1/2}, \quad (3.15)$$

(*isotropic m.c.v.* $\mathbf{w}$); whence

$$u^i_{\text{DSR,II}}(E) = \left( b_1^2(E) + b_2^2(E) + b_3^2(E) \right)^{-1/2} |\mathbf{u}_{\text{DSR,II}}(E)|_* \quad (3.16)$$

i.e., in this subcase (unlike the previous one, see (3.13),(3.12)) one can state a proportionality relation by an overall factor (even if dependent on the metric coefficients) between $u^i_{\text{DSR,II}}(E)$ and $|\mathbf{u}_{\text{DSR,II}}(E)|_*$.

We have therefore shown that the two different procedures of directional separation lead to two different mathematical definitions of maximal causal velocity, an isotropic ($\mathbf{w}$, (3.15)) and an anisotropic ($\mathbf{u}$, (3.13)) one.[3] The choice between them must be done on a physical basis (see Sect. 3.3.6).

Moreover, it must be stressed the basic physical difference with respect to the SR case. In the standard relativistic framework, the light speed $c$ must be regarded as an *absolute* maximal causal speed, i.e., it is the same for all interactions and for all values of energy exchanges. In the DSR framework, we get instead a *relative* maximal causal speed, namely a m.c.v. different for any interaction. We shall come back to this point in Sect. 3.4.

---

[2]Of course, in the case of space isotropy, we get for $u$, too, an isotropic maximal causal velocity given by

$$u^i_{\text{iso}}(E) = u^i_{\text{DSR,I}}(E)\Big|_{b_i(E)=b(E)} = c\frac{b_0(E)}{b(E)} \qquad \forall i = 1,2,3;$$
$$|\mathbf{u}_{\text{iso}}(E)| = \left( \sum_{i=1}^{3} \left( u^i_{\text{iso}}(E) \right)^2 \right)^{1/2} = \sqrt{3}c\frac{b_0(E)}{b(E)}.$$

[3]Clearly, in both cases, the light speed in vacuum, $c$, does merely play the role of a phenomenological parameter on which the values of $u$ and $w$ depend.

## 3.3 Boosts in DSR

### 3.3.1 *Deformed Lorentz Transformations along a Coordinate Axis*

It follows from the postulates (i)and (ii) of DSR that the transformation equations connecting topical "inertial" frames, i.e., the generalized Lorentz transformations, are those which leave invariant the deformed metric when passing from a topical "inertial" frame $K$, to another frame $K'$, moving with constant velocity with respect to $K$. Then, physical laws are to be covariant with respect to such generalized transformations.

In other words, the generalized Lorentz transformations are the isometries of the deformed Minkowski space $\widetilde{M}$. We shall refer to them in the following as deformed Lorentz transformations (DLT). If $X$ denotes a column four-vector, a DLT is therefore a $4 \times 4$ matrix $\Lambda_{\text{DSR}}$ connecting two topical inertial frames $K$, $K'$

$$X' = \Lambda_{\text{DSR}}(E)X \tag{3.17}$$

and leaving the deformed interval (2.17c) invariant, namely

$$\Lambda_{\text{DSR}}^{\text{T}}(E)g_{\text{DSR}}(E)\Lambda_{\text{DSR}}(E) = g_{\text{DSR}}(E). \tag{3.18}$$

(cf. (3.3)). Equation (3.18) means that, unlike the case of a standard LT, a deformed Lorentz transformation generates a similarity transformation which leaves the deformed metric tensor invariant. Let us also notice the explicit dependence of $\Lambda_{\text{DSR}}$ on the energy $E$.

The explicit form of a pure DLT (i.e., a boost) can be derived by the same procedure followed in order to find the Lorentz boost expression in the usual Minkowski space (a more formal derivation will be given in Part II) [6].

Consider two TIF, $K$ and $K'$; by definition, the DLT's leave invariant the deformed interval (2.5), i.e.,

$$\begin{aligned} & b_0^2c^2t^2 - b_1^2x^2 - b_2^2y^2 - b_3^2z^2 \\ &= b_0^2c^2t'^2 - b_1^2x'^2 - b_2^2y'^2 - b_3^2z'^2. \end{aligned} \tag{3.19}$$

Moreover, it can be assumed, without loss of generality, that the frames $K$ and $K'$ are in standard configuration (i.e., their spatial frames coincide at $t = t' = 0$). By choosing the boost direction along $\widehat{x^1} = \widehat{x}$, we have therefore $y' = y$, $z' = z$ and (3.19) reduces to

$$b_0^2c^2t^2 - b_1^2x^2 = b_0^2c^2t'^2 - b_1^2x'^2. \tag{3.20}$$

From space–time homogeneity it follows that the functional relations between the two sets of coordinates $\{x, y, z, t\}$ and $\{x', y', z', t'\}$

must be linear. Then, in general, the deformed coordinate transformations are to be searched in the form

$$\begin{cases} x' &= A_{11}x + A_{14}t, \\ y' &= y, \\ z' &= z, \\ t' &= A_{41}x + A_{44}t, \end{cases} \tag{3.21}$$

where the coefficients $A_{11}, A_{14}, A_{41}, A_{44}$ depend *a priori* in general on $\mathbf{v}$ and $\widehat{x}$ (and, parametrically, on the energy).

Notice that the origin $O'$ of TIF $K'$ must move in $K$ with velocity $\mathbf{v} = v^1\widehat{x}$, and therefore:

$$x' = 0,\ x = vt \Leftrightarrow A_{14} = -vA_{11} \Leftrightarrow x' = A_{11}(x - vt). \tag{3.22}$$

Replacing (3.21), (3.22) in (3.20) yields

$$b_0^2c^2t^2 - b_1^2x^2 = b_0^2c^2(A_{41}x + A_{44}t)^2 - A_{11}^2b_1^2x^2(x - vt)^2, \tag{3.23}$$

which entails the following $3 \times 3$ quadratic system:

$$\begin{cases} c^2 &= c^2A_{44}^2 - \left(\frac{b_1}{b_0}\right)^2 A_{11}^2v^2 \\ -1 &= c^2\left(\frac{b_0}{b_1}\right)^2 A_{41}^2 - A_{11}^2 \\ 0 &= c^2\left(\frac{b_0}{b_1}\right)^2 A_{41}A_{44} + A_{11}^2v \end{cases} \tag{3.24}$$

with general solution

$$A_{11} = A_{44} = \pm\left(1 - \left(\frac{vb_1}{cb_0}\right)^2\right)^{-1/2}; \tag{3.25}$$

$$A_{41} = \mp\left(\frac{vb_1^2}{c^2b_0^2}\right)\left(1 - \left(\frac{vb_1}{cb_0}\right)^2\right)^{-1/2} = -\left(\frac{vb_1^2}{c^2b_0^2}\right)A_{11}. \tag{3.26}$$

The final result is

$$\begin{cases} x' &= \widetilde{\gamma}(x - vt) = \widetilde{\gamma}\left(x - \widetilde{\beta}\frac{b_0}{b_1}ct\right), \\ y' &= y, \\ z' &= z, \\ t' &= \widetilde{\gamma}\left(t - \frac{vb_1^2}{c^2b_0^2}x\right) = \widetilde{\gamma}\left(t - \frac{\widetilde{\beta}^2}{v}x\right), \end{cases} \tag{3.27}$$

where $v$ is the relative speed of the reference frames, and[4]

$$\widetilde{\beta} = \frac{v}{u} , \tag{3.28}$$

$$\widetilde{\gamma} = (1 - \widetilde{\beta}^2)^{-1/2} , \tag{3.29}$$

are the *deformed velocity parameter* and *deformed relativistic $\gamma$-factor*, respectively. Quantity $u$ is the anisotropic maximal causal speed defined by (3.12). Notice that parametrizing the boost (3.27) in terms of $u$ does not imply any a priori choice between the two m.c.v. $u$ and $w$. The final choice will be done in Sect. 3.3.6 on a physical basis.

For a boost in the direction $\widehat{x}_i$, we have

$$\begin{cases} x^{i\prime} &= \widetilde{\gamma}^i \left( x^i - \widetilde{\beta}^i \dfrac{b_0}{b_i} ct \right) \\ x^{k \neq i\prime} &= x^{k \neq i} \\ t' &= \widetilde{\gamma}^i \left( t - \dfrac{\left(\widetilde{\beta}^i\right)^2}{v^i} x^i \right) \end{cases} , \tag{3.27a}$$

where

$$\widetilde{\beta}^i = \frac{v^i}{u} ; \tag{3.28 a}$$

$$\widetilde{\gamma}^i = \left(1 - \left(\widetilde{\beta}^i\right)^2\right)^{-1/2} = \left(1 - \left(\frac{v^i b_i}{c b_0}\right)^2\right)^{-1/2} \quad \text{(ESC off)}. \tag{3.29 a}$$

It must be carefully noted that, like the metric, also the generalized Lorentz transformations depend on the energy. This means that one gets different transformation laws for different values of $E$, but still with the same functional dependence on the energy, so that the invariance of the deformed interval (2.17c) is always ensured (provided that the process considered does always occur via the same interaction).

Indeed, the energy $E$ can be considered fixed also because, from a quantum point of view, energy can be transferred only by finite amounts. Differentiating (3.27), we get therefore

---

[4]Transformations (3.27) do formally coincide with the isotopic Lorentz transformations. However, in the present context their physical meaning is different, as it is easily seen e.g., by the identification of the maximal causal speed $u$ with the speed characteristic of the quanta of a given interaction (see Sect. 3.3.6). In particular, the parametrization (3.28) of the deformed velocity parameter $\widetilde{\beta}$ in terms of $u$ immediately shows that is always $\tilde{\beta} < 1$, so that $\tilde{\gamma}$ never takes imaginary values (contrarily to the isotopic case). Moreover, no reference at all is made, in this framework, to the existence of an underlying "medium."

$$\begin{cases} u\mathrm{d}t' + t'\mathrm{d}u = \widetilde{\gamma}(u\,\mathrm{d}t - \widetilde{\beta}\mathrm{d}x) + [\mathrm{d}\widetilde{\gamma}(ut - \widetilde{\beta}x) + \widetilde{\gamma}(t\,\mathrm{d}u - x\mathrm{d}\widetilde{\beta}]; \\ \quad \mathrm{d}x' = \widetilde{\gamma}(\mathrm{d}x - \widetilde{\beta}u\,\mathrm{d}t) + [\mathrm{d}\widetilde{\gamma}(x - \widetilde{\beta}ut) - \widetilde{\gamma}(t\widetilde{\beta}\mathrm{d}u + t\,\mathrm{d}\widetilde{\beta})], \end{cases} \tag{3.30}$$

where, by the earlier argument, $\mathrm{d}E = 0$ and therefore $\mathrm{d}\widetilde{\gamma} = \mathrm{d}\widetilde{\beta} = \mathrm{d}u = 0$. Squaring (3.30) and subtracting, we find

$$\mathrm{d}x'^2 - u^2\mathrm{d}t'^2 = \widetilde{\gamma}^2[(\mathrm{d}x - \widetilde{\beta}\mathrm{d}t)^2 - (u\,\mathrm{d}t - \widetilde{\beta}\mathrm{d}x)^2] = \mathrm{d}x^2 - u^2\mathrm{d}t^2 \tag{3.31}$$

where in the last step use has been made of (3.20). Exploiting the explicit expression of $u$, (3.12), one has finally

$$\mathrm{d}s'^2 = \mathrm{d}s^2, \tag{3.32}$$

i.e., the DLT (3.27) are actually (some of) the isometries of the deformed Minkowski space $\widetilde{M}$, in spite of their dependence on the energy.

Notice that the DLT reduce to the standard ones of SR in the limit $g_{\mathrm{DSR}} \longrightarrow g_{\mathrm{SR}}$. The set of the DLT's does therefore represent a *covering* of the usual LT's, just like the set of LT's is a relativistic covering of the Galilei–Newton transformations (at group level, the latter are obtained from the LT's by an Inönü-Wigner procedure of group contraction).

### 3.3.2 Boost in a Generic Direction

In this case, the relative velocity is $\mathbf{v} = v^1\widehat{x} + v^2\widehat{y} + v^3\widehat{z}$ , and we have to suitably generalize definitions (3.28), (3.29) as follows:

$$\widetilde{\boldsymbol{\beta}} \equiv \frac{\mathbf{v}}{\mathbf{u}} \equiv \left(\frac{v^1 b_1(E)}{cb_0(E)}, \frac{v^2 b_2(E)}{cb_0(E)}, \frac{v^3 b_3(E)}{cb_0(E)}\right); \tag{3.33}$$

$$\widetilde{\gamma} \equiv \left(1 - \left|\widetilde{\boldsymbol{\beta}}\right|^2\right)^{-1/2}, \tag{3.34}$$

where (cf. (3.12))

$$\mathbf{u} = \left(c\frac{b_0(E)}{b_1(E)}, c\frac{b_0(E)}{b_2(E)}, c\frac{b_0(E)}{b_3(E)}\right). \tag{3.35}$$

Notice that

$$\widetilde{\boldsymbol{\beta}} \equiv \frac{\mathbf{v}}{\mathbf{u}} \neq \frac{\mathbf{v}}{u}.$$

This follows from the anisotropy of the three-vector $\mathbf{u}$, and it has to be compared with the SR case, where

$$\boldsymbol{\beta} \equiv \frac{\mathbf{v}}{\mathbf{u}} = \frac{\mathbf{v}}{c}.$$

In general, it is possible to state that

$$\frac{\mathbf{m}}{\mathbf{n}} = \frac{1}{n}\mathbf{m} \Leftrightarrow \mathbf{n} = (n, n, n)$$

i.e., iff $\mathbf{n}$ is a spatially isotropic three-vector.

In order to derive the expression of the deformed boost in a generic direction, it is possible to use the same method of the previous case. However, it is simpler to take advantage of the scalar-product properties of three-vectors in $\widetilde{M}(E)$. Namely, we consider the physical 3D space $\widetilde{E_3}(E)$ embedded in $\widetilde{M}(E)$ (see (2.7)), and decompose the space vector $\mathbf{x}$ in two components, $\mathbf{x}_{\|}$ and $\mathbf{x}_{\perp}$, parallel and orthogonal, respectively, to the boost direction $\widehat{v}$:

$$\mathbf{x} = \mathbf{x}_{\|} + \mathbf{x}_{\perp}; \tag{3.36}$$

$$\begin{aligned}
\mathbf{x}_{\|} \equiv \widehat{v}(\widehat{v} * \mathbf{x}) &= \frac{\mathbf{v}}{|\mathbf{v}|_*^2}(\mathbf{v} * \mathbf{x}) = \frac{\mathbf{v}}{\mathbf{v} * \mathbf{v}}(\mathbf{v} * \mathbf{x}) \\
&= \frac{\sum_{i=1}^3 b_i^2(E) v^i x^i}{\sum_{i=1}^3 b_i^2(E)\left(v^i\right)^2}\mathbf{v} \neq \widehat{\widetilde{\beta}}(\widehat{\widetilde{\beta}} * \mathbf{x}) \\
&= \frac{\widetilde{\boldsymbol{\beta}}}{\left|\widetilde{\boldsymbol{\beta}}\right|_*^2}(\widetilde{\boldsymbol{\beta}} * \mathbf{x}) = \frac{\widetilde{\boldsymbol{\beta}}}{\widetilde{\boldsymbol{\beta}} * \widetilde{\boldsymbol{\beta}}}(\widetilde{\boldsymbol{\beta}} * \mathbf{x}) \\
&= \frac{\sum_{i=1}^3 b_i^2(E)\widetilde{\beta}^i x^i}{\sum_{i=1}^3 b_i^2(E)\left(\widetilde{\beta}^i\right)^2}\widetilde{\boldsymbol{\beta}};
\end{aligned} \tag{3.37}$$

$$x_{\|}^i \equiv \frac{\sum_{k=1}^3 b_k^2(E) v^k x^k}{\sum_{k=1}^3 b_k^2(E)\left(v^k\right)^2} v^i \neq \frac{\sum_{k=1}^3 b_k^2(E)\widetilde{\beta}^k x^k}{\sum_{k=1}^3 b_k^2(E)\left(\widetilde{\beta}^k\right)^2}\widetilde{\beta}^i; \tag{3.38}$$

$$\begin{aligned}
\mathbf{x}_{\perp} &\equiv \mathbf{x} - \mathbf{x}_{\|} = \mathbf{x} - \frac{\sum_{i=1}^3 b_i^2(E) v^i x^i}{\sum_{i=1}^3 b_i^2(E)\left(v^i\right)^2}\mathbf{v} \\
&\neq \mathbf{x} - \frac{\sum_{i=1}^3 b_i^2(E)\widetilde{\beta}^i x^i}{\sum_{i=1}^3 b_i^2(E)\left(\widetilde{\beta}^i\right)^2}\widetilde{\boldsymbol{\beta}};
\end{aligned} \tag{3.39}$$

$$\begin{aligned}
x_{\perp}^i &\equiv x^i - \frac{\sum_{k=1}^3 b_k^2(E) v^k x^k}{\sum_{k=1}^3 b_k^2(E)\left(v^k\right)^2} v^i \\
&\neq x^i - \frac{\sum_{k=1}^3 b_k^2(E)\widetilde{\beta}^k x^k}{\sum_{k=1}^3 b_k^2(E)\left(\widetilde{\beta}^k\right)^2}\widetilde{\beta}^i.
\end{aligned} \tag{3.40}$$

It is easily checked that indeed

$$\mathbf{x} * \mathbf{v} = \sum_{i=1}^{3} b_i^2(E) x^i v^i = \frac{\sum_{i=1}^{3} b_i^2(E) x^i v^i}{\sum_{i=1}^{3} b_i^2(E) (v^i)^2} \sum_{k=1}^{3} b_k^2(E) \left(v^k\right)^2$$
$$= \frac{\sum_{i=1}^{3} b_i^2(E) x^i v^i}{\sum_{i=1}^{3} b_i^2(E) (v^i)^2} \mathbf{v} * \mathbf{v} = \mathbf{x}_{\|} * \mathbf{v} = \left|\mathbf{x}_{\|}\right|_* |\mathbf{v}|_* ; \tag{3.41}$$

$$\mathbf{x}_{\perp} * \mathbf{v} = \mathbf{x} * \mathbf{v} - \mathbf{x}_{\|} * \mathbf{v} = 0. \tag{3.42}$$

Then, applying boost (3.27a) to $\mathbf{x}_{\|}$ and $\mathbf{x}_{\perp}$ yields

$$\begin{cases} \mathbf{x}'_{\|} = \widetilde{\gamma}(\mathbf{x}_{\|} - \mathbf{v}t) \\ \mathbf{x}'_{\perp} = \mathbf{x}_{\perp} \\ t' = \widetilde{\gamma}\left(t - \sum_{i=1}^{3} \frac{v^i b_i^2(E)}{c^2 b_0^2(E)} x^i\right) = \widetilde{\gamma}(t - \widetilde{\mathbf{B}} \cdot \mathbf{x}) = \widetilde{\gamma}\left(t - \widetilde{\mathbf{B}}^{(*)} * \mathbf{x}\right) \end{cases}, \tag{3.43}$$

where we put[5]

$$\widetilde{\gamma} \equiv (1 - \widetilde{\boldsymbol{\beta}} \cdot \widetilde{\boldsymbol{\beta}})^{-1/2} = (1 - \widetilde{\boldsymbol{\beta}}^{(*)} * \widetilde{\boldsymbol{\beta}}^{(*)})^{-1/2} = \left(1 - \sum_{i=1}^{3} \frac{v^i b_i^2(E)}{c^2 b_0^2(E)}\right)^{-1/2}$$
$$= \left[1 - \left(\frac{v^1 b_1(E)}{c b_0(E)}\right)^2 - \left(\frac{v^2 b_2(E)}{c b_0(E)}\right)^2 - \left(\frac{v^3 b_3(E)}{c b_0(E)}\right)^2\right]^{-1/2} ; \tag{3.44}$$

$$\widetilde{\boldsymbol{\beta}}^{(*)} \equiv \frac{\mathbf{v}}{\mathbf{w}} = \left(\frac{v^1}{c b_0(E)}, \frac{v^2}{c b_0(E)}, \frac{v^3}{c b_0(E)}\right) = \frac{1}{c b_0(E)} \mathbf{v} ; \tag{3.45}$$

$$\mathbf{w} \equiv (c b_0(E), c b_0(E), c b_0(E)); \tag{3.46}$$

$$\widetilde{\mathbf{B}} \equiv \frac{\mathbf{v}}{\mathbf{u}^2} = \left(\frac{v^1 b_1^2(E)}{c^2 b_0^2(E)}, \frac{v^2 b_2^2(E)}{c^2 b_0^2(E)}, \frac{v^3 b_3^2(E)}{c^2 b_0^2(E)}\right) ; \tag{3.47}$$

$$\widetilde{\mathbf{B}}^{(*)} \equiv \frac{\mathbf{v}}{\mathbf{w}^2} = \frac{1}{c^2 b_0^2(E)} \mathbf{v} \quad . \tag{3.48}$$

---

[5]Care must be exercized in not confusing $\widetilde{\gamma}$ (given by (3.44)) with $\widetilde{\gamma^i}$ (3.29a). Indeed, it is $\widetilde{\gamma} \neq \widetilde{\gamma^i}$ $\forall i = 1, 2, 3$. Moreover, these quantities are related in a nontrivial way by

$$\widetilde{\gamma^i} = \widetilde{\gamma}\big|_{\mathbf{v} = v^i \widehat{x^i}}$$

namely, $\widetilde{\gamma}$ reduces to $\widetilde{\gamma^i}$ when the deformed boost with velocity $\mathbf{v}$ in the generic space direction $\widehat{v}$ reduces to the deformed boost with velocity $v^i$ along $\widehat{x^i}$.

It follows therefore that the deformed boosts admit a double treatment, either:

(I) In terms of the Euclidean scalar product $\cdot$, of the (anisotropic) m.c.v. $\mathbf{u}$ and of the related "rapidities" $\widetilde{\boldsymbol{\beta}}$ and $\widetilde{\mathbf{B}}$, or

(II) In terms of the deformed product $*$, of the (isotropic) m.c.v. $\mathbf{w}$ and of the related quantities $\widetilde{\boldsymbol{\beta}}^{(*)}$ and $\widetilde{\mathbf{B}}^{(*)}$.[6]

Then, the space vector transforms as:

$$\begin{aligned} \mathbf{x}' &= \mathbf{x}'_{\|} + \mathbf{x}'_{\perp} = \widetilde{\gamma}(\mathbf{x}_{\|} - \mathbf{v}t) + \mathbf{x}_{\perp} \\ &= \mathbf{x} + (\widetilde{\gamma} - 1)\widehat{v}(\widehat{v} * \mathbf{x}) - \widetilde{\gamma}\mathbf{v}t \\ &= \mathbf{x} + (\widetilde{\gamma} - 1)\frac{\mathbf{v}}{|\mathbf{v}|_*^2}(\mathbf{v} * \mathbf{x}) - \widetilde{\gamma}\mathbf{v}t \end{aligned} \tag{3.49}$$

and we eventually find the expression of the deformed boost in a generic direction[7]:

$$\begin{cases} \mathbf{x}' = \mathbf{x} + (\widetilde{\gamma} - 1)\dfrac{\mathbf{v}}{|\mathbf{v}|_*^2}(\mathbf{v} * \mathbf{x}) - \widetilde{\gamma}\mathbf{v}t. \\ t' = \widetilde{\gamma}\left(t - \widetilde{\mathbf{B}} \cdot \mathbf{x}\right) = \widetilde{\gamma}\left(t - \widetilde{\mathbf{B}}^{(*)} * \mathbf{x}\right). \end{cases} \tag{3.50}$$

---

[6] It is possible to show that, in this case, more equivalent forms of the deformed boost (3.43) exist. As is easily seen, this is due to the fact that, in general, $\widehat{\widetilde{\beta}} \neq \widehat{v}$ and $\widehat{\widetilde{B}} \neq \widehat{v}$, whereas $\widehat{\widetilde{\beta}^{(*)}} = \widehat{v} = \widehat{\widetilde{B}^{(*)}}$.

[7] Notice that, from the definitions (3.33) and (3.45) of $\widetilde{\boldsymbol{\beta}}(g)$ and $\widetilde{\boldsymbol{\beta}}^{(*)}(g)$, it follows:

$$|\mathbf{v}(g)|_*^2 \equiv \sum_{k=1}^{3} b_i^2(x^5)\left(v^i(g)\right)^2 = c^2 b_0^2(x^5)\left|\widetilde{\boldsymbol{\beta}}(g)\right|^2;$$

$$|\mathbf{v}(g)|_*^2 \equiv \sum_{k=1}^{3} b_i^2(x^5)\left(v^i(g)\right)^2 = c^2 b_0^2(x^5)\left|\widetilde{\boldsymbol{\beta}}^{(*)}(g)\right|_*^2.$$

on account of the fact that (as it is easy to verify)

$$\left|\widetilde{\boldsymbol{\beta}}^{(*)}(g)\right|_*^2 = \left|\widetilde{\boldsymbol{\beta}}(g)\right|^2$$

and

$$\begin{aligned} \left|\mathbf{w}(x^5)\right|_*^2 &= \sum_{k=1}^{3} b_i^2(x^5)\left(w^i(x^5)\right)^2 = c^2 b_0^2(x^5)\sum_{k=1}^{3} b_i^2(x^5); \\ \left|\mathbf{u}(x^5)\right|^2 &= \sum_{k=1}^{3}\left(u^i(x^5)\right)^2 = c^2 b_0^2(x^5)\sum_{k=1}^{3} b_i^{-2}(x^5), \end{aligned}$$

so that, in general

$$\left|\mathbf{w}(x^5)\right|_*^2 \neq \left|\mathbf{u}(x^5)\right|^2.$$

### 3.3.3 Symmetrization of Deformed Boosts

As in the case of standard SR, it is possible to symmetrize the expression of boosts in DSR by introducing suitable time coordinates.

Let us first consider a deformed boost along $\widehat{x^i}$ ($i = 1, 2, 3$); the symmetrization transformation (a "dimensionally homogenizing dilato-contraction") of $t$ is given by

$$x^0 \equiv u^i t = c\frac{b_0(E)}{b_i(E)}t; \quad x^{i\prime} \equiv x^i. \tag{3.51}$$

The deformed metric tensor in the new "primed" coordinates, $\{x^{\mu\prime}\} = \{x^0, x, y, z\}$, reads:

$$\begin{aligned} g_{\mu\nu,\mathrm{DSR}}(E) &\overset{\text{ESC on}}{=} g_{\alpha\beta,\mathrm{DSR}}(E)\frac{\partial x^\alpha}{\partial x^{\mu\prime}}\frac{\partial x^\beta}{\partial x^{\nu\prime}} \\ &= \mathrm{diag}(b_i^2(E), -b_1^2(E), -b_2^2(E), -b_3^2(E)) \\ &\overset{\text{ESC off}}{=} \delta_{\mu\nu}\left[b_i^2(E)\delta_{\mu 0} - b_1^2(E)\delta_{\mu 1} - b_2^2(E)\delta_{\mu 2} - b_3^2(E)\delta_{\mu 3}\right]. \end{aligned} \tag{3.52}$$

Equation (3.27a) takes therefore the symmetric form in $x^i$ and $x^0$ (ESC off):

$$\begin{cases} x^{i\prime} &= \widetilde{\gamma}(x^i - \widetilde{\beta}^i x^0), \\ x^{k\neq i\prime} &= x^{k\neq i}, \\ x^{0\prime} &= \widetilde{\gamma}(x^0 - \widetilde{\beta}^i x^i). \end{cases} \tag{3.53}$$

Transformation (3.51) does not symmetrize the deformed boost in a generic direction (unlike the case of SR, where the same transformation $x^0 = ct$ symmetrizes both boosts). In this case, the symmetrization is possible only if the treatment II (based on the deformed scalar product $*$) is used.

In fact, by using the proportionality (see (3.45), (3.48)) among $\widetilde{\boldsymbol{\beta}}^{(*)}$, $\widetilde{\mathbf{B}}^{(*)}$ and $\mathbf{v}$, the following transformation on $t$ (see (3.46))

$$x^0 \equiv c\,b_0(E)t = w^k t \;\; (\forall k = 1, 2, 3) \;; \; x^{i\prime} \equiv x^i \qquad (\forall i = 1, 2, 3) \tag{3.54}$$

does symmetrize (3.43) in $\mathbf{x}_{\|}$ e $x^0$:

$$\begin{cases} \mathbf{x}'_{\|} = \left(1-\widetilde{\boldsymbol{\beta}}^{(*)} * \widetilde{\boldsymbol{\beta}}^{(*)}\right)^{-1/2}\left(\mathbf{x}_{\|}-\widetilde{\boldsymbol{\beta}}^{(*)}x^0\right) \\ \mathbf{x}'_{\perp} = \mathbf{x}_{\perp} \\ x^{0\prime} = \begin{cases} \left(1-\widetilde{\boldsymbol{\beta}}^{(*)} * \widetilde{\boldsymbol{\beta}}^{(*)}\right)^{-1/2}\left(x^0-\widetilde{\boldsymbol{\beta}}^{(*)} * \mathbf{x}\right) \\ = \left(1-\widetilde{\boldsymbol{\beta}}^{(*)} * \widetilde{\boldsymbol{\beta}}^{(*)}\right)^{-1/2}\left(x^0-\widetilde{\boldsymbol{\beta}}^{(*)} * \mathbf{x}_{\|}\right) \end{cases} \end{cases} . \quad (3.55)$$

Under transformation (3.54), the metric tensor becomes:

$$\begin{aligned} g'_{\mu\nu,\mathrm{DSR}}(E) &\overset{\text{ESC on}}{=} g_{\alpha\beta,\mathrm{DSR}}(E)\frac{\partial x^\alpha}{\partial x^{\mu\prime}}\frac{\partial x^\beta}{\partial x^{\nu\prime}} \\ &= \mathrm{diag}(1,-b_1^2(E),-b_2^2(E),-b_3^2(E)) \\ &\overset{\text{ESC off}}{=} \delta_{\mu\nu}\left[\delta_{\mu 0}-b_1^2(E)\delta_{\mu 1}-b_2^2(E)\delta_{\mu 2}-b_3^2(E)\delta_{\mu 3}\right] . \end{aligned} \quad (3.56)$$

Therefore the symmetrization of the deformed boost in a generic direction makes the 4D metric isochronous, since $g'_{00,\mathrm{DSR}}=1$.

Let us finally notice that, like in the SR case, the boost in generic direction expressed in terms of $\mathbf{x}$ e $t$ (3.50) cannot in general be symmetrized.

As to the boost along a coordinate axis (3.27), it is asymmetrical in the behavior of $x'$ and $t'$, unlike the usual Lorentz transformations, which are fully symmetric when putting $x^0=ct$. However, such asymmetry is only formal. It can be removed by introducing a time coordinate defined in terms of the anisotropic maximal causal speed $u$ in the generalized Minkowski space considered (see (3.12)):

$$x^0 = ut = \left(\frac{b_0}{b}c\right)t \quad (3.57)$$

and changing the metric tensor $g_{\mathrm{DSR}}$ into

$$g'_{\mathrm{DSR}} = \mathrm{diag}(b^2,-b^2,-b^2,-b^2) = b^2 g_{\mathrm{SR}} . \quad (3.58)$$

Then, the generalized Lorentz transformations in $\widetilde{M}'$ take the symmetrical form

$$\begin{cases} x^{0\prime} = \widetilde{\gamma}(x^0-\widetilde{\beta}x^1) \\ x^{1\prime} = \widetilde{\gamma}(x^1-\widetilde{\beta}x^0) \\ x^{2\prime} = x^2 \\ x^{3\prime} = x^3 \end{cases} . \quad (3.59)$$

It is easily seen that the deformed Minkowski spaces $\widetilde{M}$ and $\widetilde{M}'$, with metrics (2.2) and (3.58), respectively, are isometric, because they have the same interval (2.5). They are therefore fully equivalent in every respect, and it is therefore possible to use indifferently either transformation (3.27) or (3.59). The main advantage of the latter ones is that, due to relation (3.57), the formulae holding for $\widetilde{M}'$ are immediately got from those of the standard special relativity by simply replacing everywhere $c$ by $u$.

### 3.3.4 Choosing the Boost Direction in DSR

We want now to remark a difficulty arising in the context of DSR, due to the space anisotropy.

Indeed, the space anisotropy (reflected in the physical anisotropic m.c.v. $\mathbf{u}$) produces a triple indetermination in the process of identifying the motion axis with any of the space coordinate axes, since now – unlike the SR case – the space dimensions are no longer equivalent.

However, this indeterminacy can be removed (at least in principle) by means of the following *Gedankenexperiment*. Consider three particles (ruled by one and the same interaction) in general different but able to move at the maximal causal velocity $u^i(E)$. Suppose they are moving in the 3D Euclidean space along mutually independent (orthogonal) spatial directions. Assigning an arbitrary labeling to the particle motion directions, we can fix an orthogonal, left-handed frame of axes. Since by assumption we know the interaction which the particles are subjected to, we know the deformed metric and therefore the metric coefficients as functions of the energy, $b_\mu^2(E)$. Then, a measurement of the particle velocities allows us to determine the right labeling of the spatial frame.

This implies that in the context of DSR, too, it is always possible, at physical level, to let one of the three space axes to coincide with the direction of motion of a physical object, and therefore apply the suitable deformed boost.

### 3.3.5 Recovering Lorentz Invariance in DSR

We have stressed in Sect. 3.1 that the deformed interval in $\widetilde{M}$ is not preserved by standard Lorentz transformations. In this sense, we can state indeed that DSR is strictly related to the breakdown of Lorentz invariance (LI), since the deformed metric is no longer kept invariant by the standard Lorentz transformations. However, by construction the deformed Lorentz boosts (and, in general, the DLT, i.e., the isometries of $\widetilde{M}$: see Part II) *do preserve* the generalized metric and interval (2.17a, c). Therefore, Lorentz invariance, broken by the energy-dependent deformation of the space time *in its usual sense*, namely as a special-relativistic symmetry property of the interactions and/or the physical systems, *is recovered*, in the framework of DSR, in a generalized, wider meaning. We shall name *deformed Lorentz invariance (DLI)*this extended LI.

The mathematical formulation of DLI is provided by (3.18), which we rewrite here for reader's convenience, by emphasizing the dependence of the DLT on the interaction considered:

$$\Lambda^{\mathrm{T}}_{\mathrm{DSR,int.}}(E) g_{\mathrm{DSR,int.}}(E) \Lambda_{\mathrm{DSR,int.}}(E) = g_{\mathrm{DSR,int.}}(E). \tag{3.18a}$$

It can be read as follows:

– *For every physical interaction, which affects the space–time geometry by deforming it in a way described by the metric tensor* $g_{\mathrm{DSR.int.}}$, *it is always possible to find DLT* $\Lambda_{\mathrm{DSR,int.}}$ *preserving the deformed geometrical structure of space–time for the interaction considered,* namely (from a mathematical point of view) generating similarity transformations which leave the deformed metric tensor invariant.

Then, we can state that DSR not only permits to deal with LI breakdown on a physical basis, but allows one to recover Lorentz invariance as an extended, higher symmetry of physics, valid for systems and/or interactions violating LI according to the usual Special Relativity, in the usual Minkowski space–time.

### *3.3.6 Velocity Composition Law in $\widetilde{M}$ and the Invariant Maximal Speed*

We have seen in Sect. 3.2 that the directionally separating approach (mandatory in the deformed case) yields two different *mathematical* definitions $\mathbf{u}$ (3.12) and $\mathbf{w}$ (3.14) of maximal causal velocity in DSR. The choice between them must be done on a physical basis, by checking their actual invariance under deformed boosts.

To this aim, one has to derive the generalized velocity composition law valid in $\widetilde{M}$. For a deformed boost in the direction $\widehat{x^i}$, differentiating the inverse of (3.27a) yields (on account of the fact that $\mathrm{d}E = 0$ in DSR) (ESC off):

$$\left\{\begin{array}{ccc} \mathrm{d}x^i & = & \widetilde{\gamma}(\mathrm{d}x^{i\prime} + v^i \mathrm{d}t') \\ \mathrm{d}x^{k\neq i} & = & \mathrm{d}x^{k\neq i\prime} \\ \mathrm{d}t & = & \widetilde{\gamma}\left(\mathrm{d}t' + \dfrac{v^i b_i^2(E)}{c^2 b_0^2(E)} \mathrm{d}x^{i\prime}\right) \end{array}\right. , \tag{3.60}$$

with $\widetilde{\gamma}$ given by (3.29). Since

$$\frac{\mathrm{d}x^i}{\mathrm{d}t} = v^i, \quad \frac{\mathrm{d}x^{i\prime}}{\mathrm{d}t'} = v^{i\prime}, \quad \frac{\mathrm{d}x^{k\neq i}}{\mathrm{d}t} = v^{k\neq i}, \quad \frac{\mathrm{d}x^{k\neq i\prime}}{\mathrm{d}t'} = v^{k\neq i\prime} \tag{3.61}$$

one gets the *deformed velocity composition law* (in compact notation, ESC off)

$$v^k = \frac{v^{k\prime} + \delta_{ik} v^i}{\left[1 + \left(\frac{b_i(E)}{b_0(E)}\right)^2 \frac{v^i v^{i\prime}}{c^2}\right] \{\widetilde{\gamma}(E) + \delta_{ik}\left[1 - \widetilde{\gamma}(E)\right]\}}. \tag{3.62}$$

This relation can be expressed in terms of the standard 3D scalar product · (and therefore of the anisotropic maximal velocity $\mathbf{u}$) (approach I) as

$$\begin{aligned} v^k &= \frac{v^{k\prime} + \delta_{ik} v^i}{\left[1 + \frac{\mathbf{v} \cdot \mathbf{v}'}{(u^i(E))^2}\right] \{\widetilde{\gamma}(E) + \delta_{ik}\left[1 - \widetilde{\gamma}(E)\right]\}} \\ &= \frac{v^{k\prime} + \delta_{ik} v^i}{\left[1 + \frac{\widetilde{\boldsymbol{\beta}} \cdot \mathbf{v}'}{u^i(E)}\right] \{\widetilde{\gamma}(E) + \delta_{ik}\left[1 - \widetilde{\gamma}(E)\right]\}}, \end{aligned} \tag{3.63}$$

where

$$\widetilde{\beta}^i(E) = \frac{v^i}{u^i(E)} \quad ; \; \widetilde{\gamma}(E) = \left(1 - \widetilde{\boldsymbol{\beta}}(E) \cdot \widetilde{\boldsymbol{\beta}}(E)\right)^{-1/2}. \tag{3.64}$$

Alternatively, we can use approach II, based on the deformed scalar product $*$ (and therefore the isotropic maximal velocity $\mathbf{w}$) and write (3.63) as:

$$\begin{aligned} v^k &= \frac{v^{k\prime} + \delta_{ik} v^i}{\left[1 + \frac{\mathbf{v} * \mathbf{v}'}{(w^i(E))^2}\right] \{\widetilde{\gamma}(E) + \delta_{ik}\left[1 - \widetilde{\gamma}(E)\right]\}} \\ &= \frac{v^{k\prime} + \delta_{ik} v^i}{\left[1 + \frac{\widetilde{\boldsymbol{\beta}}^{(*)} * \mathbf{v}'}{w^i(E)}\right] \{\widetilde{\gamma}(E) + \delta_{ik}\left[1 - \widetilde{\gamma}(E)\right]\}}, \end{aligned} \tag{3.65}$$

with

$$\widetilde{\beta}^{(*)i}(E) = \frac{v^i}{w^i(E)} \quad ; \; \widetilde{\gamma}(E) = \left(1 - \widetilde{\boldsymbol{\beta}}^{(*)}(E) * \widetilde{\boldsymbol{\beta}}^{(*)}(E)\right)^{-1/2}. \tag{3.66}$$

It is now an easy task to check the truly maximal character of the two velocities. Indeed, if $v^{i\prime} = u^i(E)$, one gets, from (3.63)

$$v^i = \frac{u^i(E) + v^i}{1 + \frac{v^i}{u^i(E)}} = u^i(E), \tag{3.67}$$

whereas, for $v^{i\prime} = w^i(E)$, (3.65) yields

$$v^i = \frac{w^i(E) + v^i}{1 + \frac{(b_i(E))^2 v^i}{w^i(E)}} \neq w^i(E). \tag{3.68}$$

We can therefore conclude, on a physical basis, that $u$ *is the maximal, invariant causal velocity in DSR, and it plays in the deformed Minkowski space* $\widetilde{M}$ *the role of the light speed in standard SR.*

It is now easy to see why – although approach (II) looks at first sight more rigorous mathematically, because it permits to connect the peculiar features of spatial anisotropy of DSR to the deformed product $*$, "naturally induced" from the metric of $\widetilde{M}(E)$ – actually it is approach (I) which yields the physically relevant result. Indeed, the velocity $\mathbf{u}$ is just defined as $d\mathbf{x}/dt$, and it therefore represents the physically measured velocity, for a particle moving in the usual, physical Euclidean 3D space. On the other hand, this result clearly shows that the space anisotropy introduced by the deformed metric is not a mere mathematical artifact, but it reflects itself in the physical properties (imposed by the interaction involved) of the phenomenon described by the deformed space–time.

It is actually a strict consequence of the spatial anisotropy of the space–time region considered that, in a given Minkowski space with deformed metric, there exist infinitely many different, maximal causal velocities, corresponding to the different possible directions of motion (although, of course, only three of them are independent).

Let us remark that $\mathbf{u}$ depends explicitly on the metric parameters $b_\mu$, which are a priori different for every physical system. However, since the deformation of the metric represents, on average, the effects of the nonlocal interactions involved, it is expected that physical systems with the same kind of interactions (besides the electromagnetic ones) are described by metric parameters of the same order of magnitude (or, at least, this holds true for the ratio $b_0/b$). In this sense it is possible to refer to $u$ as a "speed of interaction," rather than "speed of the physical system" considered (of course, at the same energy scale). It is worth noticing that a similar result (namely, a "maximum attainable speed," a priori different for different physical processes) was also obtained by Coleman and Glashow [15], in the framework of a discussion of possible effects breaking Lorentz invariance (essentially on a local scale).

The comparison of the deformed boost expression (3.27) with the corresponding ones of the standard Lorentz boosts shows clearly that the transition from SR (based on $M$) to DSR (based on $\widetilde{M}$) is simply carried out by letting

$$\mathbf{u}_{\mathrm{SR}} = (c, c, c) \longrightarrow \mathbf{u}_{\mathrm{DSR}}(E) = \left( \frac{cb_0(E)}{b_1(E)}, \frac{cb_0(E)}{b_2(E)}, \frac{cb_0(E)}{b_3(E)} \right). \tag{3.69}$$

In other words, the difference between $M$ and $\widetilde{M}(E)$ (at least as far as the finite coordinate transformations are concerned) is completely embodied in the three-vector m.c.v. $\mathbf{u}$. Equation (3.69) amounts to a passage from an *absolute* (namely, interaction and energy *independent*) m.c.v. $u_{\mathrm{SR}} = c$ to a *relative* (i.e., interaction and energy *dependent*) m.c.v. $u_{\mathrm{DSR}} = u(E)$ in passing from SR to DSR. However, $u(E)$ remains invariant – for fixed energy values – under generalized Lorentz transformations from a given reference frame to another.

In the case of an isotropic 3D space (i.e., $b_1^2(E) = b_2^2(E) = b_3^2(E) \equiv b^2(E)$), the corresponding deformed metric is given by (2.14), (2.19), and one gets, for any component of $\mathbf{u}$:

$$u = \frac{b_0(E)}{b(E)}c. \tag{3.70}$$

In (3.70), the value of $u$ is parametrized in terms of $c$, and depends on the physical system (and its interactions). Moreover, it is

$$u \gtreqless c \Leftrightarrow \frac{b_0(E)}{b(E)} \gtreqless 1. \tag{3.71}$$

In other words, there may be maximal causal speeds either *subluminal* or *superluminal*, depending on the interaction considered.

An example is obtained by applying (3.70) to the (spatially isotropic) Einstein metric (2.21). In this case,

$$b_0 = \sqrt{1 + \frac{2\phi}{c^2}},$$

and therefore the maximal causal speed is given (in the limit of weak gravitational field $\phi$, so that $\phi/c^2 \ll 1$) by

$$u = \frac{b_0(E)}{b(E)}c = c\sqrt{1 + \frac{2\phi}{c^2}} \backsimeq c\left(1 + \frac{\phi}{c^2}\right) = c\left(1 - \frac{2Gm}{c^2 r}\right), \tag{3.72}$$

with $G$ being the Newton constant. This amounts to a modified light speed $u < c$. Such a modification of the light speed in a gravitational field can be thought of as propagation in a medium endowed with a refractive index $n = \left(1 - \dfrac{2Gm}{c^2 r}\right)^{-1}$. We recall that such an interpretation was pioneered by Levi-Civita [16].

The light-like world-lines in $\widetilde{M}$, given by (3.8), read in the isotropic case

$$\begin{aligned} d\tilde{s}^2(E) = 0 &\Leftrightarrow b_0^2(E)c^2 dt^2 - b^2(E)\left[(dx^1)^2 + (dx^2)^2 + (dx^3)^2\right] = 0 \\ &\Leftrightarrow u^2 dt^2 - dx^2 - dy^2 - dz^2 = 0. \end{aligned} \tag{3.73}$$

Such an equation represents a *Deformed Light Cone.* Precisely, it is a *Super-Light Cone*[8] or a *Sub-Light Cone* according to $u \gtrless 1$. It constitutes the counterpart of the usual light cone in standard SR, and provides a geometrical representation of the maximal causal role played by $u$ for the interaction considered.

The maximal causal speed $u$ can be therefore interpreted, from a physical standpoint, as the speed of the quanta of the interaction which requires a representation in terms of a deformed Minkowski space. Since these quanta are associated to the Deformed Light Cone in $\widetilde{M}$ (see (3.8)), they must be zero-mass particles (with respect to the interaction considered), in analogy with photons (with respect to the e.m. interaction) in the usual SR.

Let us clarify the latter statement. The carriers of a given interaction propagating with the speed $u$ typical of that interaction *are expected to be strictly massless only inside the space whose metric is determined by the interaction considered.* A priori, nothing forbids that such "deformed photons" may acquire a nonvanishing mass in a deformed Minkowski space related to a different interaction.

This might be the case of the massive bosons $W^+$, $W^-$ and $Z^0$, carriers of the weak interaction, which would therefore be massless in the space $\widetilde{M}(g_{DSR,weak}(E))$ related to the weak interaction, but would acquire a mass when considered in the standard Minkowski space $M$ of SR (that, as already stressed, is strictly connected to the electromagnetic interaction, ruling the operation of the measuring devices). In this framework, therefore, it is not necessary to postulate a "symmetry breaking" mechanism (like the Goldstone one in gauge theories) to let particles acquire mass.[9] Mass itself would assume a *relative nature*, related not only to the interaction concerned, but also to the metric background where one measures the energy of the physical system considered. This can be seen if one takes into account the fact in general, for relativistic particles, mass is the invariant norm of 4-momentum, and what is usually measured is not the value of such an invariant, but of the related energy. It is possible indeed, in this framework, to give a geometrical meaning to the electron mass, and relate it to the breakdown of local Lorentz invariance [6].

### *3.3.7 DSR and Lorentzian Relativity*

The exposition of the foundations and properties of DSR we did up to now allow us to establish some connections between DSR and *Lorentzian Relativity* (LR).

---

[8]The term "Super Light Cone" was coined by D. Mugnai, in connection with the superluminal propagation of X-waves.

[9]On the contrary, if one could build up measuring devices based on interactions different from the e.m. one, the photon might acquire a mass with respect to such a non-e.m. background.

Let us recall that LR is the pre-Einsteinian version of relativity theory, according to Lorentz [17] and Poincaré [18]. It represents an evolution of the ether theory, due to Lorentz himself (LET). Like Einsteinian Relativity (ER), the Lorentzian one is based on the relativity principle, and has the Lorentz transformations as its main mathematical tools. However, these latter have a different meaning in the two frameworks. Actually, in SR Lorentz transformations involve time, space and mass, whereas in LR they affect only matter, i.e., the clocks and the meter sticks used to measure time, space and momentum. This has to be compared with the effects of temperature, whose increase causes a clock to slow and a meter stick to increase its length, yet this does not mean that temperature affects time or space [19]. In LR, unlike SR, time and space are simply dimensions (concepts), and cannot be changed by motion. Larmor time dilation and Lorentz–Fitzgerald length contraction in LR are therefore *real effects,* not kinematical appearances. As a matter of fact, all the experiments supporting SR, being based on Lorentz transformations, do support LR, too.

To the present purposes, let us stress the following two basic points of departure of LR from standard Einstein SR:

(1) In LR the speed of light is not invariant, but depends on the observer.

(2) In SR, all motions are relative, and no preferred frame exists. In LR, the motion is relative to a preferred reference frame.

The preferred frame of reference of LR was originally the ether. Such a point of view subsequently changed. Let us recall that, in the last twenty years, there was a revival of the problem of the existence of an absolute frame $\Sigma_0$ [20]. Possible candidates for $\Sigma_0$ are:

(a) The frame where the 2.7 $^\circ K$ background thermal radiation is isotropic for all the velocities of light

(b) The Hubble frame, where an observer would see all galaxies receding away with the Hubble expansion velocity

(c) The frame tied to the moving arm of our Galaxy

(d) The frame of the stochastic background gravitational radiation

A reanalysis of classical and modern ether-drift experiments has been also carried out [21], seemingly showing positive evidence for such a preferred frame on the basis of the old data by Michelson, Morley and Miller.

However, in the further developments of LR [22] the absolute reference frame is identified with the local gravitational field (as pioneered in [23]), which is of course a different frame from place to place. The possible existence of a preferred frame is nothing, in today language, but the possibility of breakdown of Lorentz invariance(see Chap. 10).

It is easy to see that the earlier two features of LR, namely, the non-invariance of light speed and the existence of a preferred frame, are also present in DSR.

The first point is evident from the discussion of Sect. 3.2 and Sect. 3.3.6. The maximal causal speed $u_{\text{int.}}$ depends on the interaction considered; within a given interaction, the speed varies with the energy of the process, and is a priori different for different space directions (see (3.35)). It follows that there is no longer a unique invariant velocity, exactly as in LR.

The local gravitational field, which is the substitute of ether as preferred frame in the modern view to LR, is nothing but a Lorentz frame, where space–time is Minkowskian. Let us recall that the idea of the flat Minkowski space as replacing the ether can be traced back to Einstein himself, who, after abandoning the idea of ether in the old sense, identified it, within General Relativity, as a substratum without mechanical and kinematical properties, but able to codetermine mechanical and electromagnetic events.

We have stressed more and more (see e.g., Sect. 2.2) the leading role played by Minkowski space in DSR. Indeed, all the physical measurements, in particular that of the energy which determines the metric, and therefore the space–time deformation, are carried out in Minkowskian conditions *via* the electromagnetic interaction. Every local deformation of space–time described by the DSR formalism always implies an underlying Minkowskian frame of reference, which, in this sense, does play the role of (local) preferred frame.

Finally, let us recall that in Lorentzian Relativity it is understood that Lorentz transformations only apply to electromagnetic phenomena (so classical Galilei transformations still hold for mechanical laws). In other words, different coordinate transformations correspond to different classes of physical phenomena (see the comments to the second postulate of SR, Sect. 1.2). This is exactly what happens in DSR.

Due to these common features of DSR and LR, in the following physical phenomena at variance with Special Relativity, which can be ascribed to them, will be just called *Lorentzian effects*.

## 3.4 Kinematics and Wave Propagation in a Deformed Minkowski Space

### *3.4.1 Dynamic Definition of Proper Time*

#### *Proper Time in SR*

We recall that, in general, a definition of (infinitesimal) proper time can be given, in SR, as the space–time path parameter entering the time-like, unit geodesic equation

$$\mathrm{d}\tau : g_{\mu\nu}\frac{\mathrm{d}x^\mu}{c\mathrm{d}\tau}\frac{\mathrm{d}x^\nu}{c\mathrm{d}\tau} = 1 \Longleftrightarrow \mathrm{d}\tau^2 = \frac{\mathrm{d}s^2}{c^2}. \tag{3.74}$$

Notice that such a definition holds true irrespective of the number of dimensions and of the nature of the metric tensor $g$.

Equation (3.74) can be interpreted as a normalization condition on the four-velocity $V^\mu \equiv \mathrm{d}x^\mu/\mathrm{d}\tau$:

$$g_{\mu\nu}V^\mu V^\nu = c^2. \tag{3.75}$$

Since $\mathrm{d}\tau$ is an invariant, it is possible to choose a suitable inertial frame in order to simplify its expression. As is well known, one takes the rest frame of the particle, i.e.,

$$\mathrm{d}x^1 = \mathrm{d}x^2 = \mathrm{d}x^3 = 0 \left(\text{SR natural frame} : \left(x^0, x^1, x^2, x^3\right)_{\text{SR,nat}} = \left(x^0, \overline{x^1}, \overline{x^2}, \overline{x^3}\right)\right).$$

Therefore

$$\mathrm{d}\tau_{\text{SR}} = \mathrm{d}t_{\text{nat}} = \mathrm{d}t_0 \tag{3.76}$$

(where we denoted by $t_0$ the time $t_{\text{nat}}$ in the natural frame) or, in integral form

$$\tau_{\text{SR}} = t_0. \tag{3.77}$$

Such a definition of proper time in SR must be consistent with the dynamics in Minkowski space, namely with the geodetic equations

$$\frac{\mathrm{d}^2x^\mu(\tau_{\text{SR}})}{\mathrm{d}\tau_{\text{SR}}^2} = 0 \Longleftrightarrow x^\mu(\tau_{\text{SR}}) = \alpha_{\mu1}\tau_{\text{SR}} + \alpha_{\mu2} \tag{3.78}$$

with $\alpha_{\mu1}$, $\alpha_{\mu2}$ real constants. The expressions (3.76), (3.77) of the proper time in SR are therefore compatible with the following class of solutions of the geodesic equations

$$x^\mu(\tau_{\text{SR}}) \overset{\text{ESC on}}{=} \delta^{\mu0}\alpha_{\mu1}\tau_{\text{SR}} + \alpha_{\mu2}, \quad \alpha_{01} = c, \quad \alpha_{02} = 0, \quad \alpha_{i2} = \overline{x^i}, i=1,2,3. \tag{3.79}$$

Proper Time in DSR

The same procedure can be followed to define proper time in DSR. We have

$$\begin{aligned}(\mathrm{d}\tau_{\text{DSR}}(E))^2 &= \frac{1}{c^2}(\mathrm{d}s_{\text{DSR}}(E))^2 \\ &= \frac{1}{c^2}\Big[b_0^2(E)c^2\mathrm{d}t^2 - b_1^2(E)\left(\mathrm{d}x^1\right)^2 \\ &\quad -b_2^2(E)\left(\mathrm{d}x^2\right)^2 - b_3^2(E)\left(\mathrm{d}x^3\right)^2\Big].\end{aligned} \tag{3.80}$$

The natural topical reference frame corresponds in this case to the frame where the particle is at rest *with fixed energy* $\overline{E}$, namely with $\mathrm{d}x^1 = \mathrm{d}x^2 = \mathrm{d}x^3 = \mathrm{d}E = 0$, so that the *DSR natural frame* is characterized by

$$\left(x^0, x^1, x^2, x^3\right)_{\text{DSR,nat}} = \left(x^0, \overline{x^1}, \overline{x^2}, \overline{x^3}\right)_{E=\overline{E}}. \tag{3.81}$$

Then

$$\left(\mathrm{d}\tau_{\text{DSR}}(\overline{E})\right)^2 = \frac{1}{c^2}\left(\mathrm{d}s_{\text{DSR}}(\overline{E})\right)^2\Big|_{\text{nat}} = b_0^2(\overline{E}))\mathrm{d}t_0^2. \tag{3.82}$$

Finally, omitting the dependence on $\overline{E}$, one gets for the infinitesimal proper time in DSR

$$\mathrm{d}\tau_{\text{DSR}} = b_0 \mathrm{d}t_0 \tag{3.83}$$

or, in finite form:

$$\tau_{\text{DSR}} = b_0 t_0. \tag{3.84}$$

Such relations are analogous to those between proper time and coordinate time found in General Relativity ($\mathrm{d}\tau_{\text{GR}} = \sqrt{g_{00}}\mathrm{d}t$).[10] In the DSR case, too, the proper time *does not coincide* with the time measured by the local observer in the particle frame.[11] Like in GR, therefore, one has to distinguish between the *real (proper) time* $\tau$ and the coordinate (or *universe*) time $t$. As is well known, such a distinction is fundamental, within GR, for the analysis of gravitational phenomena (like gravitational collapse). Something analogous occurs in the DSR framework for interactions described by asynchronous metrics (like the strong one: see Sect. 4.1), for which this fact may have deep physical implications.

As for SR, the definition of proper time must be consistent with the dynamics of free particles. The geodesic equations in the deformed Minkowski space $\widetilde{M}$ are:

$$\frac{\mathrm{d}^2 x^\mu(\tau_{\text{DSR}})}{\mathrm{d}\tau_{\text{DSR}}^2} = 0 \Longleftrightarrow x^\mu(\tau_{\text{DSR}}) = \alpha_{\mu 1}\tau_{\text{DSR}} + \alpha_{\mu 2} \tag{3.85}$$

($\alpha_{\mu 1}$, $\alpha_{\mu 2}$ real). Therefore, at fixed energy ($E = \overline{E}$), the compatibility condition in DSR between proper time definition and dynamics is expressed by

$$x^\mu(\tau_{\text{DSR}}) \overset{\text{ESC on}}{=} \delta^{\mu 0}\alpha_{\mu 1}\tau_{\text{DSR}} + \alpha_{\mu 2}, \quad \alpha_{01} = \frac{c}{b_0}, \quad \alpha_{02} = 0, \quad \alpha_{i2} = \overline{x^i}, \quad i = 1, 2, 3, \tag{3.86}$$

that can be regarded as a dynamic definition of natural, topical reference frame.

---

[10] Indeed, since $g_{\text{DSR}00} = b_0^2$, the general-relativistic relation for $\tau$ becomes exactly the DSR relation (3.83).

[11] An exception is provided by the case of the symmetrizing transformation $x^0 \equiv cb_0(E)t$. Then, one gets $g_{\text{DSR}00} = 1$, so that $\tau = t$ (namely proper time coincides with coordinate time).

### 3.4.2 Deformed Relativistic Kinematics

From the knowledge of the generalized Lorentz transformations it is easy to derive the main kinematical and dynamic laws valid in DSR [6]. In this section, we shall merely list those which are useful to phenomenological purposes.

Velocity Composition Law (cf. (3.62))

$$V_{\text{tot}} = \frac{v_1 + v_2}{1 + \dfrac{v_1 v_2}{u^2}}, \tag{3.87}$$

which obviously for, say, $v_1 = u$ yields $V_{\text{tot}} = u$.

If the condition of spatial isotropy is given up, the composition law for motion, say, along the $x_k$-axis, becomes

$$V = \frac{v_1 + v_2}{1 + \dfrac{v_1 v_2}{u_k^2}}; \quad u_k = \frac{cb_0}{b_k} \tag{3.88}$$

and we recover the result (already derived in Sect. 3.3.6) that the invariant velocity is

$$u_k = \frac{cb_0}{b_k}. \tag{3.89}$$

Time dilation

$$\Delta t = \tilde{\gamma}(E)\Delta t_0; \tag{3.90}$$

Length contraction

$$\Delta L = \tilde{\gamma}^{-1}(E)\Delta L_0; \tag{3.91}$$

Four-velocity

$$V^\mu(E) = \frac{\mathrm{d}x^\mu}{\mathrm{d}\tau(E)} = \tilde{\gamma}(E)\frac{\mathrm{d}x^\mu}{b_0(E)\mathrm{d}t}, \tag{3.92}$$

(where (3.83), (3.90) have been taken into account). One gets explicitly, for the spatial part of $V^\mu$:

$$V^k(E) = \tilde{\gamma}(E)\frac{v^k}{b_0(E)} = \tilde{\gamma}(E)\tilde{v}^k, \tag{3.93}$$

with $\upsilon^k = dx^k/dt$ being the components of the standard three-vector velocity $\mathbf{v}$, and we introduced the *deformed 3-velocity*

$$\tilde{\upsilon}^k \equiv \frac{\upsilon^k}{b_0(E)}. \tag{3.94}$$

Notice that such a definition is analogous to that used in General Relativity (in agreement with the definition of proper time in DSR). As to the time component $V^0(E)$, its expression depends on the dimensional-conversion factor used to define $x^0$. We get:

(a) $x^0 = ct$:

$$V^0(E) = \tilde{\gamma}(E)\frac{c}{b_0(E)}; \tag{3.95}$$

(b) $x^0 = cb_0(E)t = w^i t$ (where $\mathbf{w}$ is the isotropic m.c.v. for the interaction considered):

$$V^0(E) = \tilde{\gamma}(E)c; \tag{3.96}$$

(c)

$$x^0 = u^i(E)t = c\frac{b_0(E)}{b(E)}t,$$

(where $\mathbf{u}$ is the anisotropic m.c.v. in the case of spatial isotropy):

$$V^0(E) = \tilde{\gamma}(E)\frac{u^i(E)}{b_0(E)}. \tag{3.97}$$

Therefore, the generalized expression of the momentum four-vector is

$$p^\mu(E) = mV^\mu(E) = \frac{m\tilde{\gamma}(E)}{b_0(E)}\begin{cases}(c, \mathbf{v})\\ (w^i(E), \mathbf{v})\\ (u^i(E), \mathbf{v})\end{cases}, \tag{3.98}$$

where $m$ is the rest mass.

As is well known, in SR the energy is the time component of $p^\mu$. However, in the general-relativistic case, the conserved quantity is the *covariant* component of the four-momentum vector.[12] Accordingly, in the general case, we take as deformed relativistic energy, for a particle subjected to a given

[12] Indeed, the energy for a particle in a stationary gravitational field (covariant component $p_0$ of the four-momentum) is

$$E_0 = mc^2\gamma\sqrt{g_{00}}, \tag{*}$$

where the speed $v$ in $\gamma$ is defined with respect to the proper time $\tau$. In the Newtonian limit of weak field, $g_{00} \cong 1 + 2\phi/c^2$ (with $\phi$ being the gravitational potential), so that

$$E_0 \simeq mc^2\gamma\left(1 + \frac{\phi}{c^2}\right),$$

interaction and moving along $\widehat{x^i}$, the quantity:

$$E_{\rm DEF} = mc\widetilde{\gamma}(E)b_0(E)\left\{\begin{array}{c} c \\ w^i(E) \\ u^i(E) \end{array}\right. . \tag{3.99}$$

Let us stress that $E_{\rm DEF}$ is *the energy the system under consideration possess in the deformed space–time, and is a consequence of the deformation of the metric.* It must not be confused with $E$, that is instead *the interaction energy* measured in Minkowskian conditions. The difference between the two energies can be understood by considering the internal dynamics (of geometrical origin) inherent in the mathematical structure of the space $\widetilde{M}$ (see Part II)[13].

Let us notice that (3.99) implies *a generalized dispersion relation* between energy and momentum in the deformed space–time. We have in fact

$$E_{\rm DEF}^2 = \frac{1}{b_0^2(E)}\left[m^2c^4 - b_i^2(E)(cp^i)^2\right]. \tag{3.100}$$

As is well known, generalized dispersion relations of such a kind are a characteristic feature of theories allowing for violation of local Lorentz invariance [24]. In particular, they arise in the context of multidimensional space–time theories [25]. As we shall see in Parts IV and V, DSR admits a natural embedding in a 5D Riemannian space. Equation (3.100) constitutes, in a sense, a manifestation of this for now "hidden" extra dimension.

Lastly, let us consider a plane wave propagating with speed $u$ (e.g., in the $xy$ plane, at angles $\theta$, $\theta'$ in frames $K$, $K'$) with dispersion relation $u = \lambda\nu = \lambda'\nu'$, where $\nu$, $\nu'$ are the wave frequencies in $K$, $K'$. Applying the DLT, it is easy to get the following laws:

Doppler effect

$$\nu = \widetilde{\gamma}(E)\nu'(1 + \widetilde{\beta}(E)\cos\theta'); \tag{3.101}$$

Aberration law

$$tg\theta = \frac{\sin\theta'}{\widetilde{\gamma}(E)(\widetilde{\beta}(E) + \cos\theta')}. \tag{3.102}$$

---

which is fully analogous to the expression of the relativistic energy of a charged particle in a constant electromegnetic field. Moreover, in the nonrelativistic limit, one gets

$$E_0 \simeq mc^2 + \frac{1}{2}mv^2 + m\phi,$$

just showing that (*) is the correct general-relativistic definition of the energy for a particle in a gravitational field.

[13] Precisely, in the internal fields arising from its structure of Generalized Lagrange Space (see Chap. 9).

TABLE 3.1. Kinematical laws in standard and deformed Minkowski space

| Minkowski space | deformed Minkowski space |
|---|---|
| $V_{\text{tot}} = \dfrac{v_1 + v_2}{1 + \dfrac{v_1 v_2}{c^2}}$ | $V_{\text{tot}} = \dfrac{v_1 + v_2}{1 + \left(\dfrac{b}{b_0}\right)^2 \dfrac{v_1 v_2}{c^2}}$ |
| $\Delta t = \dfrac{\Delta t_0}{\left(1 - \dfrac{v^2}{c^2}\right)^{1/2}}$ | $\Delta t = \dfrac{\Delta t_0}{\left[1 - \left(\dfrac{b}{b_0}\right)^2 \dfrac{v^2}{c^2}\right]^{1/2}}$ |
| $\Delta L = \Delta L_0 \left(1 - \dfrac{v^2}{c^2}\right)^{1/2}$ | $\Delta L = \Delta L_0 \left[1 - \left(\dfrac{b}{b_0}\right)^2 \dfrac{v^2}{c^2}\right]^{1/2}$ |

We want now to provide a comparison between the main kinematical laws in the usual Minkowski space $M$ and in the deformed one $\widetilde{M}$ (in the hypothesis of spatial isotropy), because their different behaviors may help one to understand the peculiar features of leptonic, hadronic (and gravitational) interactions with respect to the electromagnetic one. Such laws are summed up in Table 3.1, where the maximal speed $u$ has been expressed in terms of $c$, in order to emphasize the dependence of the deformed laws on the parameter ratio $b/b_0$ and exhibit their scale invariance.

In the limiting case $v = c$, one gets explicitly

$$v_1 = c \Longrightarrow V_{\text{tot}} = \frac{c + v_2}{1 + \left(\dfrac{b}{b_o}\right)^2 \dfrac{v_2}{c}}; \tag{3.103}$$

$$v = c \Longrightarrow \Delta t = \frac{\Delta t_0}{\left[1 - \left(\dfrac{b}{b_o}\right)^2\right]^{1/2}}; \tag{3.104}$$

$$v = c \Longrightarrow \Delta L = \Delta L_0 \left[1 - \left(\frac{b}{b_0}\right)^2\right]^{1/2}. \tag{3.105}$$

Remember that, in this framework, $c$ has lost its meaning of maximal causal speed, by preserving the mere role of maximal causal speed for electromagnetic phenomena in $M$.

To the purpose of experimental verification, it is worth to express the deformed kinematical laws of time dilation and length contraction for a particle of rest mass $m$ in terms of the *usual* energy $E$. Clearly, for $E \gg mc^2$, $E$ can be considered the total energy of the particle, measured by electromagnetic methods in the usual Minkowski space. We report such laws in Table 3.2 (in comparison with the standard, Einsteinian ones):

TABLE 3.2. Time dilation and length contraction as functions of energy

| Minkowski space | deformed Minkowski space |
|---|---|
| $\Delta t = \Delta t_0 \dfrac{E}{m}$ | $\Delta t = \Delta t_0 \left[1 - \left(\dfrac{b}{b_o}\right)^2 + \left(\dfrac{b}{b_o}\right)^2 \left(\dfrac{m}{E}\right)^2\right]^{-1/2}$ |
| $\Delta L = \Delta L_0 \dfrac{m}{E}$ | $\Delta L = \Delta L_0 \left[1 - \left(\dfrac{b}{b_o}\right)^2 + \left(\dfrac{b}{b_o}\right)^2 \left(\dfrac{m}{E}\right)^2\right]^{1/2}$ |

It is easily seen that, in the case of the time-dilation law, the main difference is the loss of linearity in the dependence on the energy of the deformed law, as compared to the Minkowskian one. Such a different behavior is therefore a clear signature of the presence of nonlocal effects in the interaction considered, or of LI breakdown.

### *3.4.3 Wave Propagation in a Deformed Space–Time*

*Deformed Helmholtz Equation*

We want now to approach the problem of wave propagation in a deformed Minkowski space–time produced by an interaction whose metric has a threshold behavior in energy,[14] namely it is non-Minkowskian for, say, $E < E_0$ [6]. To this end, let us introduce the generalized D'Alembert operator $\tilde{\Box}$, defined by means of the scalar product $*$ in $\widetilde{M}$ (see (2.5)):

$$\tilde{\Box} \equiv \partial * \partial = g_{\mu\nu,\mathrm{DSR}}\partial^\mu \partial^\nu = \frac{b_0^2}{c^2}\partial_t^2 - (b_1^2\partial_x^2 + b_2^2\partial_y^2 + b_3^2\partial_z^2). \tag{3.106}$$

Therefore, the deformed Helmholtz–D'Alembert wave equation is given by

$$\tilde{\Box} f = 0, \tag{3.107}$$

with $f$ being any component of the field associated to the wave considered. For instance, the field of such a wave propagating in the Minkowski space $\widetilde{M}$ can be written as:

$$\mathbf{f}(x) = \mathbf{A}(\mathbf{x})\mathrm{e}^{\mathrm{i}k*x}, \tag{3.108}$$

where $k$ is the wavevector and $\mathrm{e}^{\mathrm{i}k*x}$ is the generalized phase.

By assuming a spatially isotropic deformed metric (see (2.19)), in the corresponding deformed space–time the generalized phase takes the "Minkowskian-like" form $\mathrm{e}^{\mathrm{i}\tilde{k}\cdot x}$ (where the dot denotes the usual scalar product in the Minkowski space), with

$$\tilde{k}^\mu = \left(\frac{2\pi\nu}{c}, \tilde{k}_x, \tilde{k}_y, \tilde{k}_z\right) \tag{3.109}$$

[14] As it will be seen in Chap. 4, this is just the case of the phenomenological metrics of all the four fundamental interactions.

and $\nu$ is the frequency measured in the ordinary Minkowski space–time. Then, (3.108) becomes

$$\mathbf{f}(x) = \mathbf{A}(\mathbf{x})\mathrm{e}^{\mathrm{i}\tilde{k}\cdot x}. \tag{3.110}$$

A wave propagating in a deformed Minkowski space–time (whatever the interaction ruling it) will be referred to in the following as a *Lorentzian wave*.

*Deformation Tensor*

A scalar wave propagating in a finite region (of size $L$) of a deformed space–time (along the $z$-direction) can be expanded in a Fourier series as

$$f(z) = \sum_{n=-\infty}^{\infty} c_n \mathrm{e}^{\mathrm{i}n2\pi z/L}. \tag{3.111}$$

In particular, for an evanescent wave $\mathrm{e}^{-\chi z}$, with

$$\chi = \frac{2\pi}{c}\sqrt{\nu_c^2 - \nu^2} \tag{3.112}$$

(where $\nu_c$ is a critical frequency), one has (for $\chi L \gg 1$)

$$c_n = \frac{(1/L)}{\chi + in\frac{2\pi}{L}}. \tag{3.113}$$

It is easily seen that each Fourier component of the Lorentzian wave propagates in a different deformed Minkowski space–time. This is clearly related to the energy (and momentum) dependence of the parameters of the deformed metric. If $g_{\mu\nu,\mathrm{DSR}}^{(n)}$ is the deformed metric "seen" by the $n$th Fourier component of the wave, with metric coefficients $b_\mu^{(n)}$, we can build an *effective metric tensor* $g_{\mu\nu,\mathrm{DSR}}$ for the wave as follows:[15]

$$g_{\mu\nu,\mathrm{DSR}}(c_n) = \frac{\sum_n |c_n|^2 g_{\mu\nu,\mathrm{DSR}}^{(n)}}{\sum_n |c_n|^2}. \tag{3.114}$$

Clearly, outside the region where the space–time is deformed, all the Fourier waves propagate in a Minkowskian space–time, and definition (3.114) reduces to the usual Minkowski metric tensor $g_{\mu\nu}$. In fact, in such a region we are in full Minkowskian conditions (according to the assumption we made, this means that the energy of the process is higher than the threshold energy, $E > E_0$), i.e., $b_\mu^{(n)2} = 1$, $\mu = 0, 1, 2, 3$, and therefore $g_{\mu\nu\mathrm{DSR}}^{(n)} = g_{\mu\nu} \forall n$.

[15] Analogous results hold in the case of a growing wave, or when both an evanescent and a growing wave are present in the space–time deformation region.

Notice that tensor (3.114) is analogous to the Cauchy stress tensor of a continuous medium. In fact, let us consider, in orthogonal Cartesian coordinates, an infinitesimal tetrahedron with edges parallel to the coordinate axes and the oblique face $S$ opposite to the vertex $O$, origin of the Cartesian frame. If the tetrahedron is a part of a continuous body, the stress vector across $S$ in the point $O$ is given by

$$\boldsymbol{\phi}_{\mathbf{a}}(O) = \frac{\sum_i \boldsymbol{\phi}_i(O) a_i}{\sum_i |a_i|^2}, \tag{3.115}$$

where $\mathbf{a}$ is a vector normal to $S$ and $\boldsymbol{\phi}_i(O)$ $(i = 1, 2, 3)$ is the stress vector on the face of the tetrahedron orthogonal to the i-th axis. The nine components of the three vectors $\boldsymbol{\phi}_i(O)$ do just constitute the rank-two, symmetric Cauchy tensor.

The tensor $g_{\mathrm{DSR}}$ can be therefore regarded as the average tensor representing the space–time deformation inside the deformed region (corresponding to the energy $E < E_0$) globally "seen" by the wave (3.110). So, we can name it *average tensor of the space–time deformation (related to the interaction considered)*, $g_{\mathrm{DSR,int.}}$. It can be stated in full generality that the Minkowski space is always subjected to a stress, whenever crossed by a wavepacket. Such a stress is related to the deformation of the space–time, which may be described by the tensor $g_{\mathrm{DSR}} = g_{\mathrm{SR}}$ (*ineffectual deformation*) or by a tensor $g_{\mathrm{DSR}} \neq g_{\mathrm{SR}}$ (*effectual deformation*). The two cases $g_{\mathrm{DSR}} = g_{\mathrm{SR}}$, $g_{\mathrm{DSR}} \neq g_{\mathrm{SR}}$ are obviously determined by the interaction ruling the wavepacket propagation and by the energy of the wavepacket components.

It was shown that a description in terms of propagation in a deformed space–time holds for the tunneling of particles and photons through a barrier. In such a case, the propagation is known to be superluminal (the so-called *Hartmann–Fletcher effect* [26,27], namely the tunneling time becomes independent of the barrier width $d$ for sufficiently large $d$). The approach to faster-than-light propagation in terms of a deformation tensor is similar, in some respects, to that where superluminal propagation (e.g., of light between parallel mirrors) is connected to vacuum effects [28]. In this case, the influence of the (structured) vacuum is described in an effective way in terms of a refractive index $\mathcal{N}$ (as pioneered by Sommerfeld). The speed of propagation of electromagnetic signals, $u_{\mathrm{e.m.}}$, is related to the "speed of light" parameter $c$ (which appears in standard Lorentz transformations) by $u_{\mathrm{e.m.}} = c/\mathcal{N}$ (of course, this allows one to single out a preferred reference frame, namely the one where light propagates isotropically with speed $u_{\mathrm{e.m.}}$: see Sect. 3.3.7). Something analogous happens in General Relativity, too: many gravitational phenomena (like the bending of light rays near a massive body), usually described in terms of the curved (Riemannian) structure of space–time, can be treated by considering (in an Euclideanspace) the vacuum as a polarizable medium endowed with an effective refraction index (see (3.72)), in which light propagates [16,29]. In some cases, such a propagation – due to the influence of the gravitational vacuum – turns out to be superluminal (the refractive index is less than one, $\mathcal{N} < 1$) [30].[16] The deformed metric approach can be therefore regarded as *dual* to the general relativistic one, in which the space–time curvature for electromagnetic signals is replaced by a refractive index. In the DSR formalism, the vacuum or nonlocal effects which affect propagation inside the barrier are described in terms of a space–time deformation (and the role of the refractive index is played by the deformation tensor).

## 3.5 Field Deformation

We want now to show that the deformation of space–time, expressed by the metric $g_{\text{DSR}}$ (2.17), does affect also the external fields applied to the physical system considered.

Let us consider for instance the case of a physical process ruled by the electromagnetic interaction. Therefore, the Minkowski space $M$ is endowed with the electromagnetic tensor $F_{\mu\nu}(x)$ (external e.m. field) acting on the system. Of course $F^{\mu}_{\nu}(x) = g^{\mu\rho}_{\text{SR}} F_{\rho\nu}(x)$.

In the deformed Minkowski space $\widetilde{M}$, the covariant components of the electromagnetic tensor read (ESC on)

$$\widetilde{F}_{\mu\nu} = g_{\mu\rho\text{DSR}} F^{\rho}_{\nu} = g_{\mu\rho\text{DSR}} g^{\mu\sigma}_{\text{SR}} F_{\sigma\nu}, \tag{3.116}$$

where (ESC off)

$$(g_{\mu\rho\text{DSR}} g^{\mu\sigma}_{\text{SR}}) = \text{diag}(b_0^2, b_1^2, b_2^2, b_3^2) = (b_\sigma^2 \delta^\sigma_\rho). \tag{3.117}$$

We have therefore

$$\widetilde{F}_{0\nu} = b_0^2 F_{0\nu}; \quad \widetilde{F}_{1\nu} = b_1^2 F_{1\nu}; \quad \widetilde{F}_{2\nu} = b_2^2 F_{2\nu}; \quad \widetilde{F}_{3\nu} = b_3^2 F_{3\nu}, \tag{3.118}$$

or (ESC off)

$$\widetilde{F}_{\mu\nu} = b_\mu^2 F_{\mu\nu}, \qquad \mu, \nu = 0, 1, 2, 3. \tag{3.119}$$

It follows that the tensor $\widetilde{F}_{\mu\nu}$ is not antisymmetric:

$$\widetilde{F}_{\mu\nu} \neq -\widetilde{F}_{\nu\mu}. \tag{3.120}$$

[16] Let us recall that the evidence for such a superluminality in the gravitational field – strictly related to the dependence of the light speed $c$ on the gravitational potential $\phi$ – has been recognized by Einstein himself, in the early years of General Relativity. However, he later preferred to adopt the point of view by Nordström, in which the dependence of $c$ on $\phi$ (and therefore the superluminality) is removed and rather ascribed to the mass.

The result shown here for the electromagnetic interaction can be generalized to other fundamental interactions described by tensor fields.

On account of the well-known identification

$$\widetilde{F}_{0i} = \widetilde{E_i}, \quad \widetilde{F}_{12} = -\widetilde{B_3}, \quad \widetilde{F}_{23} = -\widetilde{B_1}, \quad \widetilde{F}_{31} = -\widetilde{B_2} \tag{3.121}$$

(and analogously for $F_{\mu\nu}$), we can write, for the energy density $\widetilde{\mathcal{E}}$ of the deformed electromagnetic field:

$$\widetilde{\mathcal{E}} = \frac{\widetilde{\mathbf{E}}^2 + \widetilde{\mathbf{B}}^2}{8\pi} = \frac{b_0^4 \mathbf{E}^2 + b_1^4 B_3^2 + b_2^4 B_1^2 + b_3^4 B_2^2}{8\pi}, \tag{3.122}$$

to be compared with the standard expression for the e.m. field $\mathbf{E}$, $\mathbf{B}$:

$$\mathcal{E} = \frac{\mathbf{E}^2 + \mathbf{B}^2}{8\pi}. \tag{3.123}$$

*There is therefore a difference in the energy associated to the electromagnetic field in the deformed space–time region.* We have, for the energy density

$$\Delta\mathcal{E} = \mathcal{E} - \widetilde{\mathcal{E}}. \tag{3.124}$$

We can state that the difference $\Delta\mathcal{E}$ represents *the energy spent by the interaction in order to deform the space–time geometry.*

We can therefore conclude that *the deformation of space–time does affect the field itself that deforms the geometry of the space.* There is therefore a feedback between space and interaction which fully implements the Solidarity Principle.

# 4

# Metric Description of Interactions

## 4.1 Review of Phenomenological Metrics

We want now to review the results obtained for the deformed metrics, describing the four fundamental interactions – electromagnetic, weak, strong and gravitational –, from the phenomenological analysis of the experimental data. A more detailed treatment can be found in [6].

The explicit functional form of the metric

$$g_{\mathrm{DSR,int.}}(E) = \mathrm{diag}(b_{0,\mathrm{int.}}^2(E), -b_{1,\mathrm{int.}}^2(E), -b_{2,\mathrm{int.}}^2(E), -b_{3,\mathrm{int.}}^2(E)) \quad (4.1)$$

for the four interactions (in its *covariant* form) is as follows.

### *4.1.1 Electromagnetic Interaction*

The experiments considered were those on the superluminal propagation of e.m. waves in conducting waveguides with variable section (first observed at Cologne [31–33] and Florence [34,35]). The introduction, in this framework, of a deformed Minkowski space was motivated by ascribing the superluminal speed of the signals to some nonlocal e.m. effect, inside the narrower part of the waveguide, which can be described in terms of an effective deformation of space–time inside the barrier region. Since one is dealing with electromagnetic forces (which are usually described by the Minkowskian metric), it is possible to assume $b_0^2 = 1$ (this is also justified by the fact that all the relevant deformed quantities depend actually on the ratio $b/b_0$). Assuming moreover an isotropically deformed three-space ($b_1 = b_2 = b_3 = b$),[1] the fit to the experimental data yields

$$g_{\text{DSR,e.m.}}(E) = \text{diag}\left(1, -b^2_{\text{e.m.}}(E), -b^2_{\text{e.m.}}(E), -b^2_{\text{e.m.}}(E)\right); \quad (4.2)$$

$$\begin{aligned} b^2_{\text{e.m.}}(E) &= \begin{cases} (E/E_{0\text{e.m.}})^{1/3}, & 0 \leq E < E_{0\text{e.m.}} \\ 1, & E_{0\text{e.m.}} < E \end{cases} \\ &= 1 + \Theta(E_{0\text{e.m.}} - E)\left[\left(\frac{E}{E_{0\text{e.m.}}}\right)^{1/3} - 1\right], E > 0 \end{aligned} \quad (4.3)$$

(where $\Theta(x)$ is the Heaviside theta function). The threshold energy $E_{0\text{e.m.}}$ is the energy value at which the metric parameters attain a constant value, i.e., the metric becomes Minkowskian. The value obtained by the fit is

$$E_{0,\text{e.m.}} = (4.5 \pm 0.2)\ \mu\text{eV}. \quad (4.4)$$

Notice that the value obtained for $E_{0,\text{e.m.}}$ is of the order of the energy $E_{\text{coh}}$ corresponding to the coherence length of a photon for radio-optical waves (used in waveguide experiments). Indeed, for such waves it is $\nu \approx 100\,\text{MHz}$ (i.e., $10^{-8}$ s), and therefore $\lambda \approx 10^2$ cm so that

$$E_{\text{coh}} \equiv \frac{hc}{\lambda} \simeq 1\ \mu\text{eV}. \quad (4.5)$$

Moreover, it must be noted that, since – as stressed in Sect. 3.3.5 – the departure of the metric from the Minkowskian one is a signature of the Lorentz invariance breakdown (in its usual meaning), breaking LI in electromagnetic processes is expected to occur at *low* (not high, as commonly thought) energies. As we shall see in Part III, from an experimental analysis, this seems to be indeed the case. This seemingly strange fact[2] can be understood by recourse to the following analogy. Consider the surface of liquid (or an elastic film, like a soap film). If one puts quite gently on it a very light object, for instance a needle, by virtue of the surface tension

[1]Notice that the assumption of spatial isotropy for the electromagnetic interaction in the waveguide propagation is only a matter of convenience, since waveguide experiments do not provide any physical information on space directions different from the propagation one (the axis of the waveguide). An analogous consideration holds true for the weak case, too.

[2]For a similar "strange" fact in physics – with a phenomenon occurring at low, rather than high, energy, contrarily to what supposed –, let us quote the erroneous belief – however widely spread in the physicist community at the beginning of the twentieth century – that only fast, highly energetic neutrons could be able to induce radioactivity. Such a prejudice was overcome only by the experimental evidence obtained by Fermi and coworkers in 1935 for the effectiveness of slow neutrons in producing nuclear reactions, and the related mechanism fully understood after a deep comprehension of nuclear forces.

the liquid surface bends but the needle does not sink. As is well known, this occurs because the weight of the needle is so light that the tension forces are able to balance it. On the contrary, a more massive body, like a stone, does break the surface film, and the body sinks. The surface felt by the needle is a curved, deformed one; the stone is unable to feel the curvature. In an analogous way, a particle with energy much greater than the electromagnetic energy threshold $E_{0,\text{e.m.}}$ cannot "feel" the space–time deformation.

### *4.1.2 Weak Interaction*

The experimental input was provided by the data on the pure leptonic decay of the meson $K_s^0$, whose lifetime $T$ is known in a wide energy range (30/350 GeV) (an almost unique case) [36, 37]. Use has been made of the deformed law of time dilation as a function of the energy, which reads (cf. Table 3.2)

$$T = \frac{T_0}{\left[1 - \left(\frac{b}{b_0}\right)^2 + \left(\frac{b}{b_0}\right)^2 \left(\frac{m}{E}\right)^2\right]^{1/2}}. \tag{4.6}$$

As in the electromagnetic case, an isotropic three-space was assumed, whereas the isochrony with the usual Minkowski metric (i.e., $b_0^2 = 1$) was *derived* by the fit of (4.6) to the experimental data. The corresponding metric is therefore given by

$$g_{\text{DSR,weak}}(E) = \text{diag}\left(1, -b^2_{\text{weak.}}(E), -b^2_{\text{weak.}}(E), -b^2_{\text{weak.}}(E)\right); \tag{4.7}$$

$$\begin{aligned} b^2_{\text{weak.}}(E) &= \begin{cases} (E/E_{0\,\text{weak}})^{1/3}, & 0 \le E < E_{0\,\text{weak}} \\ 1 & E_{0\,\text{weak}} < E \end{cases} \\ &= 1 + \Theta(E_{0\,\text{weak.}} - E)\left[\left(\frac{E}{E_{0\,\text{weak}}}\right)^{1/3} - 1\right], E > 0, \end{aligned} \tag{4.8}$$

with

$$E_{0,\,\text{weak}} = (80.4 \pm 0.2)\ \text{GeV}. \tag{4.9}$$

Two points are worth stressing. First, the value of $E_{0\,\text{weak}}$ – i.e., the energy value at which the weak metric becomes Minkowskian – corresponds to the mass of the $W$-boson, through which the $K_s^0$-decay occurs. Moreover, the leptonic metric (4.7)–(4.9) has the same form of the electromagnetic metric (4.2)–(4.4). Therefore, one recovers, by the DSR formalism, the well-known result of the Glashow–Weinberg–Salam model that, at the energy scale $E_{0\,\text{weak}}$, the weak interaction and the electromagnetic one are mixed. We want also to notice that, in both the electromagnetic and the weak case, the metric parameter exhibits a "sub-Minkowskian" behavior, i.e., $b(E)$ approaches 1 from below ($E < E_0$) as energy increases.

### 4.1.3 Strong Interaction

The phenomenon considered was the so-called Bose–Einstein (BE) effect in the strong production of identical bosons in high-energy collisions, which consists in an enhancement of their correlation probability. The DSR formalism permits to derive a generalized BE correlation function, depending on all the four metric parameters $b_\mu(E)$. By using the experimental data on pion pair production, obtained in 1984 by the UA1 collaboration at CERN [38], one gets the following expression of the strong metric for the two-pion BE phenomenon:

$$g_{\mathrm{DSR,strong}}(E) = \mathrm{diag}\,\big(b^2_{\mathrm{strong}}(E), -b^2_{1,\mathrm{strong}}(E), -b^2_{2,\mathrm{strong}}(E), -b^2_{\mathrm{strong}}(E)\big)\,; \tag{4.10}$$

$$\begin{aligned} b^2_{\mathrm{strong}}(E) &= \begin{cases} 1, & 0 \le E < E_{0\mathrm{strong}} \\ (E/E_{0\mathrm{strong}})^2, & E_{0\mathrm{strong}} < E \end{cases} \\ &= 1 + \Theta(E - E_{0,\mathrm{strong}}) \left[ \left( \frac{E}{E_{0,\mathrm{strong}}} \right)^2 - 1 \right], \\ & E > 0; \end{aligned} \tag{4.11}$$

$$b^2_{1,\mathrm{strong}}(E) = \left(\sqrt{2}/5\right)^2; \tag{4.12}$$

$$b^2_{2,\mathrm{strong}} = (2/5)^2, \tag{4.13}$$

with

$$E_{0,\mathrm{strong}} = (367.5 \pm 0.4)\ \mathrm{GeV}. \tag{4.14}$$

The threshold energy $E_{0,\mathrm{strong}}$ is still the value at which the metric becomes Minkowskian. Let us stress that, in this case, contrarily to the electromagnetic and the weak ones, *a deformation of the time coordinate occurs;* moreover, *the three-space is anisotropic*, with two spatial parameters constant (but different in value) and the third one variable with energy in an "over-Minkowskian" way. It is also worth to recall that the strong metric parameters $b_\mu$ admit of a sensible physical interpretation: the spatial parameters are (related to) the spatial sizes of the interaction region ("fireball") where pions are produced, whereas the time parameter is essentially the mean life of the process. We refer the reader to [6] for further details.

It is also worth noticing that – like a physical meaning can be attributed to $E_{0,\mathrm{weak}}$ as the energy scale of the intermediate vector bosons for electroweak interactions – an analogous interpretation can be given to $E_{0,\mathrm{strong}}$. A possible suggestion is that the value of $E_{0,\mathrm{strong}}$ does represent the energy scale corresponding to the upper limit of the mass of the Higgs boson, that – as is well known – breaks the gauge invariance of the electroweak mixed interactions by endowing the weak interaction carriers with mass.

Let us explicitly notice that the hadronic metric (4.10) is not always isochronous with the usual Minkowski metric ($b_0^2 = 1$). Actually, it follows from (4.11) that it is $b_0^2 \neq 1$ for $E_{0,\text{strong}} < E$.

Such a case is not new; indeed, as is well known, the same happens for the gravitational interaction, as shown e.g., by the various measurements of red or blue shifts of electromagnetic radiation in a gravitational field, or by the relative delays of atomic clocks put at different heights in presence of gravity.

Let us investigate the possible implications of such an anisochronism of the hadronic metric. We denote by $\mathrm{d}t_{\text{had}}$ the time interval taken by a certain hadronic process for a particle at rest ("hadronic clock"). The same process, when referred to a Minkowskian electromagnetic metric, will take a time $\mathrm{d}t_{\text{e.m.}}$ to happen. The former time corresponds to the real time felt by the particle in its local frame, namely to the proper time $\tau_{\text{DSR}}$ in the Minkowski space deformed by the strong interaction.[3] The latter is the coordinate time of an external observer, who looks at the process by means of his electromagnetic instruments.[4] The two times are therefore related by (3.83), so that we get

$$\frac{\mathrm{d}t_{\text{had}}}{\mathrm{d}t_{\text{e.m.}}} = b_{0,\text{strong}} \tag{4.15}$$

or, on account of the explicit form the strong metric (4.11):

$$\frac{\mathrm{d}t_{\text{had}}}{\mathrm{d}t_{\text{e.m.}}} = \begin{cases} 1, & 0 \leq E \leq E_{0,\text{strong}} \\ E_{0,\text{strong}}/E, & E_{0,\text{strong}} \leq E \end{cases}. \tag{4.16}$$

Equation (4.16) provides *the law of time deformation in a hadronic field.* It is easily seen that there is isochronism at low energies (i.e., physical processes have the same rate either when referred to a hadronic metric or to an electromagnetic one), whereas there is a time contraction at high energies. In other words, hadronic processes are faster when observed with respect to an electromagnetic metric.

By the way, such results provide an interesting representation of two fundamental features of strong interactions, i.e., asymptotic freedom and confinement. Briefly stated, these two effects can be thought of as related to the different time rates required by the electromagnetic and the strong interactions, respectively, in order to transfer the same amount of energy to the hadronic constituents. As a consequence, hadron constituents look

---

[3] The situation reminds one of the gravitational fall of a particle toward a collapsing body. In that case, the proper time is the real time measured by the particle (influenced by the body gravitational field), whereas the coordinate time is that measured by a distant observer in fully Minkowskian conditions.

[4] The interpretation of $\mathrm{d}t_{\text{e.m.}}$ as the coordinate (Minkowskian) time is supported also by the fact that the energy of an hadronic process is in general much higher than the electromagnetic energy threshold $E_{0,\text{e.m.}}$, and therefore the e.m. metric is fully Minkowskian.

"free" when electromagnetically probed. See [6] for a deeper discussion of this interesting point.

### 4.1.4 Gravitation

It is possible to show that the gravitational interaction, too (at least on a *local* scale, i.e., in a neighborhood of Earth) can be described in terms of an energy-dependent metric, whose time coefficient was derived by fitting the experimental results on the relative rates of clocks at different heights in the gravitational field of Earth, obtained by Alley in 1974–1976 [39]. No information can be derived from the experimental data about the space parameters. The energy-dependent gravitational metric was thus obtained in the form

$$g_{\mathrm{DSR,grav}}(E) = \mathrm{diag}\left(b^2_{0,\mathrm{grav}}(E), -b^2_{1,\mathrm{grav}}(E), -b^2_{2,\mathrm{grav}}(E), -b^2_{3,\mathrm{grav}}(E)\right); \tag{4.17}$$

$$\begin{aligned} b^2_{0,\mathrm{grav}}(E) &= \begin{cases} 1\,, & 0 \le E < E_{0\mathrm{grav}} \\ \frac{1}{4}(1+E/E_{0\mathrm{grav}})^2, & E_{0\mathrm{grav}} < E \end{cases} \\ &= 1 + \Theta(E - E_{0,\mathrm{grav.}})\left[\frac{1}{4}\left(1+\frac{E}{E_{0,\mathrm{grav.}}}\right)^2 - 1\right], \\ & E > 0 \end{aligned} \tag{4.18}$$

with the coefficients $b^2_{k,\mathrm{grav}}(E)$ $(k = 1,2,3)$ undetermined and

$$E_{0,\mathrm{grav}} = (20.2 \pm 0.1)\ \mu\mathrm{eV}. \tag{4.19}$$

Intriguingly enough, the value of the threshold energy for the gravitational case $E_{0,\mathrm{grav}}$ is approximately one order of magnitude less than the thermal energy corresponding to the 2.7 K cosmic microwave background radiation (CMWBR) in the Universe ($E_{\mathrm{CMWBR}} \approx 232.67\ \mu\mathrm{eV}$).

Let us notice that for the gravitational metric (unlike the e.m. and weak cases, and in analogy with the strong metric) a deformation of the time coordinate occurs. The time coefficient varies with energy in an over-Minkowskian way, namely, it approaches the Minkowskian limit from above ($E_0 < E$). The relation between proper time and coordinate time, (3.83), corresponding to the time coefficient (4.18) reads:

$$\mathrm{d}\tau = \left(1 + \frac{E}{E_0}\right)\mathrm{d}t, \tag{4.20}$$

at variance with the Einsteinian one (derived from metric (2.21))

$$\mathrm{d}\tau = \left(1 + \frac{E}{E_0}\right)^{1/2}\mathrm{d}t. \tag{4.21}$$

As to the explicit form of the spatial part of the gravitational metric (on which the experimental data do not provide any information), two possibilities are open:

1. *The spatial, 3D metric is Euclidean*, i.e., $b^2_{k,\text{grav}}(E) = 1 \; \forall k = 1, 2, 3$;
2. *The spatial metric is anisotropic and energy-dependent*, i.e., the 4D metric has a structure similar to the strong one (4.10), namely

$$g_{\text{DSR,grav}}(E) = \text{diag}\left(b^2_{\text{grav}}(E), -b^2_{1,\text{grav}}(E), -b^2_{2,\text{grav}}(E), -b^2_{\text{grav}}(E)\right); \tag{4.17a}$$

$$b^2_{\text{grav}}(E) = \begin{cases} 1\,, & 0 \leq E < E_{0,\text{grav}} \\ \frac{1}{4}(1 + E/E_{0,\text{grav}})^2, & E_{0,\text{grav}} < E \end{cases}$$

$$= 1 + \Theta(E - E_{0,\text{grav.}}) \left[\frac{1}{4}\left(1 + \frac{E}{E_{0,\text{grav.}}}\right)^2 - 1\right],$$

$$E > 0, \tag{4.18a}$$

with, in general, $b^2_1 \neq b^2_2$. Assuming a gravitational deformation of space, allowing also for spatial anisotropy, seems to be the more sound choice on physical grounds, on account of the experimental evidences (see Chap. 15). This is why in the following, unless otherwise specified, we shall mean for gravitational metric the form (4.17a), (4.18a).

Let us stress explicitly that nothing can be said from the experimental point of view about the behavior of the metrics (4.2)–(4.4), (4.7)–(4.9), (4.10) (4.14), (4.17) (4.19) at the threshold energies (however, in Part V, suitable mathematical assumptions will be made on this behavior for convenience' reasons). Moreover, notice that, formally, some metrics become degenerate for $E = 0$. However, actually it does not make sense, from a physical point of view, to consider a vanishing energy in our framework, because no physical process at all does take place at zero energy.

The general pattern of the four phenomenological metrics is shown in Fig. 4.1.

Let us remark that the non-Minkowskian behavior of the four phenomenological metrics, at variance with respect to Special Relativity, implies a violation of Lorentz invariance in its usual meaning (namely, as the metric homomorphisms of the Minkowski space–time $M$) (see Sect. 3.1). However, as already noted, the existence of the DLT permits to recover Lorentz invariance in a generalized sense in the framework of DSR, still in the form of isometries, but now of the deformed Minkowski space $\widetilde{M}$ (see Sect. 3.3.5). We referred to this new symmetry within DSR as deformed Lorentz invariance. Then, it is possible in a sense to state that physical phenomena,

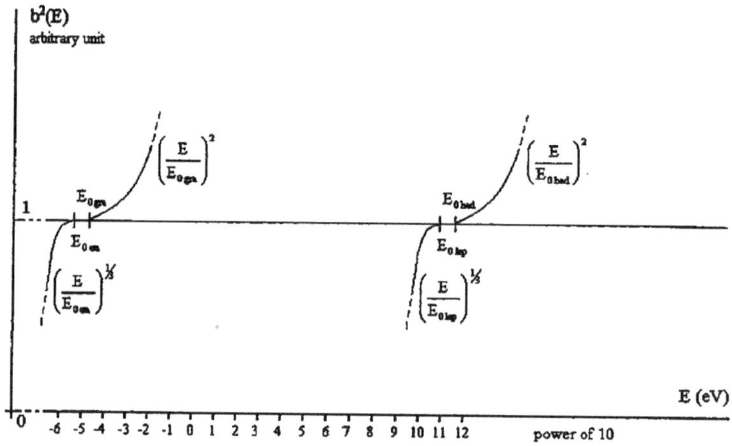


FIGURE 4.1. General pattern of the phenomenological metrics for the four fundamental interactions

exhibiting a Lorentz-violating behavior in the standard meaning, inside the standard Minkowski space $M$, actually do yield evidence (provided such a violation occurs according to the DSR conditions and modes) for deformed Lorentz invariance. For a given interaction, such an invariance is obviously dependent on the energy range considered. Such a restriction does no longer apply when embedding DSR within DR5 (its 5D counterpart, obtained by taking energy as extra metric coordinate: see Parts IV and V).

## 4.2 Threshold Energies and Recursive Metrics

A comparison among the threshold energies for the electromagnetic, weak, strong and gravitational interactions, given by (4.4), (4.9), (4.14), (4.19), yields

$$E_{0\mathrm{el}} < E_{0\mathrm{grav}} < E_{0\mathrm{w}} < E_{0\mathrm{s}}, \tag{4.22}$$

i.e., an increasing arrangement of $E_0$ from the electromagnetic to the strong interaction. Moreover

$$\frac{E_{0\mathrm{grav}}}{E_{0\mathrm{el}}} = 4.49 \pm 0.02; \frac{E_{0\mathrm{s}}}{E_{0\mathrm{w}}} = 4.57 \pm 0.01, \tag{4.23}$$

namely

$$\frac{E_{0\mathrm{grav}}}{E_{0\mathrm{el}}} \simeq \frac{E_{0\mathrm{s}}}{E_{0\mathrm{w}}}, \tag{4.24}$$

an intriguing result indeed.

A further remark concerns the possible pattern of interactions ensuing from DSR. According to the results summarized earlier, we have two pairs of interactions (i) electromagnetic and gravitational and (ii) weak and strong, ordered by the increasing arrangement of the threshold energies. Moreover, in each pair the former interaction is sub-Minkowskian, and the latter is over-Minkowskian. The first question is: Does this pattern end with the second pair, or not? If a third pair exists, we can assume that the threshold energies of the new pair, $E_{0,5}$ and $E_{0,6}$, are related to the threshold energies of the previous sub-Minkowskian and over-Minkowskian metrics according to

$$\frac{E_{0,2n}}{E_{0,2n-1}} = \frac{E_{0,2n+2}}{E_{0,2n+1}}, \quad n = 1, 2 \tag{4.25}$$

(with $E_{0\mathrm{el}} = E_{0,1}$; $E_{0\mathrm{grav}} = E_{0,2}$; $E_{0\mathrm{w}} = E_{0,3}$; $E_{0\mathrm{s}} = E_{0,4}$). In such hypothesis, with the values (4.4), (4.9), (4.14), (4.19) of the threshold energies for the known interactions, one gets

$$\begin{aligned} E_{0,5} &\simeq 1.3 \times 10^{18}\,\mathrm{GeV}; \\ E_{0,6} &\simeq 6.7 \times 10^{18}\,\mathrm{GeV}. \end{aligned} \tag{4.26}$$

A possible representation of the recursive pattern of the phenomenological metrics is illustrated in Fig. 4.2.

Such a pattern may recur again, or not, and it is of course a matter of experiment to check the real existence of these new pairs of interactions. What to be excluded is that it recurs *ad infinitum*. In this connection, we recall that it was shown that the maximum possible force in Nature is provided by the *Kosntro constant* (or *Planck force*) $K$, given by [40]

$$K = \frac{c^4}{G} = 1.211 \times 10^{44}\mathrm{N} = 7.556 \times 10^{51}\mathrm{GeV\,cm^{-1}}, \tag{4.27}$$

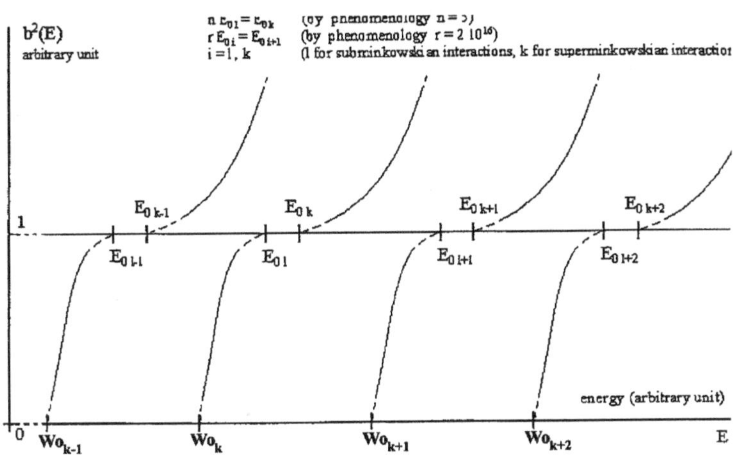


FIGURE 4.2. Qualitative behavior of recursive and asymptotic metrics

where $G$ is the gravitational constant ($G = 1.072 \times 10^{-10}\text{cm}^5\,[\text{GeV s}^4]^{-1}$). The corresponding maximum energy, i.e., the energy of the whole Universe, is therefore (assuming $R_0 \sim 10^{10}$ light-years)

$$E_{\text{max}} = KR_0 \sim 10^{79}\,\text{GeV}. \tag{4.28}$$

Either in the case of new interactions (besides the known ones) or not, we deem that the interaction pattern in the DSR scheme is bounded from above by the value $E_{\text{max}}$ (4.28) related to the Kostro limit. This holds, in particular, for the asymptotic behavior of the over-Minkowskian metrics.

Let us also stress that, in the sub-Minkowskian case, the deformed metrics can vanish, in general, for an energy value $W_0 \neq 0$, i.e., by definition

$$b_\mu^2(W_0) = 0. \tag{4.29}$$

(see Fig. 4.2). The physical meaning of the metric vanishing is that, at the energy $W_0$, the interaction is unable to measure the space–time separation. In particular, such a value of the energy corresponds to a point-like object for the interaction considered. Examples of this fact are provided by the inertial mass of the photon for an electromagnetic metric, or the neutrino masses for a leptonic metric. Indeed, for the e.m. interaction, $W_{0,\text{e.m.}}$ is essentially the upper limit for the photon inertial mass.

## 4.3 Asymptotic Metrics

Of course, the paucity of the phenomena, and the related experimental data, on which the phenomenological analysis is based, makes just preliminary and provisional the resulting metrics obtained for the electromagnetic, weak, gravitational and strong interactions.

In this connection, it is worth listing the possible functional forms of deformed metrics which might be able to describe physical interactions. We shall divide them in two main classes, according to the functional form of the parameters $b_\mu^2(E)$:

(I) First class

$$\begin{cases} b_0^2(E) = \text{const}; \\ b_k^2(E) = \left(1 - \dfrac{W_{0k}}{E}\right)^{n_k}, \end{cases} \tag{4.30}$$

with $n_k \in R$.

Coefficients with the functional form (4.30) are characteristic of sub-Minkowskian metrics, which are asymptotically Minkowskian with increasing energy. Metrics of the first class are suitable for representing the electromagnetic and leptonic interactions.

(II) Second class

$$\begin{cases} b_0^2(E) = \left(1 + \dfrac{E}{W_{00}}\right)^{n_0} ; \\ b_k^2(E) = \left(1 + \dfrac{E}{W_{0k}}\right)^{n_k} . \end{cases} \quad (4.31)$$

Metrics of class II are obviously over-Minkowskian, divergent with increasing energy, asymptotically Minkowskian with decreasing energy. They are suitable for the representation of the hadronic and the gravitational interaction.

From these results and those of Sect. 4.2 on threshold energies, it is possible to illustrate on a qualitative basis the behavior of the asymptotic metrics, as done in Fig. 4.2.

Notice that both first- and second-class metrics become spatially isotropic if constants $W_{0k}$, $n_k$ reduce to only two, $W_0$ and $n$. It is worth stressing some interesting implications of the earlier metrics.

For both classes, the constants $W_{0k}$ have the natural meaning of scale energy of the interaction described by the corresponding metrics. Moreover, in metrics of the first kind they represent the energy values at which the spatial part of the metric vanishes.

Indeed, let us recall that, applying an isotropic metric of type I to the data on e.m. wave propagation, one gets for $W_{0k,\text{e.m.}}$ values compatible with the present upper bounds on the photon inertial mass.

As a last consequence, consider the general form of the time deformation law in a hadronic field, that reads:

$$\frac{\Delta t_{\text{had}}}{\Delta t_{\text{e.m.}}} = \frac{1}{\sqrt{g_{00,\text{DSR}}}} = \left(1 + \frac{E}{W_{00}}\right)^{-n_0/2} . \quad (4.32)$$

For $n_0 \leq 2$ (on account of the results of the previous section) the earlier expression shows a behavior asymptotically decreasing toward zero as the energy $E$ increases, whereas, for $W_{00} \geq E > 0$, it goes to 1 as the energy decreases. As we have seen in Sect. 4.1, for strong interaction such a behavior can be interpreted in terms of confinement and asymptotic freedom of hadronic constituents.

As a last remark, we want to stress that there is presently no clear understanding of what happens from the metric point of view when two or more interactions are involved at the same time. Roughly speaking, one could state that the resulting space–time deformation arises from the superposition of the metrics concerned, according to their energy thresholds and their sub- or over-Minkowskian behavior. By recourse to an analogy we already exploited, one can think to the deformation of an elastic carpet ensuing from putting two or more bodies on it. However, a rigorous treatment of such a problem would require a study of the metric structure of a manifold endowed with more different metrics. As a matter of fact, there is at least an experiment whose results need recourse to two phenomenological metrics (the electromagnetic and the gravitational one) in order to be fully explained (see Chap. 14).

## 4.4 DSR as Metric Gauge Theory

It is clear from the discussion of the phenomenological metrics describing the four fundamental interactions in DSR that the Minkowski space $M$ is the space–time manifold of background of any experimental measurement and detection (namely, of any process of acquisition of information on physical reality). In particular, we can consider this Minkowski space as that associated to the electromagnetic interaction above the threshold energy $E_{0,\mathrm{e.m.}}$. Therefore, in modeling the physical phenomena, one has to take into account this fact. If one believes in the geometrical nature of interactions, i.e., assumes the validity of the Finzi principle, this means that one has to suitably *gauge* (with reference to $M$) the space–time metrics with respect to the interaction – and/or the phenomenon – under study. In other words, one needs to "adjust" suitably the local metric of space–time according to the interaction acting in the region considered. We can name such a procedure "Metric Gaugement Process" (M.G.P.). Like in usual gauge theories a different phase is chosen in different space–time *points*, in DSR different metrics are associated to different space–time *manifolds* according to the interaction acting therein. We have thus a gauge structure on the space of manifolds

$$\widetilde{\mathcal{M}} \equiv \cup_{g_{\mathrm{DSR}} \in \mathcal{P}(E)} \widetilde{M}\left(g_{\mathrm{DSR}}\right), \tag{4.33}$$

where $\mathcal{P}(E)$ is the set of the energy-dependent pseudoeuclidean metrics of the type (2.17). This is why it is possible to regard DSR as a *Metric Gauge Theory*.

However, let us notice that DSR can be considered as a metric gauge theory from another point of view, on account of the dependence of the metric coefficients on the energy. Actually, once the MGP has been applied, by selecting the suitable gauge (namely, the suitable *functional form* of the metric) according to the interaction considered (thus implementing the Finzi principle), the metric dependence on the energy implies another different gauge process. Namely, the metric is gauged according to the process under study, thus selecting the *given* metric, with the *given values* of the coefficients, suitable for the given phenomenon. The analogy of this second kind of metric gauge with the standard, non-abelian gauge theories is more evident in the framework of the 5D space–time $\Re_5$ (with energy as extra dimension) embedding $\widetilde{M}$, on which Deformed Relativity in Five Dimensions (DR5) is based (see Parts IV and V). In $\Re_5$, in fact, energy is no longer a parametric variable, like in DSR, but plays the role of fifth (metric) coordinate. The invariance under such a metric gauge, not manifest in four dimensions, is instead recovered in the form of the isometries of the 5D space–time–energy manifold $\Re_5$.

We have therefore a *double* metric gaugement, according, on one side, to the interaction ruling the physical phenomenon examined, and on the other side to its energy, in which the metric coefficients are the analogous of the gauge functions.

Another connection of DSR with *standard* gauge theory will be discussed at the end of Part II.

# Part II

# MATHEMATICS OF DEFORMED SPACE–TIME

# 5

# Generalized Minkowski Spaces and Killing Symmetries

In the first Part of this book, we discussed the physical foundations of the DSR in four dimensions. This second Part will be devoted to dealing in detail with the mathematical features and properties of DSR. In this framework, the isometries of the deformed Minkowski space $\widetilde{M}$ play a basic role. The mathematical tool needed to such a study are the Killing equations, whose solution will allow us to determine both the infinitesimal and the finite structure of the deformed chronotopical groups of symmetries [41–43]. An important result we shall report at the end of this Part – due to its physical implications – is the geometrical structure of $\widetilde{M}$ as a generalized Lagrange space [12, 13, 44].

## 5.1 Generalized Minkowski Spaces

The structure of the deformed space–time $\widetilde{M}$ of DSR can be generalized to what we shall call *generalized Minkowski space* $\widetilde{M_N}(\{x\}_{\text{n.m.}})$. We define $\widetilde{M_N}(\{x\}_{\text{n.m.}})$ as a $N$-dimensional Riemann space with a global metric structure determined by the (in general nondiagonal) metric tensor $g_{\mu\nu}(\{x\}_{\text{n.m.}})$ $(\mu, \nu = 1, 2, ..., N)$, where $\{x\}_{\text{n.m.}}$ denotes a set of $N_{\text{n.m.}}$ nonmetrical coordinates (i.e., different from the $N$ coordinates related to the dimensions of the space considered) [41]. The interval in $\widetilde{M_N}(\{x\}_{\text{n.m.}})$ therefore reads

$$\mathrm{d}s^2 = g_{\mu\nu}(\{x\}_{\text{n.m.}})\mathrm{d}x^\mu \mathrm{d}x^\nu. \tag{5.1}$$

We shall assume the signature $(T, S)$ ($T$ time-like dimensions and $S = N - T$ space-like dimensions). It follows that $\widetilde{M_N}(\{x\}_{\text{n.m.}})$ is *flat*, because all the components of the Riemann–Christoffel tensor vanish.

Of course, an example is just provided by the 4D deformed Minkowski space $\widetilde{M}(E)$. In the following, in order to comply with the notation adopted for generalized Minkowski spaces, we shall denote the DSR deformed space–time with $\widetilde{M}(x^5)$, where the coordinate $x^5$ has to be interpreted as the energy. The index 5 explicitly refers to the already mentioned fact that the deformed Minkowski space can be "naturally" embedded in a 5D (Riemannian) space (see Parts IV and V).

## 5.2 Maximal Killing Group of a $N$-Dimensional Generalized Minkowski Space

### 5.2.1 *Lie Groups, P.B.W. Theorem and the Transformation Representation*

Let us recall the essential content of the Poincaré-Birkhoff–Witt (P.B.W.) theorem and of the Lie theorems: Given a Lie group $G_{\text{L}}$ of order $M$, it is always possible to build up an exponential representative mapping of any finite element $g$ of $G_{\text{L}}$:

$$\forall g \text{finite} \in G_{\text{L}}$$
$$\Rightarrow \exists \{\alpha_i\}_{i=1\ldots M} \in R^M \ (\{\alpha_i\} = \{\alpha_i(g)\}) : g = \exp\left(\sum_{i=1}^{M} \alpha_i T^i\right), \tag{5.2}$$

where $\{T^i\}_{i=1\ldots M}$ is the generator basis of the Lie algebra of $G_{\text{L}}$ and $\{\alpha_i = \alpha_i(g)\}_{i=1\ldots M}$ is a set of $M$ real parameters (of course depending on $g \in G_{\text{L}}$).

Therefore, by a series development of the exponential

$$g = \exp\left(\sum_{i=1}^{M} \alpha_i T^i\right) = \sum_{k=0}^{\infty} \frac{1}{k!}\left(\sum_{i=1}^{M} \alpha_i(g) T^i\right)^k \quad \forall g \text{ finite} \in G_{\text{L}}, \tag{5.3}$$

we get, for an infinitesimal element $(g \to \delta g)$ $(\Leftrightarrow \{\alpha_i(g)\}_{i=1\ldots M} \in R^M \to \{\alpha_i(g)\}_{i=1\ldots M} \in I_0 \subset R^M)$:[1]

$$\delta g = 1 + \sum_{i=1}^{M} \alpha_i(g) T^i + O(\{\alpha_i^2(g)\}) \quad \forall \delta g \text{ infinitesimal} \in G_{\text{L}}. \tag{5.4}$$

[1] $I_0 \subset R^M$ is a generic neighborhood of $0 \in R^M$.

Since any Lie group admits a representation as a group of transformations acting on a $N$-dimensional manifold $S_N$ ("$N$-dimensional vector space of transformation representation," not to be confused with thegroup manifold $V_M$), given $x \in S_N$, one has, for the action of a finite and infinitesimal element of $G_{\mathrm{L}}$, respectively:

$$gx = \left[\exp\left(\sum_{i=1}^{M} \alpha_i(g)T^i\right)\right] x = \left[\sum_{k=0}^{\infty} \frac{1}{k!}\left(\sum_{i=1}^{M} \alpha_i(g)T^i\right)^k\right] x = x' \in S_N; \tag{5.5}$$

$$\left.\begin{array}{c}(\delta g)\, x = \left[1 + \sum_{i=1}^{M} \alpha_i(g)T^i\right] x = x + \left(\sum_{i=1}^{M} \alpha_i(g)T^i\right) x = x' \in S_N; \\ \delta g : S_N \ni x \to x' = x + \delta x_{(g)}(x) \in S_N\end{array}\right\} \Rightarrow$$

$$\Rightarrow \delta x_{(g)}(x) = \left(\sum_{i=1}^{M} \alpha_i(g)T^i\right) x. \tag{5.6}$$

### 5.2.2 *Killing Equations in a N-Dimensional Generalized Minkowski Space*

In general $S_N$ is endowed with a metric structure we shall assume in the following to be at most Riemannian. The interval in $S_N$ is therefore:

$$\varphi(x) \equiv \mathrm{d}s^2(x) = g_{\mu\nu}(x)\mathrm{d}x^\mu \mathrm{d}x^\nu, \tag{5.7}$$

with $g_{\mu\nu}(x)$ being the symmetric, rank-two metric tensor. By carrying out an infinitesimal transformation of the type

$$x^{\mu\prime} = x^\mu + \xi^\mu(x), \tag{5.8}$$

one has:

$$\begin{aligned} \delta \mathrm{d}x^\mu(x) \overset{[\delta,d]=0}{=} \mathrm{d}\delta x^\mu(x) &= \frac{\partial \xi^\mu(x)}{\partial x^\gamma}\mathrm{d}x^\gamma; \\ \delta g_{\mu\nu}(x) &= \frac{\partial g_{\mu\nu}(x)}{\partial x^\beta}\zeta^\beta(x), \end{aligned} \tag{5.9}$$

and therefore

$$\begin{aligned} \delta\varphi(x) &\equiv \delta \mathrm{d}s^2(x) = \delta(g_{\mu\nu}(x)\mathrm{d}x^\mu \mathrm{d}x^\nu) \\ &= (\delta g_{\mu\nu}(x))\, \mathrm{d}x^\mu \mathrm{d}x^\nu + g_{\mu\nu}(x)\, (\delta \mathrm{d}x^\mu(x))\, \mathrm{d}x^\nu + g_{\mu\nu}(x)\mathrm{d}x^\mu\, (\delta \mathrm{d}x^\nu(x)) \end{aligned}$$

$$
\begin{aligned}
&= \left(\frac{\partial g_{\mu\nu}(x)}{\partial x^\beta}\zeta^\beta(x)\right)\mathrm{d}x^\mu \mathrm{d}x^\nu + g_{\mu\nu}(x)\left(\frac{\partial \xi^\mu(x)}{\partial x^\gamma}\mathrm{d}x^\gamma\right)\mathrm{d}x^\nu \\
&\quad + g_{\mu\nu}(x)\mathrm{d}x^\mu\left(\frac{\partial \xi^\nu(x)}{\partial x^\chi}\mathrm{d}x^\chi\right) \\
&= \frac{\partial g_{\mu\nu}(x)}{\partial x^\beta}\zeta^\beta(x)\mathrm{d}x^\mu \mathrm{d}x^\nu + g_{\nu\beta}(x)\frac{\partial \xi^\beta(x)}{\partial x^\mu}\mathrm{d}x^\mu \mathrm{d}x^\nu \\
&\quad + g_{\mu\beta}(x)\frac{\partial \xi^\beta(x)}{\partial x^\nu}\mathrm{d}x^\mu \mathrm{d}x^\nu \\
&= \left(\frac{\partial g_{\mu\nu}(x)}{\partial x^\beta}\zeta^\beta(x) + g_{\nu\beta}(x)\frac{\partial \xi^\beta(x)}{\partial x^\mu} + g_{\mu\beta}(x)\frac{\partial \xi^\beta(x)}{\partial x^\nu}\right)\mathrm{d}x^\mu \mathrm{d}x^\nu
\end{aligned}
\tag{5.10}
$$

The invariance of the infinitesimal interval under transformation (5.8) requires therefore

$$
\delta \mathrm{d}s^2(x) = 0 \Leftrightarrow \left(\frac{\partial g_{\mu\nu}(x)}{\partial x^\beta}\zeta^\beta(x) + g_{\nu\beta}(x)\frac{\partial \xi^\beta(x)}{\partial x^\mu} + g_{\mu\beta}(x)\frac{\partial \xi^\beta(x)}{\partial x^\nu}\right) = 0; \tag{5.11}
$$

$$
a_\mu(x) \equiv g_{\mu\eta}(x)a^\eta(x) \overset{g_{\mu\eta}(x)g^{\eta\chi}(x)=\delta^\chi_\mu \ \ \forall x\in S_N}{\Leftrightarrow} a^\eta(x) = g^{\eta\mu}(x)a_\mu(x); \tag{5.12}
$$

$$
\begin{aligned}
\frac{\partial a^\eta(x)}{\partial x^\nu} &\equiv a^\eta(x)_{,\nu} \\
&= \frac{\partial\left(g^{\eta\mu}(x)a_\mu(x)\right)}{\partial x^\nu} = \frac{\partial g^{\eta\mu}(x)}{\partial x^\nu}a_\mu(x) + g^{\eta\mu}(x)\frac{\partial a_\mu(x)}{\partial x^\nu}.
\end{aligned}
\tag{5.13}
$$

Let us introduce the covariant derivative on $S_N$, defined by

$$
a_\mu(x)_{;\nu} \equiv a_\mu(x)_{,\nu} - \Gamma^\lambda_{\mu\nu}(x)a_\lambda(x) \tag{5.14}
$$

with $\Gamma^\lambda_{\mu\nu}(x)$ being the affine connection

$$
\Gamma^\lambda_{\mu\nu}(x) = \frac{1}{2}g^{\rho\lambda}(x)\left(\frac{\partial g_{\nu\rho}(x)}{\partial x^\mu} + \frac{\partial g_{\mu\rho}(x)}{\partial x^\nu} - \frac{\partial g_{\nu\mu}(x)}{\partial x^\rho}\right). \tag{5.15}
$$

Since the covariant derivative of the metric tensor vanishes ($g_{\mu\eta;\rho}(x) = 0$), it is possible to rewrite (5.11) as:

$$
\delta \mathrm{d}s^2(x) = 0 \Leftrightarrow \xi_\mu(x)_{;\nu} + \xi_\nu(x)_{;\mu} = 0, \tag{5.16}
$$

or, in compact form:

$$
\xi_{[\mu}(x)_{;\nu]} = 0, \tag{5.17}
$$

where the bracket [..] means symmetrization with respect to the enclosed indices.

As is well known, the $N(N+1)/2$ (5.17) in the $N$ components of the covariant $N$-vector $\xi_\mu(x)$ are *the Killing equations* of the space $S_N$. As is clearly seen from their derivation, the contravariant Killing vectors correspond to directions along which the infinitesimal interval – and therefore the metric tensor – remains unchanged. Then, they determine the infinitesimal isometries of $S_N$. Another very useful property of Killing vectors is that they are associated to constants of motion, namely to quantities which keep their value along any geodesic. Any $N$-dimensional Riemannian space admits a Killing group with at most $N(N+1)/2$ parameters; in this latter case, the space is called "maximally symmetric." It can be shown that a Riemann space is maximally symmetric *iff* its scalar curvature $R$ is constant.

In index notation, (5.6) can be written as:

$$\delta x^\mu_{(g)}(x) = \left[\left(\sum_{i=1}^{M} \alpha_i(g) T^i\right) x\right]^\mu, \mu = 1, ..., N. \tag{5.18}$$

Let us denote simply by $\alpha$ the parametric $M$-vector $\{\alpha_i\}_{i=1...M}$ of the representation (5.2) of the element $g \in G_\mathrm{L}$. Then, from (5.6) and (5.8) one gets

$$\delta x^\mu_{(g)}(x, \alpha) = \xi^\mu_{(g)}(x, \alpha); \tag{5.19}$$

$$\xi^\mu_{(g)}(x, \alpha) = \left[\left(\sum_{i=1}^{M} \alpha_i(g) T^i\right) x\right]^\mu, \tag{5.20}$$

namely $\delta x^\mu_{(g)}(x, \alpha)$ is the contravariant $N$-vector of the infinitesimal transformation associated – in the transformation representation of the Lie group $G_\mathrm{L}$ – to the infinitesimal element $\delta g$.

We can now define the mixed second-rank $N$-tensor $\delta\omega^\mu_\nu(g)$ of an infinitesimal transformation (associated to $\delta g \in G_\mathrm{L}$) as:

$$\delta x^\mu_{(g)}(x) = \left[\left(\sum_{i=1}^{M} \alpha_i(g) T^i\right) x\right]^\mu \equiv \delta\omega^\mu_\nu(g) x^\nu. \tag{5.21}$$

The number of independent components of the tensor $\delta\omega^\mu_\nu(g)$ is equal to the order $M$ of the Lie group; in general, nothing can be said about its symmetry properties. Notice that, in the context of generalized $N$-dimensional Minkowski spaces, the infinitesimal mixed tensor depends in general on the set of nonmetric variables, i.e., $\delta\omega^\mu_\nu = \delta\omega^\mu_\nu(g, \{x\}_{\mathrm{n.m.}})$. From (5.18)–(5.20) it follows

$$\xi^\mu_{(g)}(x) = \delta\omega^\mu_\nu(g) x^\nu, \tag{5.22}$$

showing that $\delta\omega^\mu_\nu(g)$ is *the tensor of the "rotation" parameters* in $S_N$. Let us stress that (5.20)–(5.21), associating the *global* tensor $\delta\omega^\mu_\nu(g)$ ($\neq \delta\omega^\mu_\nu(g, x)$) to $\delta x^\mu_{(g)}(x)$ (in general *local*), imply a reductive assumption on the possible

Lie groups considered. Actually, as is easily seen, the introduction of $\delta\omega^{\mu}_{\nu}(g)$ (independent of $x$) is possible only *iff* $\delta x^{\mu}_{(g)}(x)$ is a *linear and homogeneous* function of $x$. Of course, this imposes severe restrictions on the possible types of Lie groups under consideration.

Indeed, let us stress that the transformation representation of the $M$-order Lie group $G_{\rm L}$ we considered above *is not a group representation of* $G_{\rm L}$ (in the usual meaning of the term). Indeed, although $G_{\rm L}$ can be interpreted as a suitable transformation group acting on $S_N$ (with $M \neq N$ in general), such coordinate transformations *are not necessarily linear.* Otherwise speaking, $G_{\rm L}$ *does not admit, in general, a N-order matrix representation.* In other words, its $M$ infinitesimal generators $T^i$ $(i = 1, ..., M)$ *cannot in general be represented by* $N \times N$ *matrices acting on* $S_N$. Although (5.3) for $g$ in terms of the generators $\{T^i\}$ can be linearized with respect to the group parameters $\{\alpha_i\}$ (by means of a "parametric MacLaurin development" in the neighborhood of the null $M$-vector of parameters), thus getting the infinitesimal element $\delta g$ (5.4), $\delta gx$ *is not necessarily linear in* $x$, due to the possible dependence of some of the $T^i$'s on $x$.

Therefore, introducing the tensor $\delta\omega^{\mu}_{\nu}(g)$ amounts to consider only those Lie groups admitting a $N \times N$ matrix representation on $S_N$ (corresponding to linear and homogeneous coordinate transformations).

### 5.2.3 Maximal Killing Group of $\widetilde{M_N}$

To our present aims, we have to impose two further conditions. First, we assume that the Lie groups under consideration, in the related transformation representation, are Killing groups of $S_N$ (not necessarily maximal), namely (from (5.17), (5.18), and (5.20)):

$$\begin{aligned}
&\xi_{(g)\mu}(x)_{;\nu} + \xi_{(g)\nu}(x)_{;\mu} = 0 \Leftrightarrow \delta x_{(g)\mu}(x)_{;\nu} + \delta x_{(g)\nu}(x)_{;\mu} = 0 \\
&\Leftrightarrow \left[\left(\sum_{i=1}^{M} \alpha_i(g) T^i\right) x\right]_{\mu;\nu} + \left[\left(\sum_{i=1}^{M} \alpha_i(g) T^i\right) x\right]_{\nu;\mu} = 0 \\
&\Leftrightarrow \left(\delta\omega_{\mu\rho}(g) x^{\rho}\right)_{;\nu} + \left(\delta\omega_{\nu\rho}(g) x^{\rho}\right)_{;\mu} = 0 \\
&\Leftrightarrow \left(\delta\omega_{[\mu\rho}(g) x^{\rho}\right)_{;\nu]} = 0
\end{aligned} \tag{5.23}$$

Last (5.23) can be derived from the first one on account of (5.22) and of

$$\xi_{(g)\mu}(x) = g_{\mu\nu}\xi^{\nu}_{(g)}(x) = g_{\mu\nu}\delta\omega^{\nu}_{\rho}(g)x^{\rho} = \delta\omega_{\mu\rho}(g)x^{\rho}. \tag{5.24}$$

Moreover, $S_N$ is assumed to be endowed with a *global* metric structure, independent of $x \in S_N$ (but dependent, in general, on a set $\{x\}_{\rm n.m.}$ of nonmetric coordinates), i.e.,:

$$g_{\mu\nu}(\{x\}_{\rm n.m.}) \neq g_{\mu\nu}(x). \tag{5.25}$$

This second requirement entails that in $S_N$ all components of the Riemann–Christoffel tensor vanish (so that the covariant derivative reduces to the ordinary derivative), and then it is a flat manifold. In other words, we are assuming that $S_N$ is a *N-dimensional generalized Minkowski space*, as defined in Sect. 5.1. Therefore, we shall henceforth use the notation $\widetilde{M_N}$ instead of $S_N$.

Notice that although, in general, $\delta\omega^\mu_\nu(g, \{x\}_{\text{n.m.}})$ does depend on possible nonmetric variables, its completely covariant form *does not*, due to the dependence of $g_{\mu\nu}$ on $\{x\}_{\text{n.m.}}$:

$$\delta\omega_{\mu\rho}(g) = g_{\mu\sigma}(\{x\}_{\text{n.m.}})\delta\omega^\sigma_\rho(g, \{x\}_{\text{n.m.}}) \neq \delta\omega_{\mu\rho}(\{x\}_{\text{n.m.}}). \tag{5.26}$$

On the contrary, its completely contravariant form does depend on $\{x\}_{\text{n.m.}}$:

$$\delta\omega^{\mu\rho}(g, \{x\}_{\text{n.m.}}) = g^{\mu\sigma}(\{x\}_{\text{n.m.}})\delta\omega^\rho_\sigma(g, \{x\}_{\text{n.m.}}). \tag{5.27}$$

We can therefore state that, in a generalized Minkowski space, any form of the $N$-tensor $\delta\omega(g)$ is global (i.e., independent of all metric variables), but its completely covariant expression is independent of possible nonmetric variables, too. This independence is related to the fact that (as it will be seen in the following: see (5.26)) $\delta\omega_{\alpha\beta}(g)$ is nothing but the antisymmetric tensor of the space–time rotation parameters. Thus the dependence of $\delta\omega_{\alpha\beta}$ on the physical theory concerned is reducible to its very dependence on the element $g$ of the space–time rotation group of the $N$-d generalized Minkowski space under consideration. That is why, in the following, parentheses will be sometime used in the covariant components of $\delta\omega$ (e.g., in the form $\delta\omega_{\alpha\beta,(\text{DSR})}(g)$ or $\delta\omega_{\text{cov.},(\text{DSR})}(g)$).

In $\widetilde{M_N}$, the following formulae hold $\forall\delta g$ infinitesimal $\in G_{\text{L}}$:

$$\delta g : \widetilde{M_N} \ni x \to x'(\{x\}_{m.}, \{x\}_{\text{n.m.}}) = x + x_{(g)}(\{x\}_{m.}, \{x\}_{\text{n.m.}}) \in \widetilde{M_N}; \tag{5.28}$$

$$\begin{aligned} \delta x^\mu_{(g)}(x, \{x\}_{\text{n.m.}}) &= \left[\left(\sum_{i=1}^{M} \alpha_i(g) T^i(\{x\}_{\text{n.m.}})\right) x\right]^\mu \\ &= \delta\omega^\mu_\nu(g, \{x\}_{\text{n.m.}})x^\nu = \xi^\mu_{(g)}(x, \{x\}_{\text{n.m.}}); \end{aligned} \tag{5.29}$$

$$\xi_{(g)\mu}(x)_{,\nu} + \xi_{(g)\nu}(x)_{,\mu} = 0, \tag{5.30}$$

where "$,\mu$" denotes ordinary derivation with respect to $x^\mu$. From (5.21) and (5.27) it follows also:

$$\begin{aligned} &\xi_{(g)\mu}(x)_{,\nu} + \xi_{(g)\nu}(x)_{,\mu} = 0 \\ \Leftrightarrow &\left(\delta\omega_{\mu\nu}(g)x^\rho\right)_{,\nu} + \left(\delta\omega_{\nu\rho}(g)x^\rho\right)_{,\mu} = 0. \end{aligned} \tag{5.31}$$

The last equation entails the antisymmetry of $\delta\omega_{\mu\nu}(g)$:

$$\delta\omega_{\mu\nu}(g) + \delta\omega_{\nu\mu}(g) = 0, \tag{5.32}$$

which therefore has $N(N-1)/2$ independent components (such a number, as stressed earlier, is also equal to to the order $M$ of $G_L$, $M = N(N-1)/2$), i.e., the (rotation) transformation group related to the tensor $\delta\omega_{\mu\nu}(g)$ is a $N(N-1)/2$-parameter Killing group.

Let us observe that a $N$-dimensional, generalized Minkowski space, being (as noted earlier) a special case of a Riemann space with constant curvature, admits a maximal Killing group with $N(N+1)/2$ parameters. Since the (rotation) transformation group related to the tensor $\delta\omega_{\mu\nu}(g)$ is a $N(N-1)/2$-parameter Killing group, we have still to find another $N$-parameter Killing group of $\widetilde{M_N}$ (because $N + N(N-1)/2 = N(N+1)/2$).

This is easily done by noting that the $N(N+1)/2$ Killing equations in such a space

$$\xi_\mu(x)_{,\nu} + \xi_\nu(x)_{,\mu} = 0 \equiv \frac{\partial \xi_\mu(x)}{\partial x^\nu} + \frac{\partial \xi_\nu(x)}{\partial x^\mu} = 0 \tag{5.33}$$

are trivially satisfied by constant covariant $N$-vectors $\xi_\mu \neq \xi_\mu(x)$, to which there corresponds the infinitesimal transformation

$$\delta g : x^\mu \to x^{\mu\prime}(x, \{x\}_{\text{n.m.}}) = x^\mu + \delta x^\mu_{(g)}(\{x\}_{\text{n.m.}}) = x^\mu + \xi^\mu_{(g)}(\{x\}_{\text{n.m.}}) \tag{5.34}$$

with $\delta x^\mu_{(g)}(\{x\}_{\text{n.m.}})$ , $\xi^\mu_{(g)}(\{x\}_{\text{n.m.}})$ constant (with respect to $x^\mu$).

In conclusion, a $N$-d generalized Minkowski space $\widetilde{M_N}(\{x\}_{\text{n.m.}})$ admits a maximal Killing group which is the (semidirect) product of the Lie group of the $N$-dimensional space–time rotations (or $N$-d generalized, homogeneous Lorentz group $\mathrm{SO}(T,S)^{N(N-1)/2}_{\text{GEN.}}$) with $N(N-1)/2$ parameters, and of the Lie group of the $N$-dimensional space–time translations $\mathrm{Tr.}(T,S)^N_{\text{GEN.}}$ with $N$ parameters:

$$P(T,S)^{N(N+1)/2}_{\text{GEN.}} = \mathrm{SO}(T,S)^{N(N-1)/2}_{\text{GEN.}} \otimes_s \mathrm{Tr.}(T,S)^N_{\text{GEN.}} \tag{5.35}$$

The semidirect nature of the group product is due to the fact that, as it shall be explicitly derived (in the case $N = 4, T = 1, S = 3$ of DSR, without loss of generality) in Chap. 7, in general we have that

$$\begin{gathered} \exists \text{ at least } 1 \ \ (\mu,\nu,\rho) \in \{1,2,...,N\}^3 : \\ : [I^{\mu\nu}_{\text{GEN.}}(\{x\}_{\text{n.m.}}), \Upsilon^\rho_{\text{GEN.}}(\{x\}_{\text{n.m.}})] \neq 0, \ \forall \{x\}_{\text{n.m.}} , \end{gathered} \tag{5.36}$$

where $I^{\mu\nu}_{\text{GEN.}}(\{x\}_{\text{n.m.}})$ and $\Upsilon^\rho_{\text{GEN.}}(\{x\}_{\text{n.m.}})$ are the infinitesimal generators of $\mathrm{SO}(T,S)^{N(N-1)/2}_{\text{GEN.}}$ and $\mathrm{Tr.}(T,S)^N_{\text{GEN.}}$, respectively. We will refer to $P(S,T)^{N(N+1)/2}_{\text{GEN.}}$ as the *generalized* (or *inhomogeneous Lorentz*) *group* .

### 5.2.4 Solution of Killing Equations in a 4D Generalized Minkowski Space

We want now to solve the Killing equations in a 4D generalized Minkowski space $\widetilde{M}(\{x\}_{\text{n.m.}})$ ($S \leq 4, T = 4 - S$). A covariant Killing four-vector $\xi_\mu(x)$

must satisfy (5.17), which explicitly amounts to the system:

$$\begin{cases} \text{(I)} & \xi_0(x)_{,0} = 0; \\ \text{(II)} & \xi_0(x)_{,1} + \xi_1(x)_{,0} = 0; \\ \text{(III)} & \xi_0(x)_{,2} + \xi_2(x)_{,0} = 0; \\ \text{(IV)} & \xi_0(x)_{,3} + \xi_3(x)_{,0} = 0; \\ \text{(V)} & \xi_1(x)_{,1} = 0; \\ \text{(VI)} & \xi_1(x)_{,2} + \xi_2(x)_{,1} = 0; \\ \text{(VII)} & \xi_1(x)_{,3} + \xi_3(x)_{,1} = 0; \\ \text{(VIII)} & \xi_2(x)_{,2} = 0; \\ \text{(IX)} & \xi_2(x)_{,3} + \xi_3(x)_{,2} = 0; \\ \text{(X)} & \xi_3(x)_{,3} = 0. \end{cases} \tag{5.37}$$

From equations (5.37) (I ,V,VII, and X) one gets:

$$\begin{cases} \xi_0 = \xi_0(x^1, x^2, x^3); \\ \xi_1 = \xi_1(x^0, x^2, x^3); \\ \xi_2 = \xi_2(x^0, x^1, x^3); \\ \xi_3 = \xi_3(x^0, x^1, x^2). \end{cases} \tag{5.38}$$

Solving system (5.37) is cumbersome but straightforward [41]. The final result is:

$$\begin{cases} \xi_0(x) = -\zeta^1 x^1 - \zeta^2 x^2 - \zeta^3 x^3 + T^0; \\ \xi_1(x) = \zeta^1 x^0 + \theta^2 x^3 - \theta^3 x^2 - T^1; \\ \xi_2(x) = \zeta^2 x^0 - \theta^1 x^3 + \theta^3 x^1 - T^2; \\ \xi_3(x) = \zeta^3 x^0 + \theta^1 x^2 - \theta^2 x^1 - T^3, \end{cases} \tag{5.39}$$

where $\zeta^i$, $\theta^i$ $(i = 1, 2, 3)$ and $T^\mu$ $(\mu = 0, 1, 2, 3)$ are real coefficients.

We can draw the following conclusions:

1. In spite of the fact that no assumption was made on the functional form of the Killing vector, we got a dependence at most linear (inhomogeneous) on the metric coordinates for all components of $\xi_\mu(x)$. Therefore, in order to determine the (maximal) Killing group of a generalized Minkowski space,[2] one can, without loss of generality, consider only groups whose transformation representation is implemented by transformations at most linear in the coordinates.

2. In general, $\xi_\mu \neq \xi_\mu(\{x\}_{\text{n.m.}})$, i.e., the covariant Killing vector does not depend on possible nonmetric variables.[3] On the contrary, the *contravariant* Killing four-vector *does indeed*, due to the dependence of the fully contravariant metric tensor on $\{x\}_{\text{n.m.}}$:

$$\xi^\mu(x, \{x\}_{\text{n.m.}}) = g^{\mu\nu}(\{x\}_{\text{n.m.}})\xi_\nu(x). \tag{5.40}$$

   Such a result is consistent with the fact that $\delta\omega_{\mu\nu}(g)$, unlike $\delta\omega^\mu_\nu(g, \{x\}_{\text{n.m.}})$, is independent of $\{x\}_{\text{n.m.}}$ (cf.(5.28),(5.29)).

3. Solution (5.39) *does not depend on the metric tensor.* This entails that all 4D generalized Minkowski spaces admit the same *covariant* Killing four-vector. *It corresponds to the covariant four-vector of infinitesimal transformation of the space–time rototranslational group of* $\widetilde{M}(\{x\}_{\text{n.m.}})$. Therefore, assuming the signature $(+,-,-,-)$ (i.e., $S = 3$, $T = 1$), in a basis of "length-dimensional" coordinates, we can state that:

   (a) $\boldsymbol{\zeta} = (\zeta^1, \zeta^2, \zeta^3)$ is the three-vector of the dimensionless parameters ("rapidity") of a generalized 3D boost

   (b) $\boldsymbol{\theta} = (\theta^1, \theta^2, \theta^3)$ is the three-vector of the dimensionless parameters (angles) of a generalized 3D rotation

   (c) $T_\mu = (T^0, -T^1, -T^2, -T^3)$ is the covariant four-vector of the ("length-dimensional") parameters of a generalized 4D translation

---

[2] In fact, although we discussed explicitly the 4D case, the extension to the generic $N$-d case is straightforward.

[3] Indeed

$$\begin{aligned}\xi_{\mu(g)}(x) &= g_{\mu\nu,DSR4}(\{x\}_{\text{n.m.}})\xi^\nu_{(g)}(x, \{x\}_{\text{n.m.}}) \\ &= g_{\mu\nu,\text{DSR4}}(\{x\}_{\text{n.m.}})\delta\omega^\nu_\rho(g, \{x\}_{\text{n.m.}})x^\rho = \delta\omega_{\mu\rho}(g)x^\rho.\end{aligned}$$

# 6

# Infinitesimal Structure of Generalized Space–Time Rotation Groups

We shall now take advantage of the results derived in Chap. 5 on the Killing vectors of generalized $N$-d Minkowski spaces in order to study in detail the infinitesimal structure of their space–time rotation groups [41]. Needless to say, special emphasis will be given to the 4D case, and in particular to the deformed Minkowski space $\widetilde{M}$ of DSR.

## 6.1 Finite-Dimensional Representation of Infinitesimal Generators and Generalized Lorentz Algebra

### 6.1.1 Generalized Lorentz Algebra

As in the standard special-relativistic case, we can decompose the mixed $N$-tensor of infinitesimal transformation parameters $\delta\omega^{\mu}_{\nu}(g, \{x\}_{\text{n.m.}})$ (see (5.21)) as:[1]

$$\delta\omega^{\mu}_{\nu}(g, \{x\}_{\text{n.m.}}) = \frac{1}{2}\delta\omega_{\alpha\beta}(g)(I^{\alpha\beta})^{\mu}_{\nu}(\{x\}_{\text{n.m.}}), \tag{6.1}$$

i.e., as a linear combination of $N(N-1)/2$ matrices (independent of the group element $g$) $\left\{(I^{\alpha\beta})^{\mu}_{\nu}(\{x\}_{\text{n.m.}})\right\}_{\alpha,\beta=1\ldots N}$ [2] with coefficients

[1] The factor 1/2 is inserted only for further convenience.

[2] The pair of indices ($\alpha$ $\beta$ ) labels the different infinitesimal group generators, whereas – in the (N($<\infty$)-dimensional) matrix representation of the generators we are

$\{\delta\omega_{\alpha\beta}(g)\}_{\alpha,\beta=1\ldots N}$. Such matrices represent *the infinitesimal generators* of the space–time rotational component of the maximal Killing group of $\widetilde{M_N}(\{x\}_{\text{n.m.}})$. Since in this case $\delta\omega_{\mu\nu}(g)$ is antisymmetric (see (5.32)), the basis matrices $\{(I^{\alpha\beta})^\mu_\nu\}_{\alpha,\beta=1\ldots N}$, too, are antisymmetric in indices $\alpha$ and $\beta$:

$$\{(I^{\alpha\beta})^\mu_\nu(\{x\}_{\text{n.m.}})\}_{\alpha,\beta=1\ldots N} = -\{(I^{\beta\alpha})^\mu_\nu(\{x\}_{\text{n.m.}})\}_{\alpha,\beta=1\ldots N}. \tag{6.2}$$

For the fully covariant $\delta\omega_{\mu\nu}(g)$ the analogous decomposition reads

$$\begin{aligned}\delta\omega_{\mu\nu}(g) &= g_{\mu\rho}(\{x\}_{\text{n.m.}})\delta\omega^\rho_\nu(g,\{x\}_{\text{n.m.}})\\ &= \frac{1}{2}\delta\omega_{\alpha\beta}(g)g_{\mu\rho}(\{x\}_{\text{n.m.}})(I^{\alpha\beta})^\rho_\nu(\{x\}_{\text{n.m.}})\\ &= \frac{1}{2}\delta\omega_{\alpha\beta}(g)(I^{\alpha\beta})_{\mu\nu}(\{x\}_{\text{n.m.}}).\end{aligned} \tag{6.3}$$

But, since $\delta\omega_{\mu\nu}(g)$ is independent of $\{x\}_{\text{n.m.}}$, the same holds for its components $\delta\omega_{\alpha\beta}(g)$, and therefore (6.3) implies

$$(I^{\alpha\beta})_{\mu\nu} \neq (I^{\alpha\beta})_{\mu\nu}(\{x\}_{\text{n.m.}}). \tag{6.4}$$

In order to find the explicit form of the infinitesimal generators in the $N$-d matrix representation, let us exploit the antisymmetry of $\delta\omega_{\mu\nu}(g)$:

$$\begin{aligned}\delta\omega_{\mu\nu}(g) &= -\delta\omega_{\nu\mu}(g) \Leftrightarrow \delta\omega_{\mu\nu}(g)\\ &= \frac{1}{2}(\delta\omega_{\mu\nu}+\delta\omega_{\mu\nu}) = \frac{1}{2}(\delta\omega_{\mu\nu}-\delta\omega_{\nu\mu})\\ &= \frac{1}{2}g^\alpha_\mu g^\beta_\nu\delta\omega_{\alpha\beta} - \frac{1}{2}g^\beta_\mu g^\alpha_\nu\delta\omega_{\alpha\beta} = \frac{1}{2}\delta\omega_{\alpha\beta}(g^\alpha_\mu g^\beta_\nu - g^\beta_\mu g^\alpha_\nu)\\ &= \frac{1}{2}\delta\omega_{\alpha\beta}(g)(\delta^\alpha_\mu\delta^\beta_\nu - \delta^\beta_\mu\delta^\alpha_\nu).\end{aligned} \tag{6.5}$$

Comparing (6.3) and (6.5) yields:[3]

$$g_{\mu\rho}(\{x\}_{\text{n.m.}})(I^{\alpha\beta})^\rho_\nu(\{x\}_{\text{n.m.}}) \equiv (I^{\alpha\beta})_{\mu\nu} = (\delta^\alpha_\mu\delta^\beta_\nu - \delta^\beta_\mu\delta^\alpha_\nu). \tag{6.6}$$

---

considering – the controvariant (covariant) index is a row (column) index. This latter remark holds true for $\delta\omega^\mu_\nu(g,\{x\}_{\text{n.m.}})$, too.

[3]Equation (6.6) shows clearly that the factors with a nonmetric dependence in $g_{\mu\rho}(\{x\}_{\text{n.m.}})$ and $(I^{\alpha\beta})^\rho_\nu(\{x\}_{\text{n.m.}})$ cancel each other (cf. (6.4)). The same *does not* happen when both $\mu$ and $\nu$ are contravariant:

$$\begin{aligned}(I^{\alpha\beta})^{\mu\nu} &\equiv g^{\mu\rho}(\{x\}_{\text{n.m.}})g^{\nu\sigma}(\{x\}_{\text{n.m.}})(I^{\alpha\beta})_{\rho\sigma}\\ &= g^{\mu\rho}(\{x\}_{\text{n.m.}})g^{\nu\sigma}(\{x\}_{\text{n.m.}})(\delta^\alpha_\rho\delta^\beta_\sigma - \delta^\beta_\rho\delta^\alpha_\sigma)\\ &= g^{\mu\alpha}(\{x\}_{\text{n.m.}})g^{\nu\beta}(\{x\}_{\text{n.m.}}) - g^{\mu\beta}(\{x\}_{\text{n.m.}})g^{\nu\alpha}(\{x\}_{\text{n.m.}}).\end{aligned}$$

We get therefore the following explicit form for the mixed matrix representation of the generators:[4]

$$
\begin{aligned}
(I^{\alpha\beta})^{\mu}_{\nu}(\{x\}_{\text{n.m.}}) &= g^{\mu\rho}(\{x\}_{\text{n.m.}})(I^{\alpha\beta})_{\rho\nu} \\
&= g^{\mu\rho}(\{x\}_{\text{n.m.}})(\delta^{\alpha}{}_{\rho}\delta^{\beta}_{\nu} - \delta^{\beta}_{\rho}\delta^{\alpha}_{\nu}) \\
&= g^{\mu\alpha}(\{x\}_{\text{n.m.}})\delta^{\beta}_{\nu} - g^{\mu\beta}(\{x\}_{\text{n.m.}})\delta^{\alpha}_{\nu}.
\end{aligned}
\tag{6.7}
$$

Let us now find out the Lie algebra of the generators $\{(I^{\alpha\beta})^{\mu}_{\nu}(\{x\}_{\text{n.m.}})\}_{\alpha,\beta=1\ldots N}$, with product defined by the commutator $[,]$. Since such an algebra is independent of the representation, we can evaluate such a commutator for their matrix representation (6.7). One has, for the generators $(\alpha\beta)$ and $(\rho\sigma)$

$$
\left(I^{\alpha\beta}I^{\rho\sigma}\right)^{\mu}_{\nu} = \left(I^{\alpha\beta}\right)^{\mu}_{\chi}\left(I^{\rho\sigma}\right)^{\chi}_{\nu}
\tag{6.8}
$$

and therefore

$$
\begin{aligned}
[I^{\alpha\beta}, I^{\rho\sigma}]^{\mu}_{\nu} &= \left(I^{\alpha\beta}I^{\rho\sigma} - I^{\rho\sigma}I^{\alpha\beta}\right)^{\mu}_{\nu} \\
&= \left(I^{\alpha\beta}I^{\rho\sigma}\right)^{\mu}_{\nu} - \left(I^{\rho\sigma}I^{\alpha\beta}\right)^{\mu}_{\nu} \\
&= \left(I^{\alpha\beta}\right)^{\mu}_{\chi}\left(I^{\rho\sigma}\right)^{\chi}_{\nu} - \left(I^{\rho\sigma}\right)^{\mu}_{\chi}\left(I^{\alpha\beta}\right)^{\chi}_{\nu} \\
&= \left(g^{\mu\alpha}\delta^{\beta}_{\chi} - g^{\mu\beta}\delta^{\alpha}_{\chi}\right)\left(g^{\chi\rho}\delta^{\sigma}_{\nu} - g^{\chi\sigma}\delta^{\rho}_{\nu}\right) \\
&\quad - \left(g^{\mu\rho}\delta^{\sigma}_{\chi} - g^{\mu\sigma}\delta^{\rho}_{\chi}\right)\left(g^{\chi\alpha}\delta^{\beta}_{\nu} - g^{\chi\beta}\delta^{\alpha}_{\nu}\right) \\
&= \left(g^{\mu\alpha}\delta^{\beta}_{\chi}g^{\chi\rho}\delta^{\sigma}_{\nu} - g^{\mu\alpha}\delta^{\beta}_{\chi}g^{\chi\sigma}\delta^{\rho}_{\nu} - g^{\mu\beta}\delta^{\alpha}_{\chi}g^{\chi\rho}\delta^{\sigma}_{\nu} + g^{\mu\beta}\delta^{\alpha}_{\chi}g^{\chi\sigma}\delta^{\rho}_{\nu}\right) \\
&\quad - \left(g^{\mu\rho}\delta^{\sigma}_{\chi}g^{\chi\alpha}\delta^{\beta}_{\nu} - g^{\mu\rho}\delta^{\sigma}_{\chi}g^{\chi\beta}\delta^{\alpha}_{\nu} - g^{\mu\sigma}\delta^{\rho}_{\chi}g^{\chi\alpha}\delta^{\beta}_{\nu} + g^{\mu\sigma}\delta^{\rho}_{\chi}g^{\chi\beta}\delta^{\alpha}_{\nu}\right) \\
&= \left(g^{\mu\alpha}g^{\beta\rho}\delta^{\sigma}_{\nu} - g^{\mu\alpha}g^{\beta\sigma}\delta^{\rho}_{\nu} - g^{\mu\beta}g^{\alpha\rho}\delta^{\sigma}_{\nu} + g^{\mu\beta}g^{\alpha\sigma}\delta^{\rho}_{\nu}\right) \\
&\quad - \left(g^{\mu\rho}g^{\sigma\alpha}\delta^{\beta}_{\nu} - g^{\mu\rho}g^{\sigma\beta}\delta^{\alpha}_{\nu} - g^{\mu\sigma}g^{\rho\alpha}\delta^{\beta}_{\nu} + g^{\mu\sigma}g^{\rho\beta}\delta^{\alpha}_{\nu}\right) \\
&= g^{\alpha\sigma}\left(g^{\beta\mu}\delta^{\rho}_{\nu} - g^{\rho\mu}\delta^{\beta}_{\nu}\right) + g^{\beta\rho}\left(g^{\alpha\mu}\delta^{\sigma}_{\nu} - g^{\sigma\mu}\delta^{\alpha}_{\nu}\right) \\
&\quad - g^{\alpha\rho}\left(g^{\beta\mu}\delta^{\sigma}_{\nu} - g^{\sigma\mu}\delta^{\beta}_{\nu}\right) - g^{\beta\sigma}\left(g^{\alpha\mu}\delta^{\rho}_{\nu} - g^{\rho\mu}\delta^{\alpha}_{\nu}\right) \\
&= g^{\alpha\sigma}\left(I^{\beta\rho}\right)^{\mu}_{\nu} + g^{\beta\rho}\left(I^{\alpha\sigma}\right)^{\mu}_{\nu} - g^{\alpha\rho}\left(I^{\beta\sigma}\right)^{\mu}_{\nu} - g^{\beta\sigma}\left(I^{\alpha\rho}\right)^{\mu}_{\nu},
\end{aligned}
\tag{6.9}
$$

[4]We have analogously

$$
\begin{aligned}
(I^{\alpha\beta})^{\nu}_{\mu}(\{x\}_{\text{n.m.}}) &= g^{\nu\rho}(\{x\}_{\text{n.m.}})(I^{\alpha\beta})_{\mu\rho} \\
&= g^{\nu\rho}(\{x\}_{\text{n.m.}})(\delta^{\alpha}_{\mu}\delta^{\beta}_{\rho} - \delta^{\beta}_{\mu}\delta^{\alpha}_{\rho}) \\
&= g^{\nu\beta}(\{x\}_{\text{n.m.}})\delta^{\alpha}_{\mu} - g^{\nu\alpha}(\{x\}_{\text{n.m.}})\delta^{\beta}_{\mu}.
\end{aligned}
$$

namely

$$\begin{aligned}
&[I^{\alpha\beta}(\{x\}_{\text{n.m.}}), I^{\rho\sigma}(\{x\}_{\text{n.m.}})]^{\mu}_{\nu} \\
&\quad = g^{\alpha\sigma}(\{x\}_{\text{n.m.}}) \left(I^{\beta\rho}\right)^{\mu}_{\nu}(\{x\}_{\text{n.m.}}) + g^{\beta\rho}(\{x\}_{\text{n.m.}}) \left(I^{\alpha\sigma}\right)^{\mu}_{\nu}(\{x\}_{\text{n.m.}}) \\
&\qquad - g^{\alpha\rho}(\{x\}_{\text{n.m.}}) \left(I^{\beta\sigma}\right)^{\mu}_{\nu}(\{x\}_{\text{n.m.}}) - g^{\beta\sigma}(\{x\}_{\text{n.m.}}) \left(I^{\alpha\rho}\right)^{\mu}_{\nu}(\{x\}_{\text{n.m.}}).
\end{aligned} \tag{6.10}$$

By omitting row and column indices, we get eventually

$$\begin{aligned}
&[I^{\alpha\beta}(\{x\}_{\text{n.m.}}), I^{\rho\sigma}(\{x\}_{\text{n.m.}})] \\
&\quad = g^{\alpha\sigma}(\{x\}_{\text{n.m.}}) I^{\beta\rho}(\{x\}_{\text{n.m.}}) + g^{\beta\rho}(\{x\}_{\text{n.m.}}) I^{\alpha\sigma}(\{x\}_{\text{n.m.}}) \\
&\qquad - g^{\alpha\rho}(\{x\}_{\text{n.m.}}) I^{\beta\sigma}(\{x\}_{\text{n.m.}}) - g^{\beta\sigma}(\{x\}_{\text{n.m.}}) I^{\alpha\rho}(\{x\}_{\text{n.m.}}).
\end{aligned} \tag{6.11}$$

Equation (6.11) defines the *generalized Lorentz algebra*, associated to the generalized, homogeneous Lorentz group $SO(T,S)^{N(N-1)/2}_{\text{GEN.}}$ of the $N$-dimensional generalized Minkowski space $\widetilde{M_N}(\{x\}_{\text{n.m.}})$.

### *6.1.2 Dependence of the Transformation Commutativity on the Parametric Level*

Infinitesimal transformations do commute in $\widetilde{M_N}(\{x\}_{\text{n.m.}})$. In fact, let us consider a generic product of infinitesimal coordinate transformations of $SO(T = N - S, S)_{\text{GEN.}}$, expressed in terms of $\delta\omega_{\alpha\beta}(g)$ and $I^{\alpha\beta}$ (Greek indices have a cardinality $N$, denoted by $\{[N]\}$) (ESC off on $\alpha$ and $\beta$):[5]

$$x^{\mu\prime}_{(g)} = \left\{ \prod_{\alpha,\beta\in\{[N]\},\alpha>\beta} \left[1 + \delta\omega_{\alpha\beta}(g) I^{\alpha\beta}(\{x\}_{\text{n.m.}}\right] \right\}^{\mu}_{\nu} x^{\nu}. \tag{6.12}$$

By neglecting terms of order $\geq 2$ in the parametric tensor, one gets (ESC on):

$$\begin{aligned}
x^{\mu\prime}_{(g)} &\cong \left[1 + \delta\omega_{\alpha\beta}(g) I^{\alpha\beta}(\{x\}_{\text{n.m.}})\right]^{\mu}_{\nu} x^{\nu} \\
&= \left[\delta^{\mu}_{\nu} + \delta\omega_{\alpha\beta}(g) \left(I^{\alpha\beta}\right)^{\mu}_{\nu}(\{x\}_{\text{n.m.}})\right] x^{\nu} \\
&= x^{\mu} + \delta\omega_{\alpha\beta}(g) \left(I^{\alpha\beta}\right)^{\mu}_{\nu}(\{x\}_{\text{n.m.}}) x^{\nu}.
\end{aligned} \tag{6.13}$$

Equation (6.13) contains a matrix sum of generators, which entails commutativity of the infinitesimal transformations.

---

[5]The generator product in (6.12) is to be meant as that defined on the Lie group considered. In particular, in the $N$-dimensional matrix representation, it coincides with the usual product of square matrices.

The earlier properties hold for any Lie group $G_{\mathrm{L}}$ in the transformation representation, and for the related space $S_N$. Indeed, let us consider the product of infinitesimal transformations (see (5.17)–(5.20))

$$x^{\mu\prime}_{(g)} = \left(\left(\prod_{i=1}^{M}\left(1+\alpha_i(g)T^i\right)\right)x\right)^{\mu}, \tag{6.14}$$

where $x$ is a contravariant $N$-vector of $S_N$. Although $G_{\mathrm{L}}$ does not admit, in general, a matrix representation on $\widetilde{M_N}(\{x\}_{\mathrm{n.m.}})$, it is still possible to neglect terms of order $\geqslant 2$ in the infinitesimal transformation parameters and write

$$\begin{aligned} x^{\mu\prime}_{(g)} &= \left[\left(1+\sum_{i=1}^{N}\alpha_i(g)T^i\right)x\right]^{\mu} = x + \left[\left(\sum_{i=1}^{N}\alpha_i(g)T^i\right)x\right]^{\mu} \\ &= x^{\mu} + \left[\left(\sum_{i=1}^{N}\alpha_i(g)T^i\right)x\right]^{\mu} = x^{\mu} + \sum_{i=1}^{N}\alpha_i(g)\left(T^i x\right)^{\mu}, \end{aligned} \tag{6.15}$$

which entails again that infinitesimal transformations do commute.

The nontrivial nature of the algebraic structure of the Lie groups (not necessarily abelian) reveals itself at the finite level of their realization. Indeed, in the transformation representation, we have, by the P.B.W. and Lie theorems (cf. (5.2)):

$$S_N \ni gx = \left[\exp\left(\sum_{i=1}^{M}\alpha_i(g)T^i\right)\right]x. \tag{6.16}$$

Consider two arbitrary finite elements $g'$, $g'' \in G_{\mathrm{L}}$ such that

$$\begin{cases} \alpha_i(g') = \delta_{ii'}\alpha_{i'}(g'); \\ \alpha_i(g'') = \delta_{ii''}\alpha_{i''}(g''), \end{cases} \tag{6.17}$$

and their products (where $\circ$ denotes the product in $G_{\mathrm{L}}$)

$$g' \circ g'' = \widetilde{g} \in G_{\mathrm{L}} \tag{6.18}$$

$$g'' \circ g' = \widetilde{\widetilde{g}} \in G_{\mathrm{L}} \tag{6.19}$$

(with $\widetilde{g} \neq \widetilde{\widetilde{g}}$, because $\circ$ is in general non-abelian). The finite coordinate transformations associated to $\widetilde{g}$ and $\widetilde{\widetilde{g}}$ are, respectively, (ESC off):

$$S_N \ni \widetilde{g}x = \left[\exp\left(\alpha_{i'}(g)T^{i'}\right)\exp\left(\alpha_{i''}(g)T^{i''}\right)\right]x; \tag{6.20}$$

$$S_N \ni \widetilde{\widetilde{g}}x = \left[\exp\left(\alpha_{i''}(g)T^{i''}\right)\exp\left(\alpha_{i'}(g)T^{i'}\right)\right]x. \tag{6.21}$$

The condition to be satisfied in order to have $\widetilde{g}x = \widetilde{\widetilde{g}}x$ is

$$\begin{aligned}
\widetilde{g}x &= \widetilde{\widetilde{g}}x \Leftrightarrow \exp\left(\alpha_{i'}(g)T^{i'}\right)\exp\left(\alpha_{i''}(g)T^{i''}\right) \\
&\stackrel{\star}{=} \exp\left(\alpha_{i'}(g)T^{i'} + \alpha_{i''}(g)T^{i''}\right) \\
&= \exp\left(\alpha_{i''}(g)T^{i''} + \alpha_{i'}(g)T^{i'}\right) \\
&\stackrel{\star}{=} \exp\left(\alpha_{i''}(g)T^{i''}\right)\exp\left(\alpha_{i'}(g)T^{i'}\right).
\end{aligned} \tag{6.22}$$

Exploiting the Baker–Campbell–Hausdorff (BCH) formula , we get

$$\begin{gathered}
\exp\left(\alpha_{i'}(g)T^{i'}\right)\exp\left(\alpha_{i''}(g)T^{i''}\right) = \exp\alpha_{i'}(g)T^{i'} + \alpha_{i''}(g)T^{i''} \\
\Leftrightarrow 0 = [\alpha_{i'}(g)T^{i'}, \alpha_{i''}(g)T^{i''}] = \alpha_{i'}(g)\alpha_{i''}(g)[T^{i'}, T^{i''}] \Leftrightarrow [T^{i'}, T^{i''}] = 0,
\end{gathered} \tag{6.23}$$

namely the equalities of (6.22) marked with "$\star$" do hold *iff* the commutator of $T^{i'}$ and $T^{i''}$ vanish for any pair of generators, which is not true in general due to the non-abelian nature of $G_{\mathrm{L}}$. This implies that finite coordinate transformations in the transformation representation of a non-abelian Lie group do not commute.

The results of the present section apply in particular to the Killing groups of the generalized $N$-dimensional Minkowski spaces $\widetilde{M_N}(\{x\}_{\text{n.m.}})$, i.e., to the generalized Poincaré groups $P(T = N-S, S)_{\text{GEN.}}^{N(N+1)/2.}$ and their space–time rotation components (namely the generalized homogeneous Lorentz groups $\mathrm{SO}(T = N - S, S)_{\text{GEN.}}^{N(N-1)/2}$).

Let us notice that the (parametric) dependence of the metric of a generalized Minkowski space on the set $\{x\}_{\text{n.m.}}$ of nonmetrical coordinates reflects itself also at the group level. In particular, such a dependence shows up in:

1. The $N \times N$ matrix representation of the infinitesimal generators
2. The infinitesimal group transformations
3. The structure constants of the Lie algebra of generators

Moreover, since to any fixed value $\{\bar{x}\}_{\text{n.m.}}$ of $\{x\}_{\text{n.m.}}$ there corresponds a generalized Minkowski space $\widetilde{M_N}(\{\bar{x}\}_{\text{n.m.}})$, we have a family of $N$-d generalized Minkowski spaces

$$\left\{\widetilde{M_N}(\{x\}_{\text{n.m.}})\right\}_{\{x\}_{\text{n.m.}} \in R_{\{x\}_{\text{n.m.}}}}, \tag{6.24}$$

where $R_{\{x\}_{\text{n.m.}}}$ is the range of the set $\{x\}_{\text{n.m.}}$; if the cardinality of the range of each element of the set $\{x\}_{\text{n.m.}}$ is infinite, the cardinality of $R_{\{x\}_{\text{n.m.}}}$

(and of the family (6.24)) is $\infty^{N_{\text{n.m.}}}$. In correspondence, one gets a family of generalized Poincaré groups

$$\left\{ P(T,S)_{\text{GEN.}}^{N(N+1)/2}(\{x\}_{\text{n.m.}}) \right\}_{\{x\}_{\text{n.m.}} \in R_{\{x\}_{\text{n.m.}}}} \tag{6.25}$$

with the same cardinality structure as (6.24).

This can be symbolically summarized as:

$$\left.\begin{array}{c} \textit{(Hyper)spatial level of } N\textit{-d generalized Minkowski spaces:} \\ 1) \left\{ \widetilde{M_N}(\{x\}_{\text{n.m.}}) \right\}_{\{x\}_{\text{n.m.}} \in R_{\{x\}_{\text{n.m.}}}} \\ 2) \widetilde{M_N}(\{x\}_{\text{n.m.}}) \equiv \widetilde{M_N}(\{\bar{x}\}_{\text{n.m.}}) \end{array}\right\} \Leftrightarrow$$

$$\Leftrightarrow \left\{ \begin{array}{l} \qquad \textit{Group level of related maximal Killing groups:} \\ \qquad 1) \left\{ P(T,S)_{\text{GEN.}}^{N(N+1)/2}(\{x\}_{\text{n.m.}}) \right\}_{\{x\}_{\text{n.m.}} \in R_{\{x\}_{\text{n.m.}}}} = \\ = \left\{ \text{SO}(T,S)_{\text{GEN.}}^{N(N-1)/2}(\{x\}_{\text{n.m.}}) \otimes_s \text{Tr.}(T,S)_{\text{GEN.}}^{N}(\{x\}_{\text{n.m.}}) \right\}_{\{x\}_{\text{n.m.}} \in R_{\{x\}_{\text{n.m.}}}} \\ \qquad 2)\ P(T,S)_{\text{GEN.}}^{N(N+1)/2}(\{x\}_{\text{n.m.}}) \equiv P(T,S)_{\text{GEN.}}^{N(N+1)/2}(\{\bar{x}\}_{\text{n.m.}}) \end{array} \right. \tag{6.26}$$

## 6.2 The Case of a 4D Generalized Minkowski Space

### 6.2.1 *Self-Representation of the Infinitesimal Generators*

Let us specialize the results of Sect. 6.1.2 to a 4D generalized Minkowski space. Assuming therefore that Greek indices take the values $\{0,1,2,3\}$, and a signature ($S \leqslant 4.T = 4 - S$), we can write explicitly the generator $(\alpha\beta)$ of $\text{SO}(T = 4 - S, S)_{\text{GEN}}$ as the antisymmetric matrix:

$$I^{\alpha\beta}(\{x\}_{\text{n.m.}})$$

$$= \begin{pmatrix} 0 & I^{01}(\{x\}_{\text{n.m.}}) & I^{02}(\{x\}_{\text{n.m.}}) & I^{03}(\{x\}_{\text{n.m.}}) \\ -I^{01}(\{x\}_{\text{n.m.}}) & 0 & I^{12}(\{x\}_{\text{n.m.}}) & I^{13}(\{x\}_{\text{n.m.}}) \\ -I^{02}(\{x\}_{\text{n.m.}}) & -I^{12}(\{x\}_{\text{n.m.}}) & 0 & I^{23}(\{x\}_{\text{n.m.}}) \\ -I^{03}(\{x\}_{\text{n.m.}}) & -I^{13}(\{x\}_{\text{n.m.}}) & -I^{23}(\{x\}_{\text{n.m.}}) & 0 \end{pmatrix}. \tag{6.27}$$

Like any rank-2, antisymmetric 4-tensor, $I^{\alpha\beta}(\{x\}_{\text{n.m.}})$ can be expressed in terms of an axial and a polar three-vector. By introducing the following infinitesimal generators ($i, j, k = 1, 2, 3$, ESC on throughout)

$$S^i(\{x\}_{\text{n.m.}}) \equiv \frac{1}{2}\epsilon^i{}_{jk} I^{jk}(\{x\}_{\text{n.m.}}); \tag{6.28}$$

$$K^i(\{x\}_{\text{n.m.}}) \equiv I^{0i}(\{x\}_{\text{n.m.}}), \tag{6.29}$$

(where $\epsilon_{ijk}$ is the rank-3, fully antisymmetric Levi-Civita 3-tensor with $\epsilon_{123} \equiv 1$), components of the axial three-vector

$$\mathbf{S}(\{x\}_{\text{n.m.}}) \equiv (I^{23}(\{x\}_{\text{n.m.}}), I^{31}(\{x\}_{\text{n.m.}}), I^{12}(\{x\}_{\text{n.m.}})) \tag{6.30}$$

and of the polar one

$$\mathbf{K}(\{x\}_{\text{n.m.}}) \equiv (I^{01}(\{x\}_{\text{n.m.}}), I^{02}(\{x\}_{\text{n.m.}}), I^{03}(\{x\}_{\text{n.m.}})). \tag{6.31}$$

$I^{\alpha\beta}(\{x\}_{\text{n.m.}})$ can be rewritten as:

$$I^{\alpha\beta}(\{x\}_{\text{n.m.}})$$
$$= \begin{pmatrix} 0 & K^1(\{x\}_{\text{n.m.}}) & K^2(\{x\}_{\text{n.m.}}) & K^3(\{x\}_{\text{n.m.}}) \\ -K^1(\{x\}_{\text{n.m.}}) & 0 & S^3(\{x\}_{\text{n.m.}}) & -S^2(\{x\}_{\text{n.m.}}) \\ -K^2(\{x\}_{\text{n.m.}}) & -S^3(\{x\}_{\text{n.m.}}) & 0 & S^1(\{x\}_{\text{n.m.}}) \\ -K^3(\{x\}_{\text{n.m.}}) & S^2(\{x\}_{\text{n.m.}}) & -S^1(\{x\}_{\text{n.m.}}) & 0 \end{pmatrix}. \tag{6.32}$$

The set of generators $\mathbf{S}(\{x\}_{\text{n.m.}})$, $\mathbf{K}(\{x\}_{\text{n.m.}})$ constitutes the *self-representation basis* for $\mathrm{SO}(T = 4 - S, S)_{\text{GEN.}}$. Unlike the case of standard SR – where **S**, **K** do represent the rotation and boost generators, respectively – one cannot give them a precise physical meaning, because this latter depends on both the number $S$ of space-like dimensions and the assignment of dimensional labeling (we left here unspecified).

### 6.2.2 *Decomposition of the Parametric 4-Tensor $\delta\omega_{\mu\nu}(g)$*

We can now exploit the self-representation form of the infinitesimal generators of $\mathrm{SO}(T = 4 - S, S)_{\text{GEN.}}$ ($S \leq 4$) to decompose the infinitesimal parametric 4-tensor $\delta\omega_{\mu\nu}(g)$.

Equation (5.29), on account of (6.3), can be written as:

$$\begin{aligned} \delta x^\mu_{(g)}(x, \{x\}_{\text{n.m.}}) &= \delta\omega^\mu_\nu(g, \{x\}_{\text{n.m.}}) x^\nu \\ &= \frac{1}{2}\delta\omega_{\alpha\beta}(g)(I^{\alpha\beta})^\mu_\nu(\{x\}_{\text{n.m.}}) x^\nu, \end{aligned} \tag{6.33}$$

which is valid in the general case of $\mathrm{SO}(T = N - S, S)_{\text{GEN.}}$.

In the case $N = 4$, we have by (6.32):

$$\begin{aligned}\delta x^{\mu}_{(g)}(x,\{x\}_{\text{n.m.}}) &= \delta\omega^{\mu}_{\nu}(g,\{x\}_{\text{n.m.}})x^{\nu} = \frac{1}{2}\delta\omega_{\alpha\beta}(g)(I^{\alpha\beta})^{\mu}_{\nu}(\{x\}_{\text{n.m.}})x^{\nu} \\ &= \frac{1}{2}\delta\omega_{ij}(g)(I^{ij})^{\mu}_{\nu}(\{x\}_{\text{n.m.}})x^{\nu} + \delta\omega_{0i}(g)(I^{0i})^{\mu}_{\nu}(\{x\}_{\text{n.m.}})x^{\nu} \\ &= \frac{1}{2}\delta\omega_{ij}(g)(I^{ij})^{\mu}_{\nu}(\{x\}_{\text{n.m.}})x^{\nu} + \delta\omega_{0i}(g)(K^{i})^{\mu}_{\nu}(\{x\}_{\text{n.m.}})x^{\nu}.\end{aligned} \tag{6.34}$$

Moreover, from (6.28) it follows:[6]

$$S^{i}(\{x\}_{\text{n.m.}}) \equiv \frac{1}{2}\epsilon^{i}{}_{jk}I^{jk}(\{x\}_{\text{n.m.}}) \Leftrightarrow I^{jk}(\{x\}_{\text{n.m.}}) = \epsilon^{jk}{}_{l}S^{l}(\{x\}_{\text{n.m.}}). \tag{6.35}$$

Replacing (6.35) in (6.34) one has:

$$\begin{aligned}\delta x^{\mu}_{(g)}(x,\{x\}_{\text{n.m.}}) &= \frac{1}{2}\delta\omega_{ij}(g)(I^{ij})^{\mu}_{\nu}(\{x\}_{\text{n.m.}})x^{\nu} + \delta\omega_{0i}(g)(K^{i})^{\mu}_{\nu}(\{x\}_{\text{n.m.}})x^{\nu} \\ &= \frac{1}{2}\delta\omega_{ij}(g)(\epsilon^{ij}{}_{l}S^{l}(\{x\}_{\text{n.m.}}))^{\mu}_{\nu}x^{\nu} + \delta\omega_{0i}(g)(K^{i})^{\mu}_{\nu}(\{x\}_{\text{n.m.}})x^{\nu} \\ &= \frac{1}{2}\epsilon^{ij}{}_{l}\delta\omega_{ij}(g)(S^{l})^{\mu}_{\nu}(\{x\}_{\text{n.m.}})x^{\nu} + \delta\omega_{0i}(g)(K^{i})^{\mu}_{\nu}(\{x\}_{\text{n.m.}})x^{\nu}\end{aligned} \tag{6.36}$$

(where in the last step the change of notation $(\epsilon^{ij}{}_{l}S^{l}(\{x\}_{\text{n.m.}}))^{\mu}_{\nu} = \epsilon^{ij}{}_{l}(S^{l})^{\mu}_{\nu}(\{x\}_{\text{n.m.}})$ has been made).

We can now define *an axial and a polar parametric three-vector* by

$$\theta_i(g) \equiv -\frac{1}{2}\epsilon_{i}{}^{jk}\delta\omega_{jk}(g); \tag{6.37}$$

$$\zeta_i(g) \equiv -\delta\omega_{0i}(g), \tag{6.38}$$

namely

$$\boldsymbol{\theta}(g) \equiv (-\delta\omega_{23}(g), -\delta\omega_{31}(g), -\delta\omega_{12}(g)); \tag{6.39}$$

$$\boldsymbol{\zeta}(g) \equiv (-\delta\omega_{01}(g), -\delta\omega_{02}(g), -\delta\omega_{03}(g)). \tag{6.40}$$

---

[6] One has indeed

$$\begin{aligned}S^{i}(\{x\}_{\text{n.m.}}) &\equiv \frac{1}{2}\epsilon^{i}_{jk}I^{jk}(\{x\}_{\text{n.m.}}) \\ &= \frac{1}{2}\epsilon^{i}_{jk}\epsilon^{jk}_{l}S^{l}(\{x\}_{\text{n.m.}}) = \delta^{i}_{l}S^{l}(\{x\}_{\text{n.m.}}) = S^{i}(\{x\}_{\text{n.m.}}),\end{aligned}$$

where use has been made of the formula

$$\epsilon_{ijk}\epsilon^{k}_{lm} = \delta_{il}\delta_{jm} - \delta_{im}\delta_{jl},$$

whence

$$\epsilon^{i}_{jk}\epsilon^{jk}_{l} = \epsilon^{i}_{jk}\epsilon^{jk}_{l} = \delta^{i}_{l}\delta^{j}_{j} - \delta^{i}_{j}\delta^{j}_{l} = 3\delta^{i}_{l} - \delta^{i}_{l} = 2\delta^{i}_{l}.$$

Therefore, $\delta\omega_{\alpha\beta}(g)$ can be written in matrix form as:

$$\delta\omega_{\alpha\beta}(g) = \begin{pmatrix} 0 & -\zeta^1(g) & -\zeta^2(g) & -\zeta^3(g) \\ \zeta^1(g) & 0 & -\theta^3(g) & \theta^2(g) \\ \zeta^2(g) & \theta^3(g) & 0 & -\theta^1(g) \\ \zeta^3(g) & -\theta^2(g) & \theta^1(g) & 0 \end{pmatrix}. \tag{6.41}$$

Having left the number $S$ of space-like dimensions and the dimensional labeling unspecified, we cannot attribute a physical meaning to the parametric three-vectors (6.37)–(6.40) (unlike the case of standard SR, where $\boldsymbol{\theta}(g)$ and $\boldsymbol{\zeta}(g)$ are the space rotation and boost parameters, respectively).

Equation (6.36) can be rewritten in terms of the 3D Euclidean scalar product $\cdot$ as:

$$\begin{aligned} \delta x^\mu_{(g)}(x, \{x\}_{\text{n.m.}}) &= -\theta_l(g)(S^l)^\mu_\nu(\{x\}_{\text{n.m.}})x^\nu - \zeta_i(g)(K^i)^\mu_\nu\{x\}_{\text{n.m.}})x^\nu \\ &= \left[-\theta_l(g)S^l(\{x\}_{\text{n.m.}}) - \zeta_i(g)K^i(\{x\}_{\text{n.m.}})\right]^\mu_\nu x^\nu \\ &= \left[-\boldsymbol{\theta}(g)\cdot\mathbf{S}(\{x\}_{\text{n.m.}}) - \boldsymbol{\zeta}(g)\cdot\mathbf{K}(\{x\}_{\text{n.m.}})\right]^\mu_\nu x^\nu. \end{aligned} \tag{6.42}$$

Let us stress that with the signature $S$=3, $T$=1, in the limit

$$g_{\mu\nu}(\{x\}_{\text{n.m.}}) \to g_{\mu\nu,\text{SR}}, \tag{6.43}$$

all results valid at group-transformation level in a 4D generalized Minkowski space reduce to the standard ones in SR.

## 6.3 Space–Time Rotations in a 4D Deformed Minkowski Space

### *6.3.1 Deformed Lorentz Group and Self-Representation Basis of Infinitesimal Generators*

We want now to specialize the results obtained to the case of DSR, i.e., considering a 4D deformed Minkowski space $\widetilde{M}(x^5)$.

Let us recall that the $N$-dimensional representation of the infinitesimal generators of the Killing group in a $N$-d generalized Minkowski space is determined (by means of (6.7)) by the mere knowledge of its metric tensor. In the DSR case we have therefore

$$\begin{aligned} &(I^{\alpha\beta})^\mu_{\nu,\text{DSR}}(x^5) = g^{\mu\rho}_{\text{DSR}}(x^5)(I^{\alpha\beta})_{\rho\nu,\text{DSR}} \\ &= g^{\mu\rho}_{\text{DSR}}(x^5)(\delta^\alpha{}_\rho\delta^\beta_\nu - \delta^\beta_\rho\delta^\alpha_\nu) = g^{\mu\alpha}_{\text{DSR}}(x^5)\delta^\beta_\nu - g^{\mu\beta}_{\text{DSR}}(x^5)\delta^\alpha_\nu \\ &\overset{\text{ESC off}}{=} \delta^{\mu\alpha}\left(b_0^{-2}(x^5)\delta^{\mu0} - b_1^{-2}(x^5)\delta^{\mu1} - b_2^{-2}(x^5)\delta^{\mu2} - b_3^{-2}(x^5)\delta^{\mu3}\right)\delta^\beta_\nu \\ &\quad -\delta^{\mu\beta}\left(b_0^{-2}(x^5)\delta^{\mu0} - b_1^{-2}(x^5)\delta^{\mu1} - b_2^{-2}(x^5)\delta^{\mu2} - b_3^{-2}(x^5)\delta^{\mu3}\right)\delta^\alpha_\nu. \end{aligned} \tag{6.44}$$

From (6.44) we get the following $4 \times 4$ matrix representation of the infinitesimal generator of the deformed homogeneous Lorentz group $\mathrm{SO}(1,3)_{\mathrm{DEF.}}$ (space–time rotation component of the deformed Poincaré group $P(1,3)_{\mathrm{DEF.}}$):

$$I^{10}_{\mathrm{DSR}}(x^5) = \begin{pmatrix} 0 & -b_0^{-2}(x^5) & 0 & 0 \\ -b_1^{-2}(x^5) & 0 & 0 & 0 \\ 0 & 0 & 0 & 0 \\ 0 & 0 & 0 & 0 \end{pmatrix}; \tag{6.45}$$

$$I^{20}_{\mathrm{DSR}}(x^5) = \begin{pmatrix} 0 & 0 & -b_0^{-2}(x^5) & 0 \\ 0 & 0 & 0 & 0 \\ -b_2^{-2}(x^5) & 0 & 0 & 0 \\ 0 & 0 & 0 & 0 \end{pmatrix}; \tag{6.46}$$

$$I^{30}_{\mathrm{DSR}}(x^5) = \begin{pmatrix} 0 & 0 & 0 & -b_0^{-2}(x^5) \\ 0 & 0 & 0 & 0 \\ 0 & 0 & 0 & 0 \\ -b_3^{-2}(x^5) & 0 & 0 & 0 \end{pmatrix}; \tag{6.47}$$

$$I^{12}_{\mathrm{DSR}}(x^5) = \begin{pmatrix} 0 & 0 & 0 & 0 \\ 0 & 0 & -b_1^{-2}(x^5) & 0 \\ 0 & b_2^{-2}(x^5) & 0 & 0 \\ 0 & 0 & 0 & 0 \end{pmatrix}; \tag{6.48}$$

$$I^{23}_{\mathrm{DSR}}(x^5) = \begin{pmatrix} 0 & 0 & 0 & 0 \\ 0 & 0 & 0 & 0 \\ 0 & 0 & 0 & -b_2^{-2}(x^5) \\ 0 & 0 & b_3^{-2}(x^5) & 0 \end{pmatrix}; \tag{6.49}$$

$$I^{31}_{\mathrm{DSR}}(x^5) = \begin{pmatrix} 0 & 0 & 0 & 0 \\ 0 & 0 & 0 & b_1^{-2}(x^5) \\ 0 & 0 & 0 & 0 \\ 0 & -b_3^{-2}(x^5) & 0 & 0 \end{pmatrix}. \tag{6.50}$$

Comparing (6.45)–(6.50) with the 4D matrix representation of the infinitesimal generators of the standard homogeneous Lorentz group $\mathrm{SO}(1,3)$ shows that *deforming the metric structure implies the loss of symmetry of the boost generators and of antisymmetry of space-rotation generators.*

The antisymmetry of the generators in the labeling indices $(\alpha\beta)$ still holds:

$$\left\{(I^{\alpha\beta})^{\mu}_{\nu,\mathrm{DSR}}(x^5)\right\}_{\alpha,\beta=0,1,2,3} = -\left\{(I^{\beta\alpha})^{\mu}_{\nu,\mathrm{DSR}}(x^5)\right\}_{\alpha,\beta=0,1,2,3} \tag{6.51}$$

or

$$I^{\alpha\beta}_{\mathrm{DSR}}(x^5) = -I^{\beta\alpha}_{\mathrm{DSR}}(x^5) \quad (\alpha,\beta = 0,1,2,3). \tag{6.52}$$

Therefore, there are only six independent generators. In matrix form:

$$I^{\alpha\beta}_{\mathrm{DSR}}(x^5) = \begin{pmatrix} 0 & I^{01}_{\mathrm{DSR}}(x^5) & I^{02}_{\mathrm{DSR}}(x^5) & I^{03}_{\mathrm{DSR}}(x^5) \\ -I^{01}_{\mathrm{DSR}}(x^5) & 0 & I^{12}_{\mathrm{DSR}}(x^5) & I^{13}_{\mathrm{DSR}}(x^5) \\ -I^{02}_{\mathrm{DSR}}(x^5) & -I^{12}_{\mathrm{DSR}}(x^5) & 0 & I^{23}_{\mathrm{DSR}}(x^5) \\ -I^{03}_{\mathrm{DSR}}(x^5) & -I^{13}_{\mathrm{DSR}}(x^5) & -I^{23}_{\mathrm{DSR}}(x^5) & 0 \end{pmatrix} \tag{6.53}$$

We can now pass to the self-representation basis of the generators of $SO(1,3)_{\mathrm{DEF.}}$ by introducing the following axial and polar three-vectors by means of the Levi-Civita tensor

$$S^i_{\mathrm{DSR}}(x^5) \equiv \frac{1}{2}\epsilon^i{}_{jk} I^{jk}_{\mathrm{DSR}}(x^5); \tag{6.54}$$

$$K^i_{\mathrm{DSR}}(x^5) \equiv I^{0i}_{\mathrm{DSR}}(x^5), \tag{6.55}$$

or

$$\mathbf{S}_{\mathrm{DSR}}(x^5) \equiv (I^{23}_{\mathrm{DSR}}(x^5), I^{31}_{\mathrm{DSR}}(x^5), I^{12}_{\mathrm{DSR}}(x^5))\ ; \tag{6.56}$$

$$\mathbf{K}_{\mathrm{DSR}}(x^5) \equiv (I^{01}_{\mathrm{DSR}}(x^5), I^{02}_{\mathrm{DSR}}(x^5), I^{03}_{\mathrm{DSR}}(x^5)). \tag{6.57}$$

Then, $I^{\alpha\beta}_{\mathrm{DSR}}(x^5)$ can be written as:

$$I^{\alpha\beta}_{\mathrm{DSR}}(x^5) = \begin{pmatrix} 0 & K^1_{\mathrm{DSR}}(x^5) & K^2_{\mathrm{DSR}}(x^5) & K^3_{\mathrm{DSR}}(x^5) \\ -K^1_{\mathrm{DSR}}(x^5) & 0 & S^3_{\mathrm{DSR}}(x^5) & -S^2_{\mathrm{DSR}}(x^5) \\ -K^2_{\mathrm{DSR}}(x^5) & -S^3_{\mathrm{DSR}}(x^5) & 0 & S^1_{\mathrm{DSR}}(x^5) \\ -K^3_{\mathrm{DSR}}(x^5) & S^2_{\mathrm{DSR}}(x^5) & -S^1_{\mathrm{DSR}}(x^5) & 0 \end{pmatrix}. \tag{6.58}$$

In DSR, like in the SR case, we can identify (apart from a sign) $S^i_{\mathrm{DSR}}(x^5)$ with the infinitesimal generator of the deformed 3D space rotation around $\widehat{x^i}$, and $K^i_{\mathrm{DSR}}(x^5)$ with the infinitesimal generator of the deformed Lorentz boost with motion direction along $\widehat{x^i}$.

### 6.3.2 Decomposition of the Parametric 4-Tensor in DSR

We can now specialize (6.1) (expressing the infinitesimal variation of the contravariant four-vector $x^\mu$ in the self-representation) to the DSR case, getting (ESC on):

$$\begin{aligned} \delta x^\mu_{(g),\mathrm{DSR}}(x,x^5) = \delta\omega^\mu_{\nu,\mathrm{DSR}}(g,x^5)x^\nu &= \frac{1}{2}\delta\omega_{\alpha\beta,(\mathrm{DSR})}(g)(I^{\alpha\beta})^\mu_{\nu,\mathrm{DSR}}(x^5)x^\nu \\ &- \frac{1}{2}\epsilon_l^{ij}\delta\omega_{ij,(\mathrm{DSR})}(g)(S^l)^\mu_{\nu,\mathrm{DSR}}(x^5)x^\nu \\ &+\delta\omega_{0i,(\mathrm{DSR})}(g)(K^i)^\mu_{\nu,\mathrm{DSR}}(x^5)x^\nu. \end{aligned} \tag{6.59}$$

It follows from (6.54):[7]

$$S^i_{\text{DSR}}(x^5) \equiv \frac{1}{2}\epsilon^i{}_{jk} I^{jk}_{\text{DSR}}(x^5) \Leftrightarrow I^{jk}_{\text{DSR}}(x^5) = \epsilon^{jk}{}_l S^l_{\text{DSR}}(x^5). \tag{6.60}$$

Replacing (6.60) in (6.59) we get:

$$\begin{aligned}
\delta x^\mu_{(g),\text{DSR}}(x, x^5) &= \frac{1}{2}\delta\omega_{ij}(g)(I^{ij})^\mu_{\nu,\text{DSR}}(x^5)x^\nu + \delta\omega_{0i}(g)(K^i)^\mu_{\nu,\text{DSR}}(x^5)x^\nu \\
&= \frac{1}{2}\delta\omega_{ij}(g)(\epsilon^{ij}{}_l S^l_{\text{DSR}}(x^5))^\mu_\nu x^\nu + \delta\omega_{0i}(g)(K^i)^\mu_{\nu,\text{DSR}}(x^5)x^\nu \\
&\overset{(\epsilon^{ij}{}_l S^l_{\text{DSR}}(x^5))^\mu_\nu = \epsilon^{ij}{}_l (S^l_{\text{DSR}})^\mu_\nu (x^5)}{=} \frac{1}{2}\epsilon^{ij}{}_l \delta\omega_{ij}(g)(S^l_{\text{DSR}})^\mu_\nu(x^5)x^\nu + \delta\omega_{0i}(g)(K^i)^\mu_{\nu,\text{DSR}}(x^5)x^\nu \\
&= \frac{1}{2}\epsilon^{\ ij}_{\ l}\delta\omega_{ij}(g)(S^l)^\mu_{\nu,\text{DSR}}(x^5)x^\nu + \delta\omega_{0i}(g)(K^i)^\mu_{\nu,\text{DSR}}(x^5)x^\nu.
\end{aligned} \tag{6.61}$$

Therefore, the DSR parametric 4-tensor $\delta\omega_{\alpha\beta}(g)$ can be written in the form (6.41)

$$\delta\omega_{\alpha\beta}(g) = \begin{pmatrix} 0 & -\zeta^1(g) & -\zeta^2(g) & -\zeta^3(g) \\ \zeta^1(g) & 0 & -\theta^3(g) & \theta^2(g) \\ \zeta^2(g) & \theta^3(g) & 0 & -\theta^1(g) \\ \zeta^3(g) & -\theta^2(g) & \theta^1(g) & 0 \end{pmatrix}, \tag{6.62}$$

where the axial three-vector $\boldsymbol{\theta}(g)$ and the polar three-vector $\boldsymbol{\zeta}(g)$ are still defined by

$$\begin{aligned}
\boldsymbol{\theta}(g) &= (\theta_i(g)) \equiv \left(-\frac{1}{2}\epsilon_i{}^{jk}\delta\omega_{jk}(g)\right) \\
&= (-\delta\omega_{23}(g), -\delta\omega_{31}(g), -\delta\omega_{12}(g));
\end{aligned} \tag{6.63}$$

[7] In fact one gets, from (5.54) and (6.60):

$$\begin{aligned}
S^i_{\text{DSR}}(x^5) &\equiv \frac{1}{2}\epsilon^i{}_{jk} I^{jk}_{\text{DSR}}(x^5) \\
&= \frac{1}{2}\epsilon^i{}_{jk}\epsilon^{jk}{}_l S^l_{\text{DSR}}(x^5) = \delta^i_l S^l_{\text{DSR}}(x^5) = S^i_{\text{DSR}}(x^5),
\end{aligned}$$

on account of the relations

$$\epsilon_{ijk}\epsilon_{lm}{}^k = \delta_{il}\delta_{jm} - \delta_{im}\delta_{jl}$$

$$\epsilon^i{}_{jk}\epsilon^{jk}{}_l = \epsilon^i{}_{jk}\epsilon_l{}^{jk} = \delta^i_l\delta^j_j - \delta^i_j\delta^j_l = 3\delta^i_l - \delta^i_l = 2\delta^i_l.$$

$$\begin{aligned}\boldsymbol{\zeta}(g) &= \zeta_i(g) \equiv (-\delta\omega_{0i}(g)) \\ &= (-\delta\omega_{01}(g), -\delta\omega_{02}(g), -\delta\omega_{03}(g)),\end{aligned} \tag{6.64}$$

(cf. (6.37)–(6.40)), but now they correspond to a true deformed rotation and to a deformed boost, respectively, as in the standard SR case.

### *6.3.3 Infinitesimal Transformations of the 4D Deformed Lorentz Group*

We can utilize the results of the previous two sections to write (6.42) as:

$$\begin{aligned}\delta x^{\mu}_{(g),\text{DSR}}(x,x^5) &= -\theta_l(g)(S^l)^{\mu}_{\ \nu,\text{DSR}}(x^5)x^{\nu} - \zeta_i(g)(K^i)^{\mu}_{\ \nu,\text{DSR}}(x^5)x^{\nu} \\ &= \left(-\boldsymbol{\theta}(g)\cdot\mathbf{S}_{\text{DSR}}(x^5) - \boldsymbol{\zeta}(g)\cdot\mathbf{K}_{\text{DSR}}(x^5)\right)^{\mu}_{\ \nu} x^{\nu}.\end{aligned} \tag{6.65}$$

Therefore, the infinitesimal space–time rotation transformation in the deformed Minkowski space $\widetilde{M}(x^5)$ corresponding to the element $g$ of $\text{SO}(1,3)_{\text{DEF.}}$, can be expressed as:

$$\begin{aligned}\delta g : x^{\mu} \rightarrow x^{\mu\prime}_{(g)}(x^5) &= x^{\mu} + \delta x^{\mu}_{(g),\text{DSR}}(x,x^5) \\ &= (1 - \theta_1(g)S^1_{\text{DSR}}(x^5) - \theta_2(g)S^2_{\text{DSR}}(x^5) \\ &\quad -\theta_3(g)S^3_{\text{DSR}}(x^5) - \zeta_1(g)K^1_{\text{DSR}}(x^5) \\ &\quad -\zeta_2(g)K^2_{\text{DSR}}(x^5) - \zeta_3(g)K^3_{\text{DSR}}(x^5))^{\mu}_{\ \nu}x^{\nu},\end{aligned} \tag{6.66}$$

where 1 is the identity of $\text{SO}(1,3)_{\text{DEF.}}$.[8]

Then, on account of the physical meaning of the 3D parameter and generator vectors, $\boldsymbol{\theta}(g)$, $\boldsymbol{\zeta}(g)$, and $\mathbf{S}_{\text{DSR}}(x^5)$, $\mathbf{K}_{\text{DSR}}(x^5)$, we can get, from the matrix representation of the $\text{SO}(1,3)_{\text{DEF.}}$ generators, the explicit expressions of all the different kinds of infinitesimal transformations of the deformed Lorentz group, namely:

1. *3D deformed space (true) rotations* (parameters $\boldsymbol{\theta}(g)$ and generators $\mathbf{S}_{\text{DSR}}(x^5)$), which constitute the group $\text{SO}(3)_{\text{DEF.}}$ of rotations in a deformed 3D space, non-abelian, noninvariant proper subgroup of $\text{SO}(1,3)_{\text{DEF.}}$:

   – *(Clockwise) infinitesimal rotation by an angle $\theta_1(g)$ around $\widehat{x^1}$:*

[8] Corresponding to the origin of the deformed Lorentz algebra $su(2)_{\text{DEF.}} \otimes su(2)_{\text{DEF.}}$ (see Sect. 6.3.4).

$$x^{\mu\prime}_{(g)}(x^5) = (1 - \theta_1(g)S^1_{\mathrm{DSR}})^\mu_\nu(x^5)x^\nu$$

$$\Leftrightarrow \begin{pmatrix} x^{0\prime}_{(g),\mathrm{DSR}}(x^5) \\ x^{1\prime}_{(g),\mathrm{DSR}}(x^5) \\ x^{2\prime}_{(g),\mathrm{DSR}}(x^5) \\ x^{3\prime}_{(g),\mathrm{DSR}}(x^5) \end{pmatrix} = \begin{pmatrix} 1 & 0 & 0 & 0 \\ 0 & 1 & 0 & 0 \\ 0 & 0 & 1 & \theta_1(g)b_2^{-2}(x^5) \\ 0 & 0 & -\theta_1(g)b_3^{-2}(x^5) & 1 \end{pmatrix} \begin{pmatrix} x^0 \\ x^1 \\ x^2 \\ x^3 \end{pmatrix}$$

$$= \begin{pmatrix} x^0 \\ x^1 \\ x^2 + \theta_1(g)b_2^{-2}(x^5)x^3 \\ -\theta_1(g)b_3^{-2}(x^5)x^2 + x^3 \end{pmatrix}; \tag{6.67}$$

– *(Clockwise) infinitesimal rotation by an angle* $\theta_2(g)$ *around* $\widehat{x^2}$ :

$$x^{\mu\prime}_{(g)}(x^5) = (1 - \theta_2(g)S^2_{\mathrm{DSR}})^\mu_\nu(x^5)x^\nu$$

$$\Leftrightarrow \begin{pmatrix} x^{0\prime}_{(g),\mathrm{DSR}}(x^5) \\ x^{1\prime}_{(g),\mathrm{DSR}}(x^5) \\ x^{2\prime}_{(g),\mathrm{DSR}}(x^5) \\ x^{3\prime}_{(g),\mathrm{DSR}}(x^5) \end{pmatrix} = \begin{pmatrix} 1 & 0 & 0 & 0 \\ 0 & 1 & 0 & -\theta_2(g)b_1^{-2}(x^5) \\ 0 & 0 & 1 & 0 \\ 0 & \theta_2(g)b_3^{-2}(x^5) & 0 & 1 \end{pmatrix} \begin{pmatrix} x^0 \\ x^1 \\ x^2 \\ x^3 \end{pmatrix}$$

$$= \begin{pmatrix} x^0 \\ x^1 - \theta_2(g)b_1^{-2}(x^5)x^3 \\ x^2 \\ \theta_2(g)b_3^{-2}(x^5)x^1 + x^3 \end{pmatrix}; \tag{6.68}$$

– *(Clockwise) infinitesimal rotation by an angle* $\theta_3(g)$ *around* $\widehat{x^3}$:

$$x^{\mu\prime}_{(g)}(x^5) = (1 - \theta_3(g)S^3_{\mathrm{DSR}})^\mu_\nu(x^5)x^\nu$$

$$\Leftrightarrow \begin{pmatrix} x^{0\prime}_{(g),\mathrm{DSR}}(x^5) \\ x^{1\prime}_{(g),\mathrm{DSR}}(x^5) \\ x^{2\prime}_{(g),\mathrm{DSR}}(x^5) \\ x^{3\prime}_{(g),\mathrm{DSR}}(x^5) \end{pmatrix} = \begin{pmatrix} 1 & 0 & 0 & 0 \\ 0 & 1 & \theta_3(g)b_1^{-2}(x^5) & 0 \\ 0 & -\theta_3(g)b_2^{-2}(x^5) & 1 & 0 \\ 0 & 0 & 0 & 1 \end{pmatrix} \begin{pmatrix} x^0 \\ x^1 \\ x^2 \\ x^3 \end{pmatrix}$$

$$= \begin{pmatrix} x^0 \\ x^1 + \theta_3(g)b_1^{-2}(x^5)x^2 \\ -\theta_3(g)b_2^{-2}(x^5)x^1 + x^2 \\ x^3 \end{pmatrix}. \tag{6.69}$$

2. *3D deformed space–time (pseudo) rotations, or deformed Lorentz boosts* (parameters $\boldsymbol{\zeta}(g)$ and generators $\mathbf{K}_{\mathrm{DSR}}(x^5)$); they do not form a group:

– *Infinitesimal boost with rapidity* $\zeta_1(g)$ *along* $\widehat{x^1}$*:*

$$x^{\mu\prime}_{(g)}(x^5) = (1 - \zeta_1(g) K^1_{\mathrm{DSR}})^{\mu}_{\nu}(x^5) x^{\nu}$$

$$\Leftrightarrow \begin{pmatrix} x^{0\prime}_{(g),\mathrm{DSR}}(x^5) \\ x^{1\prime}_{(g),\mathrm{DSR}}(x^5) \\ x^{2\prime}_{(g),\mathrm{DSR}}(x^5) \\ x^{3\prime}_{(g),\mathrm{DSR}}(x^5) \end{pmatrix} = \begin{pmatrix} 1 & -\zeta_1(g) b_0^{-2}(x^5) & 0 & 0 \\ -\zeta_1(g) b_1^{-2}(x^5) & 1 & 0 & 0 \\ 0 & 0 & 1 & 0 \\ 0 & 0 & 0 & 1 \end{pmatrix} \begin{pmatrix} x^0 \\ x^1 \\ x^2 \\ x^3 \end{pmatrix}$$

$$= \begin{pmatrix} x^0 - \zeta_1(g) b_0^{-2}(x^5) x^1 \\ -\zeta_1(g) b_1^{-2}(x^5) x^0 + x^1 \\ x^2 \\ x^3 \end{pmatrix}; \tag{6.70}$$

– *Infinitesimal boost with rapidity* $\zeta_2(g)$ *along* $\widehat{x^2}$*:*

$$x^{\mu\prime}_{(g)}(x^5) = (1 - \zeta_2(g) K^2_{\mathrm{DSR}})^{\mu}_{\nu}(x^5) x^{\nu}$$

$$\Leftrightarrow \begin{pmatrix} x^{0\prime}_{(g),\mathrm{DSR}}(x^5) \\ x^{1\prime}_{(g),\mathrm{DSR}}(x^5) \\ x^{2\prime}_{(g),\mathrm{DSR}}(x^5) \\ x^{3\prime}_{(g),\mathrm{DSR}}(x^5) \end{pmatrix} = \begin{pmatrix} 1 & 0 & -\zeta_2(g) b_0^{-2}(x^5) & 0 \\ 0 & 1 & 0 & 0 \\ -\zeta_2(g) b_2^{-2}(x^5) & 0 & 1 & 0 \\ 0 & 0 & 0 & 1 \end{pmatrix} \begin{pmatrix} x^0 \\ x^1 \\ x^2 \\ x^3 \end{pmatrix}$$

$$= \begin{pmatrix} x^0 - \zeta_2(g) b_0^{-2}(x^5) x^2 \\ x^1 \\ -\zeta_2(g) b_2^{-2}(x^5) x^0 + x^2 \\ x^3 \end{pmatrix}; \tag{6.71}$$

– *Infinitesimal boost with rapidity* $\zeta_3(g)$ *along* $\widehat{x^3}$*:*

$$x^{\mu\prime}_{(g)}(x^5) = (1 - \zeta_3(g) K^3_{\mathrm{DSR}})^{\mu}_{\nu}(x^5) x^{\nu} \Leftrightarrow$$

$$\Leftrightarrow \begin{pmatrix} x^{0\prime}_{(g),\mathrm{DSR}}(x^5) \\ x^{1\prime}_{(g),\mathrm{DSR}}(x^5) \\ x^{2\prime}_{(g),\mathrm{DSR}}(x^5) \\ x^{3\prime}_{(g),\mathrm{DSR}}(x^5) \end{pmatrix} = \begin{pmatrix} 1 & 0 & 0 & -\zeta_3(g) b_0^{-2}(x^5) \\ 0 & 1 & 0 & 0 \\ 0 & 0 & 1 & 0 \\ -\zeta_3(g) b_3^{-2}(x^5) & 0 & 0 & 1 \end{pmatrix} \begin{pmatrix} x^0 \\ x^1 \\ x^2 \\ x^3 \end{pmatrix}$$

$$= \begin{pmatrix} x^0 - \zeta_3(g) b_0^{-2}(x^5) x^3 \\ x^1 \\ x^2 \\ -\zeta_3(g) b_3^{-2}(x^5) x^0 + x^3 \end{pmatrix}. \tag{6.72}$$

The explicit form of the infinitesimal contravariant four-vector $\delta x^{\mu}_{(g),\mathrm{DSR}}(x, x^5)$, corresponding to an element $g \in \mathrm{SO}(1,3)_{\mathrm{DEF.}}$, is therefore

$$\begin{cases} \delta x^0_{(g),\mathrm{DSR}}(x,x^5) = -\zeta_1(g)b_0^{-2}(x^5)x^1 - \zeta_2(g)b_0^{-2}(x^5)x^2 - \zeta_3(g)b_0^{-2}(x^5)x^3 \\ \qquad = b_0^{-2}(x^5)(-\zeta_1(g)x^1 - \zeta_2(g)x^2 - \zeta_3(g)x^3); \\ \delta x^1_{(g),\mathrm{DSR}}(x,x^5) = -\zeta_1(g)b_1^{-2}(x^5)x^0 + \theta_3(g)b_1^{-2}(x^5)x^2 - \theta_2(g)b_1^{-2}(x^5)x^3 \\ \qquad = -b_1^{-2}(x^5)(\zeta_1(g)x^0 - \theta_3(g)x^2 + \theta_2(g)x^3); \\ \delta x^2_{(g),\mathrm{DSR}}(x,x^5) = -\zeta_2(g)b_2^{-2}(x^5)x^0 - \theta_3(g)b_2^{-2}(x^5)x^1 + \theta_1(g)b_2^{-2}(x^5)x^3 \\ \qquad = -b_2^{-2}(x^5)(\zeta_2(g)x^0 + \theta_3(g)x^1 - \theta_1(g)x^3); \\ \delta x^3_{(g),\mathrm{DSR}}(x,x^5) = -\zeta_3(g)b_3^{-2}(x^5)x^0 + \theta_2(g)b_3^{-2}(x^5)x^1 - \theta_1(g)b_3^{-2}(x^5)x^2 \\ \qquad = -b_3^{-2}(x^5)(\zeta_3(g)x^0 - \theta_2(g)x^1 + \theta_1(g)x^2). \end{cases} \tag{6.73}$$

The covariant components of such a four-vector are

$$\begin{cases} \delta x_{0(g),\mathrm{DSR}}(x) = -\zeta_1(g)x^1 - \zeta_2(g)x^2 - \zeta_3(g)x^3; \\ \delta x_{1(g),\mathrm{DSR}}(x) = \zeta_1(g)x^0 - \theta_3(g)x^2 + \theta_2(g)x^3; \\ \delta x_{2(g),\mathrm{DSR}}(x) = \zeta_2(g)x^0 + \theta_3(g)x^1 - \theta_1(g)x^3; \\ \delta x_{3(g),\mathrm{DSR}}(x) = \zeta_3(g)x^0 - \theta_2(g)x^1 + \theta_1(g)x^2. \end{cases} \tag{6.74}$$

Comparing (6.74) with the expression (5.39) of the covariant Killing vector, we see the perfect correspondence between the space–time rotational component of $\xi_\mu(x)$ (unique for all the 4D generalized Minkowski spaces) and the covariant four-vector $\delta x_{\mu(g),\mathrm{DSR}}(x)$ related to $\mathrm{SO}(1,3)_{\mathrm{DEF.}}$ (see point 3 of Sect. 5.2.4).

### 6.3.4 4D Deformed Lorentz Algebra

Let us specialize (6.10) to the DSR case, in order to derive the 4D deformed Lorentz algebra, i.e., the Lie algebra of the 4D deformed, homogeneous Lorentz group $\mathrm{SO}(1,3)^6_{\mathrm{DEF.}}$. We get

$$\begin{aligned} [I^{\alpha\beta}_{\mathrm{DSR}}(x^5), I^{\rho\sigma}_{\mathrm{DSR}}(x^5)] &= g^{\alpha\sigma}_{\mathrm{DSR}}(x^5)I^{\beta\rho}_{\mathrm{DSR}}(x^5) + g^{\beta\rho}_{\mathrm{DSR}}(x^5)I^{\alpha\sigma}_{\mathrm{DSR}}(x^5) \\ &\quad - g^{\alpha\rho}_{\mathrm{DSR}}(x^5)I^{\beta\sigma}_{\mathrm{DSR}}(x^5) - g^{\beta\sigma}_{\mathrm{DSR}}(x^5)I^{\alpha\rho}_{\mathrm{DSR}}(x^5) \\ = \delta^{\alpha\sigma}&\left(b_0^{-2}(x^5)\delta^{\alpha0} - b_1^{-2}(x^5)\delta^{\alpha1} - b_2^{-2}(x^5)\delta^{\alpha2} - b_3^{-2}(x^5)\delta^{\alpha3}\right) I^{\beta\rho}_{\mathrm{DSR}}(x^5) \\ +\delta^{\beta\rho}&\left(\delta^{\beta0}b_0^{-2}(x^5) - \delta^{\beta1}b_1^{-2}(x^5) - \delta^{\beta2}b_2^{-2}(x^5) - \delta^{\beta3}b_3^{-2}(x^5)\right) I^{\alpha\sigma}_{\mathrm{DSR}}(x^5) \\ -\delta^{\alpha\rho}&\left(\delta^{\alpha0}b_0^{-2}(x^5) - \delta^{\alpha1}b_1^{-2}(x^5) - \delta^{\alpha2}b_2^{-2}(x^5) - \delta^{\alpha3}b_3^{-2}(x^5)\right) I^{\beta\sigma}_{\mathrm{DSR}}(x^5) \\ -\delta^{\beta\sigma}&\left(\delta^{\beta0}b_0^{-2}(x^5) - \delta^{\beta1}b_1^{-2}(x^5) - \delta^{\beta2}b_2^{-2}(x^5) - \delta^{\beta3}b_3^{-2}(x^5)\right) I^{\alpha\rho}_{\mathrm{DSR}}(x^5). \end{aligned} \tag{6.75}$$

On account of the physical interpretation of the infinitesimal generators, one has therefore the following kinds of commutators:

1. *Commutator of generators of 3D deformed space rotations:*

$$\begin{aligned}
&[I^{ij}_{\mathrm{DSR}}(x^5), I^{lm}_{\mathrm{DSR}}(x^5)]\\
&= \delta^{im}\left(\delta^{i0}b_0^{-2}(x^5) - \delta^{i1}b_1^{-2}(x^5) - \delta^{i2}b_2^{-2}(x^5) - \delta^{i3}b_3^{-2}(x^5)\right) I^{jl}_{\mathrm{DSR}}(x^5)\\
&+\delta^{jl}\left(\delta^{j0}b_0^{-2}(x^5) - \delta^{j1}b_1^{-2}(x^5) - \delta^{j2}b_2^{-2}(x^5) - \delta^{j3}b_3^{-2}(x^5)\right) I^{im}_{\mathrm{DSR}}(x^5)\\
&-\delta^{il}\left(\delta^{i0}b_0^{-2}(x^5) - \delta^{i1}b_1^{-2}(x^5) - \delta^{i2}b_2^{-2}(x^5) - \delta^{i3}b_3^{-2}(x^5)\right) I^{jm}_{\mathrm{DSR}}(x^5)\\
&-\delta^{jm}\left(\delta^{j0}b_0^{-2}(x^5) - \delta^{j1}b_1^{-2}(x^5) - \delta^{j2}b_2^{-2}(x^5) - \delta^{j3}b_3^{-2}(x^5)\right) I^{il}_{\mathrm{DSR}}(x^5)\\
&\overset{\text{ESC off on } i,j}{=} -\delta^{im}b_i^{-2}(x^5)I^{jl}_{\mathrm{DSR}}(x^5) - \delta^{jl}b_j^{-2}(x^5)I^{im}_{\mathrm{DSR}}(x^5)\\
&\quad +\delta^{il}b_i^{-2}(x^5)I^{jm}_{\mathrm{DSR}}(x^5) + \delta^{jm}b_j^{-2}(x^5)I^{il}_{\mathrm{DSR}}(x^5);
\end{aligned} \tag{6.76}$$

2. *Commutator of generators of 3D deformed boosts:*

$$\begin{aligned}
&[I^{i0}_{\mathrm{DSR}}(x^5), I^{j0}_{\mathrm{DSR}}(x^5)]\\
&= \delta^{i0}\left(\delta^{i0}b_0^{-2}(x^5) - \delta^{i1}b_1^{-2}(x^5) - \delta^{i2}b_2^{-2}(x^5) - \delta^{i3}b_3^{-2}(x^5)\right) I^{0j}_{\mathrm{DSR}}(x^5)\\
&+\delta^{0j}\left(\delta^{j0}b_0^{-2}(x^5) - \delta^{j1}b_1^{-2}(x^5) - \delta^{j2}b_2^{-2}(x^5) - \delta^{j3}b_3^{-2}(x^5)\right) I^{i0}_{\mathrm{DSR}}(x^5)\\
&-\delta^{ij}\left(\delta^{i0}b_0^{-2}(x^5) - \delta^{i1}b_1^{-2}(x^5) - \delta^{i2}b_2^{-2}(x^5) - \delta^{i3}b_3^{-2}(x^5)\right) I^{00}_{\mathrm{DSR}}(x^5)\\
&-\delta^{00}\left(\delta^{00}b_0^{-2}(x^5) - \delta^{01}b_1^{-2}(x^5) - \delta^{02}b_2^{-2}(x^5) - \delta^{03}b_3^{-2}(x^5)\right) I^{ij}_{\mathrm{DSR}}(x^5)\\
&= -b_0^{-2}(x^5)I^{ij}_{\mathrm{DSR}}(x^5);
\end{aligned} \tag{6.77}$$

3. *"Mixed" commutator of 3D deformed space and boost generators:*

$$\begin{aligned}
&[I^{ij}_{\mathrm{DSR}}(x^5), I^{k0}_{\mathrm{DSR}}(x^5)]\\
&= \delta^{i0}\left(\delta^{i0}b_0^{-2}(x^5) - \delta^{i1}b_1^{-2}(x^5) - \delta^{i2}b_2^{-2}(x^5) - \delta^{i3}b_3^{-2}(x^5)\right) I^{jk}_{\mathrm{DSR}}(x^5)\\
&+\delta^{jk}\left(\delta^{j0}b_0^{-2}(x^5) - \delta^{j1}b_1^{-2}(x^5) - \delta^{j2}b_2^{-2}(x^5) - \delta^{j3}b_3^{-2}(x^5)\right) I^{i0}_{\mathrm{DSR}}(x^5)\\
&-\delta^{ik}\left(\delta^{i0}b_0^{-2}(x^5) - \delta^{i1}b_1^{-2}(x^5) - \delta^{i2}b_2^{-2}(x^5) - \delta^{i3}b_3^{-2}(x^5)\right) I^{j0}_{\mathrm{DSR}}(x^5)\\
&-\delta^{j0}\left(\delta^{j0}b_0^{-2}(x^5) - \delta^{j1}b_1^{-2}(x^5) - \delta^{j2}b_2^{-2}(x^5) - \delta^{j3}b_3^{-2}(x^5)\right) I^{ik}_{\mathrm{DSR}}(x^5)\\
&\overset{\text{ESC off on } i,j}{=} -\delta^{jk}b_j^{-2}(x^5)I^{i0}_{\mathrm{DSR}}(x^5) + \delta^{ik}b_i^{-2}(x^5)I^{j0}_{\mathrm{DSR}}(x^5).
\end{aligned} \tag{6.78}$$

In the "self-representation" basis of $SO(1,3)^{6.}_{\mathrm{DEF.}}$, it is easy to show that commutation relations (6.75)–(6.78) read:

$$
\left\{
\begin{aligned}
&[S^i_{\text{DSR}}(x^5), S^j_{\text{DSR}}(x^5)] \\
&\overset{\text{ESC on}}{=} \left(\textstyle\sum_{s=1}^{3}(1-\delta_{is})((1-\delta_{js})b_s^{-2}(x^5)\right) \epsilon_{ijk} S^k_{\text{DSR}}(x^5) \\
&= \epsilon_{ijk} b_k^{-2}(x^5) S^k_{\text{DSR}}(x^5); \\
&[K^i_{\text{DSR}}(x^5), K^j_{\text{DSR}}(x^5)] \\
&\overset{\text{ESC on}}{=} -b_0^{-2}(x^5)\epsilon_{ijk} S^k_{\text{DSR}}(x^5); \\
&[S^i_{\text{DSR}}(x^5), K^j_{\text{DSR}}(x^5)] \\
&\overset{\text{ESC ``on'' on } l,\text{ESC ``off'' on } j}{=} \epsilon_{ijl} K^l_{\text{DSR}}(x^5) \left(\textstyle\sum_{s=1}^{3} \delta_{js} b_s^{-2}(x^5)\right) \\
&= \epsilon_{ijl} b_j^{-2}(x^5) K^l_{\text{DSR}}(x^5),
\end{aligned}
\right.
\tag{6.79}
$$

where use has been made of the relation

$$
\begin{gathered}
\epsilon_{ims}\epsilon_{jrs} b_s^{-2}(x^5) \\
(\delta_{ij}\delta_{mr} - \delta_{ir}\delta_{mj}) \left( \sum_{k=1}^{3} (1-\delta_{ik})(1-\delta_{mk}) b_k^{-2}(x^5) \right),
\end{gathered}
\tag{6.80}
$$

which generalizes to the DSR case the well-known formula $\epsilon_{ims}\epsilon_{jrs} = \delta_{ij}\delta_{mr} - \delta_{ir}\delta_{jm}$.[9]

The commutators (6.79) define the *(4D) deformed Lorentz algebra* $su(2)_{\text{DEF.}} \otimes su(2)_{\text{DEF.}}$ of generators of $SO(1,3)_{\text{DEF.}}$, and generalize to the

---

[9] Let us prove it. One has

$$
\begin{aligned}
&\epsilon_{ims}\epsilon_{jrs} b_s^{-2}(x^5) \\
&= \epsilon_{im1}\epsilon_{jr1} b_1^{-2}(x^5) + \epsilon_{im2}\epsilon_{jr2} b_2^{-2}(x^5) + \epsilon_{im3}\epsilon_{jr3} b_3^{-2}(x^5) \\
&= \overbrace{\begin{bmatrix} \epsilon_{231}\epsilon_{231}(=1) & \epsilon_{231}\epsilon_{321}(=-1) \\ \epsilon_{321}\epsilon_{321}(=1) & \epsilon_{321}\epsilon_{231}(=-1) \end{bmatrix}}^{s=1} b_1^{-2}(x^5) \\
&+ \overbrace{\begin{bmatrix} \epsilon_{132}\epsilon_{132}(=1) & \epsilon_{132}\epsilon_{312}(=-1) \\ \epsilon_{312}\epsilon_{312}(=1) & \epsilon_{312}\epsilon_{132}(=-1) \end{bmatrix}}^{s=2} b_2^{-2}(x^5) \\
&+ \overbrace{\begin{bmatrix} \epsilon_{123}\epsilon_{123}(=1) & \epsilon_{123}\epsilon_{213}(=-1) \\ \epsilon_{213}\epsilon_{213}(=1) & \epsilon_{213}\epsilon_{123}(=-1) \end{bmatrix}}^{s=3} b_3^{-2}(x^5) \\
&= \delta_{ij}\delta_{mr} b^{-2}_{k\neq i, k\neq m}(x^5) - \delta_{ir}\delta_{mj} b^{-2}_{k\neq i, k\neq m}(x^5) \\
&= (\delta_{ij}\delta_{mr} - \delta_{ir}\delta_{mj}) \left( \sum_{k=1}^{3} (1-\delta_{ik})(1-\delta_{mk}) b_k^{-2}(x^5) \right),
\end{aligned}
$$

where the square brackets [...] do not denote matrices, but only mean a generic array enumerating the nonzero elements, arranged according to the sum index $s$.

DSR case the infinitesimal algebraic structure of the standard homogeneous Lorentz group SO(1,3). They admit interpretations wholly analogous to those of the commutators of the usual Lorentz algebra. First (6.79) expresses the closed nature of the algebra of the deformed rotation generators; consequently the 3D deformed space rotations form a three-parameter subgroup of $SO(1,3)_{\text{DEF.}}$, $SO(3)_{\text{DEF.}}$. On the contrary, the deformed boost generator algebra is not closed (according to second (6.79)), and then the deformed boosts do not form a subgroup of the deformed Lorentz group. This implies that $SO(1,3)_{\text{DEF.}}$ cannot be considered the product of two subgroups. This is further confirmed by the noncommutativity of deformed space rotations and boosts, expressed by third commutator (6.79). Moreover, first and third (6.79) show that both $\mathbf{S}_{\text{DSR}}(x^5)$ and $\mathbf{K}_{\text{DSR}}(x^5)$ behave as three-vectors under deformed spatial rotations.

Needless to say, in the limit

$$\begin{aligned} & g_{\mu\nu,\text{DSR}}(x^5) \to g_{\mu\nu,\text{SR}} \\ & \Leftrightarrow \delta_{\mu\nu}(\delta_{\mu 0} b_0^2(x^5) - \delta_{\mu 1} b_1^2(x^5) - \delta_{\mu 2} b_2^2(x^5) - \delta_{\mu 3} b_3^2(x^5)) \\ & \to \delta_{\mu\nu}(\delta_{\mu 0} - \delta_{\mu 1} - \delta_{\mu 2} - \delta_{\mu 3}) \Leftrightarrow b_\mu^2(x^5) \to 1 \;\; \forall \mu = 0,1,2,3, \end{aligned} \tag{6.81}$$

all results valid at group-transformation level in DSR reduce to the standard ones in SR. In general, this is the limit in which all the results and reasonings of DSR become the corresponding ones in SR (due to its nature of "anisotropizing deforming" generalization of SR).

# 7

# Finite Structure of Deformed Chronotopical Groups

In Chap. 6, we discussed the infinitesimal structure of the chronotopical group of generalized Minkowski spaces and, in particular, of the deformed Minkowski space $\widetilde{M}(x_5)$ of DSR. We want now to discuss the corresponding finite structure, so deriving in a rigorous way the DLT [42].

## 7.1 Space–Time Rotations in a 4D Generalized Minkowski Space

### 7.1.1 General Case

We shall consider, without loss of generality, the case of a 4D generalized Minkowski space with $S(< 4)$ space-like and $T = 4-S$ time-like dimensions. On the basis of the results of Chap. 6, in particular exploiting (6.34) which yields the infinitesimal transformation corresponding to the element $g$ of the generalized, homogeneous Lorentz group $\mathrm{SO}(T = 4 - S, S)_{\mathrm{GEN.}}$, and in the self-representation basis of the generators, $\mathbf{S}(\{x\}_{\mathrm{n.m.}})$ and $\mathbf{K}(\{x\}_{\mathrm{n.m.}})$ (6.30)–(6.31), we can write the finite space–time rotation in $\widetilde{M}(\{x\}_{\mathrm{n.m.}})$ corresponding to $g \in \mathrm{SO}(T = 4 - S, S)_{\mathrm{GEN.}}$ as:

$$\mathrm{SO}(T = 4 - S, S)_{\mathrm{GEN.}} \ni g : x^{\mu} \to x^{\mu\prime}_{(g)}(x, \{x\}_{\mathrm{n.m.}})$$

$$= \exp(-\boldsymbol{\theta}(g) \cdot \mathbf{S}(\{x\}_{\mathrm{n.m.}}) - \boldsymbol{\zeta}(g) \cdot \mathbf{K}(\{x\}_{\mathrm{n.m.}}))^{\mu}_{\ \nu} x^{\nu}$$

$$= \exp(-\theta_1(g)S^1(\{x\}_{\text{n.m.}}) - \theta_2(g)S^2(\{x\}_{\text{n.m.}}) - \theta_3(g)S^3(\{x\}_{\text{n.m.}}) - \zeta_1(g)K^1(\{x\}_{\text{n.m.}}) - \zeta_2(g)K^2(\{x\}_{\text{n.m.}}) - \zeta_3(g)K^3(\{x\}_{\text{n.m.}}))^{\mu}_{\ \nu}x^{\nu}. \quad (7.1)$$

In matrix notation, and by a series development of the exponential, we can write:

$$\Lambda(g,\{x\}_{\text{n.m.}}) = \sum_{n=0}^{\infty}\frac{1}{n!}(-\boldsymbol{\theta}(g)\cdot\mathbf{S}(\{x\}_{\text{n.m.}}) - \boldsymbol{\zeta}(g)\cdot\mathbf{K}(\{x\}_{\text{n.m.}}))^n$$
$$= \sum_{n=0}^{\infty}\frac{1}{n!}\left[-\theta_1(g)S^1(\{x\}_{\text{n.m.}}) - \theta_2(g)S^2(\{x\}_{\text{n.m.}}) - \theta_3(g)S^3(\{x\}_{\text{n.m.}}) - \zeta_1(g)K^1(\{x\}_{\text{n.m.}}) - \zeta_2(g)K^2(\{x\}_{\text{n.m.}}) - \zeta_3(g)K^3(\{x\}_{\text{n.m.}})\right]^n. \quad (7.2)$$

Here, $\boldsymbol{\theta}(g)$ and $\boldsymbol{\zeta}(g)$ are the rotation parameter axial three-vector and the boost parameter polar three-vector, respectively, defined by (6.63)–(6.64). Since the generalized Lorentz algebra (6.11) is noncommutative, the group $\text{SO}(T,S)^{N(N-1)/2}_{\text{GEN.}}$ is non-abelian, and therefore the 4D finite transformations (7.1), (7.2) do not commute. Thus one has, using the Baker–Campbell–Hausdorff formula:

$$\begin{aligned}\Lambda(g,\{x\}_{\text{n.m.}}) &= \exp(-\theta_1(g)S^1(\{x\}_{\text{n.m.}}) - \theta_2(g)S^2(\{x\}_{\text{n.m.}}) - \theta_3(g)S^3(\{x\}_{\text{n.m.}}) \\ &\quad - \zeta_1(g)K^1(\{x\}_{\text{n.m.}}) - \zeta_2(g)K^2(\{x\}_{\text{n.m.}}) - \zeta_3(g)K^3(\{x\}_{\text{n.m.}})) \\ &\neq \exp(-\theta_1(g)S^1(\{x\}_{\text{n.m.}}))\times\exp(-\theta_2(g)S^2(\{x\}_{\text{n.m.}})) \\ &\quad \times\exp(-\theta_3(g)S^3(\{x\}_{\text{n.m.}}))\times\exp(-\zeta_1(g)K^1(\{x\}_{\text{n.m.}})) \\ &\quad \times\exp(-\zeta_2(g)K^2(\{x\}_{\text{n.m.}}))\times\exp(-\zeta_3(g)K^3(\{x\}_{\text{n.m.}})), \end{aligned} \quad (7.3)$$

where $\times$ denotes matrix product.

### *7.1.2 Deformed Lorentz Group of* DSR

In the case of the deformed Minkowski space–time $\widetilde{M}(x^5)$, we know that the generalized Lorentz algebra (6.11) becomes the 4D deformed Lorentz algebra $su(2)_{\text{DEF.}}\otimes su(2)_{\text{DEF.}}$, and the infinitesimal space–time rotation (6.42) is given by (6.66). Moreover, the parameters $\boldsymbol{\theta}(g)$ and $\boldsymbol{\zeta}(g)$ are, respectively, the deformed rotation and boost Euclidean three-vectors, whereas $\mathbf{S}_{\text{DSR}}(x^5)$ and $\mathbf{K}_{\text{DSR}}(x^5)$ are the generators of the corresponding transformations of the deformed, homogeneous Lorentz group$\text{SO}(1,3)_{\text{DEF.}}$, satisfying the (4D) deformed Lorentz algebra $su(2)_{\text{DEF.}}\otimes su(2)_{\text{DEF.}}$(6.79).

Therefore, in the DSR case, (7.1), (7.2) for a finite transformation become, respectively,

$$\begin{aligned}
\mathrm{SO}(1,3)_{\mathrm{DEF.}} \ni g : \Lambda(g,x^5) &= \exp(-\boldsymbol{\theta}(g)\cdot\mathbf{S}_{\mathrm{DSR}}(x^5) - \boldsymbol{\zeta}(g)\cdot\mathbf{K}_{\mathrm{DSR}}(x^5)) \\
&= \exp(-\theta_1(g)S^1_{\mathrm{DSR}}(x^5) - \theta_2(g)S^2_{\mathrm{DSR}}(x^5) \\
&\quad - \theta_3(g)S^3_{\mathrm{DSR}}(x^5) - \zeta_1(g)K^1_{\mathrm{DSR}}(x^5) \\
&\quad - \zeta_2(g)K^2_{\mathrm{DSR}}(x^5) - \zeta_3(g)K^3_{\mathrm{DSR}}(x^5)) \\
&= \sum_{n=0}^{\infty}\frac{1}{n!}(-\boldsymbol{\theta}(g)\cdot\mathbf{S}_{\mathrm{DSR}}(x^5) - \boldsymbol{\zeta}(g)\cdot\mathbf{K}_{\mathrm{DSR}}(x^5))^n \\
&= \sum_{n=0}^{\infty}\frac{1}{n!}(-\theta_1(g)S^1_{\mathrm{DSR}}(x^5) - \theta_2(g)S^2_{\mathrm{DSR}}(x^5) \\
&\quad - \theta_3(g)S^3_{\mathrm{DSR}}(x^5) - \zeta_1(g)K^1_{\mathrm{DSR}}(x^5) \\
&\quad - \zeta_2(g)K^2_{\mathrm{DSR}}(x^5) - \zeta_3(g)K^3_{\mathrm{DSR}}(x^5))^n ; \qquad (7.4)
\end{aligned}$$

$$\begin{aligned}
\mathrm{SO}(1,3)_{\mathrm{DEF.}} \ni g : \Lambda(g,x^5) &= \exp(-\theta_1(g)S^1_{\mathrm{DSR}}(x^5) - \theta_2(g)S^2_{\mathrm{DSR}}(x^5) \\
&\quad - \theta_3(g)S^3_{\mathrm{DSR}}(x^5) - \zeta_1(g)K^1_{\mathrm{DSR}}(x^5) \\
&\quad - \zeta_2(g)K^2_{\mathrm{DSR}}(x^5) - \zeta_3(g)K^3_{\mathrm{DSR}}(x^5)) \\
&\neq \exp(-\theta_1(g)S^1_{\mathrm{DSR}}(x^5)) \times \exp(-\theta_2(g)S^2_{\mathrm{DSR}}(x^5)) \\
&\quad \times \exp(-\theta_3(g)S^3_{\mathrm{DSR}}(x^5)) \times \exp(-\zeta_1(g)K^1_{\mathrm{DSR}}(x^5)) \\
&\quad \times \exp(-\zeta_2(g)K^2_{\mathrm{DSR}}(x^5)) \times \exp(-\zeta_3(g)K^3_{\mathrm{DSR}}(x^5)).
\end{aligned} \qquad (7.5)$$

## 7.2 Finite Space–Time Rotations in $\widetilde{M}$

### *7.2.1 Infinitesimal Generators*

The explicit form of the matrices of the infinitesimal generators $I^{\alpha\beta}_{\mathrm{DSR}}(x^5)$ of the group $\mathrm{SO}(1,3)_{\mathrm{DEF.}}$ in the 4D representation have been derived in Sect. 6.3 (see (6.45)–(6.50)). In the following, we shall need the 4×4 matrices $A_i$, $B_i$ $(i = 1,2,3)$, defined by

$$A_1 \equiv \begin{pmatrix} 1 & 0 & 0 & 0 \\ 0 & 1 & 0 & 0 \\ 0 & 0 & 0 & 0 \\ 0 & 0 & 0 & 0 \end{pmatrix}; \qquad (7.6)$$

$$A_2 \equiv \begin{pmatrix} 1 & 0 & 0 & 0 \\ 0 & 0 & 0 & 0 \\ 0 & 0 & 1 & 0 \\ 0 & 0 & 0 & 0 \end{pmatrix}; \tag{7.7}$$

$$A_3 \equiv \begin{pmatrix} 1 & 0 & 0 & 0 \\ 0 & 0 & 0 & 0 \\ 0 & 0 & 0 & 0 \\ 0 & 0 & 0 & 1 \end{pmatrix}; \tag{7.8}$$

$$B_1 \equiv \begin{pmatrix} 0 & 0 & 0 & 0 \\ 0 & 0 & 0 & 0 \\ 0 & 0 & 1 & 0 \\ 0 & 0 & 0 & 1 \end{pmatrix}; \tag{7.9}$$

$$B_2 \equiv \begin{pmatrix} 0 & 0 & 0 & 0 \\ 0 & 1 & 0 & 0 \\ 0 & 0 & 0 & 0 \\ 0 & 0 & 0 & 1 \end{pmatrix}; \tag{7.10}$$

$$B_3 \equiv \begin{pmatrix} 0 & 0 & 0 & 0 \\ 0 & 1 & 0 & 0 \\ 0 & 0 & 1 & 0 \\ 0 & 0 & 0 & 0 \end{pmatrix}. \tag{7.11}$$

Let us evaluate the powers of the deformed boost generators: we have for instance, for $I^{10}_{\text{DSR}}(x^5)$:

$$\left(I^{10}_{\text{DSR}}(x^5)\right)^2 = \begin{pmatrix} b_0^{-2}(x^5)b_1^{-2}(x^5) & 0 & 0 & 0 \\ 0 & b_0^{-2}(x^5)b_1^{-2}(x^5) & 0 & 0 \\ 0 & 0 & 0 & 0 \\ 0 & 0 & 0 & 0 \end{pmatrix}; \tag{7.12}$$

$$\left(I^{10}_{\text{DSR}}(x^5)\right)^3 = \begin{pmatrix} 0 & -b_0^{-4}(x^5)b_1^{-2}(x^5) & 0 & 0 \\ -b_0^{-2}(x^5)b_1^{-4}(x^5) & 0 & 0 & 0 \\ 0 & 0 & 0 & 0 \\ 0 & 0 & 0 & 0 \end{pmatrix}; \tag{7.13}$$

$$\left(I^{10}_{\text{DSR}}(x^5)\right)^4 = \begin{pmatrix} b_0^{-4}(x^5)b_1^{-4}(x^5) & 0 & 0 & 0 \\ 0 & b_0^{-4}(x^5)b_1^{-4}(x^5) & 0 & 0 \\ 0 & 0 & 0 & 0 \\ 0 & 0 & 0 & 0 \end{pmatrix}; \tag{7.14}$$

$$\left(I^{10}_{\text{DSR}}(x^5)\right)^5 = \begin{pmatrix} 0 & -b_0^{-6}(x^5)b_1^{-4}(x^5) & 0 & 0 \\ -b_0^{-4}(x^5)b_1^{-6}(x^5) & 0 & 0 & 0 \\ 0 & 0 & 0 & 0 \\ 0 & 0 & 0 & 0 \end{pmatrix}. \tag{7.15}$$

By induction, we get the general formula

$$
\begin{cases}
\left(I^{10}_{\mathrm{DSR}}(x^5)\right)^{2n} = b_0^{-2n}(x^5)b_1^{-2n}(x^5)\begin{pmatrix} 1 & 0 & 0 & 0 \\ 0 & 1 & 0 & 0 \\ 0 & 0 & 0 & 0 \\ 0 & 0 & 0 & 0 \end{pmatrix}, n \in N; \\
\left(I^{10}_{\mathrm{DSR}}(x^5)\right)^{2n+1} = b_0^{-2n}(x^5)b_1^{-2n}(x^5)I^{10}_{\mathrm{DSR}}(x^5) \\
= b_0^{-2n}(x^5)b_1^{-2n}(x^5)\begin{pmatrix} 0 & -b_0^{-2}(x^5) & 0 & 0 \\ -b_1^{-2}(x^5) & 0 & 0 & 0 \\ 0 & 0 & 0 & 0 \\ 0 & 0 & 0 & 0 \end{pmatrix}, n \in N \cup \{0\}.
\end{cases}
\tag{7.16}
$$

The analogous relations for the other deformed generators are:

$$
\begin{cases}
\left(I^{20}_{\mathrm{DSR}}(x^5)\right)^{2n} = b_0^{-2n}(x^5)b_2^{-2n}(x^5)\begin{pmatrix} 1 & 0 & 0 & 0 \\ 0 & 0 & 0 & 0 \\ 0 & 0 & 1 & 0 \\ 0 & 0 & 0 & 0 \end{pmatrix}, n \in N; \\
\left(I^{10}_{\mathrm{DSR}}(x^5)\right)^{2n+1} = b_0^{-2n}(x^5)b_2^{-2n}(x^5)I^{20}_{\mathrm{DSR}}(x^5) \\
= b_0^{-2n}(x^5)b_2^{-2n}(x^5)\begin{pmatrix} 0 & 0 & -b_0^{-2}(x^5) & 0 \\ 0 & 0 & 0 & 0 \\ -b_2^{-2}(x^5) & 0 & 0 & 0 \\ 0 & 0 & 0 & 0 \end{pmatrix}, n \in N \cup \{0\};
\end{cases}
\tag{7.17}
$$

$$
\begin{cases}
\left(I^{30}_{\mathrm{DSR}}(x^5)\right)^{2n} = b_0^{-2n}(x^5)b_3^{-2n}(x^5)\begin{pmatrix} 1 & 0 & 0 & 0 \\ 0 & 0 & 0 & 0 \\ 0 & 0 & 0 & 0 \\ 0 & 0 & 0 & 1 \end{pmatrix}, n \in N; \\
\left(I^{30}_{\mathrm{DSR}}(x^5)\right)^{2n+1} = b_0^{-2n}(x^5)b_3^{-2n}(x^5)I^{30}_{\mathrm{DSR}}(x^5) \\
= b_0^{-2n}(x^5)b_3^{-2n}(x^5)\begin{pmatrix} 0 & 0 & 0 & -b_0^{-2}(x^5) \\ 0 & 0 & 0 & 0 \\ 0 & 0 & 0 & 0 \\ -b_3^{-2}(x^5) & 0 & 0 & 0 \end{pmatrix}, n \in N \cup \{0\}.
\end{cases}
\tag{7.18}
$$

On account of definitions (7.6)–(7.11), (7.16)–(7.18) can be summarized as (ESC off):

$$\begin{cases} \left(I^{i0}_{\text{DSR}}(x^5)\right)^0 = 1_{4D}; \\ \left(I^{i0}_{\text{DSR}}(x^5)\right)^{2n} = b_0^{-2n}(x^5)b_i^{-2n}(x^5)A_i \ , n \in N; \\ \left(I^{i0}_{\text{DSR}}(x^5)\right)^{2n+1} = b_0^{-2n}(x^5)b_i^{-2n}(x^5)I^{i0}_{\text{DSR}}(x^5) \ , n \in N \cup \{0\} \end{cases} \tag{7.19}$$

($i = 1, 2, 3$, with $1_{4D}$ being the identity 4×4 matrix).

For the rotation generators, one gets, for e.g., $I^{12}_{\text{DSR}}(x^5)$:

$$\left(I^{12}_{\text{DSR}}(x^5)\right)^2 = \begin{pmatrix} 0 & 0 & 0 & 0 \\ 0 & -b_1^{-2}(x^5)b_2^{-2}(x^5) & 0 & 0 \\ 0 & 0 & -b_1^{-2}(x^5)b_2^{-2}(x^5) & 0 \\ 0 & 0 & 0 & 0 \end{pmatrix}; \tag{7.20}$$

$$\left(I^{12}_{\text{DSR}}(x^5)\right)^3 = \begin{pmatrix} 0 & 0 & 0 & 0 \\ 0 & 0 & b_1^{-4}(x^5)b_2^{-2}(x^5) & 0 \\ 0 & -b_1^{-2}(x^5)b_2^{-4}(x^5) & 0 & 0 \\ 0 & 0 & 0 & 0 \end{pmatrix}; \tag{7.21}$$

$$\left(I^{12}_{\text{DSR}}(x^5)\right)^4 = \begin{pmatrix} 0 & 0 & 0 & 0 \\ 0 & b_1^{-4}(x^5)b_2^{-4}(x^5) & 0 & 0 \\ 0 & 0 & b_1^{-4}(x^5)b_2^{-4}(x^5) & 0 \\ 0 & 0 & 0 & 0 \end{pmatrix}; \tag{7.22}$$

$$\left(I^{12}_{\text{DSR}}(x^5)\right)^5 = \begin{pmatrix} 0 & 0 & 0 & 0 \\ 0 & 0 & -b_1^{-6}(x^5)b_2^{-4}(x^5) & 0 \\ 0 & b_1^{-4}(x^5)b_2^{-6}(x^5) & 0 & 0 \\ 0 & 0 & 0 & 0 \end{pmatrix}; \tag{7.23}$$

and so on. Therefore

$$\begin{cases} \left(I^{12}_{\text{DSR}}(x^5)\right)^{2n} = (-1)^n b_1^{-2n}(x^5)b_2^{-2n}(x^5) \begin{pmatrix} 0 & 0 & 0 & 0 \\ 0 & 1 & 0 & 0 \\ 0 & 0 & 1 & 0 \\ 0 & 0 & 0 & 0 \end{pmatrix}, n \in N; \\ \left(I^{12}_{\text{DSR}}(x^5)\right)^{2n+1} = (-1)^n b_1^{-2n}(x^5)b_2^{-2n}(x^5)I^{12}_{\text{DSR}}(x^5) \\ = (-1)^n b_1^{-2n}(x^5)b_2^{-2n}(x^5) \begin{pmatrix} 0 & 0 & 0 & 0 \\ 0 & 0 & -b_1^{-2}(x^5) & 0 \\ 0 & b_2^{-2}(x^5) & 0 & 0 \\ 0 & 0 & 0 & 0 \end{pmatrix}, n \in N \cup \{0\}. \end{cases} \tag{7.24}$$

Analogously, the powers of the other two generators are expressed by

$$\begin{cases} \left(I^{23}_{\mathrm{DSR}}(x^5)\right)^{2n} = (-1)^n b_2^{-2n}(x^5) b_3^{-2n}(x^5) \begin{pmatrix} 0 & 0 & 0 & 0 \\ 0 & 0 & 0 & 0 \\ 0 & 0 & 1 & 0 \\ 0 & 0 & 0 & 1 \end{pmatrix}, n \in N; \\ \left(I^{23}_{\mathrm{DSR}}(x^5)\right)^{2n+1} = (-1)^n b_2^{-2n}(x^5) b_3^{-2n}(x^5) I^{23}_{\mathrm{DSR}}(x^5) \\ = (-1)^n b_2^{-2n}(x^5) b_3^{-2n}(x^5) \begin{pmatrix} 0 & 0 & 0 & 0 \\ 0 & 0 & 0 & 0 \\ 0 & 0 & 0 & -b_2^{-2}(x^5) \\ 0 & 0 & b_3^{-2}(x^5) & 0 \end{pmatrix}, n \in N \cup \{0\}; \end{cases} \tag{7.25}$$

$$\begin{cases} \left(I^{31}_{\mathrm{DSR}}(x^5)\right)^{2n} = (-1)^n b_1^{-2n}(x^5) b_3^{-2n}(x^5) \begin{pmatrix} 0 & 0 & 0 & 0 \\ 0 & 1 & 0 & 0 \\ 0 & 0 & 0 & 0 \\ 0 & 0 & 0 & 1 \end{pmatrix}, n \in N; \\ \left(I^{31}_{\mathrm{DSR}}(x^5)\right)^{2n+1} = (-1)^n b_1^{-2n}(x^5) b_3^{-2n}(x^5) I^{31}_{\mathrm{DSR}}(x^5) \\ = (-1)^n b_1^{-2n}(x^5) b_3^{-2n}(x^5) \begin{pmatrix} 0 & 0 & 0 & 0 \\ 0 & 0 & 0 & b_1^{-2}(x^5) \\ 0 & 0 & 0 & 0 \\ 0 & -b_3^{-2}(x^5) & 0 & 0 \end{pmatrix}, n \in N \cup \{0\}. \end{cases} \tag{7.26}$$

Equations(7.24)–(7.26) can be summarized as (ESC off):

$$\begin{cases} \left(I^{ij}_{\mathrm{DSR}}(x^5)\right)^0 = 1_{4D}; \\ \left(I^{ij}_{\mathrm{DSR}}(x^5)\right)^{2n} = (-1)^n b_i^{-2n}(x^5) b_j^{-2n}(x^5) B_{s\neq i, s\neq j} \\ = (-1)^n b_i^{-2n}(x^5) b_j^{-2n}(x^5) \left(\sum_{s=1}^3 (1-\delta_{si})(1-\delta_{sj}) B_s\right) \\ \overset{\text{ESC ``off'' on } i \text{ and } j, \text{ ESC ``on'' on } k}{=} (-1)^n b_i^{-2n}(x^5) b_j^{-2n}(x^5) \left|\epsilon_{ijk}\right| B_k \;, n \in N\;; \\ \left(I^{ij}_{\mathrm{DSR}}(x^5)\right)^{2n+1} = (-1)^n b_i^{-2n}(x^5) b_j^{-2n}(x^5) I^{ij}_{\mathrm{DSR}}(x^5) \;, n \in N \cup \{0\}, \end{cases} \tag{7.27}$$

where $|\epsilon_{ijk}| \equiv sgn(\epsilon_{ijk})\,\epsilon_{ijk}$. Notice that (7.19) and (7.27), although obtained by utilizing a 4D representation of the infinitesimal generators of $SO(1,3)_{\text{DEF.}}$, *hold true in general at abstract group level* (i.e., they are representation independent).

### *7.2.2 Finite Deformed Boost along a Coordinate Axis*

Let us first consider a finite deformed boost transformation, with rapidity parameter $\zeta_1(g)$ along $\widehat{x^1}$. Recalling that $K^i_{\text{DSR}}(x^5) \equiv I^{0i}_{\text{DSR}}(x^5)\ \forall i = 1,2,3$, we get, from (7.5):

$$
\begin{aligned}
&SO(1,3)_{\text{DEF.}} \ni g : \Lambda_{\text{DSR}}(g, x^5)\\
&= \exp(-\zeta_1(g)K^1_{\text{DSR}}(x^5)) = \exp(\zeta_1(g)I^{10}_{\text{DSR}}(x^5))\\
&= \sum_{n=0}^{\infty} \frac{1}{n!}(\zeta_1(g)I^{10}_{\text{DSR}}(x^5))^n\\
&= \sum_{n=0}^{\infty} \frac{1}{n!}(\zeta_1(g))^n(I^{10}_{\text{DSR}}(x^5))^n\\
&= 1_{4D} + \sum_{n=1}^{\infty} \frac{1}{(2n)!}(\zeta_1(g))^{2n}(I^{10}_{\text{DSR}}(x^5))^{2n}\\
&\quad + \sum_{n=0}^{\infty} \frac{1}{(2n+1)!}(\zeta_1(g))^{2n+1}(I^{10}_{\text{DSR}}(x^5))^{2n+1}\\
&= 1_{4D} + A_1 \sum_{n=1}^{\infty} \frac{1}{(2n)!} b_0^{-2n}(x^5) b_1^{-2n}(x^5)(\zeta_1(g))^{2n}\\
&\quad + I^{10}_{\text{DSR}}(x^5) \sum_{n=0}^{\infty} \frac{1}{(2n+1)!} b_0^{-2n}(x^5) b_1^{-2n}(x^5)(\zeta_1(g))^{2n+1}\\
&= 1_{4D} + A_1 \sum_{n=1}^{\infty} \frac{1}{(2n)!}(\zeta_1(g) b_0^{-1}(x^5) b_1^{-1}(x^5))^{2n}\\
&\quad + I^{10}_{\text{DSR}}(x^5) b_0(x^5) b_1(x^5) \sum_{n=0}^{\infty} \frac{1}{(2n+1)!}(\zeta_1(g) b_0^{-1}(x^5) b_1^{-1}(x^5))^{2n+1}\\
&= 1_{4D} + \left(\left(\cosh \zeta_1(g) b_0^{-1}(x^5) b_1^{-1}(x^5)\right) - 1\right) A_1\\
&\quad + b_0(x^5) b_1(x^5) \left(\sinh \zeta_1(g) b_0^{-1}(x^5) b_1^{-1}(x^5)\right) I^{10}_{DSR}\ (x^5),
\end{aligned}
\tag{7.28}
$$

where in the last passage the series expansions of hyperbolic functions have been used.

Let us denote by $\Lambda_{\mathrm{B,DSR},\widehat{x^1}}(g,x^5)$ the 4×4 matrix representing a boost with rapidity parameter $\zeta_1(g)$ along $\widehat{x^1}$, namely:

$$\begin{aligned}\Lambda_{\mathrm{B,DSR},\widehat{x^1}}(g,x^5) \equiv{}& 1_{4D} + \left(\left(\cosh \zeta_1(g) b_0^{-1}(x^5) b_1^{-1}(x^5)\right) - 1\right) A_1 \\ &+ b_0(x^5) b_1(x^5) \left(\sinh \zeta_1(g) b_0^{-1}(x^5) b_1^{-1}(x^5)\right) I^{10}_{\mathrm{DSR}}(x^5).\end{aligned} \tag{7.29}$$

Then, (7.28) can be rewritten as:

$$\mathrm{SO}(1,3)_{\mathrm{DEF.}} \ni g : x \rightarrow x'_{(g)}(x,x^5) = \Lambda_{\mathrm{B,DSR},\widehat{x^1}}(g,x^5)x, \tag{7.30}$$

where $x'$, $x$ have to be meant as column four-vectors.

We find therefore

$$\begin{pmatrix} x^{0\prime}_{(g)}(x,x^5) \\ x^{1\prime}_{(g)}(x,x^5) \\ x^{2\prime}_{(g)}(x,x^5) \\ x^{3\prime}_{(g)}(x,x^5) \end{pmatrix} = \Lambda_{\mathrm{B,DSR},\widehat{x^1}}(g,x^5) \times \begin{pmatrix} x^0 \\ x^1 \\ x^2 \\ x^3 \end{pmatrix}$$

$$= \begin{pmatrix} \left\{ \begin{array}{c} \left(\cosh \zeta_1(g) b_0^{-1}(x^5) b_1^{-1}(x^5)\right) x^0 + \\ -b_0^{-1}(x^5) b_1(x^5) \left(\sinh \zeta_1(g) b_0^{-1}(x^5) b_1^{-1}(x^5)\right) x^1 \end{array} \right. \\ \left\{ \begin{array}{c} -b_0(x^5) b_1^{-1}(x^5) \left(\sinh \zeta_1(g) b_0^{-1}(x^5) b_1^{-1}(x^5)\right) x^0 \\ + \left(\cosh \zeta_1(g) b_0^{-1}(x^5) b_1^{-1}(x^5)\right) x^1 \end{array} \right. \\ x^2 \\ x^3 \end{pmatrix}. \tag{7.31}$$

Analogously, one gets, respectively, for the finite deformed boosts with rapidity parameter $\zeta_2(g)$ along $\widehat{x^2}$ and with rapidity parameter $\zeta_3(g)$ along $\widehat{x^3}$:

$$\mathrm{SO}(1,3)_{\mathrm{DEF.}} \ni g : \Lambda_{\mathrm{B,DSR},\widehat{x^2}}(g, x^5)$$

$$= \exp(-\zeta_2(g) K^2_{\mathrm{DSR}}(x^5)) = \exp(\zeta_2(g) I^{20}_{\mathrm{DSR}}(x^5))$$

$$= \sum_{n=0}^{\infty} \frac{1}{n!} (\zeta_2(g) I^{20}_{\mathrm{DSR}}(x^5))^n = \sum_{n=0}^{\infty} \frac{1}{n!} (\zeta_2(g))^n (I^{20}_{\mathrm{DSR}}(x^5))^n$$

$$= 1_{4D} + \sum_{n=1}^{\infty} \frac{1}{(2n)!} (\zeta_2(g))^{2n} (I^{20}_{\mathrm{DSR}}(x^5))^{2n}$$

$$+ \sum_{n=0}^{\infty} \frac{1}{(2n+1)!} (\zeta_2(g))^{2n+1} (I^{20}_{\mathrm{DSR}}(x^5))^{2n+1}$$

$$= 1_{4D} + A_2 \sum_{n=1}^{\infty} \frac{1}{(2n)!} b_0^{-2n}(x^5) b_2^{-2n}(x^5) (\zeta_2(g))^{2n}$$

$$+ I^{20}_{\mathrm{DSR}}(x^5) \sum_{n=0}^{\infty} \frac{1}{(2n+1)!} b_0^{-2n}(x^5) b_2^{-2n}(x^5) (\zeta_2(g))^{2n+1}$$

$$= 1_{4D} + A_2 \sum_{n=1}^{\infty} \frac{1}{(2n)!} (\zeta_2(g) b_0^{-1}(x^5) b_2^{-1}(x^5))^{2n}$$

$$+ I^{20}_{\mathrm{DSR}}(x^5) b_0(x^5) b_2(x^5) \sum_{n=0}^{\infty} \frac{1}{(2n+1)!} (\zeta_2(g) b_0^{-1}(x^5) b_2^{-1}(x^5))^{2n+1}$$

$$= 1_{4D} + \left( (\cosh \zeta_2(g) b_0^{-1}(x^5) b_2^{-1}(x^5)) - 1 \right) A_2$$

$$+ b_0(x^5) b_2(x^5) \left( \sinh \zeta_2(g) b_0^{-1}(x^5) b_2^{-1}(x^5) \right) I^{20}_{DSR}\ (x^5); \qquad (7.32)$$

$$\mathrm{SO}(1,3)_{\mathrm{DEF.}} \ni g : \Lambda_{\mathrm{B,DSR},\widehat{x^3}}(g, x^5)$$

$$= \exp(-\zeta_3(g) K^3_{\mathrm{DSR}}(x^5)) = \exp(\zeta_3(g) I^{30}_{\mathrm{DSR}}(x^5))$$

$$= \sum_{n=0}^{\infty} \frac{1}{n!} (\zeta_3(g) I^{30}_{\mathrm{DSR}}(x^5))^n = \sum_{n=0}^{\infty} \frac{1}{n!} (\zeta_3(g))^n (I^{30}_{\mathrm{DSR}}(x^5))^n$$

$$= 1_{4D} + \sum_{n=1}^{\infty} \frac{1}{(2n)!} (\zeta_3(g))^{2n} (I^{30}_{\text{DSR}}(x^5))^{2n}$$

$$+ \sum_{n=0}^{\infty} \frac{1}{(2n+1)!} (\zeta_3(g))^{2n+1} (I^{30}_{\text{DSR}}(x^5))^{2n+1}$$

$$= 1_{4D} + A_3 \sum_{n=1}^{\infty} \frac{1}{(2n)!} b_0^{-2n}(x^5) b_3^{-2n}(x^5) (\zeta_3(g))^{2n}$$

$$+ I^{30}_{\text{DSR}}(x^5) \sum_{n=0}^{\infty} \frac{1}{(2n+1)!} b_0^{-2n}(x^5) b_3^{-2n}(x^5) (\zeta_3(g))^{2n+1}$$

$$= 1_{4D} + A_3 \sum_{n=1}^{\infty} \frac{1}{(2n)!} (\zeta_3(g) b_0^{-1}(x^5) b_3^{-1}(x^5))^{2n}$$

$$+ I^{30}_{\text{DSR}}(x^5) b_0(x^5) b_3(x^5) \sum_{n=0}^{\infty} \frac{1}{(2n+1)!} (\zeta_3(g) b_0^{-1}(x^5) b_3^{-1}(x^5))^{2n+1}$$

$$= 1_{4D} + \left( (\cosh \zeta_3(g) b_0^{-1}(x^5) b_3^{-1}(x^5)) - 1 \right) A_3$$

$$+ b_0(x^5) b_3(x^5) \left( \sinh \zeta_3(g) b_0^{-1}(x^5) b_3^{-1}(x^5) \right) I^{30}_{\text{DSR}} \ (x^5). \tag{7.33}$$

The boost matrices $\Lambda_{\text{B,DSR},\widehat{x^2}}(g, x^5)$, $\Lambda_{\text{B,DSR},\widehat{x^3}}(g, x^5)$ explicitly read

$$\Lambda_{\text{B,DSR},\widehat{x^2}}(g, x^5) \equiv 1_{4D} + \left( (\cosh \zeta_2(g) b_0^{-1}(x^5) b_2^{-1}(x^5)) - 1 \right) A_2$$

$$+ b_0(x^5) b_2(x^5) \left( \sinh \zeta_2(g) b_0^{-1}(x^5) b_2^{-1}(x^5) \right) I^{20}_{DSR} \ (x^5); \tag{7.34}$$

$$\Lambda_{\text{B,DSR},\widehat{x^3}}(g, x^5) \equiv 1_{4D} + \left( (\cosh \zeta_3(g) b_0^{-1}(x^5) b_3^{-1}(x^5)) - 1 \right) A_3$$

$$+ b_0(x^5) b_3(x^5) \left( \sinh \zeta_3(g) b_0^{-1}(x^5) b_3^{-1}(x^5) \right) I^{30}_{\text{DSR}} \ (x^5), \tag{7.35}$$

so that one gets from (7.32) and (7.33), in matrix notation:

$$\text{SO}(1,3)_{\text{DEF.}} \ni g : x \to x'_{(g)}(x, x^5) = \Lambda_{\text{B,DSR},\widehat{x^2}}(g, x^5) x; \tag{7.36}$$

$$\text{SO}(1,3)_{\text{DEF.}} \ni g : x \to x'_{(g)}(x, x^5) = \Lambda_{\text{B,DSR},\widehat{x^3}}(g, x^5) x. \tag{7.37}$$

The explicit form of transformations (7.36), (7.37) for the boosts along $\widehat{x^2}$ and $\widehat{x^3}$ is therefore:

$$\begin{pmatrix} x^{0\prime}_{(g)}(x,x^5) \\ x^{1\prime}_{(g)}(x,x^5) \\ x^{2\prime}_{(g)}(x,x^5) \\ x^{3\prime}_{(g)}(x,x^5) \end{pmatrix} = \Lambda_{\mathrm{B,DSR},\widehat{x^2}}(g,x^5) \times \begin{pmatrix} x^0 \\ x^1 \\ x^2 \\ x^3 \end{pmatrix}$$

$$= \begin{pmatrix} \begin{cases} \left(\cosh \zeta_2(g) b_0^{-1}(x^5) b_2^{-1}(x^5)\right) x^0 \\ -b_0^{-1}(x^5) b_2(x^5) \left(\sinh \zeta_2(g) b_0^{-1}(x^5) b_2^{-1}(x^5)\right) x^2 \end{cases} \\ x^1 \\ \begin{cases} -b_0(x^5) b_2^{-1}(x^5) \left(\sinh \zeta_2(g) b_0^{-1}(x^5) b_2^{-1}(x^5)\right) x^0 \\ + \left(\cosh \zeta_2(g) b_0^{-1}(x^5) b_2^{-1}(x^5)\right) x^2 \end{cases} \\ x^3 \end{pmatrix}; \tag{7.38}$$

$$\begin{pmatrix} x^{0\prime}_{(g)}(x,x^5) \\ x^{1\prime}_{(g)}(x,x^5) \\ x^{2\prime}_{(g)}(x,x^5) \\ x^{3\prime}_{(g)}(x,x^5) \end{pmatrix} = \Lambda_{\mathrm{B,DSR},\widehat{x^3}}(g,x^5) \times \begin{pmatrix} x^0 \\ x^1 \\ x^2 \\ x^3 \end{pmatrix}$$

$$= \begin{pmatrix} \begin{cases} \left(\cosh \zeta_3(g) b_0^{-1}(x^5) b_3^{-1}(x^5)\right) x^0 \\ -b_0^{-1}(x^5) b_3(x^5) \left(\sinh \zeta_3(g) b_0^{-1}(x^5) b_3^{-1}(x^5)\right) x^3 \end{cases} \\ x^1 \\ x^2 \\ \begin{cases} -b_0(x^5) b_3^{-1}(x^5) \left(\sinh \zeta_3(g) b_0^{-1}(x^5) b_3^{-1}(x^5)\right) x^0 \\ + \left(\cosh \zeta_3(g) b_0^{-1}(x^5) b_3^{-1}(x^5)\right) x^3 \end{cases} \end{pmatrix}. \tag{7.39}$$

Let us introduce the *(effective) deformed rapidity* $\widetilde{\boldsymbol{\zeta}}(g, x^5)$, defined by[1]

$$\widetilde{\zeta}_i(g, x^5) \equiv \zeta_i(g) b_0^{-1}(x^5) b_i^{-1}(x^5) \ \ \forall i = 1, 2, 3. \tag{7.40}$$

Then, a finite deformed boost transformation, with rapidity parameter $\zeta_i(g)$ along $\widehat{x^i}$ can be written in compact form as:

$$\begin{cases} x^{0\prime} &= \left(\cosh \widetilde{\zeta}_i(g, x^5)\right) x^0 - b_0^{-1}(x^5) b_i(x^5) \left(\sinh \widetilde{\zeta}_i(g, x^5)\right) x^i \ ; \\ x^{i\prime} &= -b_0(x^5) b_i^{-1}(x^5) \left(\sinh \widetilde{\zeta}_i(g, x^5)\right) x^0 + \left(\cosh \widetilde{\zeta}_i(g, x^5)\right) x^i \ ;. \\ x^{k \neq i\prime} &= x^{k \neq i}. \end{cases} \tag{7.41}$$

By recalling the expression of a boost in SR

$$\begin{cases} x^{0\prime} &= (\cosh \zeta_i(g))\, x^0 - (\sinh \zeta_i(g))\, x^i; \\ x^{i\prime} &= -(\sinh \zeta_i(g))\, x^0 + (\cosh \zeta_i(g))\, x^i; \\ x^{k \neq i\prime} &= x^{k \neq i}, \end{cases} , \tag{7.42}$$

it is easily seen that the deforming transition SR→DSR corresponds – at the level of group parameters – to the deforming and anisotropizing rescaling of rapidities $\zeta_i(g) \rightarrow \widetilde{\zeta}_i(g, x^5)$ $\forall i = 1, 2, 3$.

Parametric Change of Basis for a Deformed Boost along a Coordinate Axis

We recall that a deformed boost with speed parameter $v^i$ along $\widehat{x^i}$ reads (see Sect. 3.3.2) (ESC off throughout):

$$\begin{cases} x^{i\prime} &= \widetilde{\gamma}(g)(x^i - v^i(g)t) = \widetilde{\gamma}(g)\left(x^i - \widetilde{\beta}(g)\dfrac{b_0(x^5)}{b_i(x^5)}ct\right); \\ x^{k \neq i\prime} &= x^{k \neq i}; \\ t' &= \widetilde{\gamma}(g)\left(t - \dfrac{v^i(g) b_i^2(x^5)}{c^2 b_0^2(x^5)} x^i\right) = \widetilde{\gamma}(g)\left(t - \dfrac{\widetilde{\beta}^2(g)}{v^i(g)} x^i\right), \end{cases} \tag{7.43}$$

where we made explicit the dependence on the considered element $g \in \mathrm{SO}(1,3)_{\mathrm{DEF.}}$ .(the dependence on $x^5$ is omitted, but understood), and (cf. (3.28a)–(3.29a))

[1] Note that $\widetilde{\zeta}^i(g, x^5) = \widetilde{\zeta}_i(g, x^5)$ $\forall i = 1, 2, 3$, i.e., that, as $\boldsymbol{\zeta}(g)$, the (effective) deformed rapidity three-vector $\widetilde{\boldsymbol{\zeta}}(g, x^5)$ is Euclidean. This follows from definition (7.40), because $\zeta^i(g) = \zeta_i(g)$ and the position of the index $i$ in $b_i(x^5)$ is a notational convention.

$$\widetilde{\beta}(g) = \widetilde{\beta^i}(g) \equiv \frac{v^i(g)b_i(x^5)}{cb_0(x^5)} \equiv \frac{v^i(g)}{u_i}; \tag{7.44}$$

$$\widetilde{\gamma}(g) = \widetilde{\gamma^i}(g) \equiv \left(1 - \left(\widetilde{\beta^i}(g)\right)^2\right)^{-1/2} = \left(1 - \left(\frac{v^i(g)b_i(x^5)}{cb_0(x^5)}\right)^2\right)^{-1/2}. \tag{7.45}$$

Quantity $u_i$ is the maximal causal speed (along $\widehat{x^i}$) in $\widetilde{M}(x_5)$, defined by (3.12).

As already shown in Sect. 3.3, (7.43) can be put in symmetrical form with respect to time and space coordinates by introducing the dimensional coordinate $\tilde{x}^0 = u^i t$:

$$\begin{cases} x^{i\prime} = \widetilde{\gamma}(g)(x^i - \widetilde{\beta^i}(g)\tilde{x}^0); \\ x^{k\neq i\prime} = x^{k\neq i}; \\ \tilde{x}^{0\prime} = \widetilde{\gamma}(g)(\tilde{x}^0 - \widetilde{\beta^i}(g)x^i). \end{cases} \tag{7.46}$$

Such a symmetry is lost if we use the "standard" time coordinate $x^0 \equiv ct$, which is related to $\tilde{x}^0$ by

$$\tilde{x}^0 = x^0 \frac{b_0(x^5)}{b_i(x^5)}. \tag{7.47}$$

In terms of $x^0$, we have in fact

$$\begin{cases} x^{i\prime} = \widetilde{\gamma}(g)\left(x^i - \widetilde{\beta^i}(g)\dfrac{b_0(x^5)}{b_i(x^5)}x^0\right); \\ x^{k\neq i\prime} = x^{k\neq i}; \\ x^{0\prime} = \widetilde{\gamma}(g)\left(x^0 - \widetilde{\beta^i}(g)\dfrac{b_i(x^5)}{b_0(x^5)}x^i\right). \end{cases} \tag{7.48}$$

Comparing (7.48) with (7.41) allows us to get the relations connecting the dimensional parametric basis of velocities $\{v^i\}$ and the dimensionless basis of (effective) deformed rapidities $\left\{\widetilde{\zeta}^i(g, x^5)\right\}$ defined by (7.40) (the dependence on $x^5$ is now fully made explicit):

$$\left\{\begin{array}{l} I)\ \cosh\widetilde{\zeta}_i(g,x^5)=\widetilde{\gamma}(g,x^5) \\ =\left(1-\left(\widetilde{\beta}^i(g,x^5)\right)^2\right)^{-1/2}=\left(1-\left(\frac{v^i(g)b_i(x^5)}{cb_0(x^5)}\right)^2\right)^{-1/2}; \\ II)\ b_0^{-1}(x^5)b_i(x^5)\left(\sinh\widetilde{\zeta}_i(g,x^5)\right)=\widetilde{\gamma}(g,x^5)\widetilde{\beta}^i(g,x^5)\frac{b_i(x^5)}{b_0(x^5)} \\ =\frac{v^i(g)b_i^2(x^5)}{cb_0^2(x^5)}\left(1-\left(\frac{v^i(g)b_i(x^5)}{cb_0(x^5)}\right)^2\right)^{-1/2}; \\ III)\ b_0(x^5)b_i^{-1}(x^5)\left(\sinh\widetilde{\zeta}_i(g,x^5)\right)=\widetilde{\gamma}(g,x^5)\widetilde{\beta}^i(g,x^5)\frac{b_0(x^5)}{b_i(x^5)} \\ =\frac{v^i(g)}{c}\left(1-\left(\frac{v^i(g)b_i(x^5)}{cb_0(x^5)}\right)^2\right)^{-1/2}, \end{array}\right.$$

$$\forall i = 1,2,3. \tag{7.49}$$

From the earlier system one gets (ESC off):[2]

$$\left(1-\left(\frac{v^i(g)b_i(x^5)}{cb_0(x^5)}\right)^2\right)^{-1/2}=\widetilde{\gamma}(g,x^5)\overset{\diamond}{\equiv}\widetilde{\gamma}^i(g,x^5)=\cosh\widetilde{\zeta}_i(g,x^5); \tag{7.50}$$

$$\begin{aligned}\frac{v^i(g)b_i(x^5)}{cb_0(x^5)}\left(1-\left(\frac{v^i(g)b_i(x^5)}{cb_0(x^5)}\right)^2\right)^{-1/2}&=\widetilde{\gamma}(g,x^5)\widetilde{\beta}^i(g,x^5)\\ &\overset{\diamond}{\equiv}\widetilde{\gamma}^i(g,x^5)\widetilde{\beta}^i(g,x^5)\\ &=\sinh\widetilde{\zeta}_i(g,x^5).\end{aligned} \tag{7.51}$$

[2]Notice that $\widetilde{\gamma}(g,x^5)$, by the identities marked with "$\diamond$", is endowed with an (arbitrarily contravariant) index $i \in \{1,2,3\}$, on account of its dependence on $\widetilde{\beta}(g,x^5) = \widetilde{\beta}^i(g,x^5) = \dfrac{v^i(g)b_i(x^5)}{cb_0(x^5)}$.

Such relations are consistent with the properties of the hyperbolic functions, since ($\forall i = 1, 2, 3$) (ESC off)

$$\cosh^2 \widetilde{\zeta}_i(g, x^5) - \sinh^2 \widetilde{\zeta}_i(g, x^5) = 1$$

$$\Leftrightarrow \left(1 - \left(\frac{v^i(g)b_i(x^5)}{cb_0(x^5)}\right)^2\right)^{-1} - \left(\frac{v^i(g)b_i(x^5)}{cb_0(x^5)}\right)^2 \left(1 - \left(\frac{v^i(g)b_i(x^5)}{cb_0(x^5)}\right)^2\right)^{-1}$$

$$= \frac{\left(1 - \left(\frac{v^i(g)b_i(x^5)}{cb_0(x^5)}\right)^2\right)}{\left(1 - \left(\frac{v^i(g)b_i(x^5)}{cb_0(x^5)}\right)^2\right)} = 1. \tag{7.52}$$

Equations (7.50) and (7.51) reduce of course to the standard SR relations in the limit $\widetilde{M}(x^5) \longrightarrow M$, i.e., in the limit $g_{\mu\nu,\mathrm{DSR}}(x^5) \rightarrow g_{\mu\nu,\mathrm{SR}}$.

### *7.2.3 Finite Deformed Rotation about a Coordinate Axis*

Let us now consider a finite true (clockwise) deformed rotation by an angle $\theta_1(g)$ about $\widehat{x^1}$. By recalling that $\mathbf{S}_{\mathrm{DSR}}(x^5) \equiv (I^{23}_{\mathrm{DSR}}(x^5), I^{31}_{\mathrm{DSR}}(x^5), I^{12}_{\mathrm{DSR}}(x^5))$, it follows:

$$\begin{aligned}
&\mathrm{SO}(1,3)_{\mathrm{DEF.}} \ni g : x^\mu \rightarrow x^{\mu'}_{(g)}(x, x^5) \\
&= \exp(-\theta_1(g) S^1_{\mathrm{DSR}}(x^5))^\mu_{\ \nu} x^\nu = \exp(\theta_1(g) I^{32}_{\mathrm{DSR}}(x^5))^\mu_{\ \nu} x^\nu \\
&= \left(\sum_{n=0}^{\infty} \frac{1}{n!}(\theta_1(g) I^{32}_{\mathrm{DSR}}(x^5))^n\right)^\mu_{\ \nu} x^\nu = \left(\sum_{n=0}^{\infty} \frac{1}{n!}(\theta_1(g))^n (I^{32}_{\mathrm{DSR}}(x^5))^n\right)^\mu_{\ \nu} x^\nu \\
&= \left(1_{4D} + \sum_{n=1}^{\infty} \frac{1}{(2n)!}(\theta_1(g))^{2n} (I^{32}_{\mathrm{DSR}}(x^5))^{2n}\right. \\
&\qquad \left. + \sum_{n=0}^{\infty} \frac{1}{(2n+1)!}(\theta_1(g))^{2n+1} (I^{32}_{\mathrm{DSR}}(x^5))^{2n+1}\right)^\mu_{\ \nu} x^\nu \\
&= \left(1_{4D} + B_1 \sum_{n=1}^{\infty} \frac{(-1)^n}{(2n)!} b_2^{-2n}(x^5) b_3^{-2n}(x^5)(\theta_1(g))^{2n}\right. \\
&\qquad \left. + I^{32}_{\mathrm{DSR}}(x^5) \sum_{n=0}^{\infty} \frac{(-1)^n}{(2n+1)!} b_2^{-2n}(x^5) b_3^{-2n}(x^5)(\theta_1(g))^{2n+1}\right)^\mu_{\ \nu} x^\nu
\end{aligned}$$

$$= \left( 1_{4D} + B_1 \sum_{n=1}^{\infty} \frac{(-1)^n}{(2n)!} (\theta_1(g) b_2^{-1}(x^5) b_3^{-1}(x^5))^{2n} \right.$$

$$\left. + I_{\text{DSR}}^{32}(x^5) b_2(x^5) b_3(x^5) \sum_{n=0}^{\infty} \frac{(-1)^n}{(2n+1)!} (\theta_1(g) b_2^{-1}(x^5) b_3^{-1}(x^5))^{2n+1} \right)^{\mu}_{\ \nu} x^{\nu}$$

$$= \left(1_{4D} + \left( (\cos \theta_1(g) b_2^{-1}(x^5) b_3^{-1}(x^5)) - 1 \right) B_1 \right.$$

$$\left. + b_2(x^5) b_3(x^5) \left( \sin \theta_1(g) b_2^{-1}(x^5) b_3^{-1}(x^5) \right) I_{\text{DSR}}^{32}(x^5) \right)^{\mu}_{\ \nu} x^{\nu}, \tag{7.53}$$

where in the last passage the series expansions of trigonometric functions have been used.

By introducing the matrix

$$\Lambda_{\text{R,DSR},\widehat{x^1}}(g, x^5) \equiv 1_{4D} + \left( (\cos \theta_1(g) b_2^{-1}(x^5) b_3^{-1}(x^5)) - 1 \right) B_1$$

$$+ b_2(x^5) b_3(x^5) \left( \sin \theta_1(g) b_2^{-1}(x^5) b_3^{-1}(x^5) \right) I_{\text{DSR}}^{32}(x^5), \tag{7.54}$$

(7.53) can be rewritten in matrix notation as:

$$\text{SO}(1,3)_{\text{DEF.}} \ni g : x \to x'_{(g)}(x, x^5) = \Lambda_{\text{R,DSR},\widehat{x^1}}(g, x^5) x, \tag{7.55}$$

or explicitly:

$$\begin{pmatrix} x^{0\prime}_{(g)}(x, x^5) \\ x^{1\prime}_{(g)}(x, x^5) \\ x^{2\prime}_{(g)}(x, x^5) \\ x^{3\prime}_{(g)}(x, x^5) \end{pmatrix} = \Lambda_{\text{R,DSR},\widehat{x^1}}(g, x^5) \times \begin{pmatrix} x^0 \\ x^1 \\ x^2 \\ x^3 \end{pmatrix}$$

$$= \begin{pmatrix} x^0 \\ x^1 \\ \begin{cases} (\cos \theta_1(g) b_2^{-1}(x^5) b_3^{-1}(x^5)) x^2 + \\ + b_2^{-1}(x^5) b_3(x^5) \left( \sin \theta_1(g) b_2^{-1}(x^5) b_3^{-1}(x^5) \right) x^3 \end{cases} \\ \begin{cases} -b_2(x^5) b_3^{-1}(x^5) \left( \sin \theta_1(g) b_2^{-1}(x^5) b_3^{-1}(x^5) \right) x^2 + \\ + (\cos \theta_1(g) b_2^{-1}(x^5) b_3^{-1}(x^5)) x^3 \end{cases} \end{pmatrix}. \tag{7.56}$$

Analogous relations hold for the finite true (clockwise) deformed rotations by an angle $\theta_2(g)$ about $\widehat{x^2}$ and by an angle $\theta_3(g)$ about $\widehat{x^3}$:

$$
\begin{aligned}
&SO(1,3)_{\text{DEF.}} \ni g : x^\mu \to x'^\mu_{(g)}(x,x^5)\\
&= \exp(-\theta_2(g) S^2_{\text{DSR}}(x^5))^\mu_\nu x^\nu = \exp(\theta_2(g) I^{13}_{\text{DSR}}(x^5))^\mu_\nu x^\nu\\
&= \left(\sum_{n=0}^{\infty} \frac{1}{n!}(\theta_2(g) I^{13}_{\text{DSR}}(x^5))^n\right)^\mu_\nu x^\nu\\
&= \left(\sum_{n=0}^{\infty} \frac{1}{n!}(\theta_2(g))^n (I^{13}_{\text{DSR}}(x^5))^n\right)^\mu_\nu x^\nu\\
&= \left(1_{4D} + \sum_{n=1}^{\infty} \frac{1}{(2n)!}(\theta_2(g))^{2n} (I^{13}_{\text{DSR}}(x^5))^{2n}\right.\\
&\quad \left. + \sum_{n=0}^{\infty} \frac{1}{(2n+1)!}(\theta_2(g))^{2n+1} (I^{13}_{\text{DSR}}(x^5))^{2n+1}\right)^\mu_\nu x^\nu\\
&= \left(1_{4D} + B_2 \sum_{n=1}^{\infty} \frac{(-1)^n}{(2n)!} b_1^{-2n}(x^5) b_3^{-2n}(x^5)(\theta_2(g))^{2n}\right.\\
&\quad \left. + I^{13}_{\text{DSR}}(x^5) \sum_{n=0}^{\infty} \frac{(-1)^n}{(2n+1)!} b_1^{-2n}(x^5) b_3^{-2n}(x^5)(\theta_2(g))^{2n+1}\right)^\mu_\nu x^\nu\\
&= \left(1_{4D} + B_2 \sum_{n=1}^{\infty} \frac{(-1)^n}{(2n)!}(\theta_2(g) b_1^{-1}(x^5) b_3^{-1}(x^5))^{2n}\right.\\
&\quad + I^{13}_{\text{DSR}}(x^5) b_1(x^5) b_3(x^5) \sum_{n=0}^{\infty} \frac{(-1)^n}{(2n+1)!}\\
&\quad \left. \times (\theta_2(g) b_1^{-1}(x^5) b_3^{-1}(x^5))^{2n+1}\right)^\mu_\nu x^\nu\\
&= \left(1_{4D} + \left((\cos\theta_2(g) b_1^{-1}(x^5) b_3^{-1}(x^5)) - 1\right) B_2\right.\\
&\quad \left. + b_1(x^5) b_3(x^5) \left(\sin\theta_2(g) b_1^{-1}(x^5) b_3^{-1}(x^5)\right) I^{13}_{\text{DSR}}(x^5)\right)^\mu_\nu x^\nu; \qquad (7.57)
\end{aligned}
$$

$$
\begin{aligned}
&SO(1,3)_{\text{DEF.}} \ni g : x^\mu \to x'^\mu_{(g)}(x,x^5)\\
&= \exp(-\theta_3(g) S^3_{\text{DSR}}(x^5))^\mu_\nu x^\nu = \exp(\theta_3(g) I^{21}_{\text{DSR}}(x^5))^\mu_\nu x^\nu\\
&= \left(\sum_{n=0}^{\infty} \frac{1}{n!}(\theta_3(g) I^{21}_{\text{DSR}}(x^5))^n\right)^\mu_\nu x^\nu = \left(\sum_{n=0}^{\infty} \frac{1}{n!}(\theta_3(g))^n (I^{21}_{\text{DSR}}(x^5))^n\right)^\mu_\nu x^\nu
\end{aligned}
$$

$$= \left( 1_{4D} + \sum_{n=1}^{\infty} \frac{1}{(2n)!} (\theta_3(g))^{2n} (I_{\mathrm{DSR}}^{21}(x^5))^{2n} \right.$$

$$\left. + \sum_{n=0}^{\infty} \frac{1}{(2n+1)!} (\theta_3(g))^{2n+1} (I_{\mathrm{DSR}}^{21}(x^5))^{2n+1} \right)^{\mu}_{\ \nu} x^{\nu}$$

$$= \left( 1_{4D} + B_3 \sum_{n=1}^{\infty} \frac{(-1)^n}{(2n)!} b_1^{-2n}(x^5) b_2^{-2n}(x^5) (\theta_3(g))^{2n} \right.$$

$$\left. + I_{\mathrm{DSR}}^{21}(x^5) \sum_{n=0}^{\infty} \frac{(-1)^n}{(2n+1)!} b_1^{-2n}(x^5) b_2^{-2n}(x^5) (\theta_3(g))^{2n+1} \right)^{\mu}_{\ \nu} x^{\nu}$$

$$= \left( 1_{4D} + B_3 \sum_{n=1}^{\infty} \frac{(-1)^n}{(2n)!} (\theta_3(g) b_1^{-1}(x^5) b_2^{-1}(x^5))^{2n} \right.$$

$$\left. + I_{\mathrm{DSR}}^{21}(x^5) b_1(x^5) b_2(x^5) \sum_{n=0}^{\infty} \frac{(-1)^n}{(2n+1)!} (\theta_3(g) b_1^{-1}(x^5) b_2^{-1}(x^5))^{2n+1} \right)^{\mu}_{\ \nu} x^{\nu}$$

$$= \left( 1_{4D} + \left( \left( \cos \theta_3(g) b_1^{-1}(x^5) b_2^{-1}(x^5) \right) - 1 \right) B_3 \right.$$

$$\left. + b_1(x^5) b_2(x^5) \left( \sin \theta_3(g) b_1^{-1}(x^5) b_2^{-1}(x^5) \right) I_{\mathrm{DSR}}^{21}(x^5) \right)^{\mu}_{\ \nu} x^{\nu}. \tag{7.58}$$

Equations (7.57) and (7.58) can be rewritten as:

$$\mathrm{SO}(1,3)_{\mathrm{DEF.}} \ni g : x \to x'_{(g)}(x, x^5) = \Lambda_{\mathrm{R,DSR},\widehat{x^2}}(g, x^5) x; \tag{7.59}$$

$$\mathrm{SO}(1,3)_{\mathrm{DEF.}} \ni g : x \to x'_{(g)}(x, x^5) = \Lambda_{\mathrm{R,DSR},\widehat{x^3}}(g, x^5) x, \tag{7.60}$$

where we defined the rotation 4×4 matrices

$$\Lambda_{\mathrm{R,DSR},\widehat{x^2}}(g, x^5) \equiv 1_{4D} + \left( \left( \cos \theta_2(g) b_1^{-1}(x^5) b_3^{-1}(x^5) \right) - 1 \right) B_2$$

$$+ b_1(x^5) b_3(x^5) \left( \sin \theta_2(g) b_1^{-1}(x^5) b_3^{-1}(x^5) \right) I_{\mathrm{DSR}}^{13}(x^5); \tag{7.61}$$

$$\Lambda_{\mathrm{R,DSR},\widehat{x^3}}(g, x^5) \equiv 1_{4D} + \left( \left( \cos \theta_3(g) b_1^{-1}(x^5) b_2^{-1}(x^5) \right) - 1 \right) B_3$$

$$+ b_1(x^5) b_2(x^5) \left( \sin \theta_3(g) b_1^{-1}(x^5) b_2^{-1}(x^5) \right) I_{\mathrm{DSR}}^{21}(x^5). \tag{7.62}$$

Then, the finite true (clockwise) rotations in the deformed Minkowski space $\widetilde{M}(x_5)$ along $\widehat{x^2}$ and along $\widehat{x^3}$ read, respectively:

$$
\begin{pmatrix} x^{0\prime}_{(g)}(x,x^5) \\ x^{1\prime}_{(g)}(x,x^5) \\ x^{2\prime}_{(g)}(x,x^5) \\ x^{3\prime}_{(g)}(x,x^5) \end{pmatrix} = \Lambda_{\mathrm{R,DSR},\widehat{x^2}}(g,x^5) \times \begin{pmatrix} x^0 \\ x^1 \\ x^2 \\ x^3 \end{pmatrix}
$$

$$
= \left( \begin{array}{c} x^0 \\ \left\{ \begin{array}{c} (\cos\theta_2(g) b_1^{-1}(x^5) b_3^{-1}(x^5))x^1 \\ -b_1^{-1}(x^5) b_3^{-1}(x^5) \left(\sin\theta_2(g) b_1^{-1}(x^5) b_3^{-1}(x^5)\right) x^3 \end{array} \right. \\ x^2 \\ \left\{ \begin{array}{c} b_1(x^5) b_3^{-1}(x^5) \left(\sin\theta_2(g) b_1^{-1}(x^5) b_3^{-1}(x^5)\right) x^1 + \\ +(\cos\theta_2(g) b_1^{-1}(x^5) b_3^{-1}(x^5))x^3 \end{array} \right. \end{array} \right); \tag{7.63}
$$

$$
\begin{pmatrix} x^{0\prime}_{(g)}(x,x^5) \\ x^{1\prime}_{(g)}(x,x^5) \\ x^{2\prime}_{(g)}(x,x^5) \\ x^{3\prime}_{(g)}(x,x^5) \end{pmatrix} = \Lambda_{\mathrm{R,DSR},\widehat{x^3}}(g,x^5) \times \begin{pmatrix} x^0 \\ x^1 \\ x^2 \\ x^3 \end{pmatrix}
$$

$$
= \left( \begin{array}{c} x^0 \\ \left\{ \begin{array}{c} (\cos\theta_3(g) b_1^{-1}(x^5) b_2^{-1}(x^5))x^1 \\ +b_1^{-1}(x^5) b_2(x^5) \left(\sin\theta_3(g) b_1^{-1}(x^5) b_2^{-1}(x^5)\right) x^2 \end{array} \right. \\ \left\{ \begin{array}{c} -b_1(x^5) b_2^{-1}(x^5) \left(\sin\theta_3(g) b_1^{-1}(x^5) b_2^{-1}(x^5)\right) x^1 \\ (\cos\theta_3(g) b_1^{-1}(x^5) b_2^{-1}(x^5))x^2 \end{array} \right. \\ x^3 \end{array} \right). \tag{7.64}
$$

By introducing the *(effective) deformed angles* $\widetilde{\boldsymbol{\theta}}(g)$ $(i \neq j, i \neq k, j \neq k)$[3]

$$\widetilde{\theta}_i(g, x^5) \equiv \theta_i(g) b_j^{-1}(x^5) b_k^{-1}(x^5)$$

$$\underset{=}{\text{ESC "off" on i, ESC "on" on j and k}} \frac{1}{2}\theta_i(g)\,|\epsilon_{ijk}|\,b_j^{-1}(x^5)b_k^{-1}(x^5),\ \forall i = 1,2,3, \tag{7.65}$$

Equations (7.56), (7.63), (7.64) can be written in compact form, respectively, as

$$\begin{pmatrix} x^{0\prime} \\ x^{1\prime} \\ x^{2\prime} \\ x^{3\prime} \end{pmatrix} = \begin{pmatrix} x^0 \\ x^1 \\ (\cos\widetilde{\theta}_1(g,x^5))x^2 + b_2^{-1}(x^5)b_3(x^5)\left(\sin\widetilde{\theta}_1(g,x^5)\right)x^3 \\ -b_2(x^5)b_3^{-1}(x^5)\left(\sin\widetilde{\theta}_1(g,x^5)\right)x^2 + (\cos\widetilde{\theta}_1(g,x^5))x^3 \end{pmatrix}; \tag{7.66}$$

$$\begin{pmatrix} x^{0\prime} \\ x^{1\prime} \\ x^{2\prime} \\ x^{3\prime} \end{pmatrix} = \begin{pmatrix} x^0 \\ (\cos\widetilde{\theta}_2(g,x^5))x^1 - b_1^{-1}(x^5)b_3(x^5)\left(\sin\widetilde{\theta}_2(g,x^5)\right)x^3 \\ x^2 \\ -b_2(x^5)b_3^{-1}(x^5)\left(\sin\widetilde{\theta}_1(g,x^5)\right)x^2 + (\cos\widetilde{\theta}_1(g,x^5))x^3 \end{pmatrix}; \tag{7.67}$$

$$\begin{pmatrix} x^{0\prime} \\ x^{1\prime} \\ x^{2\prime} \\ x^{3\prime} \end{pmatrix} = \begin{pmatrix} x^0 \\ (\cos\widetilde{\theta}_3(g,x^5))x^1 + b_1^{-1}(x^5)b_2(x^5)\left(\sin\widetilde{\theta}_3(g,x^5)\right)x^2 \\ -b_1(x^5)b_2^{-1}(x^5)\left(\sin\widetilde{\theta}_3(g,x^5)\right)x^1 + (\cos\widetilde{\theta}_3(g,x^5))x^2 \\ x^3 \end{pmatrix}. \tag{7.68}$$

By comparing the finite true (clockwise) rotations by an angle $\theta^i(g)$ about $\widehat{x^i}$ in SR

$$\begin{pmatrix} x^{0\prime} \\ x^{1\prime} \\ x^{2\prime} \\ x^{3\prime} \end{pmatrix} = \begin{pmatrix} x^0 \\ x^1 \\ (\cos\theta_1(g))\,x^2 + (\sin\theta_1(g))\,x^3 \\ -(\sin\theta_1(g))\,x^2 + (\cos\theta_1(g))\,x^3 \end{pmatrix}; \tag{7.69}$$

---

[3]Definition (7.65) of $\widetilde{\boldsymbol{\theta}}(g)$ does only formally coincides with that of the isotopic angles introduced by Santilli.

Note also that $\widetilde{\theta}^i(g, x^5) = \widetilde{\theta}_i(g, x^5)$ $\forall i = 1,2,3$, i.e., that, as $\boldsymbol{\theta}(g)$, the (effective) deformed angle three-vector $\widetilde{\boldsymbol{\theta}}(g, x^5)$ is Euclidean. This follows from (7.65), because $\theta^i(g) = \theta_i(g)$ and, as mentioned before for the deformed rapidity, the position of the index $i$ in $b_i(x^5)$ is a notational convention.

$$\begin{pmatrix} x^{0\prime} \\ x^{1\prime} \\ x^{2\prime} \\ x^{3\prime} \end{pmatrix} = \begin{pmatrix} x^0 \\ (\cos\theta_2(g))\,x^1 - (\sin\theta_2(g))\,x^3 \\ x^2 \\ (\sin\theta_2(g))\,x^1 + (\cos\theta_2(g))\,x^3 \end{pmatrix}; \tag{7.70}$$

$$\begin{pmatrix} x^{0\prime} \\ x^{1\prime} \\ x^{2\prime} \\ x^{3\prime} \end{pmatrix} = \begin{pmatrix} x^0 \\ (\cos\theta_3(g))\,x^1 + (\sin\theta_3(g))\,x^2 \\ -(\sin\theta_3(g))\,x^1 + (\cos\theta_3(g))\,x^2 \\ x^3 \end{pmatrix} \tag{7.71}$$

with those in DSR, it is easily seen that the deforming transition SR→DSR corresponds – at the level of group parameters – to the deforming and anisotropizing rescaling of angles $\theta_i(g) \to \widetilde{\theta}_i(g, x^5)$ $\forall i = 1, 2, 3$.

### *7.2.4 Antisymmetric Tensor of Deformed Rotation Parameters*

We have seen in Sect. 5.3 that, for a generalized Minkowski space, all forms of the tensor $\delta\omega$ are global, i.e., independent of the set of metric variables $x = \{x^0, x^1, x^2, x^3\}$, but only $\delta\omega_{\alpha\beta}(g)$ is a priori independent of possible nonmetric variables $\{x\}_{\text{n.m.}}$. For instance, consider the completely contravariant form of $\delta\omega$:

$$\delta\omega^{\alpha\beta}(g, \{x\}_{\text{n.m.}}) \equiv g^{\alpha\gamma}(\{x\}_{\text{n.m.}})g^{\beta\delta}(\{x\}_{\text{n.m.}})\delta\omega_{\alpha\beta}(g) \tag{7.72}$$

or, in matrix form:

$$\delta\omega_{\text{contrav.}}(g, \{x\}_{\text{n.m.}}) \equiv g^{\text{T}}_{\text{contrav.}}(\{x\}_{\text{n.m.}}) \times \delta\omega_{\text{cov.}}(g) \times g_{\text{contrav.}}(\{x\}_{\text{n.m.}}). \tag{7.73}$$

In the DSR case, the completely contravariant metric tensor reads

$$g^{\mu\nu}_{\text{DSR}}(x^5) = \text{diag}(b_0^{-2}(x^5), -b_1^{-2}(x^5), -b_2^{-2}(x^5), -b_3^{-2}(x^5))$$

$$\stackrel{\text{ESC off}}{=} \delta^{\mu\nu}(\delta^{\mu 0}b_0^{-2}(x^5) - \delta^{\mu 1}b_1^{-2}(x^5) - \delta^{\mu 2}b_2^{-2}(x^5) - \delta^{\mu 3}b_3^{-2}(x^5)) \tag{7.74}$$

or

$$g_{\text{contrav.DSR}}(x^5) = \begin{pmatrix} b_0^{-2}(x^5) & 0 & 0 & 0 \\ 0 & -b_1^{-2}(x^5) & 0 & 0 \\ 0 & 0 & -b_2^{-2}(x^5) & 0 \\ 0 & 0 & 0 & -b_3^{-2}(x^5) \end{pmatrix}. \tag{7.75}$$

Therefore

$$\begin{aligned}
\delta\omega^{\alpha\beta}_{\text{DSR}}(g, x^5) &\equiv \delta\omega_{\text{contrav.,DSR}}(g, x^5) \\
&\equiv g^{\text{T}}_{\text{contrav.DSR}}(x^5) \times \delta\omega_{\text{cov.,(DSR)}}(g) \times g_{\text{contrav.DSR}}(x^5) \\
&= \begin{pmatrix} b_0^{-2}(x^5) & 0 & 0 & 0 \\ 0 & -b_1^{-2}(x^5) & 0 & 0 \\ 0 & 0 & -b_2^{-2}(x^5) & 0 \\ 0 & 0 & 0 & -b_3^{-2}(x^5) \end{pmatrix} \\
&\times \begin{pmatrix} 0 & -\zeta^1(g) & -\zeta^2(g) & -\zeta^3(g) \\ \zeta^1(g) & 0 & -\theta^3(g) & \theta^2(g) \\ \zeta^2(g) & \theta^3(g) & 0 & -\theta^1(g) \\ \zeta^3(g) & -\theta^2(g) & \theta^1(g) & 0 \end{pmatrix} \\
&\times \begin{pmatrix} b_0^{-2}(x^5) & 0 & 0 & 0 \\ 0 & -b_1^{-2}(x^5) & 0 & 0 \\ 0 & 0 & -b_2^{-2}(x^5) & 0 \\ 0 & 0 & 0 & -b_3^{-2}(x^5) \end{pmatrix} \\
&= \begin{pmatrix} b_0^{-2}(x^5) & 0 & 0 & 0 \\ 0 & -b_1^{-2}(x^5) & 0 & 0 \\ 0 & 0 & -b_2^{-2}(x^5) & 0 \\ 0 & 0 & 0 & -b_3^{-2}(x^5) \end{pmatrix} \\
&\times \begin{pmatrix} 0 & \zeta^1(g)b_1^{-2}(x^5) & \zeta^2(g)b_2^{-2}(x^5) & \zeta^3(g)b_3^{-2}(x^5) \\ \zeta^1(g)b_0^{-2}(x^5) & 0 & \theta^3(g)b_2^{-2}(x^5) & -\theta^2(g)b_3^{-2}(x^5) \\ \zeta^2(g)b_0^{-2}(x^5) & -\theta^3(g)b_1^{-2}(x^5) & 0 & \theta^1(g)b_3^{-2}(x^5) \\ \zeta^3(g)b_0^{-2}(x^5) & \theta^2(g)b_1^{-2}(x^5) & -\theta^1(g)b_2^{-2}(x^5) & 0 \end{pmatrix},
\end{aligned} \tag{7.76}$$

whence, by recalling definitions (7.40) and (7.65) of the (effective) deformed rapidity and angle (Euclidean) three-vectors, we obtain:

$$\begin{cases} \delta\omega^{0i}_{\text{DSR}}(g,x^5) = \zeta^i(g)b_0^{-2}(x^5)b_i^{-2}(x^5) \\ = \widetilde{\zeta}^i(g,x^5)b_0^{-1}(x^5)b_i^{-1}(x^5), \ \forall\, i=1,2,3; \\ \left. \begin{array}{l} \delta\omega^{12}_{\text{DSR}}(g,x^5) = -\theta^3(g)b_1^{-2}(x^5)b_2^{-2}(x^5) = -\widetilde{\theta}^3(g,x^5)b_1^{-1}(x^5)b_2^{-1}(x^5) \\ \delta\omega^{13}_{\text{DSR}}(g,x^5) = \theta^2(g)b_1^{-2}(x^5)b_3^{-2}(x^5) = \widetilde{\theta}^2(g,x^5)b_1^{-1}(x^5)b_3^{-1}(x^5) \\ \delta\omega^{23}_{\text{DSR}}(g,x^5) = -\theta^1(g)b_2^{-2}(x^5)b_3^{-2}(x^5) = -\widetilde{\theta}^1(g,x^5)b_2^{-1}(x^5)b_3^{-1}(x^5) \end{array} \right\} \\ \Rightarrow \delta\omega^{jk}_{\text{DSR}}(g,x^5) \overset{\text{ESC on on } i,\text{ ESC off on } j \text{ and } k}{=} \epsilon_{ikj}\widetilde{\theta}^i(g,x^5)b_j^{-1}(x^5)b_k^{-1}(x^5), \\ \forall\,(j,k) \in (1,2,3)\,. \end{cases} \tag{7.77}$$

The earlier relation between the components of $\delta\omega_{\alpha\beta,(\text{DSR})}(g)$ and those of $\delta\omega^{\alpha\beta}_{\text{DSR}}(g,x^5)$ can be therefore written as $(i,j=1,2,3)$ (ESC off):

$$\begin{cases} \delta\omega_{ij,(\text{DSR})}(g) = b_i^2(x^5)b_j^2(x^5)\delta\omega^{ij}_{\text{DSR}}(g,x^5); \\ \delta\omega_{0i,(\text{DSR})}(g) = -b_0^{-2}(x^5)b_i^{-2}(x^5)\delta\omega^{0i}_{\text{DSR}}(g,x^5). \end{cases} \tag{7.78}$$

By recalling the expressions (7.40) and (7.65) of the deformed rapidity and angle three-vectors, we can therefore state that the formal "anisotropizing deforming" transition SR→DSR is summarized, at the group parameter level, by the passage from the antisymmetric, parametric *covariant* tensor $\delta\omega_{\alpha\beta,(\text{SR})}(g)$[4] to the antisymmetric *contravariant* tensor $\delta\omega^{\alpha\beta}_{\text{DSR}}(g,x^5)$ of the (effective) deformed parameters of $SO(1,3)_{\text{DEF.}}$ .

Let us notice that the same conclusion (namely, the characterization of $\delta\omega^{\alpha\beta}_{\text{DSR}}(g,x^5)$ as tensor of the effective deformed parameters) does *not* hold at the infinitesimal level. Indeed, in DSR (see Part I) – and in SR as well – one cannot rescale parameters in a unique way, since in any fixed infinitesimal transformation typology the related group parameter is rescaled in a different way, depending on the coordinate concerned. Effective deformed group parameters can be therefore defined only at the finite level

[4]Notice that formally

$$\delta\omega_{\alpha\beta,(\text{SR})}(g) = \delta\omega_{\alpha\beta,(\text{DSR})}(g), \tag{$\circ$}$$

but the $g$'s belong to *different* space–time rotation groups. In the l.h.s of ($\circ$) $g \in SO(1,3)_{\text{STD.}}$ (homogeneous Lorentz group), while in the r.h.s. of ($\circ$) $g$ belong to the "deformed" counterpart, i.e., $g \in SO(1,3)_{\text{DEF.}}$ (homogeneous "deformed" Lorentz group).

of the space–time rotation component $SO(1,3)_{\text{DEF.}}$ of the maximal Killing group $P(1,3)_{\text{DEF.}}$ of the deformed 4D Minkowski space $\widetilde{M}(x^5)$.

The possibility of generalizing such results to the generalized 4D (S,T=4-S) (and possibly to the $N$-d (S,T=N–S)) Minkowskian spaces depends strongly on the explicit form of the metric tensor. For instance, for a non-diagonal $N$-d metric tensor, the parametric tensor $\delta\omega^{\alpha\beta}(g, \{x\}_{\text{n.m.}})$ in general has no definite symmetry property, and therefore it is more difficult (or even impossible) to identify its components with possible (effective) generalized parameters.

### *7.2.5 Parameter Range and Group Compactness*

As is by now well known, in general, in the generalized 4D (S,T=4-S) Minkowski spaces, the rank-2, completely covariant, antisymmetric tensor $\delta\omega_{\mu\nu}(g)$ is – apart from a sign – the tensor of the $4(4-1)/2 = 6$ dimensionless parameters of the rotational component $SO(T=4-S,S)^{6}_{\text{GEN.}}$ of the (maximal) Killing group $P(T=4-S,S)^{10}_{\text{GEN.}}$ of a generalized 4D Minkowski space. Moreover, due to its antisymmetry, it can always be decomposed in an axial three-vector $\boldsymbol{\theta}(g)$ and a polar three-vector $\boldsymbol{\zeta}(g)$. The physical interpretation of such (Euclidean) three-vectors is strictly related to the number of space-like (or time-like) dimensions and on their labeling.

In the standard case (including SR and DSR) of $S = 3, T = 1$, with usual index range $\{0,1,2,3\}$ and dimensional labeling, the components of the axial (Euclidean) three-vector $\boldsymbol{\theta}(g)$ are the rotation angles about the three space-like directions (i.e., the physical space in SR and in DSR), and therefore:

$$\theta^{i}(g) \in [0, 2\pi] \quad \forall i = 1,2,3. \tag{7.79}$$

On the other hand, the components of the polar (Euclidean) three-vector $\boldsymbol{\zeta}(g)$ are the dimensionless parameters (rapidities) of the (pseudo-)rotations in a mixed (space-like–time-like and space–time in SR and in DSR) plane. Since they enter the expressions of finite pseudorotations as arguments of hyperbolic functions, each component of the rapidity three-vector has a noncompact range (the whole real line):

$$\zeta^{i}(g) \in R \equiv (-\infty, +\infty) \quad \forall i = 1,2,3. \tag{7.80}$$

Then, the chronotopical rotation group $SO(1,3)_{\text{GEN.}}$ of a generalized 4D (3,1) Minkowski space is *noncompact*. This is obviously related to the existence of at least one time-like dimension, namely of pseudorotations (or boost transformations).

Such a conclusion holds true, in general, for the $N$-dimensional case. The presence of time-like dimensions (i.e., $T > 0$), and therefore of true space–time mixing, entails the lack of compactness of the chronotopical rotation group of the $N$-d generalized Minkowski space, i.e., of the homogeneous component of the corresponding (maximal) Killing group $P(T,S)_{\text{GEN.}}$.

An example supporting the earlier statement is provided by the transition from the 4D Euclidean space $E_4(T = 0, S = 4$, with metric tensor $g_{\mu\nu,E_4} = -\delta_{\mu\nu})$ to the standard 4D Minkowski space $M$ $(T = 1, S = 3$, with metric tensor $g_{\mu\nu,\mathrm{SR}} \stackrel{\mathrm{ESCoff}}{=} \delta_{\mu\nu}(\delta_{\mu 0} - \delta_{\mu 1} - \delta_{\mu 2} - \delta_{\mu 3}))$. The transition $E_4 \to M$ amounts to a signature change of a space-like dimension (which becomes time-like). By keeping the Euclidean metric unchanged, such a change can be expressed by an "inverse Wick rotation," i.e., by a formal complexification of the dimension considered:[5]

$$x^0_{E_4} = ix^0_M. \tag{7.81}$$

Whereas the chronotopical rotation group $\mathrm{SO}(4)_{\mathrm{STD.}}$ of $E_4$ is compact – since *all* its transformations are true rotations between space-like dimensions, and therefore their parameters can be interpreted as real rotation angles –, the corresponding rotation group $\mathrm{SO}(1,3)_{\mathrm{STD.}}$ of $M$ is *not* compact, due to the time-like dimension (parameterized by $x^0$).

In SR, the light speed $c$ being the maximal causal velocity (m.c.v.) implies that the range of the dimensional boost parameter $v^i(g)$ is the real, noncompact (since bounded but open) interval $(-c, +c)$ $\forall g \in \mathrm{SO}(1,3)_{\mathrm{STD.}}$(namely, "luminal" boosts are not allowed).[6] Analogously, in DSR, where the m.c.v. is given by $u^i_{\mathrm{DSR}} = \frac{b_0(x^5)}{b_i(x^5)}c$, the range of the dimensional velocity parameter $v^i(g)$ of the deformed boost is the real, noncompact interval:

$$\left(-u^i_{\mathrm{DSR}} = -c\frac{b_0(x^5)}{b_i(x^5)}, +u^i_{\mathrm{DSR}} = +c\frac{b_0(x^5)}{b_i(x^5)}\right)$$

$$v^i(g) \in (-u^i_{\mathrm{DSR}}, +u^i_{\mathrm{DSR}})\ \forall i = 1,2,3,\ \ \forall g \in \mathrm{SO}(1,3)_{\mathrm{DEF.}} \tag{7.82}$$

since

$$\widetilde{\beta}^i(g)|_{v^i(g)=\pm u^i_{\mathrm{DSR}}} = \pm\frac{u^i_{\mathrm{DSR}}}{u^i_{\mathrm{DSR}}} = \pm 1; \tag{7.83}$$

$$\begin{aligned}\widetilde{\gamma}^i(g)|_{v^i(g)=\pm u^i_{\mathrm{DSR}}} &= \left[\left(1 - \left(\widetilde{\beta}^i(g)\right)^2\right)^{-1/2}\right]|_{v^i(g)=\pm u^i_{\mathrm{DSR}}} \\ &= 1 - \left[\left(\widetilde{\beta}^i(g)|_{v^i(g)=\pm u^i_{\mathrm{DSR}}}\right)^2\right]^{-1/2} = \infty.\end{aligned} \tag{7.84}$$

---

[5]Notice the merely formal meaning of the complexification procedure of $x^0$. The inverse Wick rotation (7.81) implies only the change of signature of the time dimension and not the change of range of $x^0$ (which is in general real and unbounded both in $E_4$ and in $M$).

[6]As is well known, this amounts to say that no rest frame exists for a massless particle.

### 7.2.6 Deformed Boosts as Pseudorotations

As in the SR case, for deformed boosts, too, it is possible to make some considerations essentially related to the time-like nature of $x^0$.

Firstly, it is easily seen that the finite expression (7.41) of a deformed boost along $\widehat{x^i}$ in the parametric basis of effective deformed rapidities $\left\{\widetilde{\zeta}^i(g)\right\}$ looks very much (except for some signs) like a true deformed (clockwise) rotation by an angle $\widetilde{\zeta}^i(g)$ in the 2D deformed plane $\Pi_{(x^0,x^i)\text{def.}} \subset \widetilde{M}(x^5)$ (endowed with the metric structure determined by $g_{ab,\text{DSR}} \overset{\text{ESCoff}}{=} \delta_{ab}(\delta_{a0}b_0^2(x^5) - \delta_{ai}b_i^2(x^5))$, with $a, b \in \{0, i\}$), except for the replacement of the trigonometric functions sin and cos with the hyperbolic functions sinh and cosh. As in the SR case, this can be regarded as an inverse Wick rotation on the effective rotation angle $\widetilde{\zeta}^i(g)$:

$$\widetilde{\zeta}^i(g) \to i\widetilde{\zeta}^i(g) \ \forall i = 1, 2, 3. \tag{7.85}$$

Such a result can be actually extended to an arbitrary, $N$-d generalized Minkowski space (with $T$ time-like and $S$ space-like dimensions), and therefore boost transformations can be regarded as deformed pseudorotations (by an imaginary angle, in the Wick sense specified earlier), mixing time-like and space-like dimensions.

Like in the SR case, it is possible to further highlight the pseudorotational character of a deformed boost by trying to represent it as a true rotation by a *real* angle $\theta^i(g)$ (instead of an imaginary angle $\widetilde{\zeta}^i(g)$), on a deformed 2D plane $\Pi_{(x^0,x^i)def} \subset \widetilde{M}(x^5)$. To this aim, consider the expression (7.43) of a deformed boost along $\widehat{x^i}$ in the parametric basis of velocities $\{v^i(g)\}$, we can rewrite as:

$$\begin{cases} x'^0 = \dfrac{1}{\sqrt{1-\left(\widetilde{\beta}^i(g)\right)^2}}\left(x^0 - \widetilde{\beta}^i(g)\dfrac{b_i(x^5)}{b_0(x^5)}x^i\right); \\ x^{i\prime} = \dfrac{1}{\sqrt{1-\left(\widetilde{\beta}^i(g)\right)^2}}\left(x^i - \widetilde{\beta}^i(g)\dfrac{b_0(x^5)}{b_i(x^5)}x^0\right) \end{cases} \tag{7.86}$$

(we omitted the components orthogonal to the boost direction, which are unaffected by the transformation). Dividing and multiplying the r.h.s. by quantity $\sqrt{1+\left(\widetilde{\beta}^i(g)\right)^2}$, one gets

$$\begin{cases} x'^0 = \sqrt{\dfrac{1+\left(\widetilde{\beta^i}(g)\right)^2}{1-\left(\widetilde{\beta^i}(g)\right)^2}} \left[ \dfrac{x^0 - \widetilde{\beta^i}(g)\dfrac{b_i(x^5)}{b_0(x^5)}x^i}{\sqrt{1+\left(\widetilde{\beta^i}(g)\right)^2}} \right]; \\ x^{i\prime} = \sqrt{\dfrac{1+\left(\widetilde{\beta^i}(g)\right)^2}{1-\left(\widetilde{\beta^i}(g)\right)^2}} \left[ \dfrac{x^i - \widetilde{\beta^i}(g)\dfrac{b_0(x^5)}{b_i(x^5)}x^0}{\sqrt{1+\left(\widetilde{\beta^i}(g)\right)^2}} \right]. \end{cases} \tag{7.87}$$

Defining the deformed angular parameters[7]

$$\theta^i_{\text{def}}(g) = \theta^i(g) b_0^{-1}(x^5) b_i^{-1}(x^5) \equiv \text{arctg}\left(\widetilde{\beta^i}(g)\right) \Leftrightarrow \widetilde{\beta^i}(g) \equiv \text{tg}\left(\theta^i_{\text{def}}(g)\right), \tag{7.88}$$

one gets

$$\frac{1}{\sqrt{1+\left(\widetilde{\beta^i}(g)\right)^2}} = \cos\left(\theta^i_{\text{def}}(g)\right); \tag{7.89}$$

$$\frac{\widetilde{\beta^i}(g)}{\sqrt{1+\left(\widetilde{\beta^i}(g)\right)^2}} = \sin\left(\theta^i_{\text{def}}(g)\right). \tag{7.90}$$

Moreover, by putting

$$\alpha^i_{\text{DSR}}(g) \equiv \sqrt{\frac{1+\left(\widetilde{\beta^i}(g)\right)^2}{1-\left(\widetilde{\beta^i}(g)\right)^2}}, \tag{7.91}$$

(7.86) becomes

$$\begin{cases} x'^0 = \alpha^i_{\text{DSR}}(g) x^0 \cos\left(\theta^i_{\text{def}}(g)\right) - b_0^{-1}(x^5) b_i(x^5) x^i \sin\left(\theta^i_{\text{def}}(g)\right); \\ x^{i\prime} = \alpha^i_{\text{DSR}}(g) x^i \cos\left(\theta^i_{\text{def}}(g)\right) - b_0(x^5) b_i^{-1}(x^5) x^0 \sin\left(\theta^i_{\text{def}}(g)\right). \end{cases} \tag{7.92}$$

Apart from the "DSR length-deformation parameter" $\alpha^i_{\text{DSR}}(g)$, (7.92) does not represent a true 3D-deformed rotation of axes $\widehat{x^0}$ and $\widehat{x^i}$. In fact, it is easily seen (from the parity properties of functions sin and cos) that $\forall i \in \{1,2,3\}$ axes $\widehat{x^0}$ and $\widehat{x^i}$ rotate by the same deformed angle $\theta^i_{\text{def}}(g) =$

[7]Not to be confused with the effective deformed angles $\widetilde{\theta}_i$ given by (7.65)!

$\theta^i(g)b_0^{-1}(x^5)b_i^{-1}(x^5)$ but in opposite verses. In other words, when represented as a deformed rotation by a real angle on the deformed plane $\Pi_{(x^0,x^i),\text{def.}} \subset \widetilde{M}(x^5)$, a deformed boost implies not only a length deformation, but also a transition from a pair $(\widehat{x^0}, \widehat{x^i})$ of orthogonal axes[8] to a pair $(\widehat{x^{0\prime}}, \widehat{x^{i\prime}})$ of oblique axes (in strict analogy with the SR case).

Needless to say, the earlier results, obtained in the finite transformational case, hold true at the infinitesimal level, too. This is easily seen by recalling the expression of a deformed infinitesimal boost along $\widehat{x^i}$ in the rapidity basis (see (7.41))

$$\left\{\begin{array}{c} x^{0\prime}_{(g)}(x,x^5) = x^0 - \zeta^i(g)b_0^{-2}(x^5)x^i; \\ x^{i\prime}_{(g)}(x,x^5) = -\zeta^i(g)b_i^{-2}(x^5)x^0 + x^i : \\ x^{k\neq i\prime}_{(g)}(x,x^5) = x^{k\neq i} \end{array}\right. \tag{7.93}$$

and comparing it with a true deformed infinitesimal rotation by an angle $\theta^i(g)$ about $\widehat{x^i}$, say $i = 1$ (see (7.56)):

$$\begin{pmatrix} x^{0\prime}_{(g)}(x,x^5) \\ x^{1\prime}_{(g)}(x,x^5) \\ x^{2\prime}_{(g)}(x,x^5) \\ x^{3\prime}_{(g)}(x,x^5) \end{pmatrix} = \begin{pmatrix} x^0 \\ x^1 \\ x^2 + \theta_1(g)b_2^{-2}(x^5)x^3 \\ -\theta_1(g)b_3^{-2}(x^5)x^2 + x^3 \end{pmatrix} \tag{7.94}$$

(now, of course, the role of the real deformed angle $\theta^i(g)$ is played by $\zeta^i(g)$).

## 7.3 Deformed True Rotation about a Generic Axis

### 7.3.1 *Parametric Decomposition*

We want now to derive the finite, deformed true (clockwise) rotations by a generic angle $\varphi(g)$ about a generic axis $\widehat{\varepsilon}(g)$ in the physical 3D space $\widetilde{E_3}(x^5)$ embedded in $\widetilde{M}(x^5)$, by exploiting the form of the infinitesimal generators of the DSR chronotopical group $\mathrm{SO}(1,3)_{\text{DEF.}}$ derived in Sect. 6.3.1. The unit vector of the rotation axis is $\widehat{\varepsilon}(g) = \varepsilon^1(g)\widehat{x^1} + \varepsilon^2(g)\widehat{x^2} + \varepsilon^3(g)\widehat{x^3}$, with $|\widehat{\varepsilon}(g)|_*^2 \equiv \sum_{i=1}^3 b_i^2(x^5)\left(\varepsilon^i(g)\right)^2 = 1$ (remember that $|\cdot|_*$ is the 3D norm associated to $-g_{ij,\text{DSR}}(x^5)$).

In general, by the very basic properties of a group, it is always possible to find a (not unique) axial-parametric decomposition which transfers – once fixed the three coordinate axes in the space considered – all the

[8] Here orthogonality must be obviously meant in the deformed Minkowski space $\widetilde{M}(x^5)$, namely according to the deformed metric $g_{\text{DSR}}$ (see Sect. 3.1).

dependence on the group element $g$ only to the transformation parameters $\{\theta^i(g)\}_{i=1,2,3}$[9]

$$(\widehat{\varepsilon}(g),\varphi(g)) \to (\widehat{x^1},\theta^1(g))(\widehat{x^2},\theta^2(g))(\widehat{x^3},\theta^3(g)). \tag{7.95}$$

At the infinitesimal (i.e., algebraic) level, such a decomposition is independent of the order, due to the commutativity of the infinitesimal elements (i.e., of the transformations at an algebraic level) of any Lie group of transformations. This of course does no longer hold at a finite (i.e., group) level – due to the non-abelian nature of Lie groups of chronotopical transformations – and therefore the order on the r.h.s. of (7.95) is fixed for a given pair $(\widehat{\varepsilon}(g),\varphi(g))$.

Needless to say, at the algebraic and group level the rotation angles $\varphi(g)$ and $\{\theta^i(g)\}_{i=1,2,3}$ are infinitesimal and finite, respectively. In the treated case of finite deformed true (clockwise) rotation with generic finite angle $\varphi(g)$ around the axis $\widehat{\varepsilon}(g)$ of $\widetilde{E_3}(x^5)$, it is possible to write down a possible explicit dependence of the assigned couple $(\varphi(g),\widehat{\varepsilon}(g))$ on the set $\{\theta^i(g)\}_{i=1,2,3}$, i.e., on the components of the finite deformed angle Euclidean three-vector $\boldsymbol{\theta}(g)$; this corresponds to the following possible axial-parametric decomposition:

$$(\widehat{\varepsilon}(g),\varphi(g)) \to (\widehat{x^1_{\mathrm{DSR}}},\theta^1(g))(\widehat{x^2_{\mathrm{DSR}}},\theta^2(g))(\widehat{x^3_{\mathrm{DSR}}},\theta^3(g)). \tag{7.96}$$

We proceed in the following way. Let us define a deformed three-vector

$$\begin{aligned}\boldsymbol{\psi}(g) &\equiv \varphi(g)\widehat{\varepsilon}(g)\\ &= \varphi(g)\varepsilon^1(g)\widehat{x^1_{\mathrm{DSR}}} + \varphi(g)\varepsilon^2(g)\widehat{x^2_{\mathrm{DSR}}} + \varphi(g)\varepsilon^3(g)\widehat{x^3_{\mathrm{DSR}}}.\end{aligned} \tag{7.97}$$

Then, the set $\{\psi^i(g) \equiv \varphi(g)\varepsilon^i(g)\}_{i=1,2,3}$, when understood as the set of components of an Euclidean three-vector $\boldsymbol{\psi}_{\mathrm{Eucl}}(g)$, can be identified with $\{\theta^i(g)\}_{i=1,2,3}$; that is, we have:

$$\begin{aligned}\boldsymbol{\theta}(g) &\equiv \boldsymbol{\psi}_{\mathrm{Eucl}}(g)\\ &= \varphi(g)\varepsilon^1(g)\widehat{x^1_{\mathrm{SR}}} + \varphi(g)\varepsilon^2(g)\widehat{x^2_{\mathrm{SR}}} + \varphi(g)\varepsilon^3(g)\widehat{x^3_{\mathrm{SR}}}.\end{aligned} \tag{7.98}$$

In this sense, we have "axially-parametrically decomposed" the expression $-\varphi(g)\widehat{\varepsilon}(g)\cdot\mathbf{S}_{\mathrm{DSR}}(x^5)$ in (7.4)–(7.5).

As mentioned before, let us stress that the axial-parametric decomposition considered is not unique; this is due to the fact that (infinitesimal) finite deformed true rotations do form a proper (subalgebra)

---

[9]In general, with $\left\{\widehat{x^1},\widehat{x^2},\widehat{x^3}\right\}$ fixed in $\widetilde{E_3}(x^5)$, $\theta^i = \theta^i\left(\widehat{\varepsilon}(g),\varphi(g)\right) = \overline{\theta^i}(g)$, but, for simplicity's sake, the short notation $\theta^i(g)$ will be used $\forall i = 1,2,3$.

non-abelian subgroup of the larger embedding deformed space–time rotation (i.e., deformed Lorentz homogeneous) group, namely ($su(2)_{\text{DEF.}} \subset su(2)_{\text{DEF.}} \otimes su(2)_{\text{DEF.}}$) $\text{SO}(3)_{\text{DEF.}} \subset \text{SO}(1,3)_{\text{DEF.}}$ ·

We have therefore, in DSR (· is the Euclidean scalar product):

$$\begin{aligned}
&-\varphi(g)\widehat{\varepsilon}(g)\cdot \mathbf{S}_{\text{DSR}}(x^5)\\
&\to -\theta^1(g)S^1_{\text{DSR}}(x^5) - \theta^2(g)S^2_{\text{DSR}}(x^5) - \theta^3(g)S^3_{\text{DSR}}(x^5)\\
&= -\sum_{i=1}^{3}\theta^i(g)S^i_{\text{DSR}}(x^5)
\end{aligned} \tag{7.99}$$

(infinitesimal level, any composition order);[10]

$$\begin{aligned}
&\exp\left(-\varphi(g)\widehat{\varepsilon}(g)\cdot \mathbf{S}_{\text{DSR}}(x^5)\right)\\
&\to \exp\left(-\theta^1(g)S^1_{\text{DSR}}(x^5)\right)\times \exp\left(-\theta^2(g)S^2_{\text{DSR}}(x^5)\right)\times \exp\left(-\theta^3(g)S^3_{\text{DSR}}(x^5)\right)
\end{aligned} \tag{7.100}$$

(finite level, fixed composition order).

A different axial-parametric decomposition utilizes the Euler angles $\left\{\theta_1^{\overline{i}}(g),\, \theta^{\overline{j}\neq\overline{i}}(g),\, \theta_2^{\overline{i}}(g)\right\}$:

$$\begin{aligned}
(\widehat{\varepsilon}(g),\varphi(g)) &\to (\widehat{x^{\overline{i}}},\theta_1^{\overline{i}}(g))(\widehat{x^{\overline{j}}},\theta^{\overline{j}}(g))(\widehat{x^{\overline{i}}},\theta_2^{\overline{i}}(g));\\
\overline{i},\overline{j} &\in \{1,2,3\},\overline{i}\neq\overline{j}.
\end{aligned} \tag{7.101}$$

In this case, only two coordinate space axes, $\widehat{x^{\overline{i}}}$ and $\widehat{x^{\overline{j}\neq\overline{i}}}$, are used, with a special composition order. At infinitesimal level, one has

$$\begin{aligned}
&(\widehat{\varepsilon}(g),\varphi(g))\\
&\to (\widehat{x^{\overline{i}}},\theta_1^{\overline{i}}(g))(\widehat{x^{\overline{j}\neq\overline{i}}},\theta^{\overline{j}\neq\overline{i}}(g))(\widehat{x^{\overline{i}}},\theta_2^{\overline{i}}(g)) \equiv (\widehat{x^{\overline{i}}},\theta_1^{\overline{i}}(g)+\theta_2^{\overline{i}}(g))(\widehat{x^{\overline{i}}},\theta_2^{\overline{i}}(g)).
\end{aligned} \tag{7.102}$$

For DSR, we have

$$\begin{aligned}
&-\varphi(g)\widehat{\varepsilon}(g)\cdot \mathbf{S}_{\text{DSR}}(x^5)\\
&\longrightarrow -\theta_1^{\overline{i}}(g)S^{\overline{i}}_{\text{DSR}}(x^5) - \theta^{\overline{j}\neq\overline{i}}(g)S^{\overline{j}\neq\overline{i}}_{\text{DSR}}(x^5) - \theta_2^{\overline{i}}(g)S^{\overline{i}}_{\text{DSR}}(x^5)\\
&= -\left(\theta_1^{\overline{i}}(g)+\theta_2^{\overline{i}}(g)\right)S^{\overline{i}}_{\text{DSR}}(x^5) - \theta^{\overline{j}\neq\overline{i}}(g)S^{\overline{j}\neq\overline{i}}_{\text{DSR}}(x^5)
\end{aligned} \tag{7.103}$$

[10] For precision's sake, in (7.99) – and in general for infinitesimal deformed true rotation transformations – $g \in \text{SO}(3)_{\text{DEF.}}$ should be replaced by $\delta g$, where $\delta g \in su(2)_{\text{DEF.}}$, i.e., it is an element of the deformed true rotation algebra. But, for simplicity's sake, we will omit, but mean, this cumbersome notation.

(infinitesimal level, any composition order);

$$\begin{aligned}&\exp\left(-\varphi(g)\widehat{\varepsilon}(g)\cdot\mathbf{S}_{\mathrm{DSR}}(x^5)\right)\\ \to&\exp\left(-\theta_1^{\bar{i}}(g)S^{\bar{i}}_{\mathrm{DSR}}(x^5)\right)\times\exp\left(-\theta^{\bar{j}\neq\bar{i}}(g)S^{\bar{j}\neq\bar{i}}_{\mathrm{DSR}}(x^5)\right)\\ &\times\exp\left(-\theta_2^{\bar{i}}(g)S^{\bar{i}}_{\mathrm{DSR}}(x^5)\right)\end{aligned}\tag{7.104}$$

(finite level, fixed composition order).

It is also possible to find out by direct computation (namely, by integration on the group parameters) the 4×4 matrix representative of the finite element $g$ of the rotation group $\mathrm{SO}(1,3)_{\mathrm{DEF.}}$ in DSR corresponding to a rotation by $\varphi(g)$ about $\widehat{\varepsilon}(g)$. We have

$$\begin{aligned}&-\varphi(g)\widehat{\varepsilon}(g)\cdot\mathbf{S}_{\mathrm{DSR}}(x^5)\\ =&-\varphi(g)\sum_{i=1}^{3}\varepsilon^i(g)S^i_{\mathrm{DSR}}(x^5)\ \overset{\text{integration on group parameters}}{\longrightarrow}\\ \to&\exp\left(-\varphi(g)\widehat{\varepsilon}(g)\cdot\mathbf{S}_{\mathrm{DSR}}(x^5)\right)=\exp\left(-\varphi(g)\sum_{i=1}^{3}\varepsilon^i(g)S^i_{\mathrm{DSR}}(x^5)\right)\\ &\neq\prod_{i=1}^{3}\exp\left(-\varphi(g)\varepsilon^i(g)S^i_{\mathrm{DSR}}(x^5)\right),\end{aligned}\tag{7.105}$$

where in the last passage the Baker–Campbell–Hausdorff formula is exploited, the non-abelian nature of $\mathrm{SO}(3)_{\mathrm{DEF.}}\subset\mathrm{SO}(1,3)_{\mathrm{DEF.}}$ is used, and matrix product is understood in $\prod_{i=1}^{3}$.

Let us notice an interesting fact. Define the following functions:

$$\Gamma_{(\text{algebraic})}:su(2)_{\mathrm{DEF.}}\ni\delta g\to((\widehat{\varepsilon}(g),\varphi(g)))\in\left(\widetilde{M}(x^5)\supset\right)\widetilde{E_3}(x^5)\times I_\epsilon(0);\tag{7.106}$$

$$\Gamma_{(\text{group})}:\mathrm{SO}(3)_{\mathrm{DEF.}}\ni g\to((\widehat{\varepsilon}(g),\varphi(g)))\in\left(\widetilde{M}(x^5)\supset\right)\widetilde{E_3}(x^5)\times[-2\pi,2\pi];\tag{7.107}$$

$$\begin{aligned}\Delta_{(\text{algebraic})}\ :&\ (\text{boost component of})\ su(2)_{\mathrm{DEF.}}\otimes su(2)_{\mathrm{DEF.}}\ni\delta g\\ \to&\ ((\widehat{\varepsilon}(g),\rho(g)))\in\widetilde{M}(x^5)\left(\supset\widetilde{E_3}(x^5)\right)\times I_\epsilon(0);\end{aligned}\tag{7.108}$$

$$\begin{aligned}&\Delta_{(\text{group})}:(\text{boost component of})\ \mathrm{SO}(1,3)_{\mathrm{DEF.}}\ni g\\ &\to((\widehat{\varepsilon}(g),\rho(g)))\in\widetilde{M}(x^5)\left(\supset\widetilde{E_3}(x^5)\right)\times R;\end{aligned}\tag{7.109}$$

where $R \supset I_\epsilon(0) \equiv (-\epsilon, \epsilon)$ is a generic neighborhood centered in 0 with radius $\epsilon$ and the notation $\widetilde{M}(x^5) \left(\supset \widetilde{E_3}(x^5)\right)$ means that $\widehat{\varepsilon}(g) \in \widetilde{E_3}(x^5)$, but deformed boost transformations do affect the time-like time coordinate, also.

It can be seen, on a physically-grounded basis, that these "group-mapping" functions have a periodicity $Z_2$, i.e., they identify "diametrically opposed" points of their codomain manifolds; therefore,. the correct defining expressions are:

$$\Gamma_{(\text{algebraic})} : su(2)_{\text{DEF.}} \ni \delta g$$
$$\to ((\widehat{\varepsilon}(g), \varphi(g))) \in \left(\left(\widetilde{M}(x^5) \supset\right) \widetilde{E_3}(x^5) \times I_\epsilon(0)\right) / Z_2 = \widetilde{E_3}(x^5) \times I_\epsilon^+(0); \tag{7.110}$$

$$\Gamma_{(\text{group})} : \text{SO}(3)_{\text{DEF.}} \ni g$$
$$\to \left(((\widehat{\varepsilon}(g), \varphi(g))) \in \left(\widetilde{M}(x^5) \supset\right) \widetilde{E_3}(x^5) \times [-2\pi, 2\pi]\right) / Z_2 = \widetilde{E_3}(x^5) \times [0, 2\pi]; \tag{7.111}$$

$$\Delta_{(\text{algebraic})} : (\text{boost component of})\ su(2)_{\text{DEF.}} \otimes su(2)_{\text{DEF.}} \ni \delta g$$
$$\to ((\widehat{\varepsilon}(g), \rho(g))) \in \left(\widetilde{M}(x^5) \left(\supset \widetilde{E_3}(x^5)\right) \times I_\epsilon(0)\right) / Z_2; \tag{7.112}$$

$$\Delta_{(\text{group})} : (\text{boost component of})\ \text{SO}(1,3)_{\text{DEF.}} \ni g$$
$$\to \left(((\widehat{\varepsilon}(g), \rho(g))) \in \widetilde{M}(x^5) \left(\supset \widetilde{E_3}(x^5)\right) \times R\right) / Z_2, \tag{7.113}$$

where $R \supset I_\epsilon(0) \equiv [0, \epsilon)$ is a generic right neighborhood of 0 with radius $\epsilon$.

In fact, as it is intuitively reasonable, it can be argued on physical grounds that the same (algebraic) group element $(\delta)\, g$ is equivalently mapped in the following ways, respectively:

$$(\delta)\, g \to \Gamma\left((\delta)\, g\right) = \begin{cases} (\widehat{\varepsilon}(g), \varphi(g)) \\ (-\widehat{\varepsilon}(g), -\varphi(g)) \end{cases}$$

$$\Leftrightarrow (\delta)\, g^{-1} \to \Gamma\left((\delta)\, g^{-1}\right) = \begin{cases} (\widehat{\varepsilon}(g), -\varphi(g)) \\ (-\widehat{\varepsilon}(g), \varphi(g)) \end{cases}; \tag{7.114}$$

$$(\delta)\, g \to \Delta\left((\delta)\, g\right) = \begin{cases} (\widehat{\varepsilon}(g), \rho(g)) \\ (-\widehat{\varepsilon}(g), -\rho(g)) \end{cases}$$

$$\Leftrightarrow (\delta)\, g^{-1} \to \Delta\left((\delta)\, g^{-1}\right) = \begin{cases} (\widehat{\varepsilon}(g), -\rho(g)) \\ (-\widehat{\varepsilon}(g), \rho(g)) \end{cases} . \tag{7.115}$$

In other words, the codomain manifolds of the "group-mapping" functions $\Gamma$ and $\Delta$ have a discrete "antipodal" symmetry group $Z_2$. Whence the physical identification of algebraic and group elements must be made modulo a parity space-symmetry.

In the case of true rotations, this topologically means that

$$\Pi_0\left(O(3)_{\mathrm{DEF.}}\right) (= \Pi_0\left(O(3)_{\mathrm{STD.}}\right)) = Z_2, \tag{7.116}$$

i.e., that the zeroth homotopy group of the – standard or deformed – orthogonal Lie group of order three is $Z_2$.

### *7.3.2 Exponentiating the Deformed Infinitesimal Rotation*

Needless to say, all the parametrization procedures we discussed earlier are physically equivalent (although they may yield different formal results). We shall exploit the first one. Let us denote by $R_{(\varphi(g),\widehat{\varepsilon}(g)),\mathrm{DSR}}(x^5)$ the 4×4 matrix corresponding to an infinitesimal (clockwise) rotation by an (infinitesimal) angle $\varphi(g)$ about the axis $\widehat{\varepsilon}(g)$ (matrix belonging a 4D representation of the algebra of SO(3)$_{\mathrm{DEF.}}$, namely $su(2)_{\mathrm{DEF.}}$):

$$R_{(\varphi(g),\widehat{\varepsilon}(g)),\mathrm{DSR}}(x^5) \equiv$$

$$\equiv -\varphi(g)\widehat{\varepsilon}(g) \cdot \mathbf{S}_{\mathrm{DSR}}(x^5) \quad \begin{matrix}\text{axial-parametric decomposition}\\ \text{(algebraic level)}\end{matrix} \tag{7.117}$$

$$\to \left(-\theta^1(g) S^1_{\mathrm{DSR}}(x^5) - \theta^2(g) S^2_{\mathrm{DSR}}(x^5) - \theta^3(g) S^3_{\mathrm{DSR}}(x^5)\right)$$

passage to finite (i.e., group) level:

$$\overset{\text{Exponentiation}}{\longrightarrow} \quad \exp(R_{(\varphi(g),\widehat{\varepsilon}(g)),\mathrm{DSR}}(x^5))$$

$$= \exp\left(-\theta^1(g) S^1_{\mathrm{DSR}}(x^5) - \theta^2(g) S^2_{\mathrm{DSR}}(x^5) - \theta^3(g) S^3_{\mathrm{DSR}}(x^5)\right)$$

$$= \sum_{n=0}^{\infty} \frac{1}{n!} \left(-\theta^1(g) S^1_{\mathrm{DSR}}(x^5) - \theta^2(g) S^2_{\mathrm{DSR}}(x^5) - \theta^3(g) S^3_{\mathrm{DSR}}(x^5)\right)^n , \tag{7.118}$$

where $\exp(R_{(\varphi(g),\widehat{\varepsilon}(g)),\mathrm{DSR}}(x^5))$ is the 4×4 matrix corresponding to a finite (clockwise) rotation by a (finite) angle $\varphi(g)$ about the axis $\widehat{\varepsilon}(g)$, belonging to a 4D representation of $\mathrm{SO}(3)_{\mathrm{DEF.}}$, of course.

On account of the explicit form of $\mathbf{S}_{\mathrm{DSR}}(x^5)$ and of the deformed rotation generators (see Sect. 7.2.1), we have

$$R_{(\varphi(g),\widehat{\varepsilon}(g)),\mathrm{DSR}}(x^5)$$

$$= \begin{pmatrix} 0 & 0 & 0 & 0 \\ 0 & 0 & \theta^3(g)b_1^{-2}(x^5) & -\theta^2(g)b_1^{-2}(x^5) \\ 0 & -\theta^3(g)b_2^{-2}(x^5) & 0 & \theta^1(g)b_2^{-2}(x^5) \\ 0 & \theta^2(g)b_3^{-2}(x^5) & -\theta^1(g)b_3^{-2}(x^5) & 0 \end{pmatrix} \tag{7.119}$$

and therefore the corresponding finite form is given by the matrix

$$\Lambda_{R(\varphi(g),\widehat{\varepsilon}(g)),\mathrm{DSR}}(x^5) \equiv \exp(R_{(\varphi(g),\widehat{\varepsilon}(g)),\mathrm{DSR}}(x^5))$$

$$= \exp \begin{pmatrix} 0 & 0 & 0 & 0 \\ 0 & 0 & \theta^3(g)b_1^{-2}(x^5) & -\theta^2(g)b_1^{-2}(x^5) \\ 0 & -\theta^3(g)b_2^{-2}(x^5) & 0 & \theta^1(g)b_2^{-2}(x^5) \\ 0 & \theta^2(g)b_3^{-2}(x^5) & -\theta^1(g)b_3^{-2}(x^5) & 0 \end{pmatrix}. \tag{7.120}$$

By calculating the first powers of the matrix $R_{(\varphi(g),\widehat{\varepsilon}(g)),\mathrm{DSR}}(x^5)$, one gets the following recursive relations (with $1_{4D}$ being the 4D unit matrix):[11]

---

[11] $\left|\tilde{\boldsymbol{\theta}}(g,x^5)\right|^2$ is the Euclidean norm of the (effective) deformed rotation angle (Euclidean) three-vector $\tilde{\boldsymbol{\theta}}(g,x^5)$ defined by (7.65):

$$\left|\tilde{\boldsymbol{\theta}}(g,x^5)\right|^2 =$$
$$= \left(\theta^1(g)\right)^2 b_2^{-2}(x^5)b_3^{-2}(x^5) + \left(\theta^2(g)\right)^2 b_1^{-2}(x^5)b_3^{-2}(x^5) + \left(\theta^3(g)\right)^2 b_1^{-2}(x^5)b_2^{-2}(x^5).$$

$$
\begin{cases}
(1)\ \left(R_{(\varphi(g),\widehat{\varepsilon}(g)),\mathrm{DSR}}(x^5)\right)^0 = 1_{4D}; \\
(2)\ \left(R_{(\varphi(g),\widehat{\varepsilon}(g)),\mathrm{DSR}}(x^5)\right)^{2n} \\
= (-1)^{n-1} \left|\widetilde{\boldsymbol{\theta}}(g,x^5)\right|^{2(n-1)} \left(R_{(\varphi(g),\widehat{\varepsilon}(g)),\mathrm{DSR}}(x^5)\right)^2 \\
= -\left|\widetilde{\boldsymbol{\theta}}(g,x^5)\right|^{-2} (-1)^n \left|\widetilde{\boldsymbol{\theta}}(g,x^5)\right|^{2n} \left(R_{(\varphi(g),\widehat{\varepsilon}(g)),\mathrm{DSR}}(x^5)\right)^2, \ n \in N; \\
(3)\ \left(R_{(\varphi(g),\widehat{\varepsilon}(g)),\mathrm{DSR}}(x^5)\right)^{2n+1} \\
= (-1)^n \left|\widetilde{\boldsymbol{\theta}}(g,x^5)\right|^{2n} R_{(\varphi(g),\widehat{\varepsilon}(g)),\mathrm{DSR}}(x^5), \ n \in N \cup \{0\}.
\end{cases}
\tag{7.121}
$$

By comparing relations (7.121) with the corresponding "undeformed" ones valid in SR, and recalling the explicit forms of $R_{(\varphi(g),\widehat{\varepsilon}(g)),\mathrm{DSR}}(x^5)$ (see (7.117)), of $\left(R_{(\varphi(g),\widehat{\varepsilon}(g)),\mathrm{DSR}}(x^5)\right)^2$ and of the corresponding "undeformed" counterparts in SR, it can be immediately seen that the (local) generalizing "anisotropizing deforming" transition SR$\rightarrow$DSR entails the loss of symmetry of all the powers of $R_{(\varphi(g),\widehat{\varepsilon}(g)),\mathrm{DSR}}(x^5)$ (this is due to the loss of symmetry of the 4D representation of the infinitesimal generators of $\mathrm{SO}(1,3)_{\mathrm{DEF.}}$: see Sect. 6.3.1).

Then, it is possible to evaluate the exponential of $R_{(\varphi(g),\widehat{\varepsilon}(g)),\mathrm{DSR}}(x^5)$:

$$
\begin{aligned}
&\exp(R_{(\varphi(g),\widehat{\varepsilon}(g)),\mathrm{DSR}}(x^5)) = \sum_{n=0}^{\infty} \frac{1}{n!} \left(R_{(\varphi(g),\widehat{\varepsilon}(g)),\mathrm{DSR}}(x^5)\right)^n \\
&= 1_{4D} + \sum_{n=1}^{\infty} \frac{1}{(2n)!} \left(R_{(\varphi(g),\widehat{\varepsilon}(g)),\mathrm{DSR}}(x^5)\right)^{2n} \\
&\quad + \sum_{n=0}^{\infty} \frac{1}{(2n+1)!} \left(R_{(\varphi(g),\widehat{\varepsilon}(g)),\mathrm{DSR}}(x^5)\right)^{2n+1} \\
&= 1_{4D} - \left|\widetilde{\boldsymbol{\theta}}(g,x^5)\right|^{-2} \left(R_{(\varphi(g),\widehat{\varepsilon}(g)),\mathrm{DSR}}(x^5)\right)^2 \sum_{n=1}^{\infty} (-1)^n \frac{\left|\widetilde{\boldsymbol{\theta}}(g,x^5)\right|^{2n}}{(2n)!} \\
&\quad + R_{(\varphi(g),\widehat{\varepsilon}(g)),\mathrm{DSR}}(x^5) \sum_{n=0}^{\infty} (-1)^n \frac{\left|\widetilde{\boldsymbol{\theta}}(g,x^5)\right|^{2n}}{(2n+1)!} \\
&= 1_{4D} - \left|\widetilde{\boldsymbol{\theta}}(g,x^5)\right|^{-2} \left(R_{(\varphi(g),\widehat{\varepsilon}(g)),\mathrm{DSR}}(x^5)\right)^2 \sum_{n=1}^{\infty} (-1)^n \frac{\left|\widetilde{\boldsymbol{\theta}}(g,x^5)\right|^{2n}}{(2n)!} \\
&\quad + \left|\widetilde{\boldsymbol{\theta}}(g,x^5)\right|^{-1} R_{(\varphi(g),\widehat{\varepsilon}(g)),\mathrm{DSR}}(x^5) \sum_{n=0}^{\infty} (-1)^n \frac{\left|\widetilde{\boldsymbol{\theta}}(g,x^5)\right|^{2n+1}}{(2n+1)!}
\end{aligned}
$$

$$= 1_{4D} - \left|\widetilde{\boldsymbol{\theta}}(g,x^5)\right|^{-2} \left(\cos\left|\widetilde{\boldsymbol{\theta}}(g,x^5)\right| - 1\right) \left(R_{(\varphi(g),\widehat{\varepsilon}(g)),\mathrm{DSR}}(x^5)\right)^2 + \left|\widetilde{\boldsymbol{\theta}}(g,x^5)\right|^{-1} \sin\left|\widetilde{\boldsymbol{\theta}}(g,x^5)\right| R_{(\varphi(g),\widehat{\varepsilon}(g)),\mathrm{DSR}}(x^5). \tag{7.122}$$

In conclusion, we have, for the finite, deformed (clockwise) true rotation by an angle $\varphi(g)$ about a generic axis $\widehat{\varepsilon}(g)$ of $\widetilde{E_3}(x^5)$ $\left(\subset \widetilde{M}(x^5)\right)$:[12]

$$\begin{pmatrix} x^{0\prime}(g,x,x^5) \\ x^{1\prime}(g,x,x^5) \\ x^{2\prime}(g,x,x^5) \\ x^{3\prime}(g,x,x^5) \end{pmatrix} = \Lambda_{R(\varphi(g),\widehat{\varepsilon}(g)),\mathrm{DSR}}(x^5) \begin{pmatrix} x^0 \\ x^1 \\ x^2 \\ x^3 \end{pmatrix}, \tag{7.123}$$

with explicit components (where $\times$ denotes the usual algebraic multiplication)

$$x^{0\prime}(g,x,x^5) = \sum_{\mu=0}^{3} \left(\Lambda_{R(\varphi(g),\widehat{\varepsilon}(g)),\mathrm{DSR}}(x^5)\right)^0_{\ \mu} x^\mu = x^0; \tag{7.124}$$

$$x^{1\prime}(g,x,x^5) = \sum_{\mu=0}^{3} \left(\Lambda_{R(\varphi(g),\widehat{\varepsilon}(g)),\mathrm{DSR}}(x^5)\right)^1_{\ \mu} x^\mu$$

$$= \left[1 + \frac{\left(\cos\left|\widetilde{\boldsymbol{\theta}}(g,x^5)\right| - 1\right)\left(\left|\widetilde{\boldsymbol{\theta}}(g,x^5)\right|^2 - \left(\theta^1(g)\right)^2 b_2^{-2}(x^5) b_3^{-2}(x^5)\right)}{\left|\widetilde{\boldsymbol{\theta}}(g,x^5)\right|^2}\right] x^1$$

$$+ \frac{\theta^3(g) b_1^{-2}(x^5) \left|\widetilde{\boldsymbol{\theta}}(g,x^5)\right| \sin\left|\widetilde{\boldsymbol{\theta}}(g,x^5)\right|}{\left|\widetilde{\boldsymbol{\theta}}(g,x^5)\right|^2} x^2$$

---

[12] In general ($\forall \mu = 0,1,2,3$)

$$x^{\mu\prime}_{\mathrm{DSR}} = x^{\mu\prime}_{\mathrm{DSR}}(\varphi(g), \widehat{\varepsilon}(g), \{x\}_{m.}, x^5) = \overline{x^{\mu\prime}_{\mathrm{DSR}}}(g,x,x^5)$$

but, for simplicity's sake, the simpler notation $x^{\mu\prime}_{\mathrm{DSR}}(g,x,x^5)$ will be used.

$$
-\frac{\left(\cos\left|\widetilde{\boldsymbol{\theta}}(g,x^5)\right|-1\right)\theta^1(g)\theta^2(g)b_1^{-2}(x^5)b_3^{-2}(x^5)}{\left|\widetilde{\boldsymbol{\theta}}(g,x^5)\right|^2}x^2
$$

$$
-\frac{\theta^2(g)b_1^{-2}(x^5)\left|\widetilde{\boldsymbol{\theta}}(g,x^5)\right|\sin\left|\widetilde{\boldsymbol{\theta}}(g,x^5)\right|}{\left|\widetilde{\boldsymbol{\theta}}(g,x^5)\right|^2}x^3
$$

$$
+\frac{\left(\cos\left|\widetilde{\boldsymbol{\theta}}(g,x^5)\right|-1\right)\theta^1(g)\theta^3(g)b_1^{-2}(x^5)b_2^{-2}(x^5)}{\left|\widetilde{\boldsymbol{\theta}}(g,x^5)\right|^2}x^3
$$

$$
\begin{aligned}
&=\frac{1}{\left|\widetilde{\boldsymbol{\theta}}(g,x^5)\right|^2}\left\{\left[\left|\widetilde{\boldsymbol{\theta}}(g,x^5)\right|^2\right.\right. \qquad (7.125)\\
&+\left(\cos\left|\widetilde{\boldsymbol{\theta}}(g,x^5)\right|-1\right)\left(\left|\widetilde{\boldsymbol{\theta}}(g,x^5)\right|^2-\left(\theta^1(g)\right)^2b_2^{-2}(x^5)b_3^{-2}(x^5)\right)\Big]x^1\\
&+\Big[\theta^3(g)b_1^{-2}(x^5)\left|\widetilde{\boldsymbol{\theta}}(g,x^5)\right|\sin\left|\widetilde{\boldsymbol{\theta}}(g,x^5)\right|\\
&-\left(\cos\left|\widetilde{\boldsymbol{\theta}}(g,x^5)\right|-1\right)\theta^1(g)\theta^2(g)b_1^{-2}(x^5)b_3^{-2}(x^5)\Big]x^2\\
&-\Big[\theta^2(g)b_1^{-2}(x^5)\left|\widetilde{\boldsymbol{\theta}}(g,x^5)\right|\sin\left|\widetilde{\boldsymbol{\theta}}(g,x^5)\right|\\
&\left.+\left(\cos\left|\widetilde{\boldsymbol{\theta}}(g,x^5)\right|-1\right)\theta^1(g)\theta^3(g)b_1^{-2}(x^5)b_2^{-2}(x^5)\Big]x^3\right\};
\end{aligned}
$$

$$
x^{2\prime}(g,x,x^5)=\sum_{\mu=0}^{3}\left(\exp(R_{(\varphi(g),\widehat{\varepsilon}(g)),\mathrm{DSR}}(x^5))\right)^2_{\mu}x^{\mu}
$$

$$
=-\frac{\theta^3(g)b_2^{-2}(x^5)\left|\widetilde{\boldsymbol{\theta}}(g,x^5)\right|\sin\left|\widetilde{\boldsymbol{\theta}}(g,x^5)\right|}{\left|\widetilde{\boldsymbol{\theta}}(g,x^5)\right|^2}x^1
$$

$$
+\frac{\left(\cos\left|\widetilde{\boldsymbol{\theta}}(g,x^5)\right|-1\right)\theta^1(g)\theta^2(g)b_2^{-2}(x^5)b_3^{-2}(x^5)}{\left|\widetilde{\boldsymbol{\theta}}(g,x^5)\right|^2}x^1
$$

$$
+\left[1+\frac{\left(\cos\left|\widetilde{\boldsymbol{\theta}}(g,x^5)\right|-1\right)\left(\left|\widetilde{\boldsymbol{\theta}}(g,x^5)\right|^2-\left(\theta^2(g)\right)^2b_1^{-2}(x^5)b_3^{-2}(x^5)\right)}{\left|\widetilde{\boldsymbol{\theta}}(g,x^5)\right|^2}\right]x^2
$$

$$+\frac{\theta^1(g)b_2^{-2}(x^5)\left|\widetilde{\boldsymbol{\theta}}(g,x^5)\right|\sin\left|\widetilde{\boldsymbol{\theta}}(g,x^5)\right|}{\left|\widetilde{\boldsymbol{\theta}}(g,x^5)\right|^2}x^3$$

$$-\frac{\left(\cos\left|\widetilde{\boldsymbol{\theta}}(g,x^5)\right|-1\right)\theta^2(g)\theta^3(g)b_1^{-2}(x^5)b_2^{-2}(x^5)}{\left|\widetilde{\boldsymbol{\theta}}(g,x^5)\right|^2}x^3$$

$$\begin{aligned}
&=\frac{1}{\left|\widetilde{\boldsymbol{\theta}}(g,x^5)\right|^2}\left\{-\left[\theta^3(g)b_2^{-2}(x^5)\left|\widetilde{\boldsymbol{\theta}}(g,x^5)\right|\sin\left|\widetilde{\boldsymbol{\theta}}(g,x^5)\right|\right.\right.\\
&\left.+\left(\cos\left|\widetilde{\boldsymbol{\theta}}(g,x^5)\right|-1\right)\theta^1(g)\theta^2(g)b_2^{-2}(x^5)b_3^{-2}(x^5)\right]x^1\\
&+\left[\left|\widetilde{\boldsymbol{\theta}}(g,x^5)\right|^2+\left(\cos\left|\widetilde{\boldsymbol{\theta}}(g,x^5)\right|-1\right)\right.\\
&\left.\times\left(\left|\widetilde{\boldsymbol{\theta}}(g,x^5)\right|^2-\left(\theta^2(g)\right)^2b_1^{-2}(x^5)b_3^{-2}(x^5)\right)\right]x^2\\
&+\left[\theta^1(g)b_2^{-2}(x^5)\left|\widetilde{\boldsymbol{\theta}}(g,x^5)\right|\sin\left|\widetilde{\boldsymbol{\theta}}(g,x^5)\right|\right.\\
&\left.\left.-\left(\cos\left|\widetilde{\boldsymbol{\theta}}(g,x^5)\right|-1\right)\theta^2(g)\theta^3(g)b_1^{-2}(x^5)b_2^{-2}(x^5)\right]x^3\right\};
\end{aligned}\tag{7.126}$$

$$x^{3\prime}(g,x,x^5)=\sum_{\mu=0}^{3}\left(\Lambda_{R(\varphi(g),\widehat{\varepsilon}(g)),\mathrm{DSR}}(x^5)\right)^3_{\ \mu}x^\mu$$

$$=\frac{\theta^2(g)b_3^{-2}(x^5)\left|\widetilde{\boldsymbol{\theta}}(g,x^5)\right|\sin\left|\widetilde{\boldsymbol{\theta}}(g,x^5)\right|}{\left|\widetilde{\boldsymbol{\theta}}(g,x^5)\right|^2}x^1$$

$$-\frac{\left(\cos\left|\widetilde{\boldsymbol{\theta}}(g,x^5)\right|-1\right)\theta^1(g)\theta^3(g)b_2^{-2}(x^5)b_3^{-2}(x^5)}{\left|\widetilde{\boldsymbol{\theta}}(g,x^5)\right|^2}x^1$$

$$-\frac{\theta^1(g)b_3^{-2}(x^5)\left|\widetilde{\boldsymbol{\theta}}(g,x^5)\right|\sin\left|\widetilde{\boldsymbol{\theta}}(g,x^5)\right|}{\left|\widetilde{\boldsymbol{\theta}}(g,x^5)\right|^2}x^2$$

$$+\frac{\left(\cos\left|\widetilde{\boldsymbol{\theta}}(g,x^5)\right|-1\right)\theta^2(g)\theta^3(g)b_1^{-2}(x^5)b_3^{-2}(x^5)}{\left|\widetilde{\boldsymbol{\theta}}(g,x^5)\right|^2}x^2$$

$$+\left[1+\frac{\left(\cos\left|\widetilde{\boldsymbol{\theta}}(g,x^5)\right|-1\right)\left(\left|\widetilde{\boldsymbol{\theta}}(g,x^5)\right|^2-\left(\theta^3(g)\right)^2 b_1^{-2}(x^5)b_2^{-2}(x^5)\right)}{\left|\widetilde{\boldsymbol{\theta}}(g,x^5)\right|^2}\right]x^3$$

$$\begin{aligned}
&=\frac{1}{\left|\widetilde{\boldsymbol{\theta}}(g,x^5)\right|^2}\left\{\left[\theta^2(g)b_3^{-2}(x^5)\left|\widetilde{\boldsymbol{\theta}}(g,x^5)\right|\sin\left|\widetilde{\boldsymbol{\theta}}(g,x^5)\right|\right.\right.\\
&\quad\left.-\left(\cos\left|\widetilde{\boldsymbol{\theta}}(g,x^5)\right|-1\right)\theta^1(g)\theta^3(g)b_2^{-2}(x^5)b_3^{-2}(x^5)\right]x^1\\
&\quad-\left[\theta^1(g)b_3^{-2}(x^5)\left|\widetilde{\boldsymbol{\theta}}(g,x^5)\right|\sin\left|\widetilde{\boldsymbol{\theta}}(g,x^5)\right|\right.\\
&\quad\left.+\left(\cos\left|\widetilde{\boldsymbol{\theta}}(g,x^5)\right|-1\right)\theta^2(g)\theta^3(g)b_1^{-2}(x^5)b_3^{-2}(x^5)\right]x^2\\
&\quad+\left[\left|\widetilde{\boldsymbol{\theta}}(g,x^5)\right|^2+\left(\cos\left|\widetilde{\boldsymbol{\theta}}(g,x^5)\right|-1\right)\times\right.\\
&\quad\left.\left.\times\left(\left|\widetilde{\boldsymbol{\theta}}(g,x^5)\right|^2-\left(\theta^3(g)\right)^2 b_1^{-2}(x^5)b_2^{-2}(x^5)\right)\right]x^3\right\}.
\end{aligned}\tag{7.127}$$

## 7.4 Finite 3D Deformed Boosts in a Generic Direction

### *7.4.1 Parametric Decomposition*

We want now to derive the finite, deformed pseudorotations (boosts) with generic dimensionless parameter (rapidity) $\rho(g)$ along a generic direction $\widehat{\varepsilon}(g)$ in the physical 3D space $\widetilde{E_3}(x^5) \subset \widetilde{M}(x^5)$, by exploiting the form of the infinitesimal generators of the DSR chronotopical group $\mathrm{SO}(1,3)_{\mathrm{DEF.}}$. obtained in Chap. 6.

One can proceed in a way similar to the case of finite deformed true rotations treated in Sect. 7.3.2, and write down the following (not unique) axial-parametric decomposition:

$$(\widehat{\varepsilon}(g),\rho(g)) \to (\widehat{x^1},\varsigma^1(g))(\widehat{x^2},\varsigma^2(g))(\widehat{x^3},\varsigma^3(g)).\tag{7.128}$$

In the discussed case of a finite deformed boost with generic finite rapidity $\rho\,(g)$ along the axis $\widehat{\varepsilon}\,(g)$ of $\widetilde{E_3}(x^5)$, it is possible to make explicit a possible dependence of the assigned couple $(\rho\,(g)\,,\widehat{\varepsilon}\,(g))$ on the set $\left\{\varsigma^i(g)\right\}_{i=1,2,3}$, i.e., on the components of the finite, deformed rapidity (Euclidean) three-vector $\boldsymbol{\varsigma}(g)$. This corresponds to the axial-parametric decomposition:

$$(\widehat{\varepsilon}(g),\rho(g)) \to (\widehat{x^1_{\mathrm{DSR}}},\varsigma^1(g))(\widehat{x^2_{\mathrm{DSR}}},\varsigma^2(g))(\widehat{x^3_{\mathrm{DSR}}},\varsigma^3(g)).\tag{7.129}$$

One can proceed in the following way. Let us define a deformed three-vector

$$\boldsymbol{\pi}(g) \equiv$$
$$\equiv \rho(g)\widehat{\varepsilon}(g) = \rho(g)\varepsilon^1(g)\widehat{x^1_{\mathrm{DSR}}} + \rho(g)\varepsilon^2(g)\widehat{x^2_{\mathrm{DSR}}} + \rho(g)\varepsilon^3(g)\widehat{x^3_{\mathrm{DSR}}}. \tag{7.130}$$

Then, the set $\{\pi^i(g) \equiv \rho(g)\varepsilon^i(g)\}_{i=1,2,3}$, when understood as the set of components of an Euclidean three-vector $\boldsymbol{\pi}_{\mathrm{Eucl}}(g)$, can be identified with $\{\zeta^i(g)\}_{i=1,2,3}$; that is, one has:

$$\boldsymbol{\zeta}(g) \equiv \boldsymbol{\pi}_{\mathrm{Eucl}}(g)$$
$$= \rho(g)\varepsilon^1(g)\widehat{x^1_{\mathrm{SR}}} + \rho(g)\varepsilon^2(g)\widehat{x^2_{\mathrm{SR}}} + \rho(g)\varepsilon^3(g)\widehat{x^3_{\mathrm{SR}}}. \tag{7.131}$$

In this sense, we have "axially-parametrically decomposed" the expression $-\rho(g)\,\widehat{\boldsymbol{\varepsilon}}(g) \cdot \mathbf{K}_{\mathrm{DSR}}(x^5)$ in (7.4)–(7.5):

$$(\widehat{\varepsilon}(g), \rho(g)) \to (\widehat{x^1_{\mathrm{DSR}}}, \sigma^1(g))(\widehat{x^2_{\mathrm{DSR}}}, \sigma^2(g))(\widehat{x^3_{\mathrm{DSR}}}, \sigma^3(g)). \tag{7.132}$$

As mentioned before, let us stress that the above axial-parametric decomposition is not unique, and it is due to the fact that (infinitesimal) finite deformed pseudorotations do belong to the deformed Lorentz (algebra) group $(su(2)_{\mathrm{DEF.}} \otimes su(2)_{\mathrm{DEF.}})$ $SO(1,3)_{\mathrm{DEF.}}$.

Therefore, we obtain in DSR ($\cdot$ denotes the Euclidean scalar product):

$$-\rho(g)\widehat{\varepsilon}(g) \cdot \mathbf{K}_{\mathrm{DSR}}(x^5)$$
$$\to -\zeta^1(g)K^1_{\mathrm{DSR}}(x^5) - \zeta^2(g)K^2_{\mathrm{DSR}}(x^5) - \zeta^3(g)K^3_{\mathrm{DSR}}(x^5)$$
$$= -\sum_{i=1}^{3} \zeta^i(g)K^i_{\mathrm{DSR}}(x^5) \tag{7.133}$$

(infinitesimal level, any composition order);

$$\exp\left(-\rho(g)\widehat{\varepsilon}(g) \cdot \mathbf{K}_{\mathrm{DSR}}(x^5)\right)$$
$$\to \exp\left(-\zeta^1(g)K^1_{\mathrm{DSR}}(x^5)\right) \times \exp\left(-\zeta^2(g)K^2_{\mathrm{DSR}}(x^5)\right) \times \exp\left(-\zeta^3(g)K^3_{\mathrm{DSR}}(x^5)\right) \tag{7.134}$$

(finite level, fixed composition order).

Denoting by $B_{(\rho(g),\widehat{\varepsilon}(g)),\mathrm{DSR}}(x^5)$ the 4×4 matrix corresponding to an infinitesimal boost with (infinitesimal) rapidity $\rho(g)$ about the axis $\widehat{\varepsilon}(g)$ (matrix belonging to a 4D representation of the deformed Lorentz algebra

$su(2)_{\text{DEF.}} \otimes su(2)_{\text{DEF.}}$), we have to go through the following steps:

$$B_{(\rho(g),\widehat{\varepsilon}(g)),\text{DSR}}(x^5) \equiv -\rho(g)\widehat{\varepsilon}(g) \cdot \mathbf{K}_{\text{DSR}}(x^5) \quad \begin{array}{c}\text{axial-parametric decomposition}\\ \text{(algebraic level)}\end{array}$$

$$\to \left(-\zeta^1(g)K^1_{\text{DSR}}(x^5) - \zeta^2(g)K^2_{\text{DSR}}(x^5) - \zeta^3(g)K^3_{\text{DSR}}(x^5)\right)$$

$$\begin{array}{c}\text{passage to finite (i.e., group) level:}\\ \text{Exponentiation}\\ \longrightarrow\end{array} \quad \exp(B_{(\rho(g),\widehat{\varepsilon}(g)),\text{DSR}}(x^5))$$

$$= \exp\left(-\zeta^1(g)K^1_{\text{DSR}}(x^5) - \zeta^2(g)K^2_{\text{DSR}}(x^5) - \zeta^3(g)K^3_{\text{DSR}}(x^5)\right)$$

$$= \sum_{n=0}^{\infty} \frac{1}{n!}\left(-\zeta^1(g)K^1_{\text{DSR}}(x^5) - \zeta^2(g)K^2_{\text{DSR}}(x^5) - \zeta^3(g)K^3_{\text{DSR}}(x^5)\right)^n, \tag{7.135}$$

where $\exp(B_{(\rho(g),\widehat{\varepsilon}(g)),\text{DSR}}(x^5))$ is the $4\times 4$ matrix corresponding to a finite pseudorotation with finite rapidity $\varphi(g)$ along the axis $\widehat{\varepsilon}(g)$, belonging to a 4D representation of $\text{SO}(1,3)_{\text{DEF.}}$, of course.

On account of the explicit form of $\mathbf{K}_{\text{DSR}}(x^5)$ and of the deformed boost generators (see Sects. 7.1 and 7.2), we have

$$B_{(\rho(g),\widehat{\varepsilon}(g)),\text{DSR}}(x^5) =$$

$$= \begin{pmatrix} 0 & -\zeta^1(g)b_0^{-2}(x^5) & -\zeta^2(g)b_0^{-2}(x^5) & -\zeta^3(g)b_0^{-2}(x^5) \\ -\zeta^1(g)b_1^{-2}(x^5) & 0 & 0 & 0 \\ -\zeta^2(g)b_2^{-2}(x^5) & 0 & 0 & 0 \\ -\zeta^3(g)b_3^{-2}(x^5) & 0 & 0 & 0 \end{pmatrix}, \tag{7.136}$$

and so the corresponding finite form is given by the matrix

$$\Lambda_{B(\rho(g),\widehat{\varepsilon}(g)),\text{DSR}}(x^5) \equiv \exp\left(B_{(\rho(g),\widehat{\varepsilon}(g)),\text{DSR}}(x^5)\right)$$

$$= \exp\begin{pmatrix} 0 & -\zeta^1(g)b_0^{-2}(x^5) & -\zeta^2(g)b_0^{-2}(x^5) & -\zeta^3(g)b_0^{-2}(x^5) \\ -\zeta^1(g)b_1^{-2}(x^5) & 0 & 0 & 0 \\ -\zeta^2(g)b_2^{-2}(x^5) & 0 & 0 & 0 \\ -\zeta^3(g)b_3^{-2}(x^5) & 0 & 0 & 0 \end{pmatrix}. \tag{7.137}$$

Therefore, in matrix form, the deformed boost transformation reads

$$\begin{pmatrix} x^{0\prime}_{\mathrm{DSR}}\left(g,x,x^5\right) \\ x^{1\prime}_{\mathrm{DSR}}\left(g,x,x^5\right) \\ x^{2\prime}_{\mathrm{DSR}}\left(g,x,x^5\right) \\ x^{3\prime}_{\mathrm{DSR}}\left(g,x,x^5\right) \end{pmatrix} = \Lambda_{B(\rho(g),\widehat{\varepsilon}(g)),\mathrm{DSR}}(x^5) \begin{pmatrix} x^0 \\ x^1 \\ x^2 \\ x^3 \end{pmatrix}$$

$$\Leftrightarrow \begin{pmatrix} x^{0\prime}\left(g,x,x^5\right) \\ x^{1\prime}\left(g,x,x^5\right) \\ x^{2\prime}\left(g,x,x^5\right) \\ x^{3\prime}\left(g,x,x^5\right) \end{pmatrix} = \exp \begin{pmatrix} 0 & -\zeta^1(g)b_0^{-2}(x^5) & -\zeta^2(g)b_0^{-2}(x^5) & -\zeta^3(g)b_0^{-2}(x^5) \\ -\zeta^1(g)b_1^{-2}(x^5) & 0 & 0 & 0 \\ -\zeta^2(g)b_2^{-2}(x^5) & 0 & 0 & 0 \\ -\zeta^3(g)b_3^{-2}(x^5) & 0 & 0 & 0 \end{pmatrix} \times \begin{pmatrix} x^0 \\ x^1 \\ x^2 \\ x^3 \end{pmatrix}. \tag{7.138}$$

A remark is in order. As mentioned in Sect. 7.3, finite deformed true rotations do form a proper non-abelian subgroup of the deformed homogeneous Lorentz group, namely $SO(3)_{\mathrm{DEF.}} \subset SO(1,3)_{\mathrm{DEF.}}$ (this can be seen at an algebraic level, because the algebra of the infinitesimal generators of deformed true rotations is closed, and forms a proper Lie subalgebra of the deformed Lorentz algebra $su(2)_{\mathrm{DEF.}} \subset su(2)_{\mathrm{DEF.}} \otimes su(2)_{\mathrm{DEF.}}$: see Sect. 6.3.4). Thus the axial-parametric decompositions (7.96)–(7.98) are valid, and in Sect. 7.3 just the explicit form of the infinitesimal generators of $SO(3)_{\mathrm{DEF.}}$ (and not of the larger, covering group $SO(1,3)_{\mathrm{DEF.}}$) was needed.

In general, finite deformed boosts do *not* form a proper subgroup of $SO(1,3)_{\mathrm{DEF.}}$ (this can be seen at an algebraic level, because the algebra of the infinitesimal generators of deformed boosts is *not* closed, and doesn't form a proper Lie subalgebra of the deformed Lorentz algebra $su(2)_{\mathrm{DEF.}} \otimes su(2)_{\mathrm{DEF.}}$: see Sect. 6.3.4). Thus, in order to obtain the finite, deformed pseudorotations with generic rapidity $\rho(g)$ along a generic direction $\widehat{\varepsilon}(g)$ in the physical 3D space $\widetilde{E_3}(x^5) \subset \widetilde{M}(x^5)$, the explicit form of *all* the infinitesimal generators of $SO(1,3)_{\mathrm{DEF.}}$ should in general be needed.

This means that, by the very basic properties of a group (namely $SO(1,3)_{\mathrm{DEF.}}$), it is always possible to find a (not unique) axial-parametric decomposition which transfers – once fixed the four space–time coordinate axes of $\widetilde{M}(x^5)$ – all the dependence on the group element $g$ only on the deformed true rotation angles $\{\theta^i(g)\}_{i=1,2,3}$ and on the deformed rapidities $\{\eta^i(g)\}_{i=1,2,3}$:

$$(\widehat{\varepsilon}(g), \rho(g)) \\ \longrightarrow (\widehat{x^1}, \theta^1(g))(\widehat{x^2}, \theta^2(g))(\widehat{x^3}, \theta^3(g))(\widehat{x^1}, \eta^1(g))(\widehat{x^2}, \eta^2(g))(\widehat{x^3}, \eta^3(g)). \tag{7.139}$$

In general, with $\left\{\widehat{x^0}, \widehat{x^1}, \widehat{x^2}, \widehat{x^3}\right\}$ fixed in $\widetilde{M}(x^5)$, $\theta^i = \theta^i\left(\widehat{\varepsilon}(g), \rho(g)\right) = \overline{\theta^i}\,(g)$ and $\eta^i = \eta^i\left(\widehat{\varepsilon}(g), \rho(g)\right) = \overline{\eta^i}\,(g)$, but, for simplicity's sake, the short notations $\theta^i\,(g)$ and $\eta^i\,(g)$ has been used $\forall i = 1, 2, 3$.

As already noticed in Sect. 7.3, at the algebraic level, such a decomposition is independent of the order, due to the commutativity of the infinitesimal elements (i.e., of the transformations at an algebraic level) of any Lie group of transformations. This of course does no longer hold at a finite (i.e., group) level – due to the non-abelian nature of $SO(1,3)_{\text{DEF.}}$ – and therefore the order on the r.h.s. of the expression (7.139) is fixed for a given pair $(\widehat{\varepsilon}(g), \rho(g))$. Needless to say, at the algebraic and group level parameters $\rho(g)$ and $\{\theta^i(g)\}, \{\eta^i(g)\}_{i=1,2,3}$ are infinitesimal and finite, respectively. This will be understood, and, for simplicity's sake, no notational distinction will be made.

In summary, all the earlier equations should have an "hidden" dependence on the parametric sets $\{\theta^i(g)\}$ and $\{\eta^i(g)\}_{i=1,2,3}$; that is, for instance (7.135) would actually have to read explicitly:

$$B_{(\rho(\boldsymbol{\theta}(g),\boldsymbol{\eta}(g)),\widehat{\varepsilon}(\boldsymbol{\theta}(g),\boldsymbol{\eta}(g))),\text{DSR}}(x^5) \\ \equiv -\rho(\boldsymbol{\theta}(g), \boldsymbol{\eta}(g))\widehat{\varepsilon}(\boldsymbol{\theta}(g), \boldsymbol{\eta}(g)) \cdot \mathbf{K}_{\text{DSR}}(x^5). \tag{7.135 a}$$

For simplicity's sake, we have omitted, but understood, this cumbersome notation.

### *7.4.2 Deformed Generic Boost from Velocity Decomposition*

As noticed in Sect. 7.3, all the possible procedures applicable in these frameworks are equivalent (although they may yield different formal results). Thus, in order to obtain the general form of a finite, deformed boost with generic rapidity $\rho(g)$ along a generic direction $\widehat{\varepsilon}(g)$ in $\widetilde{E_3}(x^5) \subset \widetilde{M}(x^5)$, let us exploit the approach of velocity decomposition (already used in Sect. 3.3.2), instead of explicitly calculating $\exp\left(B_{(\rho(g),\widehat{\varepsilon}(g)),\text{DSR}}(x^5)\right)$.

Consider a generic finite, deformed boost with generic velocity $\mathbf{v}(g) = v(g)\widehat{v}(g) \equiv v(g)\widehat{\varepsilon}(g)$ along a generic direction $\widehat{\varepsilon}(g)$ in $\widetilde{E_3}(x^5) \subset \widetilde{M}(x^5)$.

Let us decompose the three-vector $\mathbf{x}$ in two components $\mathbf{x}_{\|}(g)$ and $\mathbf{x}_{\perp}(g)$, respectively, parallel and orthogonal to $\mathbf{v}(g)$. Here, "parallelism" and "orthogonality" are to be meant in the deformed 3D space $\widetilde{E_3}(x^5) \subset \widetilde{M}(x^5)$ (namely according to the deformed scalar product $*$ associated to the 3D metric tensor $-g_{ij,\mathrm{DSR}}(x^5) \overset{\text{ESC off}}{=} b_i^2(x^5)\delta_{ij}$: see Sect. 3.1). We have, in the notation of Sect. 3.3.2:

$$\mathbf{x}_{\|}(g) \equiv \widehat{v}(g)(\widehat{v}(g) * \mathbf{x}) = \frac{\mathbf{v}(g)}{|\mathbf{v}(g)|_*^2}(\mathbf{v}(g) * \mathbf{x}) = \frac{\mathbf{v}(g)}{\mathbf{v}(g) * \mathbf{v}(g)}(\mathbf{v}(g) * \mathbf{x})$$

$$= \frac{\sum_{i=1}^3 b_i^2(x^5) v^i(g) x^i}{\sum_{i=1}^3 b_i^2(x^5)\left(v^i(g)\right)^2}\mathbf{v}(g) \overset{\widetilde{\boldsymbol{\beta}}(g) \equiv \left(\frac{\mathbf{v(g)}}{\mathbf{u}}\right) \neq \frac{\mathbf{v}(g)}{u}}{\neq} \widehat{\widetilde{\beta}}(g)(\widehat{\widetilde{\beta}}(g) * \mathbf{x})$$

$$= \frac{\widetilde{\boldsymbol{\beta}}(g)}{\left|\widetilde{\boldsymbol{\beta}}(g)\right|_*^2}(\widetilde{\boldsymbol{\beta}}(g) * \mathbf{x}) = \frac{\widetilde{\boldsymbol{\beta}}(g)}{\widetilde{\boldsymbol{\beta}}(g) * \widetilde{\boldsymbol{\beta}}(g)}(\widetilde{\boldsymbol{\beta}}(g) * \mathbf{x})$$

$$= \frac{\sum_{i=1}^3 b_i^2(x^5)\widetilde{\beta}^i(g) x^i}{\sum_{i=1}^3 b_i^2(x^5)\left(\widetilde{\beta}^i(g)\right)^2}\widetilde{\boldsymbol{\beta}}(g); \tag{7.140}$$

$$x_{\|}^i(g) \equiv \frac{\sum_{k=1}^3 b_k^2(x^5) v^k(g) x^k}{\sum_{k=1}^3 b_k^2(x^5)\left(v^k(g)\right)^2} v^i$$

$$\overset{\widetilde{\boldsymbol{\beta}}(g) \equiv \left(\frac{\mathbf{v(g)}}{\mathbf{u}}\right) \neq \frac{\mathbf{v}(g)}{u}}{\neq} \frac{\sum_{k=1}^3 b_k^2(x^5)\widetilde{\beta}^k(g) x^k}{\sum_{k=1}^3 b_k^2(x^5)\left(\widetilde{\beta}^k(g)\right)^2}\widetilde{\beta}^i(g), \forall i = 1,2,3; \tag{7.141}$$

$$\mathbf{x}_{\perp}(g) \equiv \mathbf{x} - \mathbf{x}_{\|}(g) = \mathbf{x} - \frac{\sum_{i=1}^3 b_i^2(x^5) v^i(g) x^i}{\sum_{i=1}^3 b_i^2(x^5)\left(v^i(g)\right)^2}\mathbf{v}(g)$$

$$\overset{\widetilde{\boldsymbol{\beta}}(g) \equiv \left(\frac{\mathbf{v(g)}}{\mathbf{u}}\right) \neq \frac{\mathbf{v}(g)}{u}}{\neq} \mathbf{x} - \frac{\sum_{i=1}^3 b_i^2(x^5)\widetilde{\beta}^i(g) x^i}{\sum_{i=1}^3 b_i^2(x^5)\left(\widetilde{\beta}^i(g)\right)^2}\widetilde{\boldsymbol{\beta}}(g); \tag{7.142}$$

$$x_{\perp}^{i}(g) \equiv x^{i} - \frac{\sum_{k=1}^{3} b_{k}^{2}(x^{5}) v^{k}(g) x^{k}}{\sum_{k=1}^{3} b_{k}^{2}(x^{5}) \left(v^{k}(g)\right)^{2}} v^{i}(g)$$

$$\overset{\widetilde{\boldsymbol{\beta}}(g) \equiv \left(\frac{\mathbf{v(g)}}{\mathbf{u}}\right) \neq \frac{\mathbf{v}(g)}{u}}{\neq} \quad x^{i} - \frac{\sum_{k=1}^{3} b_{k}^{2}(x^{5}) \widetilde{\beta}^{k}(g) x^{k}}{\sum_{k=1}^{3} b_{k}^{2}(x^{5}) \left(\widetilde{\beta}^{k}(g)\right)^{2}} \widetilde{\beta}^{i}(g), \forall i = 1, 2, 3, \qquad (7.143)$$

with $\widetilde{\boldsymbol{\beta}}(g)$ being given by (3.33).

On account of the form of a finite, deformed boost along a coordinate axis (cf. (3.27)), a finite, deformed boost with generic (finite) velocity $\mathbf{v}(g)$ in a generic direction $\widehat{v}(g)$ is therefore given by (see Sect. 3.3) ($\cdot$ denotes, as before, the Euclidean 3D scalar product)

$$\left\{ \begin{array}{lcl} \mathbf{x}_{\|}{}'(g) & = & \widetilde{\gamma}(g)(\mathbf{x}_{\|}(g) - \mathbf{v}(g)t); \\ \mathbf{x}'_{\perp}(g) & = & \mathbf{x}_{\perp}(g); \\ t' & = & \left\{ \begin{array}{l} \widetilde{\gamma}(g)\left(t - \sum_{i=1}^{3} \frac{v^{i}(g) b_{i}^{2}(x^{5})}{c^{2} b_{0}^{2}(x^{5})} x^{i}\right) = \widetilde{\gamma}(g)\left(t - \tilde{\mathbf{B}}(g) \cdot \mathbf{x}\right) \\ = \widetilde{\gamma}\left(t - \tilde{\mathbf{B}}^{(*)}(g) * \mathbf{x}\right). \end{array} \right. \end{array} \right. \qquad (7.144)$$

where $\widetilde{\gamma}(g)$, $\tilde{\mathbf{B}}(g)$, $\tilde{\mathbf{B}}^{(*)}(g)$ are are defined in (3.44), (3.47), and (3.48).

In terms of the three-vector $\mathbf{x} = \mathbf{x}_{\|}(g) + \mathbf{x}_{\perp}(g)$, the deformed boost (7.144) in a generic direction $\widehat{v}(g)$ reads therefore

$$\left\{ \begin{array}{l} \mathbf{x}'(g) = \left\{ \begin{array}{l} \mathbf{x}'_{\|}(g) + \mathbf{x}'_{\perp}(g) = \\ = \mathbf{x} + (\widetilde{\gamma}(g) - 1)\frac{\mathbf{v}(g)}{|\mathbf{v}(g)|_{*}^{2}}(\mathbf{v}(g) * \mathbf{x}) - \widetilde{\gamma}(g)\mathbf{v}(g)t = \\ = \mathbf{x} + (\widetilde{\gamma}(g) - 1)\frac{\sum_{k=1}^{3} b_{k}^{2}(x^{5}) v^{k}(g) x^{k}}{\sum_{k=1}^{3} b_{k}^{2}(x^{5}) \left(v^{k}(g)\right)^{2}} \mathbf{v}(g) - \widetilde{\gamma}(g)\mathbf{v}(g)t; \end{array} \right. \\ t'(g) = \left\{ \begin{array}{l} \widetilde{\gamma}(g)(t - \tilde{\mathbf{B}}(g) \cdot \mathbf{x}) = \widetilde{\gamma}(g)(t - \tilde{\mathbf{B}}^{(*)}(g) * \mathbf{x}) = \\ = \widetilde{\gamma}(g)\left(t - \sum_{k=1}^{3} \frac{v^{k}(g) b_{k}^{2}(x^{5})}{c^{2} b_{0}^{2}(x^{5})} x^{k}\right), \end{array} \right. \end{array} \right. \qquad (7.145)$$

or (as usual $x^0 \equiv ct$, and $\times$ now denotes usual algebraic multiplication) ($\forall i = 1, 2, 3$):

$$x^{0\prime}(g, x^5) = \begin{cases} \widetilde{\gamma}(g)\left(x^0 - \sum_{k=1}^{3} \frac{v^k(g) b_k^2(x^5)}{c b_0^2(x^5)} x^k\right) \\ = \left(1 - \sum_{k=1}^{3} \left(\frac{v^k(g) b_k^2(x^5)}{c b_0^2(x^5)}\right)^2\right)^{-1/2} \\ \times \left(x^0 - \sum_{k=1}^{3} \frac{v^k(g) b_k^2(x^5)}{c b_0^2(x^5)} x^k\right); \end{cases}$$

$$x^{i\prime}(g, x^5) = \begin{cases} x^i + (\widetilde{\gamma}(g) - 1) \frac{\sum_{k=1}^{3} b_k^2(x^5) v^k(g) x^k}{\sum_{k=1}^{3} b_k^2(x^5) \left(v^k(g)\right)^2} v^i(g) \\ -\widetilde{\gamma}(g) \frac{v^i(g)}{c} x^0 \\ = x^i + \left[\left(1 - \sum_{k=1}^{3} \left(\frac{v^k(g) b_k^2(x^5)}{c b_0^2(x^5)}\right)^2\right)^{-1/2} - 1\right] \\ \times \frac{\sum_{k=1}^{3} b_k^2(x^5) v^k(g) x^k}{\sum_{k=1}^{3} b_k^2(x^5) \left(v^k(g)\right)^2} v^i(g) \\ -\frac{1}{c}\left(1 - \sum_{k=1}^{3} \left(\frac{v^k(g) b_k^2(x^5)}{c b_0^2(x^5)}\right)^2\right)^{-1/2} x^0 v^i(g). \end{cases} \tag{7.146}$$

Different explicit forms of a finite, deformed boost with generic velocity $\mathbf{v}(g) = v(g)\widehat{v}(g) \equiv v(g)\widehat{\varepsilon}(g)$ along a generic direction $\widehat{\varepsilon}(g)$ in $\widetilde{E}_3(x^5) \subset \widetilde{M}(x^5)$, and consequently of $\Lambda_{B(\rho(g),\widehat{\varepsilon}(g)),\mathrm{DSR}}(x^5)$, can also be obtained by exploiting relations (3.44)–(3.48).

Notice that the lack of symmetry properties of the $4 \times 4$ matrices representing deformed boosts is obviously related to the "anisotropizing deforming" character of the DSR generalization of SR, i.e., to the fact that, in general

$$b_\mu(x^5) \neq b_\nu(x^5) \ \forall \mu, \nu \in \{0, 1, 2, 3\}, \mu \neq \nu, \tag{7.147}$$

where the $b_\mu(x^5)$'s are the coefficients of the deformed metric $g_{\mu\nu,\mathrm{DSR}}(x^5)$.

### 7.4.3 Parametric Change of Basis for a Deformed Boost in a Generic Direction

On account of (7.50)–(7.51), which relate, through the use of definition (7.40), the dimensionless parameter basis of deformed rapidities $\{\zeta^i(g)\}$ and the dimensional parameter basis of deformed boost velocities $\{v^i(g)\}$, one gets (ESC off)

$$\begin{cases} I)\dfrac{v^i(g)b_i(x^5)}{cb_0(x^5)} \equiv \widetilde{\beta}^i(g) = b_i(x^5)\widetilde{\beta}^{i(*)}(g) = \operatorname{tgh}\left(\zeta_i(g)b_0^{-1}(x^5)b_i^{-1}(x^5)\right); \\ II)\left(1-\dfrac{b_i^2(x^5)}{c^2b_0^2(x^5)}\left(v^i(g)\right)^2\right)^{-1/2} = (1-(\widetilde{\beta}^i(g))^2)^{-1/2} \\ = (1-b_i^2(x^5)(\widetilde{\beta}^{i(*)}(g))^2)^{-1/2} \equiv \widetilde{\gamma}^i(g) = \cosh\left(\zeta_i(g)b_0^{-1}(x^5)b_i^{-1}(x^5)\right), \end{cases} \tag{7.148}$$

and therefore

$$\begin{aligned} &\widetilde{\gamma}(g) \equiv \\ &\equiv \left(1-\frac{1}{c^2b_0^2(x^5)}\sum_{k=1}^{3}\left(v^i(g)\right)^2 b_i^2(x^5)\right)^{-1/2} = \left(1-\sum_{i=1}^{3}(\widetilde{\beta}^i(g))^2\right)^{-1/2} \\ &= \left(1-\sum_{i=1}^{3}b_i^2(x^5)(\widetilde{\beta}^{i(*)}(g))^2\right)^{-1/2} \\ &= \left(1-\sum_{i=1}^{3}(\operatorname{tgh}\left(\zeta_i(g)b_0^{-1}(x^5)b_i^{-1}(x^5)\right))^2\right)^{-1/2}; \end{aligned} \tag{7.149}$$

$$\frac{\left(\widetilde{\beta^i}(g)\right)^2}{|\widetilde{\boldsymbol{\beta}}(g)|^2} = \frac{b_i^2(x^5)\left(\widetilde{\beta^i}^{(*)}(g)\right)^2}{|\widetilde{\boldsymbol{\beta}}^{(*)}(g)|_*^2} = \frac{\left(\widetilde{\beta^i}(g)\right)^2}{\widetilde{\boldsymbol{\beta}}(g)\cdot\widetilde{\boldsymbol{\beta}}(g)}$$

$$= \frac{b_i^2(x^5)\left(\widetilde{\beta^i}^{(*)}(g)\right)^2}{\widetilde{\boldsymbol{\beta}}^{(*)}(g) * \widetilde{\boldsymbol{\beta}}^{(*)}(g)} = \frac{\left(\widetilde{\beta^i}(g)\right)^2}{\sum_{k=1}^3(\widetilde{\beta^k}(g))^2} = \frac{b_i^2(x^5)\left(\widetilde{\beta^i}^{(*)}(g)\right)^2}{\sum_{k=1}^3 b_k^2(x^5)(\widetilde{\beta^k}^{(*)}(g))^2}$$

$$= \frac{\left(\operatorname{tgh}\left(\zeta_i(g) b_0^{-1}(x^5) b_i^{-1}(x^5)\right)\right)^2}{\sum_{k=1}^3 (\operatorname{tgh}\left(\zeta_k(g) b_0^{-1}(x^5) b_k^{-1}(x^5)\right))^2}; \tag{7.150}$$

$$\frac{\widetilde{\beta^i}(g)\widetilde{\beta^j}(g)}{|\widetilde{\boldsymbol{\beta}}(g)|^2} = \frac{b_i(x^5) b_j(x^5) \widetilde{\beta^i}^{(*)}(g)\widetilde{\beta^j}^{(*)}(g)}{|\widetilde{\boldsymbol{\beta}}^{(*)}(g)|_*^2}$$

$$= \frac{\widetilde{\beta^i}(g)\widetilde{\beta^j}(g)}{\widetilde{\boldsymbol{\beta}}(g)\cdot\widetilde{\boldsymbol{\beta}}(g)} = \frac{b_i(x^5) b_j(x^5) \widetilde{\beta^i}^{(*)}(g)\widetilde{\beta^j}^{(*)}(g)}{\widetilde{\boldsymbol{\beta}}^{(*)}(g) * \widetilde{\boldsymbol{\beta}}^{(*)}(g)}$$

$$= \frac{\widetilde{\beta^i}(g)\widetilde{\beta^j}(g)}{\sum_{k=1}^3 (\widetilde{\beta}^k(g))^2} = \frac{b_i(x^5) b_j(x^5) \widetilde{\beta^i}^{(*)}(g)\widetilde{\beta^j}^{(*)}(g)}{\sum_{k=1}^3 b_k^2(x^5)(\widetilde{\beta}^{k(*)}(g))^2}$$

$$= \frac{\left(\operatorname{tgh}\left(\zeta_i(g) b_0^{-1}(x^5) b_i^{-1}(x^5)\right)\right)\left(\operatorname{tgh}\left(\zeta_j(g) b_0^{-1}(x^5) b_j^{-1}(x^5)\right)\right)}{\sum_{k=1}^3 (\operatorname{tgh}\left(\zeta_k(g) b_0^{-1}(x^5) b_k^{-1}(x^5)\right))^2}; \tag{7.151}$$

$$\widetilde{\beta}^2 \equiv |\widetilde{\boldsymbol{\beta}}|^2 = |\widetilde{\boldsymbol{\beta}}^{(*)}|_*^2 = \widetilde{\boldsymbol{\beta}}(g)\cdot\widetilde{\boldsymbol{\beta}}(g) = \widetilde{\boldsymbol{\beta}}^{(*)}(g) * \widetilde{\boldsymbol{\beta}}^{(*)}(g)$$

$$= \sum_{k=1}^3 (\widetilde{\beta}^k(g))^2 = \sum_{k=1}^3 b_k^2(x^5)(\widetilde{\beta}^{k(*)}(g))^2$$

$$= \sum_{k=1}^3 (\operatorname{tgh}\left(\zeta_k(g) b_0^{-1}(x^5) b_k^{-1}(x^5)\right))^2. \tag{7.152}$$

The earlier relations between the sets $\left\{\zeta^i(g)\right\}_{i=1,2,3}$ and $\left\{v^i(g)\right\}_{i=1,2,3}$ express the change of parametric base for deformed boosts in DSR, from the dimensional parameter basis of deformed boost velocities to the dimensionless parameter basis of deformed rapidities By means of (7.149)–(7.152) one can therefore express $\exp\left(B_{(\rho(g),\widehat{\varepsilon}(g)),\mathrm{DSR}}(x^5)\right)$ (derived in Sect. 7.4.2) in terms of rapidities. We leave this task to the interested reader.

Inverting (7.148) one gets

$$\zeta^i(g) = b_0(x^5)b_i(x^5)\,\text{arctgh}\left(\frac{v^i(g)b_i(x^5)}{cb_0(x^5)}\right). \tag{7.153}$$

Then, (7.135) can be rewritten as

$$\begin{aligned}
&B_{(\rho(g),\widehat{\varepsilon}(g)),\text{DSR}}(x^5)\\
&= -b_0(x^5)b_1(x^5)\,\text{arctgh}\left(\frac{v^1(g)b_1(x^5)}{cb_0(x^5)}\right)K^1_{\text{DSR}}(x^5)\\
&\quad -b_0(x^5)b_2(x^5)\,\text{arctgh}\left(\frac{v^2(g)b_2(x^5)}{cb_0(x^5)}\right)K^2_{\text{DSR}}(x^5)\\
&\quad -b_0(x^5)b_3(x^5)\,\text{arctgh}\left(\frac{v^3(g)b_3(x^5)}{cb_0(x^5)}\right)K^3_{\text{DSR}}(x^5)\\
&= -b_0(x^5)\sum_{i=1}^{3} b_i(x^5)\,\text{arctgh}\left(\frac{v^i(g)b_i(x^5)}{cb_0(x^5)}\right)K^i_{\text{DSR}}(x^5);
\end{aligned} \tag{7.154}$$

$$\begin{aligned}
&\exp(B_{(\rho(g),\widehat{\varepsilon}(g)),\text{DSR}}(x^5))\\
&= \exp(-b_0(x^5)b_1(x^5)\,\text{arctgh}\left(\frac{v^1(g)b_1(x^5)}{cb_0(x^5)}\right)K^1_{\text{DSR}}(x^5)\\
&\quad -b_0(x^5)b_2(x^5)\,\text{arctgh}\left(\frac{v^2(g)b_2(x^5)}{cb_0(x^5)}\right)K^2_{\text{DSR}}(x^5)\\
&\quad -b_0(x^5)b_3(x^5)\,\text{arctgh}\left(\frac{v^3(g)b_3(x^5)}{cb_0(x^5)}\right)K^3_{\text{DSR}}(x^5))\\
&= \exp\left(-b_0(x^5)\sum_{i=1}^{3} b_i(x^5)\,\text{arctgh}\left(\frac{v^i(g)b_i(x^5)}{cb_0(x^5)}\right)K^i_{\text{DSR}}(x^5)\right).
\end{aligned} \tag{7.155}$$

We have already stressed that the dimensionless parametric three-vectors $\boldsymbol{\theta}(g)$ and $\boldsymbol{\zeta}(g)$ are Euclidean in DSR (as well as in SR). On the contrary, the dimensional parametric three-vector of boost velocity, $\mathbf{v}(g)$, changes its nature according to the metric framework considered. In particular, $\mathbf{v}(g) \in E_3 \subset M$ in SR and $\mathbf{v}(g) \in \widetilde{E_3}(x^5) \subset \widetilde{M}(x^5)$ in DSR, namely $\mathbf{v}(g)$ is an Euclidean three-vector in SR and a deformed one in DSR. By (7.153), the relation between $\boldsymbol{\zeta}(g)$ and $\mathbf{v}(g)$ is therefore

$$\begin{aligned} \boldsymbol{\zeta}(g) &= b_0(x^5)b_1(x^5)\,\mathrm{arctgh}\left(\frac{v^1(g)b_1(x^5)}{cb_0(x^5)}\right)\widehat{x^1} \\ &+b_0(x^5)b_2(x^5)\,\mathrm{arctgh}\left(\frac{v^3(g)b_3(x^5)}{cb_0(x^5)}\right)\widehat{x^2} \\ &+b_0(x^5)b_3(x^5)\,\mathrm{arctgh}\left(\frac{v^3(g)b_3(x^5)}{cb_0(x^5)}\right)\widehat{x^3} \\ &= |\boldsymbol{\zeta}(g)|\,\widehat{\zeta}(g) = (\boldsymbol{\zeta}(g)\cdot\boldsymbol{\zeta}(g))^{1/2}\,\widehat{\zeta}(g) \\ &= \left(\sum_{i=1}^{3}\left(b_0(x^5)b_i(x^5)\,\mathrm{arctgh}\left(\frac{v^i(g)b_i(x^5)}{cb_0(x^5)}\right)\right)^2\right)^{1/2}\widehat{\zeta}(g), \end{aligned} \tag{7.156}$$

where, on account of the Euclidean nature of $\boldsymbol{\zeta}(g)$ in DSR, the unit vector $\widehat{\zeta}(g)$ is given by

$$\widehat{\zeta}(g) \equiv \frac{\boldsymbol{\zeta}(g)}{\left(\sum_{i=1}^{3}\left(b_0(x^5)b_i(x^5)\,\mathrm{arctgh}\left(\frac{v^i(g)b_i(x^5)}{cb_0(x^5)}\right)\right)^2\right)^{1/2}}. \tag{7.157}$$

Equation (7.154), (7.155) can be therefore rewritten as:

$$\begin{aligned} B_{(\rho(g),\widehat{\varepsilon}(g)),\mathrm{DSR}}(x^5) &= -\boldsymbol{\zeta}(g)\cdot\mathbf{K}_{\mathrm{DSR}}(x^5) \\ &= -\left(b_0(x^5)\sum_{i=1}^{3}\left(b_i(x^5)\,\mathrm{arctgh}\left(\frac{v^i(g)b_i(x^5)}{cb_0(x^5)}\right)\right)^2\right)^{1/2}\widehat{\zeta}(g)\cdot\mathbf{K}_{\mathrm{DSR}}(x^5); \end{aligned} \tag{7.158}$$

$$\begin{aligned} \Lambda_{B\left(\rho(g),\widehat{\varepsilon}(g)\right),\mathrm{DSR}}(x^5) &= \exp\left(-\boldsymbol{\zeta}(g)\cdot\mathbf{K}_{\mathrm{DSR}}(x^5)\right) \\ &= \exp\left(-\left(b_0(x^5)\sum_{i=1}^{3}\left(b_i(x^5)\,\mathrm{arctgh}\left(\frac{v^i(g)b_i(x^5)}{cb_0(x^5)}\right)\right)^2\right)^{1/2}\widehat{\zeta}(g)\cdot\mathbf{K}_{\mathrm{DSR}}(x^5)\right). \end{aligned} \tag{7.159}$$

Let us remark that care must be exercised in relating the unit vectors $\widehat{v}(g)$ and $\widehat{\zeta}(g)$ in DSR. In fact, consider the special case of a deformed boost along a coordinate axis, namely $\widehat{v}(g) \equiv \widehat{x^k}$ ($k \in \{1,2,3\}$). Then (7.153) becomes

$$\boldsymbol{\zeta}(g) = b_0(x^5)b_k(x^5)\,\mathrm{arctgh}\left(\frac{v^k(g)b_k(x^5)}{cb_0(x^5)}\right)\widehat{x^k}. \tag{7.160}$$

We have therefore

$$\widehat{\zeta}(g) = \widehat{x^k}; \tag{7.161}$$

$$\widehat{v}(g) \equiv \widehat{x^k}. \tag{7.162}$$

However, unlike the SR case, such relations *do not allow one to conclude* that $\widehat{v}(g) = \widehat{\zeta}(g)$. The reason is that actually the two unit vectors $\widehat{x^k}$ in (7.161) and (7.162) *are different*: Whereas the $\widehat{x^k}$ in (7.161) is Euclidean (due to the Euclidean nature of $\boldsymbol{\zeta}(g)$ in DSR), the $\widehat{x^k}$ in (7.162) is deformed, because $\mathbf{v}(g)$ in the DSR case is defined in $\widetilde{E}_3(x^5) \subset \widetilde{M}(x^5)$. In other words, we have

$$\begin{cases} \widetilde{M}(x^5) \supset \widetilde{E}_3(x^5) \ni \widehat{x^k_{\text{DSR}}} \equiv \widehat{v}(g) \neq \widehat{\zeta}(g) = \widehat{x^k_{\text{SR}}} \in E_3 \subset M; \\ \widehat{x^k_{\text{DSR}}} \equiv \widehat{v}(g) = b_k^{-1}(x^5)\widehat{\zeta}(g) = b_k^{-1}(x^5)\widehat{x^k_{\text{SR}}}, \end{cases} \tag{7.163}$$

where the notation $\widehat{x^k_{\text{SR}}}$, $\widehat{x^k_{\text{DSR}}}$ is used in order to stress the Euclidean or deformed nature of the coordinate unit vector. Then

$$\begin{aligned} \boldsymbol{\zeta}(g) &= b_0(x^5)b_k(x^5)\,\text{arctgh}\left(\frac{v^k(g)b_k(x^5)}{cb_0(x^5)}\right)\widehat{x^k_{\text{SR}}} \\ &= b_0(x^5)b_k^2(x^5)\,\text{arctgh}\left(\frac{|\mathbf{v}(g)|_*}{cb_0(x^5)}\right)\widehat{v}(g), \end{aligned} \tag{7.164}$$

where $|\mathbf{v}(g)|_*$ is the deformed norm of a three-vector. On account of (7.156), this equation can also been rewritten as

$$\begin{aligned} \boldsymbol{\zeta}(g) &= b_0(x^5)b_k^2(x^5)\,\text{arctgh}\left(\widetilde{\beta^k}(g)\right)\widehat{v}(g) \\ &= b_0(x^5)b_k^2(x^5)\,\text{arctgh}\left(b_k(x^5)\widetilde{\beta^k}^{(*)}(g)\right)\widehat{v}(g) \\ &= b_0(x^5)b_k^2(x^5)\,\text{arctgh}\left(b_k(x^5)\widetilde{\beta^k}^{(*)}(g)\right)\widehat{\widetilde{\beta}}^{(*)}(g) \\ &\overset{**}{=} b_0(x^5)b_k^2(x^5)\,\text{arctgh}\left(\widetilde{\beta^k}(g)\right)\widehat{\widetilde{\beta}}(g), \end{aligned} \tag{7.165}$$

where the last step (marked with "∗∗") follows from the fact that, although in general $\widehat{\widetilde{\beta}}(g) \neq \widehat{v}(g)$, in the special case $\mathbf{v}(g) = v^k(g)\widehat{x^k}$ (ESC off) it is $\widehat{\widetilde{\beta}}(g) = \widehat{v}(g)$.

For a deformed boost in a generic direction $\widehat{v}(g) \equiv \widehat{\varepsilon}(g)$, it is still $\widehat{v}(g) \neq \widehat{\zeta}(g)$ and we have explicitly[13]

$$\begin{aligned} \boldsymbol{\zeta}(g) &= b_0(x^5)b_1(x^5)\,\text{arctgh}\left(\frac{v^1(g)b_1(x^5)}{cb_0(x^5)}\right)\widehat{x^1_{\text{SR}}} \\ &+ b_0(x^5)b_2(x^5)\,\text{arctgh}\left(\frac{v^3(g)b_3(x^5)}{cb_0(x^5)}\right)\widehat{x^2_{\text{SR}}} \\ &+ b_0(x^5)b_3(x^5)\,\text{arctgh}\left(\frac{v^3(g)b_3(x^5)}{cb_0(x^5)}\right)\widehat{x^3_{\text{SR}}} \\ &= \left(\sum_{i=1}^{3}\left(b_0(x^5)b_i(x^5)\,\text{arctgh}\left(\frac{v^i(g)b_i(x^5)}{cb_0(x^5)}\right)\right)^2\right)^{1/2}\widehat{\zeta}(g), \end{aligned} \tag{7.166}$$

whence

$$\begin{aligned} \widehat{\zeta}(g) \equiv\; & \frac{b_0(x^5)b_1(x^5)\,\text{arctgh}\left(\frac{v^1(g)b_1(x^5)}{cb_0(x^5)}\right)}{\left(\sum_{i=1}^{3}\left(b_0(x^5)b_i(x^5)\,\text{arctgh}\left(\frac{v^i(g)b_i(x^5)}{cb_0(x^5)}\right)\right)^2\right)^{1/2}}\widehat{x^1_{\text{SR}}} \\ & + \frac{b_0(x^5)b_2(x^5)\,\text{arctgh}\left(\frac{v^2(g)b_2(x^5)}{cb_0(x^5)}\right)}{\left(\sum_{i=1}^{3}\left(b_0(x^5)b_i(x^5)\,\text{arctgh}\left(\frac{v^i(g)b_i(x^5)}{cb_0(x^5)}\right)\right)^2\right)^{1/2}}\widehat{x^2_{\text{SR}}} \\ & + \frac{b_0(x^5)b_3(x^5)\,\text{arctgh}\left(\frac{v^3(g)b_3(x^5)}{cb_0(x^5)}\right)}{\left(\sum_{i=1}^{3}\left(b_0(x^5)b_i(x^5)\,\text{arctgh}\left(\frac{v^i(g)b_i(x^5)}{cb_0(x^5)}\right)\right)^2\right)^{1/2}}\widehat{x^3_{\text{SR}}}; \end{aligned} \tag{7.167}$$

$$\left\{ \begin{aligned} & \mathbf{v}(g) = v^1(g)\widehat{x^1_{\text{DSR}}} + v^2(g)\widehat{x^2_{\text{DSR}}} + v^3(g)\widehat{x^3_{\text{DSR}}} = \\ & \quad = \left(\sum_{i=1}^{3} b_i^2(x^5)\left(v^i(g)\right)^2\right)^{1/2}\widehat{v}(g) \end{aligned} \right.$$

---

[13]Notice that the vector $\widehat{v}(g) \in \widetilde{E_3}(x^5) \subset \widetilde{M}(x^5)$, and therefore it has unit norm according to the 3D deformed scalar product $*$ (with metric tensor – ESC off – $\delta_{ij}b_i^2(x^5) = -g_{ij,\text{DSR}}(x^5)$), unlike $\widehat{\zeta}(g) \in E_3 \subset M$ which is a unit vector according to the Euclidean scalar product $\cdot$ (with metric tensor $\delta_{ij} = -g_{ij,\text{SR}}$).

$$\leftrightarrow \begin{cases} \widehat{v}(g) \equiv \dfrac{v^1(g)}{\left(\sum_{i=1}^{3} b_i^2(x^5)\left(v^i(g)\right)^2\right)^{1/2}} \widehat{x_{\rm DSR}^1} \\ + \dfrac{v^2(g)}{\left(\sum_{i=1}^{3} b_i^2(x^5)\left(v^i(g)\right)^2\right)^{1/2}} \widehat{x_{\rm DSR}^2} + \dfrac{v^3(g)}{\left(\sum_{i=1}^{3} b_i^2(x^5)\left(v^i(g)\right)^2\right)^{1/2}} \widehat{x_{\rm DSR}^3} \\ = \dfrac{b_1^{-1}(x^5)v^1(g)}{\left(\sum_{i=1}^{3} b_i^2(x^5)\left(v^i(g)\right)^2\right)^{1/2}} \widehat{x_{\rm SR}^1} \\ + \dfrac{b_2^{-1}(x^5)v^2(g)}{\left(\sum_{i=1}^{3} b_i^2(x^5)\left(v^i(g)\right)^2\right)^{1/2}} \widehat{x_{\rm SR}^2} + \dfrac{b_3^{-1}(x^5)v^3(g)}{\left(\sum_{i=1}^{3} b_i^2(x^5)\left(v^i(g)\right)^2\right)^{1/2}} \widehat{x_{\rm SR}^3}. \end{cases} \tag{7.168}$$

Therefore, for a deformed boost along the coordinate axis $\widehat{x_{\rm DSR}^k}$, the matrix $B_{(\rho(g),\widehat{\varepsilon}(g)),{\rm DSR}}(x^5)$ reads

$$\begin{aligned} B_{(\rho(g),\widehat{\varepsilon}(g)),{\rm DSR}}(x^5) &= -\boldsymbol{\zeta}(g)\cdot \mathbf{K}_{\rm DSR}(x^5) \\ &= -b_0(x^5)b_k^2(x^5)\,\mathrm{arctgh}\left(\widetilde{\beta^k}(g)\right)\widehat{\widetilde{\beta}}(g)\cdot \mathbf{K}_{\rm DSR}(x^5) \\ &= -b_0(x^5)b_k^2(x^5)\,\mathrm{arctgh}\left(b_k(x^5)\widetilde{\beta^{k}}^{(*)}(g)\right)\widehat{\widetilde{\beta}^{(*)}}(g)\cdot \mathbf{K}_{\rm DSR}(x^5) \\ &= -b_0(x^5)b_k^2(x^5)\,\mathrm{arctgh}\left(\widetilde{\beta^k}(g)\right)\widehat{v}(g)\cdot \mathbf{K}_{\rm DSR}(x^5) \\ &= -b_0(x^5)b_k^2(x^5)\,\mathrm{arctgh}\left(\widetilde{\beta^k}(g)\right)\widehat{x_{\rm DSR}^k}\cdot \mathbf{K}_{\rm DSR}(x^5) \\ &= -b_0(x^5)b_k(x^5)\,\mathrm{arctgh}\left(\widetilde{\beta^k}(g)\right)\widehat{x_{\rm SR}^k}\cdot \mathbf{K}_{\rm DSR}(x^5) \\ &= -b_0(x^5)b_k(x^5)\,\mathrm{arctgh}\left(\widetilde{\beta^k}(g)\right)K_{\rm DSR}^k(x^5) \\ &\overset{K_{\rm DSR}^i(x^5)=I_{\rm DSR}^{0i}(x^5)=-I_{\rm DSR}^{i0}(x^5)}{=} b_0(x^5)b_k(x^5)\,\mathrm{arctgh}\left(\widetilde{\beta^k}(g)\right)I_{\rm DSR}^{k0}(x^5), \end{aligned} \tag{7.169}$$

whence one gets finally, by exponentiation, the expression (ESC off):

$$\Lambda_{B(\rho(g),\widehat{\varepsilon}(g)),{\rm DSR}}(x^5) = \exp\left(b_0(x^5)b_k(x^5)\,\mathrm{arctgh}\left(\widetilde{\beta^k}(g)\right)I_{\rm DSR}^{k0}(x^5)\right). \tag{7.170}$$

Of course, it is only a matter of convenience to use one of the (physically equivalent) forms we derived for the deformed boosts in this chapter.

Let us remark the basic common feature of all these expressions, i.e.,their dependence on the metric of the deformed Minkowski space, and therefore on the interaction considered. Such a result was of course expected, since DLT are the isometries of $\widetilde{M}$. As already noted in Sect. 3.3.7, DSR shares with Lorentzian Relativity the property that different sets of space–time coordinate transformations correspond to different classes of physical phenomena. The adaptability of the mathematical structure of DSR allows DLT to change suitably, in such a way to fit the distortions of space–time produced by interactions different from (the Minkowskian part of) the electromagnetic one. The space–time deformations associated to non-Minkowskian interactions, which are felt as departures from the usual Minkowski metric when probed by the standard, special-relativistic Lorentz transformations – and therefore as signatures of breakdown of usual Lorentz invariance –, are in a sense "mathematically absorbed" by the flexible instrument of deformed LT's. The consequence is the physical recovering of Lorentz invariance in DSR (as discussed in Sect. 3.3.5). We shall come back to these issues in the conclusive Chap. 26.

# 8

# Deformed Space–Time Translations in Four Dimensions

## 8.1 Translations in 4D Generalized Minkowski Spaces

We want now to discuss the translational component of the generalized Poincaré group [43]. Without loss of generality, we shall consider the 4D case.

In the case $N = 4$, it was seen in Sect. 5.2.4 that the components $\xi_\mu$ of the covariant Killing four-vector of a generic 4D generalized Minkowski space $\tilde{M}(\{x\}_{\text{n.m.}})$ read (cf. (5.39))

$$\begin{cases} \xi_0(x) = -\zeta^1 x^1 - \zeta^2 x^2 - \zeta^3 x^3 + T^0; \\ \xi_1(x) = \zeta^1 x^0 + \theta^2 x^3 - \theta^3 x^2 - T^1; \\ \xi_2(x) = \zeta^2 x^0 - \theta^1 x^3 + \theta^3 x^1 - T^2; \\ \xi_3(x) = \zeta^3 x^0 + \theta^1 x^2 - \theta^2 x^1 - T^3 \end{cases} \tag{8.1}$$

(we omitted, for simplicity's sake, the dependence on the group element $g \in \text{Tr.}(T = 4 - S, S)_{\text{GEN.}} \subset P(T = 4 - S, S)_{\text{GEN.}}$). Then – as already stressed –, independently of the explicit form of the metric tensor, all 4D generalized Minkowski spaces admit the same *covariant* Killing vector. In particular, with the signature $(+, -, -, -)$ (namely $T = 1$, $S = 3$) of SR and DSR, we have shown that $\boldsymbol{\zeta} = (\zeta^1, \zeta^2, \zeta^3)$ is the (Euclidean) three-vector of the dimensionless parameters (i.e., "generalized rapidities") of a generalized 3D boost and $\boldsymbol{\theta} = (\theta^1, \theta^2, \theta^3)$ is the (Euclidean) three-vector of the dimensionless parameters (i.e., generalized angles) of a generalized 3Dtrue rotations, whereas

$$T_\mu = (T^0, -T^1, -T^2, -T^3) \tag{8.2}$$

is the covariant four-vector of the length-dimensioned parameters of a generalized 4D translation.

Notice that the "length-dimensioned," parametric contravariant four-vectors $T^\mu(g)$ are to be regarded as infinitesimal at the algebraic level and finite at the group level. This will be understood, and, for simplicity' sake, no notational distinction will be made. As it will be explicitly seen later (in the DSR case, without loss of generality), the infinitesimal or finite nature of the translation parameter $N$-vectors (such as $T^\mu_{\mathrm{DSR}}(g, x^5)$ in DSR) is in general the only difference between the algebraic and the group level in translation coordinate transformations in a generalized $N$-d Minkowski space $\widetilde{M_N}(\{x\}_{\mathrm{n.m.}})$.

The inhomogeneity of the (infinitesimal) translation transformation (8.1) obviously entails that it cannot be represented by a $4 \times 4$ matrix (at the infinitesimal, and then at the finite, level), i.e., no 4D representation of the infinitesimal generators of $\mathrm{Tr.}(1,3)_{\mathrm{GEN.}}$ exists. However, it is possible to get a matrix representation of the infinitesimal generators of the generalized translation group $\mathrm{Tr.}(T = 4 - S, S)_{\mathrm{GEN.}} \subset P(T = 4 - S, S)_{\mathrm{GEN.}}$ by introducing a fifth, constant auxiliary coordinate [45] $y = 1$, devoid of any physical or metric meaning, as it is clearly seen by the fact that its total differential is zero. This fictitious extra coordinate plays only a parametrizing role, i.e., it is introduced in order to span the "transformative degree of freedom" associated to the inhomogeneous component of the (maximal) Killing group $P(1,3)_{\mathrm{GEN.}}$ of the 4D $(3,1)$ generalized Minkowski space $\widetilde{M}(\{x\}_{\mathrm{n.m.}})$ considered. In other words, $y = 1$ has to express the translation component of the Poincaré generalized coordinate transformations of $\widetilde{M}(\{x\}_{\mathrm{n.m.}})$.

More generally, the introduction of $y$ is necessary to give an explicit $(N+1)$-dimensional (matrix) representation of the infinitesimal generators of the generalized translation group $\mathrm{Tr.}(T = N - S, S)_{\mathrm{GEN.}}$, and then to calculate the (representation-independent) $N(S,T)$ generalized "mixed" Poincaré algebra, i.e., the commutator-exploited algebraic structure between the infinitesimal generators of $\mathrm{SO}(T = N - S, S)_{\mathrm{GEN.}}$ and the infinitesimal generators of $\mathrm{Tr.}(T = N - S, S)_{\mathrm{GEN.}}$ (for the case $N = 4$, $T = 1$, $S = 3$ of DSR, see Sect. 6.3).

Then, following the notation of [45],[1] one can consider[2] the following 5D (matrix) representation of the infinitesimal generators[3] $\left\{(\Upsilon_\mu)^A_B\right\}_{\mu=0,1,2,3}$ of the group $\mathrm{Tr.}(1,3)_{\mathrm{GEN.}}$:

$$\Upsilon_0 \equiv \begin{pmatrix} 0 & 0 & 0 & 0 & 1 \\ 0 & 0 & 0 & 0 & 0 \\ 0 & 0 & 0 & 0 & 0 \\ 0 & 0 & 0 & 0 & 0 \\ 0 & 0 & 0 & 0 & 0 \end{pmatrix}; \quad \Upsilon_1 \equiv \begin{pmatrix} 0 & 0 & 0 & 0 & 0 \\ 0 & 0 & 0 & 0 & 1 \\ 0 & 0 & 0 & 0 & 0 \\ 0 & 0 & 0 & 0 & 0 \\ 0 & 0 & 0 & 0 & 0 \end{pmatrix}; \tag{8.3}$$

$$\Upsilon_2 \equiv \begin{pmatrix} 0 & 0 & 0 & 0 & 0 \\ 0 & 0 & 0 & 0 & 0 \\ 0 & 0 & 0 & 0 & 1 \\ 0 & 0 & 0 & 0 & 0 \\ 0 & 0 & 0 & 0 & 0 \end{pmatrix}; \quad \Upsilon_3 \equiv \begin{pmatrix} 0 & 0 & 0 & 0 & 0 \\ 0 & 0 & 0 & 0 & 0 \\ 0 & 0 & 0 & 0 & 0 \\ 0 & 0 & 0 & 0 & 1 \\ 0 & 0 & 0 & 0 & 0 \end{pmatrix}. \tag{8.4}$$

Namely, the only nonzero components of the above 5D representative matrices are:

$$(\Upsilon_0)^0_6 = (\Upsilon_1)^1_6 = (\Upsilon_2)^2_6 = (\Upsilon_3)^3_6 = 1 \tag{8.5}$$

or

$$(\Upsilon_\mu)^A_B = \delta^A_\mu \delta_{6B}. \tag{8.6}$$

From (8.4) and (8.5) one easily find the following properties of the above considered 5D representation of the *covariant* infinitesimal generators of $\mathrm{Tr.}(1,3)_{\mathrm{GEN.}}$ (here and in the following, $0_{5D}$ and $1_{5D}$ denote the $5 \times 5$ zero and unity matrix, respectively):

---

[1] The only difference with [45] (treating the SR case) is an overall minus sign. This is fully justifiable assuming that the parametric contravariant four-vector $\varepsilon^\mu$, used in $(6-5.35)$ of page 150 in [45], is the opposite of $T^\mu_{\mathrm{SR}}$; that is, by omitting, for simplicity's sake, the dependence on $g \in \mathrm{Tr}(1,3)_{\mathrm{STD.}} \subset P(1,3)_{\mathrm{STD.}}$:

$$\varepsilon^\mu \equiv -T^\mu_{\mathrm{SR}} = \left(-T^0, -T^1, -T^2, -T^3\right).$$

[2] This choice could now seem a bit arbitrary, but it will prove to be justified and self-consistent from the following results, obtained, without loss of generality, in the case of DSR.

[3] In the following, the upper-case Latin indices have range $\{0, 1, 2, 3, 6\}$, where the index 6 labels the auxiliary coordinate:

$$x^6 \equiv x_6 = y = 1$$

Moreover, independently of the contravariant or covariant nature of infinitesimal generators, contravariant and covariant indices in their (matrix) representations conventionally stand for row and column indices, respectively.

$$(\Upsilon_\mu)^n = 0_{5D}, \forall n \geq 2 \Rightarrow \exp(\Upsilon_\mu) = 1_{5D} + \Upsilon_\mu; \tag{8.7}$$

$$[\Upsilon_\mu, \Upsilon_\nu] = 0_{5D}, \quad \forall (\mu, \nu) \in \{0,1,2,3\}^2; \tag{8.8}$$

$$\Upsilon_\mu \neq \Upsilon_\mu \left(x, \{x\}_{\text{n.m.}}\right). \tag{8.9}$$

In the following, we shall see that, in the DSR case, properties (8.8) and (8.9) still hold for the 5D matrix representation of the *contravariant* infinitesimal deformed translation generators. Moreover, as it is clear from (8.4)–(8.7), the 5D representation of the *covariant* infinitesimal generators of $\mathrm{Tr.}(1,3)_{\mathrm{GEN.}}$ are independent of the metric tensor (namely, they are the same irrespective of the 4D generalized Minkowski space $\widetilde{M}(\{x\}_{\text{n.m.}})$ considered). On the contrary, the *contravariant* generators *do depend* on the generalized metric, since

$$\Upsilon^\mu \overset{\text{ESC on}}{=} g^{\mu\rho}(\{x\}_{\text{n.m.}})\Upsilon_\rho = \Upsilon^\mu(\{x\}_{\text{n.m.}}). \tag{8.10}$$

The assumed explicit 5D representations (8.4) and (8.5) are justifiable by the following reasoning. Equation (8.1) expresses the independence of $T_\mu(g)$ on the (geo)metric context considered; instead, its *contravariant* form will in general be "context-dependent," of course (see Footnote 5):

$$T^\mu(g, \{x\}_{\text{n.m.}}) \overset{\text{ESC on}}{\equiv} g^{\mu\nu}(\{x\}_{\text{n.m.}})T_\nu(g). \tag{8.11}$$

Therefore, since in general a 4D "context-dependent" scalar product

$$T_\mu(g)\Upsilon^\mu(\{x\}_{\text{n.m.}}) = T_\mu(g)g^{\mu\nu}(\{x\}_{\text{n.m.}})\Upsilon_\nu = T^\mu(g, \{x\}_{\text{n.m.}})\Upsilon_\mu \tag{8.12}$$

appears in the explicit form of the 5D matrix of the infinitesimal generalized translation (e.g., in DSR case, $\mathcal{T}_{T^\mu_{\mathrm{DSR}}(g,x^5),\mathrm{DSR}}(x^5)$ of (8.41)), it is clear that the set of the *covariant* infinitesimal generalized translation generators has to be "context-independent," i.e., has to be the same irrespective of the 4D generalized Minkowski space $\widetilde{M}(\{x\}_{\text{n.m.}})$ considered.

What just said is a consequence of a peculiar feature of the translation component $\mathrm{Tr.}(1,3)_{\mathrm{GEN.}}$ of the 4D $(1,3)$ generalized Poincaré group $P(1,3)_{\mathrm{GEN.}}$.

Indeed, the elements of the 4D $(1,3)$ generalized rotation group $\mathrm{SO}(1,3)_{\mathrm{GEN.}}$ correspond, at either infinitesimal and finite level, to coordinate transformations homogeneous in their arguments, i.e., in the "length-dimensioned" coordinate basis $\{x^\mu\}_{\mu=0,1,2,3}$. Therefore, the generalized parametric (Euclidean) three-vectors $\boldsymbol{\theta}(g)$ and $\boldsymbol{\zeta}(g)$ must be dimensionless. In the cases $S=3, T=1$ of SR (corresponding to $M$) and of DSR (corresponding to $\widetilde{M}\left(x^5\right)$) $\boldsymbol{\theta}(g)$ and $\boldsymbol{\zeta}(g)$ can be identified with the generalized true rotation angle and generalized "boost rapidity" three-vectors, respectively. By using the dimension-transposing constant velocity $c$, it is possible to introduce a "velocity-dimensioned" boost parametric three-vector $\mathbf{v}(g)$, that is a contravariant three-vector in the 3D physical space embedded in the 4D Minkowski space considered ($E_3 \subset M$ in SR, and $\widetilde{E_3}\left(x^5\right) \subset \widetilde{M}\left(x^5\right)$ in DSR, respectively).

Analogously, because of the fact that the elements of the 4D generalized translation group $\text{Tr.}(1,3)_{\text{GEN.}}$ correspond, both at infinitesimal and finite level, to purely inhomogeneous coordinate transformations, it is clear that the generalized translation parametric four-vector $T^{\mu}(g)$ must be "length-dimensioned" and have a "context-dependent" geometric nature. For example the translation parameters are a standard contravariant four-vector $T^{\mu}_{\text{SR}}(g)$ of $M$ in SR, and instead a deformed one $T^{\mu}_{\text{DSR}}(g)$ of $\widetilde{M}\left(x^5\right)$ in DSR.

That is why, e.g., in the DSR case, both 3D Euclidean scalar products $\boldsymbol{\theta}(g) \cdot \mathbf{S}_{\text{DSR}}(x^5)$ and $\boldsymbol{\zeta}(g) \cdot \mathbf{K}_{\text{DSR}}(x^5)$, and 4D "deformed" scalar products (ESC on)

$$T_{\mu,(\text{DSR})}(g)\Upsilon^{\mu}_{\text{DSR}}(x^5) = \Upsilon^{\mu}_{\text{DSR}}(x^5) g_{\mu\nu,\text{DSR}}(x^5) T^{\nu}_{\text{DSR}}(g, x^5) \qquad (8.13)$$

do enter into the general expression of the $5 \times 5$ matrix corresponding to a finite transformation of the 4D deformed (inhomogeneous Lorentz) Poincaré group $P(1,3)_{\text{DEF.}}$ (see (8.46)). The notation "(DSR)" in $T_{\mu,(\text{DSR})}$ has been just used to remember that, as already stressed, $T_{\mu}(g)$ is independent of the 4D metric context considered.

## 8.2 The Group of 4D Deformed Translations

### *8.2.1 5D Representation of the Infinitesimal Contravariant Generators*

Let us consider the case of $\text{Tr.}(1,3)_{\text{DEF.}}$, i.e., the deformed space–time translation group of the 4D deformed Minkowski space $\widetilde{M}(x^5)$ of DSR. Then, on account of (8.4),(8.5), and (8.11), the considered 5D matrix representation of the infinitesimal *contravariant* deformed translation generators reads

$$\Upsilon^{0}_{\text{DSR}}(x^5) \equiv \begin{pmatrix} 0 & 0 & 0 & 0 & b_0^{-2}(x^5) \\ 0 & 0 & 0 & 0 & 0 \\ 0 & 0 & 0 & 0 & 0 \\ 0 & 0 & 0 & 0 & 0 \\ 0 & 0 & 0 & 0 & 0 \end{pmatrix}; \qquad (8.14)$$

$$\Upsilon^{1}_{\text{DSR}}(x^5) \equiv \begin{pmatrix} 0 & 0 & 0 & 0 & 0 \\ 0 & 0 & 0 & 0 & -b_1^{-2}(x^5) \\ 0 & 0 & 0 & 0 & 0 \\ 0 & 0 & 0 & 0 & 0 \\ 0 & 0 & 0 & 0 & 0 \end{pmatrix}; \qquad (8.15)$$

$$\Upsilon^2_{\text{DSR}}(x^5) \equiv \begin{pmatrix} 0 & 0 & 0 & 0 & 0 \\ 0 & 0 & 0 & 0 & 0 \\ 0 & 0 & 0 & 0 & -b_2^{-2}(x^5) \\ 0 & 0 & 0 & 0 & 0 \\ 0 & 0 & 0 & 0 & 0 \end{pmatrix}; \tag{8.16}$$

$$\Upsilon^3_{\text{DSR}}(x^5) \equiv \begin{pmatrix} 0 & 0 & 0 & 0 & 0 \\ 0 & 0 & 0 & 0 & 0 \\ 0 & 0 & 0 & 0 & 0 \\ 0 & 0 & 0 & 0 & -b_3^{-2}(x^5) \\ 0 & 0 & 0 & 0 & 0 \end{pmatrix}. \tag{8.17}$$

The only nonzero components are therefore given by

$$\begin{aligned} \left(\Upsilon^0_{\text{DSR}}(x^5)\right)_6^{\ 0} &= -\frac{b_0^{-2}(x^5)}{b_1^{-2}(x^5)} \left(\Upsilon^1_{\text{DSR}}(x^5)\right)_6^{\ 1} \\ &= -\frac{b_0^{-2}(x^5)}{b_2^{-2}(x^5)} \left(\Upsilon^2_{\text{DSR}}(x^5)\right)_6^{\ 2} = -\frac{b_0^{-2}(x^5)}{b_3^{-2}(x^5)} \left(\Upsilon^3_{\text{DSR}}(x^5)\right)_6^{\ 3} = b_0^{-2}(x^5) \end{aligned} \tag{8.18}$$

or (ESC off):

$$\begin{aligned} &\left(\Upsilon^\mu_{\text{DSR}}(x^5)\right)^A_{\ \ B} \\ &= \left(b_0^{-2}(x^5)\delta^{\mu 0} - b_1^{-2}(x^5)\delta^{\mu 1} - b_2^{-2}(x^5)\delta^{\mu 2} - b_3^{-2}(x^5)\delta^{\mu 3}\right) \delta_\mu^{\ A} \delta_{6B}. \end{aligned} \tag{8.19}$$

From (8.16)–(8.19) one immediately gets the following (representation-independent) properties of the *contravariant* deformed translation infinitesimal generators in $\widetilde{M}(x^5)$:

$$\begin{aligned} &\left(\Upsilon^\mu_{\text{DSR}}(x^5)\right)^n = 0_{5D}, \forall n \geq 2 \\ &\Rightarrow \exp(\Upsilon^\mu_{\text{DSR}}(x^5)) = 1_{5D} + \Upsilon^\mu_{\text{DSR}}(x^5); \end{aligned} \tag{8.20}$$

$$[\Upsilon^\mu_{\text{DSR}}(x^5), \Upsilon^\nu_{\text{DSR}}(x^5)] = 0_{5D}, \forall \mu, \nu \in \{0,1,2,3\}; \tag{8.21}$$

$$\Upsilon^\mu_{\text{DSR}} = \Upsilon^\mu_{\text{DSR}}(x^5), \Upsilon^\mu_{\text{DSR}} \neq \Upsilon^\mu_{\text{DSR}}(x). \tag{8.22}$$

It follows from (8.21) that ($tr.(1,3)_{\text{DEF.}}$) $\text{Tr.}(1,3)_{\text{DEF.}}$ is a proper (subalgebra) abelian subgroup of the 4D deformed Poincaré (algebra) group ($su(2)_{\text{DEF.}} \otimes su(2)_{\text{DEF.}} \otimes_s \text{tr.}(1,3)_{\text{DEF.}}$) $P(1,3)_{\text{DEF.}}$, whose infinitesimal (*contravariant*) generators (by (8.22)) are independent of the metric variables of $\widetilde{M}(x^5)$, but do (parametrically) depend on the nonmetric variable $x^5$.

### 8.2.2 "Mixed" Deformed Poincaré Algebra

It is now possible to find the "mixed" algebraic structure of the 4D deformed Poincaré group $P(1,3)_{\mathrm{DEF.}}$ of DSR; this can be carried out by evaluating the commutators (which, as in the case of the 4D deformed Lorentz algebra $su(2)_{\mathrm{DEF.}} \otimes su(2)_{\mathrm{DEF.}}$, are representation-independent) among the infinitesimal generators of $\mathrm{Tr.}(1,3)_{\mathrm{DEF.}}$ and the infinitesimal generators of the deformed homogeneous Lorentz group $\mathrm{SO}(1,3)_{\mathrm{DEF.}}$. To this aim, one has to represent the infinitesimal generators of $\mathrm{SO}(1,3)_{\mathrm{DEF.}}$ as $5\times 5$ matrices in the auxiliary fictitious 5D space with $y = 1$ as extra dimension. It is easy to see that this amounts to the following trivial replacement

$$I^{\alpha\beta}_{\mathrm{DSR}}(x^5) \to \begin{pmatrix} I^{\alpha\beta}_{\mathrm{DSR}}(x^5) & 0 \\ 0 & 0 \end{pmatrix} \ \forall\,(\alpha,\beta) \in \{0,1,2,3\}^2\,, \tag{8.23}$$

where $I^{\alpha\beta}_{\mathrm{DSR}}(x^5)$ are the infinitesimal generators of the 4D deformed homogeneous Lorentz group $\mathrm{SO}(1,3)_{\mathrm{DEF.}}$ of DSR in the 4D matrix representation derived in Sect. 6.3. Needless to say, the trivial process of "dimensional embedding" (4D→5D) of the (matrix) representation of the infinitesimal generators of the group $\mathrm{SO}(1,3)_{\mathrm{GEN.}}$ (as expressed by (8.23)), does *not* change the infinitesimal-algebraic structure in any way. This is due to the fact that the matrix rows and columns corresponding to the "auxiliary coordinate" $y$ do *not* "mix" with the homogeneous components of the coordinate transformations being considered.

We have therefore

$$I^{10}_{\mathrm{DSR}}(x^5) = \begin{pmatrix} 0 & -b_0^{-2}(x^5) & 0 & 0 & 0 \\ -b_1^{-2}(x^5) & 0 & 0 & 0 & 0 \\ 0 & 0 & 0 & 0 & 0 \\ 0 & 0 & 0 & 0 & 0 \\ 0 & 0 & 0 & 0 & 0 \end{pmatrix}; \tag{8.24}$$

$$I^{20}_{\mathrm{DSR}}(x^5) = \begin{pmatrix} 0 & 0 & -b_0^{-2}(x^5) & 0 & 0 \\ 0 & 0 & 0 & 0 & 0 \\ -b_2^{-2}(x^5) & 0 & 0 & 0 & 0 \\ 0 & 0 & 0 & 0 & 0 \\ 0 & 0 & 0 & 0 & 0 \end{pmatrix}; \tag{8.25}$$

$$I^{30}_{\mathrm{DSR}}(x^5) = \begin{pmatrix} 0 & 0 & 0 & -b_0^{-2}(x^5) & 0 \\ 0 & 0 & 0 & 0 & 0 \\ 0 & 0 & 0 & 0 & 0 \\ -b_3^{-2}(x^5) & 0 & 0 & 0 & 0 \\ 0 & 0 & 0 & 0 & 0 \end{pmatrix}; \tag{8.26}$$

$$I^{12}_{\mathrm{DSR}}(x^5) = \begin{pmatrix} 0 & 0 & 0 & 0 & 0 \\ 0 & 0 & -b_1^{-2}(x^5) & 0 & 0 \\ 0 & b_2^{-2}(x^5) & 0 & 0 & 0 \\ 0 & 0 & 0 & 0 & 0 \\ 0 & 0 & 0 & 0 & 0 \end{pmatrix}; \tag{8.27}$$

$$I^{23}_{\mathrm{DSR}}(x^5) = \begin{pmatrix} 0 & 0 & 0 & 0 & 0 \\ 0 & 0 & 0 & 0 & 0 \\ 0 & 0 & 0 & -b_2^{-2}(x^5) & 0 \\ 0 & 0 & b_3^{-2}(x^5) & 0 & 0 \\ 0 & 0 & 0 & 0 & 0 \end{pmatrix}; \tag{8.28}$$

$$I^{31}_{\mathrm{DSR}}(x^5) = \begin{pmatrix} 0 & 0 & 0 & 0 & 0 \\ 0 & 0 & 0 & b_1^{-2}(x^5) & 0 \\ 0 & 0 & 0 & 0 & 0 \\ 0 & -b_3^{-2}(x^5) & 0 & 0 & 0 \\ 0 & 0 & 0 & 0 & 0 \end{pmatrix}. \tag{8.29}$$

From (8.16)–(8.19) and (8.24)–(8.29) one gets the following form of the 4D "mixed" deformed Poincaré algebra ($\forall\,(i,j,k) \in \{1,2,3\}^3$):

$$\left\{ \begin{array}{l} [I^{i0}_{\mathrm{DSR}}(x^5), \Upsilon^{0}_{\mathrm{DSR}}(x^5)] = b_0^{-2}(x^5)\Upsilon^{i}_{\mathrm{DSR}}(x^5); \\ [I^{i0}_{\mathrm{DSR}}(x^5), \Upsilon^{j}_{\mathrm{DSR}}(x^5)] \overset{\text{ESC off on i}}{=} \delta^{ij}(x^5) b_i^{-2}(x^5)\Upsilon^{0}_{\mathrm{DSR}}(x^5); \\ [I^{ij}_{\mathrm{DSR}}(x^5), \Upsilon^{0}_{\mathrm{DSR}}(x^5)] = 0; \\ [I^{ij}_{\mathrm{DSR}}(x^5), \Upsilon^{k}_{\mathrm{DSR}}(x^5)] \\ \overset{\text{ESC off on } i \text{ and } j}{=} \delta^{ik} b_i^{-2}(x^5)\Upsilon^{j}_{\mathrm{DSR}}(x^5) - \delta^{jk} b_j^{-2}(x^5\Upsilon^{i}_{\mathrm{DSR}}(x^5), \end{array} \right. \tag{8.30}$$

or, in compact form ($\forall\,(\mu,\nu,\rho) \in \{0,1,2,3\}^3$):

$$\begin{aligned} [I^{\mu\nu}_{\mathrm{DSR}}(x^5), \Upsilon^{\rho}_{\mathrm{DSR}}(x^5)] &= g^{\nu\rho}_{\mathrm{DSR}}(x^5)\Upsilon^{\mu}_{\mathrm{DSR}}(x^5) - g^{\mu\rho}_{\mathrm{DSR}}(x^5)\Upsilon^{\nu}_{\mathrm{DSR}}(x^5) \\ \overset{\text{ESC off}}{=} \delta^{\nu\rho}&\left(b_0^{-2}(x^5)\delta^{\nu 0} - b_1^{-2}(x^5)\delta^{\nu 1} - b_2^{-2}(x^5)\delta^{\nu 2} - b_3^{-2}(x^5)\delta^{\nu 3}\right)\Upsilon^{\mu}_{\mathrm{DSR}}(x^5) \\ -\delta^{\mu\rho}&\left(b_0^{-2}(x^5)\delta^{\mu 0} - b_1^{-2}(x^5)\delta^{\mu 1} - b_2^{-2}(x^5)\delta^{\mu 2} - b_3^{-2}(x^5)\delta^{\mu 3}\right)\Upsilon^{\nu}_{\mathrm{DSR}}(x^5), \end{aligned} \tag{8.31}$$

whence in general

$$\exists \text{ at least } 1\ \ (\mu,\nu,\rho) \in \{0,1,2,3\}^3 : [I^{\mu\nu}_{\mathrm{DSR}}(x^5), \Upsilon^{\rho}_{\mathrm{DSR}}(x^5)] \neq 0,\ \ \forall x^5 \in R_0^+. \tag{8.32}$$

Therefore, although ($su(2)_{\mathrm{DEF.}} \otimes su(2)_{\mathrm{DEF.}}$) $\mathrm{SO}(1,3)_{\mathrm{DEF.}}$ and (tr. $(1,3)_{\mathrm{DEF.}}$) $\mathrm{Tr.}(1,3)_{\mathrm{DEF.}}$ are proper (subalgebras) subgroups – non-abelian and abelian, respectively – of the 4D deformed Poincaré (algebra) group, they determine it only by their semidirect product.

Let us change the basis of the infinitesimal generators of $\mathrm{SO}(1,3)_{\mathrm{DEF.}}$ to the "self-representative" one of the (Euclidean) three-vector, deformed, space–time infinitesimal generators $\mathbf{S}_{\mathrm{DSR}}(x^5)$, $\mathbf{K}_{\mathrm{DSR}}(x^5)$, defined by (6.54)–(6.57). In this basis, the "mixed" part of the 4D deformed Poincaré algebra can be written as (ESC off):

$$\begin{cases} [K^i_{\mathrm{DSR}}(x^5), \Upsilon^0_{\mathrm{DSR}}(x^5)] = -b_0^{-2}(x^5)\Upsilon^i_{\mathrm{DSR}}(x^5); \\ [K^i_{\mathrm{DSR}}(x^5), \Upsilon^j_{\mathrm{DSR}}(x^5)] \overset{\text{ESC off on } i}{=} -\delta^{ij} b_i^{-2}(x^5)\Upsilon^0_{\mathrm{DSR}}(x^5); \\ [S^i_{\mathrm{DSR}}(x^5), \Upsilon^0_{\mathrm{DSR}}(x^5)] = [\frac{1}{2}\epsilon^i_{jk} I^{jk}_{\mathrm{DSR}}(x^5), \Upsilon^0_{\mathrm{DSR}}](x^5) \\ = \frac{1}{2}\epsilon^i_{jk}[I^{jk}_{\mathrm{DSR}}(x^5), \Upsilon^0_{\mathrm{DSR}}(x^5)] = 0; \\ [S^i_{\mathrm{DSR}}(x^5), \Upsilon^k_{\mathrm{DSR}}(x^5)] = [\frac{1}{2}\epsilon^i_{jl} I^{jl}_{\mathrm{DSR}}(x^5), \Upsilon^k_{\mathrm{DSR}}(x^5)] \\ = \frac{1}{2}\epsilon^i{}_{jl}[I^{jl}_{\mathrm{DSR}}(x^5), \Upsilon^k_{\mathrm{DSR}}(x^5)] \\ = \frac{1}{2}\epsilon^i{}_{jl}\left(\delta^{jk} b_j^{-2}(x^5)\Upsilon^l_{\mathrm{DSR}}(x^5) - \delta^{lk} b_l^{-2}(x^5)\Upsilon^j_{\mathrm{DSR}}(x^5)\right) \\ \overset{\text{ESC off on } k}{=} \frac{1}{2}\left(\epsilon^{ik}_l b_k^{-2}(x^5)\Upsilon^l_{\mathrm{DSR}}(x^5) - \epsilon^{ik}_j b_k^{-2}(x^5)\Upsilon^j_{\mathrm{DSR}}(x^5)\right) \\ \overset{\text{ESC off on } k}{=} \epsilon_{ikl} b_k^{-2}(x^5)\Upsilon^l_{\mathrm{DSR}}(x^5). \end{cases} \tag{8.33}$$

On account of the results obtained in Sect. 6.3 for the 4D deformed (homogeneous) Lorentz algebra $su(2)_{\mathrm{DEF.}} \otimes su(2)_{\mathrm{DEF.}}$, we can write the Lie algebra $(su(2)_{\mathrm{DEF.}} \otimes su(2)_{\mathrm{DEF.}}) \otimes_s \mathrm{tr.}(1,3)_{\mathrm{DEF.}}$ of the 4D deformed Poincaré group $P(1,3)_{\mathrm{DEF.}}$ (i.e., the algebraic-infinitesimal structure of the maximal Killing group of $\widetilde{M}(x^5)$) as:

4D Deformed space–time rotation algebra $su(2)_{\mathrm{DEF.}} \otimes su(2)_{\mathrm{DEF.}}$

$$\begin{cases} [I^{\alpha\beta}_{\mathrm{DSR}}(x^5), I^{\rho\sigma}_{\mathrm{DSR}}(x^5)] \\ = g^{\alpha\sigma}_{\mathrm{DSR}}(x^5) I^{\beta\rho}_{\mathrm{DSR}}(x^5) + g^{\beta\rho}_{\mathrm{DSR}}(x^5) I^{\alpha\sigma}_{\mathrm{DSR}}(x^5) \\ -g^{\alpha\rho}_{\mathrm{DSR}}(x^5) I^{\beta\sigma}_{\mathrm{DSR}}(x^5) - g^{\beta\sigma}_{\mathrm{DSR}}(x^5) I^{\alpha\rho}_{\mathrm{DSR}}(x^5) \\ \overset{\text{ESC off}}{=} \delta^{\alpha\sigma}(b_0^{-2}(x^5)\delta^{\alpha 0} - b_1^{-2}(x^5)\delta^{\alpha 1} \\ -b_2^{-2}(x^5)\delta^{\alpha 2} - b_3^{-2}(x^5)\delta^{\alpha 3}) I^{\beta\rho}_{\mathrm{DSR}}(x^5) \\ +\delta^{\beta\rho}(\delta^{\beta 0} b_0^{-2}(x^5) - \delta^{\beta 1} b_1^{-2}(x^5) \\ -\delta^{\beta 2} b_2^{-2}(x^5) - \delta^{\beta 3} b_3^{-2}(x^5)) I^{\alpha\sigma}_{\mathrm{DSR}}(x^5) \\ -\delta^{\alpha\rho}(\delta^{\alpha 0} b_0^{-2}(x^5) - \delta^{\alpha 1} b_1^{-2}(x^5) \\ -\delta^{\alpha 2} b_2^{-2}(x^5) - \delta^{\alpha 3} b_3^{-2}(x^5)) I^{\beta\sigma}_{\mathrm{DSR}}(x^5) \\ -\delta^{\beta\sigma}(\delta^{\beta 0} b_0^{-2}(x^5) - \delta^{\beta 1} b_1^{-2}(x^5) \\ -\delta^{\beta 2} b_2^{-2}(x^5) - \delta^{\beta 3} b_3^{-2}(x^5)) I^{\alpha\rho}_{\mathrm{DSR}}(x^5); \end{cases} \tag{8.34}$$

4D Deformed space–time translation algebra $\mathrm{tr.}(1,3)_{\mathrm{DEF.}}$

$$[\Upsilon^{\mu}_{\mathrm{DSR}}(x^5), \Upsilon^{\nu}_{\mathrm{DSR}}(x^5)] = 0; \tag{8.35}$$

4D "Mixed" deformed space–time rototranslational algebra

$$\begin{cases} [I^{\mu\nu}_{\mathrm{DSR}}(x^5), \Upsilon^{\rho}_{\mathrm{DSR}}(x^5)] \\ = g^{\nu\rho}_{\mathrm{DSR}}(x^5)\Upsilon^{\mu}_{\mathrm{DSR}} - g^{\mu\rho}_{\mathrm{DSR}}(x^5)\Upsilon^{\nu}_{\mathrm{DSR}} \\ \overset{\text{ESC off}}{=} \delta^{\nu\rho}(b_0^{-2}(x^5)\delta^{\nu 0} - b_1^{-2}(x^5)\delta^{\nu 1} \\ -b_2^{-2}(x^5)\delta^{\nu 2} - b_3^{-2}(x^5)\delta^{\nu 3})\Upsilon^{\mu}_{\mathrm{DSR}}(x^5) \\ -\delta^{\mu\rho}(b_0^{-2}(x^5)\delta^{\mu 0} - b_1^{-2}(x^5)\delta^{\mu 1} \\ -b_2^{-2}(x^5)\delta^{\mu 2} - b_3^{-2}(x^5)\delta^{\mu 3})\Upsilon^{\nu}_{\mathrm{DSR}}(x^5), \end{cases} \tag{8.36}$$

or, in the "self-representation" basis of the deformed infinitesimal generators:

4D Deformed space–time rotation algebra $su(2)_{\mathrm{DEF.}} \otimes su(2)_{\mathrm{DEF.}}$

$$\begin{cases} [S^i_{\mathrm{DSR}}(x^5), S^j_{\mathrm{DSR}}(x^5)] \overset{\text{ESC off on } i \text{ and } j}{=} \\ = \left(\sum_{s=1}^{3}(1-\delta_{is})((1-\delta_{js})b_s^{-2}(x^5)\right) \epsilon_{ijk} S^k_{\mathrm{DSR}}(x^5) \\ = \epsilon_{ijk} b_k^{-2}(x^5) S^k_{\mathrm{DSR}}(x^5); \\ [K^i_{\mathrm{DSR}}(x^5), K^j_{\mathrm{DSR}}(x^5)] \\ = -b_0^{-2}(x^5)\epsilon_{ijk} S^k_{\mathrm{DSR}}(x^5); \\ [S^i_{\mathrm{DSR}}(x^5), K^j_{\mathrm{DSR}}(x^5)] \overset{\text{ESC off on } j}{=} \\ = \epsilon_{ijl} K^l_{\mathrm{DSR}}(x^5)\left(\sum_{s=1}^{3}\delta_{js} b_s^{-2}(x^5)\right) \\ \overset{\text{ESCoff on } j}{=} \epsilon_{ijl} b_j^{-2}(x^5) K^l_{\mathrm{DSR}}(x^5); \end{cases} \tag{8.37}$$

4D Deformed space–time translation algebra $\mathrm{tr}(1,3)_{\mathrm{DEF.}}$

$$[\Upsilon^{\mu}_{\mathrm{DSR}}(x^5), \Upsilon^{\nu}_{\mathrm{DSR}}(x^5)] = 0; \tag{8.38}$$

4D "Mixed" deformed space–time rototranslational algebra

$$\begin{cases} [K^i_{\mathrm{DSR}}(x^5), \Upsilon^0_{\mathrm{DSR}}(x^5)] = -b_0^{-2}(x^5)\Upsilon^i_{\mathrm{DSR}}(x^5); \\ [K^i_{\mathrm{DSR}}(x^5), \Upsilon^j_{\mathrm{DSR}}(x^5)] \\ \overset{\text{ESC off on } i}{=} -\delta^{ij} b_i^{-2}(x^5)\Upsilon^0_{\mathrm{DSR}}(x^5); \\ [S^i_{\mathrm{DSR}}(x^5), \Upsilon^0_{\mathrm{DSR}}(x^5)] = 0; \\ [S^i_{\mathrm{DSR}}(x^5), \Upsilon^k_{\mathrm{DSR}}(x^5)] \\ \overset{\text{ESC off on } k}{=} \epsilon_{ikl} b_k^{-2}(x^5)\Upsilon^l_{\mathrm{DSR}}(x^5). \end{cases} \tag{8.39}$$

### 8.2.3 *Infinitesimal and Finite Deformed Translations in* DSR

Let us consider the infinitesimal (i.e., algebraic) element[4] $\delta g \in \mathrm{tr.}(1,3)_{\mathrm{DEF.}} \subset ((su(2)_{\mathrm{DEF.}} \otimes su(2)_{\mathrm{DEF.}}) \otimes_s \mathrm{tr.}(1,3)_{\mathrm{DEF.}})$. It corresponds, in $\widetilde{M}(x^5)$, to a deformed, infinitesimal 4D space–time translation by the parametric, length-dimensioned (infinitesimal) *contravariant* four-vector

$$\begin{aligned} T^\mu_{\mathrm{DSR}}(g, x^5) &\equiv g^{\mu\rho}_{\mathrm{DSR}}(x^5) T_\rho(g) \\ &= \left(b_0^{-2}(x^5)T^0, b_1^{-2}(x^5)T^1, b_2^{-2}(x^5)T^2, b_3^{-2}(x^5)T^3\right). \end{aligned} \tag{8.40}$$

The $5 \times 5$ matrix $\mathcal{T}_{T^\mu_{\mathrm{DSR}}(g,x^5),\mathrm{DSR}}(x^5)$ representing such a translation is defined by

$$\begin{aligned} \mathcal{T}_{T^\mu_{\mathrm{DSR}}(g,x^5),\mathrm{DSR}}(x^5) &\equiv T^\mu_{\mathrm{DSR}}(g, x^5)\Upsilon_{\mu,(\mathrm{DSR})} \\ &= g^{\mu\rho}_{\mathrm{DSR}}(x^5)T_\rho(g)\Upsilon_{\mu,(\mathrm{DSR})} = T_\mu(g)\Upsilon^\mu_{\mathrm{DSR}}(x^5) \\ &= T^0(g)\Upsilon^0_{\mathrm{DSR}}(x^5) - T^1(g)\Upsilon^1_{\mathrm{DSR}}(x^5) - T^2(g)\Upsilon^2_{\mathrm{DSR}}(x^5) - T^3(g)\Upsilon^3_{\mathrm{DSR}}(x^5), \end{aligned} \tag{8.41}$$

(where the "DSR" in brackets in $\Upsilon_{\mu,(\mathrm{DSR})}$ means again that – as already stressed – the deformed translation infinitesimal *covariant* generators are independent of the (geo)metric context considered) or, explicitly

$$\mathcal{T}_{T^\mu_{\mathrm{DSR}}(g,x^5),\mathrm{DSR}}(x^5) = \begin{pmatrix} 0 & 0 & 0 & 0 & b_0^{-2}(x^5)T^0(g) \\ 0 & 0 & 0 & 0 & b_1^{-2}(x^5)T^1(g) \\ 0 & 0 & 0 & 0 & b_2^{-2}(x^5)T^2(g) \\ 0 & 0 & 0 & 0 & b_3^{-2}(x^5)T^3(g) \\ 0 & 0 & 0 & 0 & 0 \end{pmatrix}. \tag{8.42}$$

[4] For precision's sake, at the infinitesimal transformation level $\delta g \in \mathrm{tr.}(1,3)_{\mathrm{DEF.}} \subset ((su(2)_{\mathrm{DEF.}} \otimes su(2)_{\mathrm{DEF.}}) \times_s \mathrm{tr}(1,3)_{\mathrm{DEF.}})$ should be substituted for $g \in \mathrm{Tr.}(1,3)_{\mathrm{DEF.}} \subset P(1,3)_{\mathrm{DEF.}}$. But, for simplicity's sake, we will omit, but understand, this cumbersome notation.

Therefore, the 4D deformed, infinitesimal space–time translation by the (infinitesimal) vector (8.40) in $\widetilde{M}(x^5)$ – corresponding to $\delta g \in$ tr.$(1,3)_{\mathrm{DEF.}}$ – is given by (ESC on; for simplicity's sake, the dependence on $x$ is omitted):

$$\begin{pmatrix} x^{0\prime}(g,x^5) \\ x^{1\prime}(g,x^5) \\ x^{2\prime}(g,x^5) \\ x^{3\prime}(g,x^5) \\ y' \end{pmatrix} \overset{\text{ESC on}}{=} \left(1_{5D} + \mathcal{T}_{T^{\mu}_{\mathrm{DSR}}(g,x^5),\mathrm{DSR}}(x^5)\right)^{A}_{\ B} x^{B}$$

$$= \begin{pmatrix} 1 & 0 & 0 & 0 & b_0^{-2}(x^5)T^0(g) \\ 0 & 1 & 0 & 0 & b_1^{-2}(x^5)T^1(g) \\ 0 & 0 & 1 & 0 & b_2^{-2}(x^5)T^2(g) \\ 0 & 0 & 0 & 1 & b_3^{-2}(x^5)T^3(g) \\ 0 & 0 & 0 & 0 & 1 \end{pmatrix} \begin{pmatrix} x^0 \\ x^1 \\ x^2 \\ x^3 \\ (y=)\,1 \end{pmatrix}$$

$$= \begin{pmatrix} x^0 + b_0^{-2}(x^5)T^0(g) \\ x^1 + b_1^{-2}(x^5)T^1(g) \\ x^2 + b_2^{-2}(x^5)T^2(g) \\ x^3 + b_3^{-2}(x^5)T^3(g) \\ (y=)\,1 \end{pmatrix}. \tag{8.43}$$

At the finite transformation level, one has to evaluate the exponential of the matrix $\mathcal{T}_{T^{\mu}_{\mathrm{DSR}}(g,x^5),\mathrm{DSR}}(x^5)$, i.e., the $5 \times 5$ matrix $\exp\left(\mathcal{T}_{T^{\mu}_{\mathrm{DSR}}(g,x^5),\mathrm{DSR}}(x^5)\right)$, representing the finite (i.e., group) element $g \in \mathrm{Tr.}(1,3)_{\mathrm{DEF.}} \subset \mathrm{SO}(1,3)_{\mathrm{DEF.}}$ which corresponds to a deformed, finite 4D space–time translation by a parametric, length-dimensioned (finite) *contravariant* four-vector $T^{\mu}_{\mathrm{DSR}}(g,x^5)$:

$$\begin{aligned} \exp\left(\mathcal{T}_{T^{\mu}_{\mathrm{DSR}}(g,x^5),\mathrm{DSR}}(x^5)\right) &= \sum_{n=0}^{\infty} \frac{1}{n!} \left(\mathcal{T}_{T^{\mu}_{\mathrm{DSR}}(g,x^5),\mathrm{DSR}}(x^5)\right)^n \\ &= 1_{5D} + \mathcal{T}_{T^{\mu}_{\mathrm{DSR}}(g,x^5),\mathrm{DSR}}(x^5). \end{aligned} \tag{8.44}$$

Then, as anticipated in Sect. 8.1, for translation transformations the only difference between the infinitesimal level and the finite one is provided by the infinitesimal or finite nature of the *contravariant* translation parameters $T^{\mu}_{\mathrm{DSR}}(g,x^5)$. Such a result is a peculiar feature of the space–time translation component of the 4D deformed Poincaré group $P(1,3)_{\mathrm{DEF.}}$, and in general of translation coordinate transformations in $N$-d generalized Minkowski spaces $\widetilde{M_N}\left(\{x\}_{\mathrm{n.m.}}\right)$. Still in the case of DSR, it can be recovered also by exploiting the abelian nature of $\mathrm{Tr.}(1,3)_{\mathrm{DEF.}}$ and the property of the powers of its infinitesimal generators (see (8.20)), and by using the Baker–Campbell–Hausdorff formula; one has indeed

$$
\begin{aligned}
&\exp\left(\mathcal{T}_{T^{\mu}_{\mathrm{DSR}}(g,x^5),\mathrm{DSR}}(x^5)\right) = \exp(T^0(g)\Upsilon^0_{\mathrm{DSR}}(x^5) - T^1(g)\Upsilon^1_{\mathrm{DSR}}(x^5)\\
&\quad -T^2(g)\Upsilon^2_{\mathrm{DSR}}(x^5) - T^3(g)\Upsilon^3_{\mathrm{DSR}}(x^5)) = \exp\left(T^0(g)\Upsilon^0_{\mathrm{DSR}}(x^5)\right)\\
&\quad \times \exp\left(T^1(g)\Upsilon^1_{\mathrm{DSR}}(x^5)\right) \times \exp\left(T^2(g)\Upsilon^2_{\mathrm{DSR}}(x^5)\right) \times \exp\left(T^3(g)\Upsilon^3_{\mathrm{DSR}}(x^5)\right)\\
&= \left(\sum_{n=0}^{\infty} \frac{(T^0(g))^n}{n!}\left(\Upsilon^0_{\mathrm{DSR}}(x^5)\right)^n\right) \times \left(\sum_{n=0}^{\infty} \frac{(T^1(g))^n}{n!}\left(\Upsilon^1_{\mathrm{DSR}}(x^5)\right)^n\right)\\
&\quad \times \left(\sum_{n=0}^{\infty} \frac{(T^2(g))^n}{n!}\left(\Upsilon^2_{\mathrm{DSR}}(x^5)\right)^n\right) \times \left(\sum_{n=0}^{\infty} \frac{(T^3(g))^n}{n!}\left(\Upsilon^3_{\mathrm{DSR}}(x^5)\right)^n\right)\\
&= \left(1_{5D} + T^0(g)\Upsilon^0_{\mathrm{DSR}}(x^5)\right) \times \left(1_{5D} + T^1(g)\Upsilon^1_{\mathrm{DSR}}(x^5)\right)\\
&\quad \times \left(1_{5D} + T^2(g)\Upsilon^2_{\mathrm{DSR}}(x^5)\right) \times \left(1_{5D} + T^3(g)\Upsilon^3_{\mathrm{DSR}}(x^5)\right)\\
&= 1_{5D} + T^0(g)\Upsilon^0_{\mathrm{DSR}}(x^5) + T^1(g)\Upsilon^1_{\mathrm{DSR}}(x^5) + T^2(g)\Upsilon^2_{\mathrm{DSR}}(x^5)\\
&\quad +T^3(g)\Upsilon^3_{\mathrm{DSR}}(x^5)\\
&= 1_{5D} + \mathcal{T}_{T^{\mu}_{\mathrm{DSR}}(g,x^5),\mathrm{DSR}}(x^5).
\end{aligned} \tag{8.45}
$$

On account of the noncommutativity of the infinitesimal generators of $\mathrm{Tr.}(1,3)_{\mathrm{DEF.}}$ and of $\mathrm{SO}(1,3)_{\mathrm{DEF.}}$ (see (8.30),(8.31)), and by exploiting again the BCH formula, it is possible to state the following inequality for the 5D matrix representing the finite (i.e., group) element $g \in P(1,3)_{\mathrm{DEF.}}$ corresponding to a finite, 4D deformed space–time rototranslation in $\widetilde{M}(x^5)$, of dimensionless, parametric deformed angular (Euclidean) three-vector $\boldsymbol{\theta}(g)$, dimensionless parametric deformed rapidity (Euclidean) three-vector $\boldsymbol{\zeta}(g)$ and "length-dimensioned," parametric, translational contravariant (deformed) four-vector $T^{\mu}_{\mathrm{DSR}}(g,x^5)$:[5]

$$
\begin{aligned}
&\exp\left(-\boldsymbol{\theta}(g)\cdot \mathbf{S}_{\mathrm{DSR}}(x^5) - \boldsymbol{\zeta}(g)\cdot \mathbf{K}_{\mathrm{DSR}}(x^5) + T_{\mu}(g)\Upsilon^{\mu}_{\mathrm{DSR}}(x^5)\right)\\
&\neq \exp\left(-\boldsymbol{\theta}(g)\cdot \mathbf{S}_{\mathrm{DSR}}(x^5) - \boldsymbol{\zeta}(g)\cdot \mathbf{K}_{\mathrm{DSR}}(x^5)\right) \times \exp\left(T_{\mu}(g)\Upsilon^{\mu}_{\mathrm{DSR}}(x^5)\right)\\
&= \exp\left(-\boldsymbol{\theta}(g)\cdot \mathbf{S}_{\mathrm{DSR}}(x^5) - \boldsymbol{\zeta}(g)\cdot \mathbf{K}_{\mathrm{DSR}}(x^5)\right) \times \exp\left(\mathcal{T}_{T^{\mu}_{\mathrm{DSR}}(g,x^5),\mathrm{DSR}}(x^5)\right)\\
&= \exp\left(-\boldsymbol{\theta}(g)\cdot \mathbf{S}_{\mathrm{DSR}}(x^5) - \boldsymbol{\zeta}(g)\cdot \mathbf{K}_{\mathrm{DSR}}(x^5)\right) \times \left(1_{5D} + \mathcal{T}_{T^{\mu}_{\mathrm{DSR}}(g,x^5),\mathrm{DSR}}(x^5)\right)
\end{aligned}
$$

[5] Let us instead notice that, at infinitesimal level, all the transformations of the Lie group $P(1,3)_{\mathrm{DEF.}}$ commute.

$$\neq \exp\left(-\boldsymbol{\theta}(g)\cdot \mathbf{S}_{\mathrm{DSR}}(x^5)\right) \times \exp\left(-\boldsymbol{\zeta}(g)\cdot \mathbf{K}_{\mathrm{DSR}}(x^5)\right)$$
$$\times \left(1_{5D} + \mathcal{T}_{T^{\mu}_{\mathrm{DSR}}(g,x^5),\mathrm{DSR}}(x^5)\right)$$
$$\neq \exp\left(-\theta^1(g) S^1_{\mathrm{DSR}}(x^5)\right) \times \exp\left(-\theta^2(g) S^2_{\mathrm{DSR}}(x^5)\right)$$
$$\times \exp\left(-\theta^3(g) S^3_{\mathrm{DSR}}(x^5)\right) \times \exp\left(-\zeta^1(g) K^1_{\mathrm{DSR}}(x^5)\right)$$
$$\times \exp\left(-\zeta^2(g) K^2_{\mathrm{DSR}}(x^5)\right) \times \exp\left(-\zeta^3(g) K^3_{\mathrm{DSR}}(x^5)\right)$$
$$\times \left(1_{5D} + \mathcal{T}_{T^{\mu}_{\mathrm{DSR}}(g,x^5),\mathrm{DSR}}(x^5)\right), \qquad (8.46)$$

where in the last two lines the noncommutativity of deformed true rotation and boost generators has been used (see (8.36) and (8.39)).

By comparing (8.43) with the expression of a translation in the usual Minkowski space $M$ of SR (as before ESC on and, for simplicity's sake, dependence on $\{x_{m.}\}$ is omitted)

$$\begin{pmatrix} x^{0\prime}(g) \\ x^{1\prime}(g) \\ x^{2\prime}(g) \\ x^{3\prime}(g) \\ y' \end{pmatrix} = \left(1_{5D} + \mathcal{T}_{T^{\mu}_{\mathrm{SR}}(g),\mathrm{SR}}\right)^{A}{}_{B}\, x^B$$
$$= \begin{pmatrix} 1 & 0 & 0 & 0 & T^0(g) \\ 0 & 1 & 0 & 0 & T^1(g) \\ 0 & 0 & 1 & 0 & T^3(g) \\ 0 & 0 & 0 & 1 & T^3(g) \\ 0 & 0 & 0 & 0 & 1 \end{pmatrix} \begin{pmatrix} x^0 \\ x^1 \\ x^2 \\ x^3 \\ (y=)\,1 \end{pmatrix} = \begin{pmatrix} x^0 + T^0(g) \\ x^1 + T^1(g) \\ x^2 + T^2(g) \\ x^3 + T^3(g) \\ (y=)\,1 \end{pmatrix}, \qquad (8.47)$$

it is easily seen that passing from SR to DSR – i.e., locally deforming and spatially anisotropizing $M$ – amounts to the following parameter change (as far as space–time translations are concerned)

$$\begin{aligned} T^{\mu}_{\mathrm{SR}}(g) &= \left(T^0(g), T^1(g), T^2(g), T^3(g)\right) \to T^{\mu}_{\mathrm{DSR}}(g,x^5) \\ &= \left(b_0^{-2}(x^5) T^0(g), b_0^{-2}(x^5) T^1(g), b_0^{-2}(x^5) T^2(g), b_0^{-2}(x^5) T^3(g)\right) \\ &\equiv \widetilde{T}^{\mu}_{\mathrm{DSR}}(g,x^5). \end{aligned} \qquad (8.48)$$

Then, extending the meaning of effective transformation parameters (cf. (7.40) and (7.65) for the effective rapidities and angles, respectively) to the translation ones, we can say that in the translational case the ("length-dimensioned") deformed translation parameter four-vector $T^{\mu}_{\mathrm{DSR}}(g,x^5)$ coincides with the *effective* ("length-dimensioned")[6] deformed translation

[6] The contravariant four-vector identity (8.49), *at index* – and not component – *level*, is due to the very fact that the passage SR→DSR:
$g_{\mu\nu,\mathrm{SR}} = \mathrm{diag}\,(1,-1,-1,-1) \to g_{\mu\nu,\mathrm{DSR}}(x^5) = \mathrm{diag}\left(b_0^2(x^5), -b_1^2(x^5), -b_2^2(x^5), -b_3^2(x^5)\right)$

parameter four-vector $\widetilde{T}^{\mu}_{\text{DSR}}(g, x^5)$:

$$T^{\mu}_{\text{DSR}}(g, x^5) = \widetilde{T}^{\mu}_{\text{DSR}}(g, x^5). \tag{8.49}$$

This is a peculiar feature of the translation component $\text{Tr.}(1,3)_{\text{DEF.}}$ of the 4D deformed Poincaré group $P(1,3)_{\text{DEF.}}$, and it is due to the following fact. While for 4D space–time rotations (homogeneous coordinate transformations) the deformed transformation parameter three-vectors $\boldsymbol{\theta}(g)$ and $\boldsymbol{\zeta}(g)$ are Euclidean (see Chap. 7), in the case of 4D deformed translations (inhomogeneous coordinate transformations) the translation parameter (*contravariant*) four-vector $T^{\mu}_{\text{DSR}}(g, x^5)$ is "length-dimensioned" and deformed, i.e., dependent on the metric structure under consideration (as already pointed out).

Let us finally stress that the (local) "deforming anisotropizing" generalization of SR corresponding to DSR is fully self-consistent at space–time translation level, too. Namely, all the results obtained for the Lie group $\text{Tr.}(1,3)_{\text{DEF.}}$ of space–time deformed translations in the 4D "deformed" Minkowski space $\widetilde{M}(x^5)$ reduce, in the limit $g_{\mu\nu,\text{DSR}}(x^5) \to g_{\mu\nu,\text{SR}}$ (i.e., for $b^2_{\mu}(x^5) \to 1, \forall\mu = 0,1,2,3$), to the well-known special-relativistic results concerning the Lie group $\text{Tr.}(1,3)_{\text{STD.}}$ of space–time translations in the standard 4D Minkowski space $M$.

---

preserves the diagonality of the 2-rank, symmetric metric 4-tensor, still destroying its isochrony and spatial isotropy.

# 9

# Deformed Minkowski Space as Generalized Lagrange Space

## 9.1 Generalized Lagrange Spaces

We want now to show that the deformed Minkowski space $\widetilde{M}$ of DSR does possess another well-defined geometrical structure, besides the deformed metrical one. Precisely, we will show (following [44]) that $\widetilde{M}$ is a *generalized Lagrange space*.

Let us give the definition of generalized Lagrange space [12], since usually one is not acquainted with it.

Consider a $N$-dimensional, differentiable manifold $\mathcal{M}$ and its ($N$-dimensional) tangent space in a point, $T\mathcal{M}_{\mathbf{x}}$ ($\mathbf{x} \in \mathcal{M}$). As is well known, the union

$$\bigcup_{\mathbf{x}\in\mathcal{M}} T\mathcal{M}_{\mathbf{x}} \equiv T\mathcal{M} \tag{9.1}$$

has a fiber bundle structure. Let us denote by $\mathbf{y}$ the generic element of $T\mathcal{M}_{\mathbf{x}}$, namely a vector tangent to $\mathcal{M}$ in $\mathbf{x}$. Then, an element $u \in T\mathcal{M}$ is a vector tangent to the manifold in some point $\mathbf{x} \in \mathcal{M}$. Local coordinates for $T\mathcal{M}$ are introduced by considering a local coordinate system $(x^1, x^2, ..., x^N)$ on $\mathcal{M}$ and the components of $y$ in such a coordinate system $(y^1, y^2, ..., y^N)$. The $2N$ numbers $(x^1, x^2, ..., x^N, y^1, y^2, ..., y^N)$ constitute a local coordinate system on $T\mathcal{M}$. We can write synthetically $u = (\mathbf{x}, \mathbf{y})$. $T\mathcal{M}$ is a $2N$-dimensional, differentiable manifold.

Let $\pi$ be the mapping (*natural projection*) $\pi : u = (\mathbf{x}, \mathbf{y}) \longrightarrow \mathbf{x}$. ($\mathbf{x} \in \mathcal{M}$, $\mathbf{y} \in T\mathcal{M}_{\mathbf{x}}$). Then, the tern $(T\mathcal{M}, \pi, \mathcal{M})$ is the *tangent bundle* to the base manifold $\mathcal{M}$. The image of the inverse mapping $\pi^{-1}(\mathbf{x})$ is of course the tangent space $T\mathcal{M}_{\mathbf{x}}$, which is called the *fiber corresponding to the point* $\mathbf{x}$ *in the fiber bundle* One considers also sometimes the manifold $\widetilde{T\mathcal{M}} = T\mathcal{M}/\{0\}$, where 0 is the zero section of the projection $\pi$. We do not dwell further on the theory of the fiber bundles, and refer the reader to the wide and excellent literature on the subject [46].

The natural basis of the tangent space $T_u(T\mathcal{M})$ at a point

$$u = (\mathbf{x}, \mathbf{y}) \in T\mathcal{M} \text{is} \left\{ \frac{\partial}{\partial x^i}, \frac{\partial}{\partial y^j} \right\}, i, j = 1, 2, ..., N.$$

A local coordinate transformation in the differentiable manifold $T\mathcal{M}$ reads

$$\begin{cases} x'^i = x'^i(\mathbf{x}), \qquad \det\left(\dfrac{\partial x'^i}{\partial x^j}\right) \neq 0, \\ \qquad\qquad y'^i = \dfrac{\partial x'^i}{\partial x^j} y^j. \end{cases} \tag{9.2}$$

Here, $y^i$ is *the Liouville vector field* on $T\mathcal{M}$, i.e., $y^i \dfrac{\partial}{\partial y^i}$.

On account of (9.2), the natural basis of $T\mathcal{M}_{\mathbf{x}}$ can be written as:

$$\begin{cases} \dfrac{\partial}{\partial x^i} = \dfrac{\partial x'^k}{\partial x^i} \dfrac{\partial}{\partial x'^k} + \dfrac{\partial y'^k}{\partial x^i} \dfrac{\partial}{\partial y'^k}, \\ \\ \dfrac{\partial}{\partial y^j} = \dfrac{\partial y'^k}{\partial y^j} \dfrac{\partial}{\partial y'^k}. \end{cases} \tag{9.3}$$

Second (9.3) shows therefore that the vector basis $(\partial/\partial y^j)$, $j = 1, 2, ..., N$, generates a distribution $\mathcal{V}$ defined everywhere on $T\mathcal{M}$ and integrable, too (*vertical distribution on* $T\mathcal{M}$).

If $\mathcal{H}$ is a distribution on $T\mathcal{M}$ supplementary to $\mathcal{V}$, namely

$$T_u(T\mathcal{M}) = \mathcal{H}_u \oplus \mathcal{V}_u \quad , \quad \forall u \in T\mathcal{M}, \tag{9.4}$$

then $\mathcal{H}$ is called a *horizontal distribution*, or a *nonlinear connection* on $T\mathcal{M}$. A basis for the distributions $\mathcal{H}$ and $\mathcal{V}$ are given, respectively, by $\delta/\delta x^i$ and $\partial/\partial y^j$, where the basis in $\mathcal{H}$ explicitly reads

$$\frac{\delta}{\delta x^i} = \frac{\partial}{\partial x^i} - H^j_{\ i}(\mathbf{x}, \mathbf{y}) \frac{\partial}{\partial y^j}. \tag{9.5}$$

Here, $H^j_i(\mathbf{x}, \mathbf{y})$ are the *coefficients* of the nonlinear connection $\mathcal{H}$. The basis

$$\left\{ \frac{\delta}{\delta x^i}, \frac{\partial}{\partial y^j} \right\} = \left\{ \delta_i, \dot{\partial}_j \right\}$$

is called the *adapted basis*.

The dual basis to the adapted basis is $\{dx^i, \delta y^j\}$, with

$$\delta y^j = dy^j + H^j_{\ i}(\mathbf{x}, \mathbf{y})dx^i. \tag{9.6}$$

A *distinguished tensor* (or *d-tensor*) *field of (r,s)-type* is a quantity whose components transform like a tensor under the first coordinate transformation (9.2) on $T\mathcal{M}$ (namely they change as tensor in $\mathcal{M}$). For instance, for a $d$-tensor of type (1, 2):

$$R'^i_{jk} = \frac{\partial x'^i}{\partial x^s}\frac{\partial x^r}{\partial x'^j}\frac{\partial x^p}{\partial x'^k}R^s_{rp}. \tag{9.7}$$

In particular, both $\delta/\delta x^i$ and $\partial/\partial y^j$ are $d$-(covariant) vectors, whereas $dx^i$, $\delta y^j$ are $d$-(contravariant) vectors.

A *generalized Lagrange space* is a pair $\mathcal{GL}^N$=($\mathcal{M}$, $g_{ij}(\mathbf{x}, \mathbf{y})$), with $g_{ij}(\mathbf{x}, \mathbf{y})$ being a $d$-tensor of type (0,2) (covariant) on the manifold $T\mathcal{M}$, which is symmetric, nondegenerate[1] and of constant signature.

A function

$$L : (\mathbf{x}, \mathbf{y}) \in T\mathcal{M} \rightarrow L(\mathbf{x}, \mathbf{y}) \in \mathcal{R} \tag{9.8}$$

differentiable on $\widehat{T\mathcal{M}}$ and continuous on the null section of $\pi$ is named a *regular Lagrangian* if the Hessian of $L$ with respect to the variables $y^i$ is nonsingular. A generalized Lagrange space $\mathcal{GL}^N$=($\mathcal{M}$, $g_{ij}(\mathbf{x}, \mathbf{y})$) is reducible to a *Lagrange space* $\mathcal{L}^N$ if there is a regular Lagrangian $L$ satisfying

$$g_{ij} = \frac{1}{2}\frac{\partial^2 L}{\partial y^i \partial y^j} \tag{9.9}$$

on $\widehat{T\mathcal{M}}$ . In order that $\mathcal{GL}^N$ is reducible to a Lagrange space, a necessary condition is the total symmetry of the $d$-tensor $(\partial g_{ij}/\partial y^k)$. If such a condition is satisfied, and $g_{ij}$ are 0-homogeneous in the variables $y^i$, then the function $L = g_{ij}(\mathbf{x}, \mathbf{y})y^i y^j$ is a solution of the system (9.9). In this case, the pair $(\mathcal{M}, L)$ is a *Finsler space*[2] defined for $\mathbf{x} \in \mathcal{M}$ and $\boldsymbol{\xi} \in T_{\mathbf{x}}\mathcal{M}$ such that $\Phi(\mathbf{x}, \cdot)$ is a possibly nonsymmetric norm on $T_{\mathbf{x}}\mathcal{M}$.

Notice that every Riemann manifold ($\mathcal{M}$, $\mathbf{g}$) is also a Finsler space, the norm $\Phi(\mathbf{x}, \boldsymbol{\xi})$ being the norm induced by the scalar product $\mathbf{g}(\mathbf{x})$.

A finite dimensional Banach space is another simple example of Finsler space, where $\Phi(\mathbf{x}, \boldsymbol{\xi}) \equiv \|\boldsymbol{\xi}\|$ . $(\mathcal{M}, \Phi)$, with $\Phi^2 = L$. One says that $\mathcal{GL}^N$ is reducible to a Finsler space.

Of course, $\mathcal{GL}^N$ reduces to a pseudo-Riemannian (or Riemannian) space $(\mathcal{M}, g_{ij}(\mathbf{x}))$ if the $d$-tensor $g_{ij}(\mathbf{x}, \mathbf{y})$ does not depend on $\mathbf{y}$. On the contrary, if $g_{ij}(\mathbf{x}, \mathbf{y})$ depends only on $\mathbf{y}$ (at least in preferred charts), it is a generalized Lagrange space which is locally Minkowski.

---

[1]Namely it must be $rank \|g_{ij}(\mathbf{x}, \mathbf{y})\| = N$.

[2]Let us recall that a Finsler space [7] is a couple ($\mathcal{M}$, $\Phi$), where $\mathcal{M}$ is be an $N$-dimensional differential manifold and $\Phi : T\mathcal{M} \Rightarrow \mathcal{R}$ a function $\Phi(\mathbf{x}, \boldsymbol{\xi})$

Since, in general, a generalized Lagrange space is not reducible to a Lagrange one, it cannot be studied by means of the methods of symplectic geometry, on which – as is well known – analytical mechanics is based.

A linear $\mathcal{H}-$connection on $T\mathcal{M}$ (or on $\widetilde{T\mathcal{M}}$) is defined by a couple of geometrical objects $\mathcal{C}\Gamma(\mathcal{H}) = (L^i{}_{jk}, C^i{}_{jk})$ on $T\mathcal{M}$ with different transformation properties under the coordinate transformation (9.2). Precisely, $L^i{}_{jk}(\mathbf{x}, \mathbf{y})$ transform like the coefficients of a linear connection on $\mathcal{M}$, whereas $C^i_{jk}(\mathbf{x}, \mathbf{y})$ transform like a $d$-tensor of type (1,2). $\mathcal{C}\Gamma(\mathcal{H})$ is called *the metrical canonical* $\mathcal{H}-$*connection* of the generalized Lagrange space $\mathcal{GL}^N$.

In terms of $L^i_{jk}$ and $C^i_{jk}$ one can define two kinds of covariant derivatives: a *covariant horizontal (h-) derivative*, denoted by "$_{\shortmid}$," and a *covariant vertical (v-) derivative*, denoted by "$|$." For instance, for the $d$-tensor $g_{ij}(\mathbf{x}, \mathbf{y})$ one has

$$\begin{cases} g_{ij_{\shortmid}k} = \dfrac{\delta g_{ij}}{\delta x^k} - g_{sj} L^s{}_{ik} - g_{is} L^s{}_{jk}; \\[2ex] g_{ij|k} = \dfrac{\partial g_{ij}}{\partial x^k} - g_{sj} C^s{}_{ik} - g_{is} C^s{}_{jk}. \end{cases} \tag{9.10}$$

The two derivatives $g_{ij_{\shortmid}k}$ and $g_{ij|k}$ are both $d$-tensors of type (0,3).

The coefficients of $\mathcal{C}\Gamma(\mathcal{H})$ can be expressed in terms of the following *generalized Christoffel symbols*:

$$\begin{cases} L^i{}_{jk} = \frac{1}{2} g^{is} \left( \dfrac{\delta g_{sj}}{\delta x^k} + \dfrac{\delta g_{ks}}{\delta x^j} + \dfrac{\delta g_{jk}}{\delta x^s} \right); \\[2ex] C^i{}_{jk} = \frac{1}{2} g^{is} \left( \dfrac{\partial g_{sj}}{\partial x^k} + \dfrac{\partial g_{ks}}{\partial x^j} + \dfrac{\partial g_{jk}}{\partial x^s} \right). \end{cases} \tag{9.11}$$

Moreover, by means of the connection $\mathcal{C}\Gamma(\mathcal{H})$ it is possible to define a *d-curvature* in $T\mathcal{M}$ by means of the tensors $R_j{}^i{}_{kh}$, $S_j{}^i{}_{kh}$ and $P_j{}^i{}_{kh}$ given by

$$\begin{aligned} R_j{}^i{}_{kh} &= \frac{\delta L^i{}_{jk}}{\delta x^h} - \frac{\delta L^i{}_{jh}}{\delta x^k} + L^r{}_{jk} L^i{}_{rh} - L^r{}_{jh} L^i{}_{rk} + C^i{}_{jr} R^r{}_{kh}; \\ S_j{}^i{}_{kh} &= \frac{\partial C^i{}_{jk}}{\partial y^h} - \frac{\partial C^i{}_{jh}}{\partial y^k} + C^r{}_{jk} C^i{}_{rh} - C^r{}_{jh} C^i{}_{rk}; \\ P_j{}^i{}_{kh} &= \frac{\partial L^i{}_{jk}}{\partial y^h} - C^i{}_{j_{\shortmid}h} + C^i{}_{jr} P^r{}_{kh}. \end{aligned} \tag{9.12}$$

Here, the $d$-tensor $R^i_{jh}$ is related to the bracket of the basis $\delta/\delta x^i$:

$$\left[ \frac{\delta}{\delta x^i}, \frac{\delta}{\delta x^j} \right] = R^s{}_{ij} \frac{\partial}{\partial y^s} \tag{9.13}$$

and is explicitly given by[3]

$$R^i{}_{jk} = \frac{\delta H^i{}_j}{\delta x^k} - \frac{\delta H^i{}_k}{\delta x^j}. \tag{9.14}$$

The tensor $P^i{}_{jk}$, together with $T^i{}_{jk}$, $S^i{}_{jk}$, defined by

$$\begin{aligned} P^i{}_{jk} &= \frac{\partial H^i{}_j}{\partial y^k} - L^i{}_{jk}; \\ T^i{}_{jk} &= L^i{}_{jk} - L^i{}_{kj}; \\ S^i{}_{jk} &= C^i{}_{jk} - C^i{}_{kj} \end{aligned} \tag{9.15}$$

are *the d-tensors of torsion of the metrical connection* $C\Gamma(\mathcal{H})$.

From the curvature tensors one can get the corresponding Ricci tensors of $C\Gamma(\mathcal{H})$:

$$\begin{cases} R_{ij} = R_i{}^s{}_{js}; & S_{ij} = S_i{}^s{}_{js}; \\ \overset{1}{P}_{ij} = P_i{}^s{}_{js} & \overset{2}{P}_{ij} = P_i{}^s{}_{sj}, \end{cases} \tag{9.16}$$

and the scalar curvatures

$$R = g^{ij} R_{ij}; \quad S = g^{ij} S_{ij}. \tag{9.17}$$

Finally, *the deflection d-tensors associated to the connection* $C\Gamma(\mathcal{H})$ are

$$\begin{cases} D^i{}_j = y^i_{|j} = -H^i{}_j + y^s L^i{}_{sj}; \\ d^i{}_j = y^i|_j = \delta^i{}_j + y^s C^i_{sj}, \end{cases} \tag{9.18}$$

namely the $h$- and $v$-covariant derivatives of the Liouville vector fields.

In the generalized Lagrange space $\mathcal{GL}^N$ it is possible to write the Einstein equations with respect to the canonical connection $C\Gamma(\mathcal{H})$ as follows:

$$\begin{cases} R_{ij} - \frac{1}{2} R g_{ij} = \kappa \overset{H}{T}_{ij}; & \overset{1}{P}_{ij} = \kappa \overset{1}{T}_{ij}; \\ S_{ij} - \frac{1}{2} S g_{ij} = \kappa \overset{V}{T}_{ij}; & \overset{2}{P}_{ij} = \kappa \overset{2}{T}_{ij}, \end{cases} \tag{9.19}$$

where $\kappa$ is a constant and $\overset{H}{T}_{ij}$, $\overset{V}{T}_{ij}$, $\overset{1}{T}_{ij}$, $\overset{2}{T}_{ij}$ are the components of the energy-momentum tensor.

[3] $R^i_{jk}$ plays the role of a curvature tensor of the nonlinear connection $\mathcal{H}$. The corresponding tensor of torsion is instead

$$t^i_{jk} = \frac{\partial H^i_j}{\partial y^k} - \frac{\partial H^i_k}{\partial y^j}.$$

## 9.2 Generalized Lagrangian Structure of $\widetilde{M}$

On the basis of the previous considerations, let us analyze the geometrical structure of the deformed Minkowski space of DSR $\widetilde{M}$, endowed with the by now familiar metric $g_{\mu\nu,\mathrm{DSR}}(E)$. As explained in Part I, $E$ is the energy of the process measured by the detectors in Minkowskian conditions. Therefore, $E$ is a function of the velocity components, $u^\mu = \mathrm{d}x^\mu/\mathrm{d}\tau$, where $\tau$ is the (Minkowskian) proper time:

$$E = E\left(\frac{\mathrm{d}x^\mu}{\mathrm{d}\tau}\right). \tag{9.20}$$

The derivatives $\mathrm{d}x^\mu/\mathrm{d}\tau$ define a contravariant vector tangent to $M$ at $x$, namely they belong to $TM_{\mathbf{x}}$. We shall denote this vector (according to the notation of Sect. 9.1) by $\mathbf{y} = (y^\mu)$. Then, $(\mathbf{x},\mathbf{y})$ is a point of the tangent bundle to $M$. We can therefore consider the generalized Lagrange space $\mathcal{GL}^4 = (M, g_{\mu\nu}(\mathbf{x},\mathbf{y}))$, with

$$\begin{cases} g_{\mu\nu}(\mathbf{x},\mathbf{y}) = g_{\mu\nu,\mathrm{DSR}}(E(\mathbf{x},\mathbf{y})), \\ \\ \qquad E(\mathbf{x},\mathbf{y}) = E(\mathbf{y}). \end{cases} \tag{9.21}$$

Then, it is possible to prove the following theorem:

*The pair $\mathcal{GL}^4 = (M, g_{\mu\nu,\mathrm{DSR}}(\mathbf{x},\mathbf{y})) \equiv \widetilde{M}$ is a generalized Lagrange space which is not reducible to a Riemann space, or to a Finsler space, or to a Lagrange space.*

We already proved the first statement in Sect. 2.2 of Part I, on account of the dependence of the deformed metric tensor on $E$ (and therefore on $\mathbf{y}$) only: $g_{\mu\nu,\mathrm{DSR}}(\mathbf{x},\mathbf{y}) \equiv g_{\mu\nu,\mathrm{DSR}}(\mathbf{y})$. To prove that $\mathcal{GL}^4$ is reducible neither to a Lagrange space nor to a Finsler one, it is sufficient (as stated in Sect. 9.1) to show that the $d$-tensor field $(\partial g_{\mu\nu,\mathrm{DSR}}/\partial y^\rho)$ is not totally symmetric, i.e., the equation

$$\frac{\partial g_{\mu\nu,\mathrm{DSR}}}{\partial y^\rho} = \frac{\partial g_{\mu\rho,\mathrm{DSR}}}{\partial y^\nu} \tag{9.22}$$

does not hold. Let us assume *by absurdum* that (9.22) is satisfied. Then, for $\mu = \nu \neq \rho$, one gets

$$\frac{\partial g_{\mu\mu,\mathrm{DSR}}}{\partial y^\rho} = \frac{\partial g_{\mu\rho,\mathrm{DSR}}}{\partial y^\mu} \tag{9.23}$$

whence

$$\frac{\partial g_{\mu\nu,\mathrm{DSR}}}{\partial y^\rho} = 0, \qquad \mu \neq \rho \tag{9.24}$$

(since $g_{\mathrm{DSR},\mu\nu}$ is diagonal). Equation (9.24) implies

$$\frac{\partial g_{\mu\mu,\mathrm{DSR}}}{\partial E}\frac{\partial E}{\partial y^\rho} = 0, \quad \forall \mu, \rho; \mu \neq \rho \Longrightarrow \qquad \frac{\partial E}{\partial y^\rho} = 0, \tag{9.25}$$

which is impossible. This proves the theorem. Notice that such a result is strictly related to the fact that the deformed metric tensor of DSR is diagonal, and therefore it does not hold, in general, for the generalized Minkowski spaces we defined in Chapter 5.

If an external electromagnetic field $F_{\mu\nu}$ is present in the Minkowski space $M$, in $\widetilde{M}$ the deformed electromagnetic field is given by $\widetilde{F}^{\mu}_{\ \nu}(\mathbf{x},\mathbf{y}) = g^{\mu\rho}_{\mathrm{DSR}} F_{\rho\nu}(\mathbf{x})$ (see Sect. 3.5). Such a field is a $d$-tensor and is called *the electromagnetic tensor of the generalized Lagrange space.* Then, the nonlinear connection $\mathcal{H}$ is given by

$$H^{\mu}_{\ \nu} = \left\{ \begin{array}{c} \mu \\ \nu\rho \end{array} \right\} y^{\rho} - \widetilde{F}^{\mu}_{\ \nu}(\mathbf{x},\mathbf{y}), \tag{9.26}$$

where $\left\{ \begin{array}{c} \mu \\ \nu\rho \end{array} \right\}$, the Christoffel symbols of the Minkowski metric $g_{\mu\nu}$, are zero, so that

$$H^{\mu}_{\ \nu} = -\widetilde{F}^{\mu}_{\ \nu}(\mathbf{x},\mathbf{y}). \tag{9.27}$$

The adapted basis of the distribution $\mathcal{H}$ reads therefore

$$\frac{\delta}{\delta x^{\mu}} = \frac{\partial}{\partial x^{\mu}} + \widetilde{F}^{\nu}_{\ \mu}(\mathbf{x},\mathbf{y}).\frac{\partial}{\partial y^{\nu}}. \tag{9.28}$$

The local covector field of the dual basis (cf. (9.6)) is given by

$$\delta y^{\mu} = \mathrm{d}y^{\mu} - \widetilde{F}^{\mu}_{\ \nu}(\mathbf{x},\mathbf{y})\mathrm{d}x^{\nu}. \tag{9.29}$$

## 9.3 Canonical Metric Connection of $\widetilde{M}$

The derivation operators applied to the deformed metric tensor of the space $\mathcal{GL}^4 = \widetilde{M}$ yield

$$\frac{\delta g_{\mu\nu,\mathrm{DSR}}}{\delta x^{\rho}} = \frac{\partial g_{\mu\nu,\mathrm{DSR}}}{\partial x^{\rho}} + \widetilde{F}^{\sigma}_{\ \rho}\frac{\partial g_{\mu\nu,\mathrm{DSR}}}{\partial y^{\sigma}} = \widetilde{F}^{\sigma}_{\ \rho}\frac{\partial g_{\mu\nu,\mathrm{DSR}}}{\partial E}\frac{\partial E}{\partial y^{\sigma}}, \tag{9.30}$$

$$\frac{\partial g_{\mu\nu,\mathrm{DSR}}}{\partial y^{\sigma}} = \frac{\partial g_{\mu\nu,\mathrm{DSR}}}{\partial E}\frac{\partial E}{\partial y^{\sigma}}. \tag{9.31}$$

Then, the coefficients of the canonical metric connection $C\Gamma(\mathcal{H})$ in $\widetilde{M}$ (see (9.11)) are given by

$$\left\{ \begin{array}{l} L^{\mu}_{\ \nu\rho} = \frac{1}{2} g^{\mu\sigma}_{\mathrm{DSR}} \frac{\partial E}{\partial y^{\alpha}} \left( \frac{\partial g_{\sigma\nu,\mathrm{DSR}}}{\partial E} \widetilde{F}^{\alpha}_{\ \rho} + \frac{\partial g_{\sigma\rho,\mathrm{DSR}}}{\partial E} \widetilde{F}^{\alpha}_{\ \nu} - \frac{\partial g_{\nu\rho,\mathrm{DSR}}}{\partial E} \widetilde{F}^{\alpha}_{\ \sigma} \right), \\ \\ C^{\mu}_{\ \nu\rho} = \frac{1}{2} g^{\mu\sigma}_{\mathrm{DSR}} \frac{\partial E}{\partial y^{\alpha}} \left( \frac{\partial g_{\sigma\nu,\mathrm{DSR}}}{\partial E} \delta^{\alpha}_{\ \rho} + \frac{\partial g_{\sigma\rho,\mathrm{DSR}}}{\partial E} \delta^{\alpha}_{\ \nu} - \frac{\partial g_{\nu\rho,\mathrm{DSR}}}{\partial E} \delta^{\alpha}_{\ \sigma} \right). \end{array} \right. \tag{9.32}$$

The vanishing of the electromagnetic field tensor, $F^{\alpha}_{\ \rho} = 0$, implies $L^{\mu}_{\ \nu\rho} = 0$.

One can define the deflection tensors associated to the metric connection $\mathcal{C}\Gamma(\mathcal{H})$ as follows (cf. (9.18)):

$$\begin{aligned} D^{\mu}_{\ \nu} &= y^{\mu}_{\ |\nu} = \frac{\delta y^{\mu}}{\delta x^{\nu}} + y^{\alpha} L^{\mu}_{\ \alpha\nu} = \widetilde{F^{\mu}_{\ \nu}} + y^{\alpha} L^{\mu}_{\ \alpha\nu}; \\ d^{\mu}_{\ \nu} &= y^{\mu}_{\ |\nu} = \delta^{\mu}_{\ \nu} + y^{\alpha} C^{\mu}_{\ \alpha\nu}. \end{aligned} \tag{9.33}$$

The covariant components of these tensors read

$$\begin{aligned} D_{\mu\nu} &= g_{\mu\sigma,\mathrm{DSR}} D^{\sigma}_{\ \nu} = g_{\mu\sigma,\mathrm{DSR}} \left( \widetilde{F^{\sigma}_{\ \nu}} + y^{\alpha} L^{\sigma}_{\ \alpha\nu} \right) \\ &= F_{\mu\nu}(\mathbf{x}) + \frac{1}{2} y^{\sigma} \frac{\partial E}{\partial y^{\alpha}} \left( \frac{\partial g_{\mu\sigma,\mathrm{DSR}}}{\partial E} \widetilde{F^{\alpha}_{\ \nu}} + \frac{\partial g_{\mu\nu,\mathrm{DSR}}}{\partial E} \widetilde{F^{\alpha}_{\ \sigma}} - \frac{\partial g_{\sigma\nu,\mathrm{DSR}}}{\partial E} \widetilde{F^{\alpha}_{\ \mu}} \right); \\ d_{\mu\nu} &= g_{\mu\sigma,\mathrm{DSR}} d^{\sigma}_{\ \nu} \\ &= g_{\mathrm{DSR},\mu\nu} + \frac{1}{2} y^{\sigma} \frac{\partial E}{\partial y^{\alpha}} \left( \frac{\partial g_{\mu\sigma,\mathrm{DSR}}}{\partial E} \delta^{\alpha}_{\ \nu} + \frac{\partial g_{\mu\nu,\mathrm{DSR}}}{\partial E} \delta^{\alpha}_{\ \sigma} - \frac{\partial g_{\sigma\nu,\mathrm{DSR}}}{\partial E} \delta^{\alpha}_{\ \mu} \right). \end{aligned} \tag{9.34}$$

Let us show how the formalism of the generalized Lagrange space allows one to recover some results on the phenomenological energy-dependent metrics discussed in Chap. 4.

Consider the following metric ($c = 1$):

$$ds^2 = a(E)dt^2 + (dx^2 + dy^2 + dz^2), \tag{9.35}$$

where $a(E)$ is an arbitrary function of the energy and spatial isotropy ($b^2 = 1$) has been assumed. In absence of external electromagnetic field ($F_{\mu\nu} = 0$), the nonvanishing components $C^{\mu}_{\ \nu\rho}$ of the canonical metric connection $\mathcal{C}\Gamma(\mathcal{H})$ (see (9.32)) are

$$\begin{cases} C^{0}_{00} = \dfrac{a'}{a} y^{0}, \quad C^{0}_{01} = -\dfrac{a'}{a} y^{1}, \quad C^{0}_{02} = -\dfrac{a'}{a} y^{2}, \quad C^{0}_{03} = \dfrac{a'}{a} y^{3}, \\ C^{1}_{00} = -a' y^{1}, \quad C^{2}_{00} = -a' y^{2}, \quad C^{0}_{00} = -a' y^{3}, \end{cases} \tag{9.36}$$

where the prime denotes derivative with respect to $E$: $a' = da/dE$.

According to the formalism of generalized Lagrange spaces, we can write the Einstein equations in vacuum corresponding to the metrical connection of the deformed Minkowski space (see (9.19)). It is easy to see that the independent equations are given by

$$a' = 0; \tag{9.37}$$

$$2aa'' - (a')^2 = 0. \tag{9.38}$$

The first equation has the solution $a =$ const., namely we get the Minkowski metric. Equation (9.38) has the solution

$$a(E) = \frac{1}{4}\left(a_0 + \frac{E}{E_0}\right)^2, \qquad (9.39)$$

where $a_0$ and $E_0$ are two integration constants.

This solution represents the time coefficient of an over-Minkowskian metric of the second class, (4.31), with $n_0 = 2$. For $a_0 = 0$ it coincides with (the time coefficient of) the phenomenological metric of the strong interaction, (4.11). On the other hand, by choosing $a_0 = 1$, one gets the time coefficient of the metric for gravitational interaction, (4.18).

In other words, *considering $\widetilde{M}$ as a generalized Lagrange space permits to recover (at least partially) the metrics of two interactions (strong and gravitational) derived on a phenomenological basis.*

It is also worth noticing that this result shows that a space–time deformation (of over-Minkowskian type) exists even in absence of an external electromagnetic field (remember that (9.37),(9.38) have been derived by assuming $F_{\mu\nu} = 0$).

## 9.4 Intrinsic Physical Structure of a Deformed Minkowski Space: Gauge Fields

As we have seen, the deformed Minkowski space $\widetilde{M}$, considered as a generalized Lagrange space, is endowed with a rich geometrical structure. But the important point, to our purposes, is the presence of a physical richness, intrinsic to $\widetilde{M}$. Indeed, let us introduce the following *internal electromagnetic field tensors* on $\mathcal{GL}^4 = \widetilde{M}$, defined in terms of the deflection tensors:

$$\begin{aligned}\mathcal{F}_{\mu\nu} &\equiv \frac{1}{2}\left(D_{\mu\nu} - D_{\nu\mu}\right) \\ &= F_{\mu\nu}(\mathbf{x}) + \frac{1}{2}y^{\sigma}\frac{\partial E}{\partial y^{\alpha}}\left(\frac{\partial g_{\mu\sigma,\mathrm{DSR}}}{\partial E}\widetilde{F}^{\alpha}_{\ \nu} - \frac{\partial g_{\nu\sigma,\mathrm{DSR}}}{\partial E}\widetilde{F}^{\alpha}_{\ \mu}\right) \qquad (9.40)\end{aligned}$$

(*horizontal electromagnetic internal tensor*) and

$$\begin{aligned}f_{\mu\nu} &\equiv \frac{1}{2}\left(d_{\mu\nu} - d_{\nu\mu}\right) \\ &= \frac{1}{2}y^{\sigma}\frac{\partial E}{\partial y^{\alpha}}\left(\frac{\partial g_{\mu\sigma,\mathrm{DSR}}}{\partial E}\delta^{\alpha}_{\ \nu} - \frac{\partial g_{\nu\sigma,\mathrm{DSR}}}{\partial E}\delta^{\alpha}_{\ \mu}\right) \qquad (9.41)\end{aligned}$$

(*vertical electromagnetic internal tensor*).

The internal electromagnetic $h$- and $v$-fields $\mathcal{F}_{\mu\nu}$ and $f_{\mu\nu}$ satisfy the following *generalized Maxwell equations*

$$\begin{aligned} 2\left(\mathcal{F}_{\mu\nu\shortmid\rho}+\mathcal{F}_{\nu\rho\shortmid\mu}+\mathcal{F}_{\rho\mu\shortmid\nu}\right) &= y^{\alpha}\left(R^{\beta}{}_{\mu\nu}C_{\beta\alpha\rho}+R^{\beta}{}_{\nu\rho}C_{\beta\alpha\mu}+R^{\beta}{}_{\rho\mu}C_{\beta\alpha\nu}\right), \\ R^{\beta}{}_{\mu\nu} &= g^{\beta\sigma}\frac{\partial F_{\mu\nu}}{\partial x^{\sigma}}; \end{aligned} \tag{9.42}$$

$$\mathcal{F}_{\mu\nu|\rho}+\mathcal{F}_{\nu\rho|\mu}+\mathcal{F}_{\rho\mu|\nu}=f_{\mu\nu\shortmid\rho}+f_{\nu\rho\shortmid\mu}+f_{\rho\mu\shortmid\nu}; \tag{9.43}$$

$$f_{\mu\nu|\rho}+f_{\nu\rho|\mu}+f_{\rho\mu|\nu}=0. \tag{9.44}$$

Let us stress explicitly the different nature of the two internal electromagnetic fields. In fact, the horizontal field $\mathcal{F}_{\mu\nu}$ is strictly related to the presence of the external electromagnetic field $F_{\mu\nu}$ , and vanishes if $F_{\mu\nu}=0$. On the contrary, *the vertical field $f_{\mu\nu}$ has a geometrical origin, and depends only on the deformed metric tensor $g_{\mu\nu,\mathrm{DSR}}(E(\mathbf{y}))$ of $\mathcal{GL}^4=\widetilde{M}$ and on $E(\mathbf{y})$. Therefore, it is present also in space–time regions where no external electromagnetic field occurs.* As we shall see in Part III, this fact has deep physical implications.

A few remarks are in order. First, the main results obtained for the (abelian) electromagnetic field can be probably generalized (with suitable changes) to non-abelian gauge fields. Second, the presence of the internal electromagnetic $h$- and $v$-fields $\mathcal{F}_{\mu\nu}$ and $f_{\mu\nu}$, intrinsic to the geometrical structure of $\widetilde{M}$ as a generalized Lagrange space, is the cornerstone to build up a *dynamics (of merely geometrical origin) internal to the deformed Minkowski space.*

The important point worth emphasizing is that *such an intrinsic dynamics springs from gauge fields.* Indeed, the two internal fields $\mathcal{F}_{\mu\nu}$ and $f_{\mu\nu}$ (in particular the latter one) do satisfy equations of the gauge type (cf. (9.42)–(9.44)). Then, we can conclude that *the (energy-dependent) deformation of the metric of $\widetilde{M}$, which induces its geometrical structure as generalized Lagrange space, leads in turn to the appearance of (internal) gauge fields.*

Such a fundamental result can be schematized as follows:

$$\widetilde{M}=(M,g_{\mu\nu,\mathrm{DSR}}(E))\Longrightarrow\mathcal{GL}^4=(M,g_{\mu\nu}(\mathbf{x},\mathbf{y}))\Longrightarrow\left(\widetilde{M},\mathcal{F}_{\mu\nu},f_{\mu\nu}\right) \tag{9.45}$$

(with self-explanatory meaning of the notation).

We want also to stress explicitly that this result follows by the fact that, in deforming the metric of the space–time, *we assumed the energy as the physical (nonmetric) observable on which letting the metric coefficients depend* (see Chap. 2). This is crucial in stating the generalized Lagrangian structure of $\widetilde{M}$, as shown in Sect. 9.2.

As is well known, successfully embodying gauge fields in a space–time structure is one of the basic goals of the research in theoretical physics starting from the beginning of the twentieth century. The almost unique tool to achieve such objective is increasing the number of space–time dimensions. In such a kind of theories (whose prototype is the celebrated Kaluza–Klein formalism), one preserves the usual (special-relativistic or general-relativistic) structure of 4D space–time, and gets rid of the nonobservable extra dimensions by compactifying them (for example to circles). Then the motions of the extra metric components over the standard Minkowski space satisfy identical equations to gauge fields. The gauge invariance of these fields is simply a consequence of the Lorentz invariance in the enlarged space. In this framework, gauge fields are *external* to the space–time, because they are *added* to it by the hypothesis of extra dimensions.

In the case of the DSR theory, gauge fields arise from the very geometrical, basic structure of $\widetilde{M}$, namely they are a consequence of the metric deformation. The arising gauge fields are *intrinsic and internal to the deformed space–time, and do not need to be added from the outside.* As a matter of fact, *DSR is the first theory based on a 4D space–time able to embody gauge fields in a natural way.*

Such a conventional, intrinsic gauge structure is related to a *given* deformed Minkowski space $\overline{\widetilde{M}}$, in which the deformed metric is fixed:

$$\overline{\widetilde{M}} = (M, \bar{g}_{\mu\nu,\mathrm{DSR}}(E)) \,. \tag{9.46}$$

On the contrary, with varying $g_{\mathrm{DSR}}$, we have another gauge-like structure – as already stressed in Sect. 4.4 – namely what we called a metric gauge. In the latter case, the gauge freedom amounts to choosing the metric according to the interaction considered.

The circumstance that the deformed Minkowski space $\widetilde{M}$ is endowed with the geometry of a generalized Lagrange space testifies the richness of nontrivial mathematical properties present in the seemingly so simple structure of the deformation of the Minkowski metric. This will be further supported in Part IV, where we shall show that $\widetilde{M}$ can be naturally embedded in a 5D Riemannian space.

# Part III

# EXPERIMENTS ON DEFORMED SPACE–TIME

# 10
# Lorentz and CPT Symmetries in DSR

One of the fundamental teachings of Einstein's relativity theories is that the geometry of the (4D) space–time is globally curved (Riemannian) and locally flat (Minkowskian). This implies the existence of a local frame in which Special Relativity (SR) strictly holds for nongravitational interactions. Such a property is referred to as *local Lorentz invariance (LLI)* (distinct in principle from local position invariance, LPI, i.e., the independence of such a local Lorentzian frame from space–time position). An alternative statement of LLI is that all inertial frames of reference are (locally) equivalent. However, it is a long-disputed problem whether LLI preserves its validity at any length or energy scale (far enough from the Planck scale – corresponding to the Planck length $l_{\text{Planck}} = \sqrt{G\hbar/c^3}$ –, when quantum fluctuations are expected to come into play and affect the very geometrical structure of space–time). Doubts as to the reliability of a Lorentz-invariant description of physical phenomena at subnuclear distances were e.g., put forward, in the mid of the twentieth century, even in standard (and renowned) textbooks [47]. Typically, a violation of LLI would single out an individual frame in which a preferred inertial observer is at rest.

It must also be stressed that Lorentz invariance is strictly related to CPT symmetry, namely the combined transformation of charge conjugation ($\mathcal{C}$), parity inversion ($\mathcal{P}$), and time reversal ($\mathcal{T}$). This is due to the celebrated CPT theorem of quantum field theory (however, a classical explanation of the strict connection between LLI and CPT invariance can be given by taking into account that the inversion of both the time and the space axes is a proper Lorentz transformation, by applying it also to the energy–momentum space, and taking into account the Stückelberg–Feynman–Sudarshan “reinterpretation principle” [48,2]).

However, experiments carried out till now show to exceptionally high precision that all the basic laws of nature seem to have both Lorentz and CPT symmetry. By this evidence, the problem of their possible breakdown has been completely out of the mainstream until the last decade of the twentieth century. Only at that time some physicists started questioning seriously the validity of LLI and CPT – and therefore of Special Relativity (SR) – from both the theoretical and the experimental side.

The relevance of testing Lorentz and CPT invariance is obviously due to the basic fact that a positive evidence for their breakdown would constitute a significant signature of new, unconventional physics.

This is why LLI breakdown has been investigated by means of several, quite different approaches, sometimes extending, sometimes relaxing the concepts and the formalism of SR.

As already stressed in Part I, a deep connection exists between the DSR formalism and the breakdown of LLI. In fact, the usual Lorentz transformations do not preserve the deformed interval (2.17c), as expressed by (3.3). This is also clearly seen, on a phenomenological point of view, by inspecting the very form of the metrics we derived from the analysis of the experimental data in Sect. 4.1. Indeed, they show a departure of the space–time metric from the Minkowskian one for the four fundamental interactions. This occurs below the energy thresholds for sub-Minkowskian metrics (corresponding to the electromagnetic and weak interactions) and above the energy thresholds for over-Minkowskian ones (strong and gravitational). The most straightforward meaning of such results is that (apart from the gravitational case, where of course a departure from the Minkowskian metric is expected and natural) the three phenomena considered (i.e., the superluminal propagation of evanescent e.m. waves in waveguide, the $K_S^o$ decay and the BE correlation) do show a violation of the usual Lorentz symmetry for the electromagnetic, weak and strong interaction, respectively.

In this connection, two points deserve a moment's thought. Firstly, one may wonder whether such an LLI breakdown should be considered as an *actual* or an *effective* one. Namely, two interpretations can be given to the need for a deformed Minkowski metric in order to describe the above phenomena. The first one is to state that such results correspond to an actual local deformation of the space–time geometry, induced by the interaction considered. This would constitute an evidence in favor of a *real* analogy of the e.m., weak and strong interactions with the gravitational one, and therefore of the *real* validity of the solidarity principle for all the four interactions. The second possible interpretation is inspired by the analysis of the e.m. case. Indeed, in the e.m. wave propagation in waveguides, it is perhaps more sound to understand the nonminkowskian metric as describing, in an effective way, the nonlocal e.m. effects which occur inside the waveguide and give rise to the superluminal propagation. In the same spirit, one can regard the arising of nonminkowskian metrics in the description of $K_S^o$ decay and BE correlation as due to the presence of nonlocal forces responsible for such phenomena. Otherwise speaking, the nonminkowskian metrics involved in such cases would play an *effective* role, in the sense that they would actually be a signature of the presence and effectiveness of nonpotential effects in the phenomena considered. In such a minimalist interpretation, DSR would only constitute a suitable formalism to parametrize and to account for the breakdown of local Lorentz invariance.

Another basic point worth emphasizing is related to what we stressed in the previous two parts. Namely, physical phenomena exhibiting a Lorentz-invariance breaking behavior according to standard Special Relativity, can be regarded as preserving LLI in the generalized sense of DSR, namely satisfying deformed (local) Lorentz invariance (DLLI). Then, the breakdown of LLI can be considered as evidence for DLLI, and therefore signature of a (local at least) deformation of space–time, provided the violation of LI occurs as prescribed and predicted by DSR (see Sect. 3.3.5).

In this Part, we shall discuss some experiments – most of which directly inspired by DSR – whose results are seemingly in favor of DSR as describing a *real deformation of (local) space–time geometry* for the interaction considered, and provide evidence for violation of the usual Lorentz invariance, and for its recover in the DSR, deformed sense (i.e., as DLLI). In the following, when speaking about LLI breaking, we shall always mean it in the usual, special-relativistic sense (unless otherwise specified).

# 11 Lorentz Invariance Breakdown: A Brief Survey

## 11.1 Theoretical Developments

As said in Chap. 10, serious interest in the question of the validity of LLI and SR can be traced back only to the end of the twentieth century. This did not prevent some pioneering people from facing the problem many years ago. For instance, early theoretical speculations (in the mid of the 1900) on a possible breakdown of LLI, and its experimental consequences, are due e.g., to Bjorken [49], Blokhintsev [50] and Redei [51], Phillips [52]. These works are based on the main characteristics of LLI violation. In general, it can be stated that the LLI breakdown, in its simplest formulations, is associated to an absolute object in vacuum. The existence of a preferred frame (which can be also the space–time vacuum itself) entails that the four-velocity of this preferred inertial observer is a (time-like) vector field which has almost the same value throughout space–time. This is the simplest example of an absolute vector object, the so-called *internal vector* $N_\mu$ [50, 52, 10].

Another physical feature of a Lorentz-noninvariant vacuum is provided by the concept of universal length $l_0$ [50,10]. Such a length can be thought off as the minimal limiting length for all physical distances whose measurement can be carried out relative to a large (ideally infinite) number of physically equivalent inertial frames. Clearly, the length $l_0$ acts as an absolute demarcation line between macroscopic (large) distances and microscopic (small) ones, and it too is an absolute property of a Lorentz-noninvariant vacuum.

In Bjorken's model, Lorentz invariance is spontaneously broken, but it has no physical consequences [49]. In Blokhintsev's approach, the origin of the breakdown is the existence of a fundamental length [50], which yields observable effects in the meanlife of unstable particles [51]. The Phillips proposal of a cosmological background field [52] leads to a preferred frame of reference.

In the next years, the issue was faced from a different, more basic point of view, namely by questioning the very foundations and formalism of Special Relativity (SR) – at least in its usual, Einsteinian formulation. For instance, in the years 1972–1975 Recami and one of the present authors (R.M.) – in the framework of a generalization of SR including also superluminal reference frames – carried out a critical examination of the basic principles of SR [2]. As already stressed in Chap. 1, they pointed out, among the others, that:

1. A correct use of the Relativity Principle requires to specify the class of physical phenomena to which it applies

2. A priori, for each different class of phenomena considered, a different formal (and therefore mathematical) formulation of SR is expected to hold (in particular, the usual SR is expected to be strictly valid only for processes ruled by the electromagnetic interaction)

3. Different invariant speeds correspond, in general, to the different formulations of SR

Among the most serious attempts at generalizing the SR mathematical formalism (i.e., the structure of the Minkowski space), more or less based on the above critical analysis of the SR foundations, let us recall the anisotropic theory of Bogoslowski [8] (based on a Finslerian metric [7]), and the "isotopic" SR [9] (we already quoted them as theories with prototype deformed metrics in Sect. 2.1). Moreover, a (constant) nonminkowskian metric was introduced on a phenomenological basis, for weak interactions, by Nielsen and Picek [10].

In recent times, a number of theoretical formalisms do admit for observable effects of LLI violation [17]. They can be roughly divided in two classes: unified theories and theories with modified space–times. To the former one belong e.g., Grand-Unified Theories, (Super) String/Brane theories, (Loop) Quantum Gravity, and the so-called "effective field theories." The latter include e.g., foam-like quantum space–times, space–times endowed with a nontrivial topology or with a discrete structure at the Planck length, $\kappa$-deformed Lie algebra noncommutative space–times (for instance Doubly Special Relativity [14]). Theories with a variable speed of light or variable physical constants also imply LLI breakdown.

A very interesting analysis of LLI breakdown within the framework of the Standard Model have been considered e.g., by Coleman and Glashow[15], with the proposal of new tests of SR in cosmic-ray and neutrino physics. They developed a perturbative approach whereby to deal with the observable implications of tiny departures from LLI, and carried out a thorough investigation of possible Lorentz-violating mechanisms within the Standard Model. It has also been proposed that such small departures from LLI can affect particle kinematics in such a way to remove the cosmological Greisen–Zatsepin–Kuz'min cutoff (of the order of $4 \times 10^{19}$ eV )[53]. The problem of violation of LLI and CPT by Chern–Simons terms was also considered by Jackiw [54].

Lastly, in order to take account of the LLI breaking effects, an extension of the Standard Model has been proposed by Kostelecky [55]. He essentially assumes that the breakdown of Lorentz and/or CPT invariance is due to spontaneous symmetry breaking (namely to a noninvariance of the vacuum under these symmetries). Therefore, his approach is similar in spirit to the old Bjorken one. Such a kind of CPT and LLI breakdown poses however some unsolved questions. The basic one is the nature of the Nambu–Goldstone bosons in this framework. Actually, Goldstone's theorem doesn't apply to a discrete symmetry like CPT. For global Lorentz symmetry, it implies that spontaneous breaking must be accompanied by massless bosons, which might be identified with photons. But if gravity is included then Lorentz symmetry becomes local, and the identification of the Goldstone boson with the photon becomes controversial. In spite of these problems, the Kostelecky model is able to suggest a number of new tests of LLI and CPT, and to put stringent limits on their breakdown. We refer the reader to [55] for a thorough discussion of the Standard Model Extension.

## 11.2 Experimental Tests

From the experimental side, the main classical tests of LLI can be roughly divided in three groups [56]:

(a) Michelson–Morley (MM)- type experiments, aimed at testing isotropy of the round-trip speed of light

(b) Tests of the isotropy of the one-way speed of light (based on atomic spectroscopy and atomic timekeeping)

(c) Hughes–Drever-type (HD) experiments, testing the isotropy of nuclear energy levels

All such experiments put stringent upper limits on the degree of violation of LLI.

The breakdown of standard local Lorentz invariance is expressed by the LLI breaking factor parameter $\delta$ [56]. We recall that two different kinds of LLI violation parameters exist: the isotropic (essentially obtained by means of experiments based on the propagation of e.m. waves, for instance of the Michelson–Morley type), and the anisotropic ones (obtained by experiments of the Hughes–Drever type, which test the isotropy of the nuclear levels).

In the former case, the LLI violation parameter reads [56]

$$\delta = \left(\frac{u}{c}\right)^2 - 1; \tag{11.1}$$

$$u = c + v, \tag{11.2}$$

where $c$ is, as usual, the speed of light *in vacuo*, $v$ is *a phenomenological LLI invariance breakdown speed* (for example. the speed of the preferred frame) and $u$ is the new speed of light.[1] Notice that $u$ is nothing but the "maximal causal speed" of the electromagnetic interaction, in DSR, or the "maximum attainable speed," in the words of Coleman and Glashow [15]. In the anisotropic case, there are different contributions $\delta_A$ to the anisotropy parameter from the different interactions. In the HD experiment, it is $A = S, HF, ES, W$, meaning strong, hyperfine, electrostatic and weak, respectively. These correspond to four parameters $\delta_{\mathrm{S}}$ (due to the strong interaction), $\delta_{\mathrm{ES}}$ (related to the nuclear electrostatic energy), $\delta_{\mathrm{HF}}$ (coming from the hyperfine interaction between the nuclear spins and the applied external magnetic field) and $\delta_{\mathrm{W}}$ (the weak interaction contribution).

Many other tests of LLI (different from (a)–(c) above) have been proposed in the framework of the so-called $TH\varepsilon\mu$ formalism [56] or the Robertson–Mansouri–Sexl test theory of SR [57,58]. Space experiments have also been envisaged [59,60].

Moreover, a lot of tests based on the Standard Model Extension (that provides a quantitative theoretical framework within which various experimental tests of CPT and Lorentz symmetry can be studied and compared) have been proposed by Kostelecky [55]. They include:

- Observations of neutral-meson oscillations
- Observations of neutrino oscillations
- Clock-comparison tests on Earth and in space
- Studies of the motion of a spin-polarized torsion pendulum
- Spectroscopy of hydrogen and antihydrogen
- Comparative tests of QED in Penning traps
- Determination of muon properties

[1] Needless to say, (11.1) holds true only for $u > c$, and does not includes cases with $u < c$ (well possible, like that of (3.72)).

- Measurements of cosmological birefringence
- Tests with microwave cavities and lasers
- Observation of the baryon asymmetry

To our present aims, let us stress that, on one side, DSR allows one to shed new light on some aspects of LLI breakdown (for instance, in superluminal propagation); on the other hand, it permits to explicitly design new tests (and therefore new classes of experiments) aimed at testing LLI for all four fundamental interactions. It is just the purpose of the next chapters to examine these aspects of LLI violation, namely those concerning its (theoretical and experimental) connections with DSR.

# 12
# Superluminal Propagation of Electromagnetic Waves

In the last five years of the twentieth century some experimental results in different branches of physics have provided significant evidence for phenomena involving faster-than-light (superluminal) speeds [61,62]. In particular, a number of experiments have been carried out concerning the superluminal propagation of electromagnetic signals. These include superluminal photon tunneling – both in the microwave range and in the optical domain – in experiments performed in a number of laboratories; optical experiments with total internal reflection (in which the barrier is represented by the air gap between two prisms); optical propagation in media with anomalous dispersion; and finally propagation of e.m. X-waves (i.e., nonmonochromatic, nondispersive Bessel beams) in free space. Propagation at a group velocity greater than the light velocity has been therefore experimentally demonstrated not only for evanescent (tunneling) waves, but also for nonevanescent ones (like X-shaped waves).

It must be noticed that, from a kinematical point of view, the existence of faster-than-light e.m. signals is related to the breakdown of LLI through the parameter $\delta$ defined by (11.1).[1]

One of the main problems for a theoretical treatment of the superluminal photon propagation is due to the fact that it was observed in different

[1] Actually, this is not strictly true if one takes into account the generalization of SR to superluminal inertial frames built up by Recami and one of the present authors (RM) in the early 1970s of the past century [2]. This is why the true signature of LLI breakdown in superluminal e.m. propagation is represented by the link with energy provided by DSR.

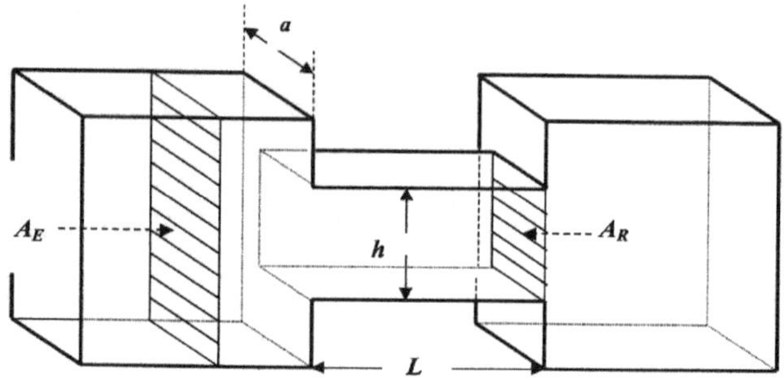


FIGURE 12.1. Rectangular waveguide with variable section used in the Cologne experiment. $L$ is the length of the narrow part of the waveguide ("barrier"); $h$ is height of the guide; $a$ is thickness of the guide; $A_{\mathrm{E}}$ is area of the large section ("emitter"); $A_{\mathrm{R}}$ is the area of the small section ("receiver")

kinds of experiments [61,62], which are not easily comparable. It is so quite impossible to state if the results of different experiments are compatible with each other.

However, it can be shown (by just using, among the others, the DSR formalism) that two of the first performed experiments on superluminal photon propagation, namely, the 1992 Cologne experiment on the tunneling of evanescent waves in an undersized waveguide (see Fig. 12.1) [32,33], and the 1993 Florence experiment on the microwave propagation in air between two not-coaxial horn antennas (see Fig. 12.2) [35], do admit a common interpretation. Precisely, both experimental devices behave as a high-pass filter [6]. We got this result by two different methods, one based on the Friis law (which yields the efficiency of a transmitting device), and the other on the deformation of the Minkowski space–time. This allowed us to set intriguing connections between these two (a priori different) classes of experiments. In particular, in either case the superluminal propagation can be described as a tunneling and is related to evanescent waves. Let us also recall that the results of the Cologne experiment allowed us to derive the explicit form (4.2), (4.3) of the DSR metric for the electromagnetic interaction, by assuming that, inside the barrier, the space–time is no longer Minkowskian but is just endowed with an energy-dependent (spatially isotropic), deformed metric, with energy threshold $E_{0,\mathrm{e.m.}} = (4.5 \pm 0.2)\ \mu\mathrm{eV}$.

We do not enter into the details of this derivation (based on (3.70) and Sect. 3.4.3) and refer the reader to [6]. Let us only stress that the analysis of the Florence experiment by the formalism of the deformed Minkowski space permits also to describe the behavior of the Florence device as a barrier, with a decaying law for the energy of the evanescent-wave type, and therefore to interpret the experiment as a genuine tunneling one. It confirmed also the value of the energy threshold for the electromagnetic interaction, $E_{0,\mathrm{e.m.}}$, originally derived by the fit to the data of the Cologne

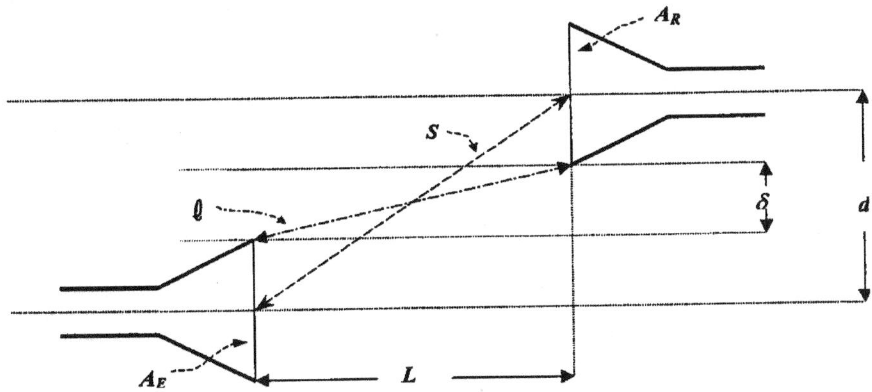


FIGURE 12.2. Schematic view of the two horn antennas used in the Florence experiment. $L$ is the distance between antennas; $S$ is distance between the centers of the antenna surfaces; $d$ is distance between the axes of antennas

tunneling experiment (see Sect. 4.1). What's more, by means of such an analysis it was possible to get the fundamental result that *the breakdown of LLI exhibits also a spatial threshold,* $\ell \simeq 9\,\mathrm{cm}$. This finding reveals itself of basic relevance in designing new electromagnetic tests of LLI and of DSR, as we shall see in Chap. 13.

# 13

# The Shadow of Light: Lorentzian Violation of Electrodynamics in Photon Systems

The experiments on superluminal group propagation, analyzed in terms of the tools of DSR, provided evidence for a breakdown of local Lorentz invariance with a threshold both in energy and space (at least for the electromagnetic interaction). In order to confirm these results, we carried out new experiments explicitly designed to test them. Let us briefly discuss these experiments, together with their implications.

## 13.1 Double-Slit-Like Experiments

The experiments we performed were optical ones, in the infrared range, of the double-slit type. We were essentially aimed at searching for a possible anomalous photon behavior, at variance with the predictions of classical and/or quantum electrodynamics, and therefore related to Lorentz invariance violation. Let us briefly report the main features and results of these three experiments, carried out at L'Aquila University [63–66].

The employed apparatus (schematically depicted in Fig. 13.1) consisted of a Plexiglas box with wooden base and lid. The box (thoroughly screened from those frequencies susceptible of affecting the measurements) contained two identical infrared (IR) LEDs, as (incoherent) sources of light, and three identical photodiodes, as detectors (A, B, C). The two sources $S_1$, $S_2$ were placed in front of a screen with three circular apertures $F_1$, $F_2$, $F_3$ on it. The apertures $F_1$ and $F_3$ were lined up with the two LEDs A and C respectively, so that each IR beam propagated perpendicularly through each of them.

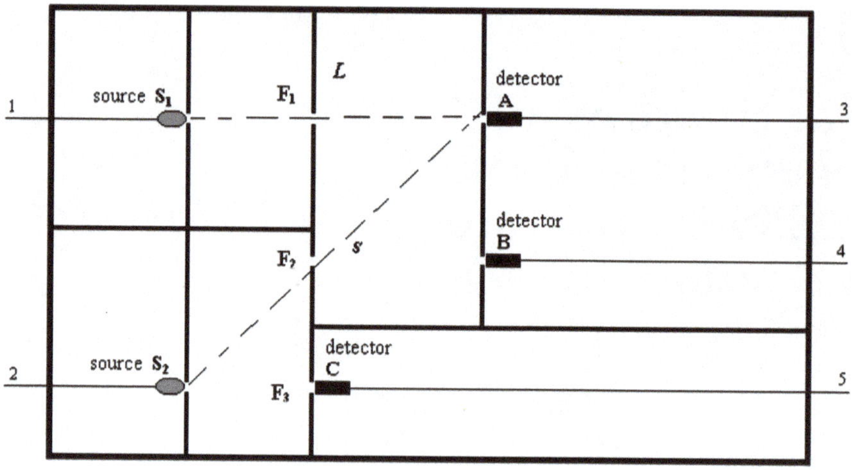


FIGURE 13.1. Above view of the experimental apparatus used in the first double-slit experiment

The geometry of this equipment was designed so that no photon could pass through aperture $F_2$ on the screen.[1] Let us stress that the dimensions of the apparatus were inferred from the geometrical size of the Florence microwave experiment [35], namely the horizontal distance between the planes of the antennas (see Fig. 12.2).

The wavelength of the two photon sources was $\lambda = 8.5 \times 10^{-5}$ cm. The apertures were circular, with a diameter of 0.5 cm, much larger than $\lambda$. We worked therefore in absence of single-slit (Fresnel) diffraction. However, the Fraunhofer diffraction was still present, and its effects have been taken into account in the background measurement.

Detector C was fixed in front of the source $S_2$; detectors A and B were placed on a common vertical, movable panel (see Fig. 13.1). This latter feature allowed us to study the space dependence of the anomalous effect, predicted by DSR.

Let us highlight the role played by the three detectors. Detector C destroyed the eigenstates of the photons emitted by $S_2$. Detector B ensured that no photon passed through the aperture $F_2$. Finally, detector A measured the photon signal from the source $S_1$.

In summary, detectors B and C played a controlling role and ensured that no spurious and instrumental effects could be mistaken for the anomalous effect which had to be revealed on detector A. The design of the

[1] In this connection, let us notice that the dotted line $S$ in Fig. 13.1 is a mere geometrical one, and does not represent any physical trajectory of photons emitted by the source $S_2$, since the aperture $F_2$ was well outside the emission cone of $S_2$ [63]. It is only to mean that the distance between $S_2$ and the detector A is the same as the distance $S$ in Fig. 12.2.

box and the measurement procedure were conceived so that detector A was not influenced by the source $S_2$ according to the known and officially accepted laws of physics governing electromagnetic phenomena: classical and/or quantum electrodynamics. In other words, with regards to detector A, all went as if the source $S_2$ was not there at all or as if it was always kept turned off.

In essence, the experiment just consisted in the measurement of the signal of detector A (aligned with the source $S_1$) in two different states of source lighting. Precisely, a single measurement on detector A consisted of two steps:

1. Sampling measurement of the signal on A with source $S_1$ switched on and source $S_2$ off
2. Sampling measurement of the signal on A with both sources $S_1$ and $S_2$ on

As already stressed, due to the geometry of the apparatus, no difference in signal on A between these two source states ought to be observed, according to either classical or quantum electrodynamics. If $A(S_1 i\ S_2 k)$ ($i, k = \text{on}, \text{off}$) denotes the value of the signal on A when source $S_1$ is in the lighting state $i$ and $S_2$ in the state $k$, a possible nonzero difference $\Delta A = A(S_1\text{"on"}\ S_2\text{"off"}) - A(S_1\text{"on"}\ S_2\text{"on"})$ in the signal measured by A when source $S_2$ was off or on (and the signal in B was strictly null) has to be considered evidence for the searched anomalous effect.

The outcomes of the first experiment were positive, namely *the differences $\Delta A$ between the measured signals on detector A in the two conditions were different from zero and below the threshold value of energy for the breakdown of local Lorentz invariance as predicted by DSR*. In particular, $\Delta A$ ranged from (2.2±0.4) to (2.3±0.5) $\mu$V, values well below the threshold $E_{0,\text{e.m.}}$=4.5 $\mu$V. Moreover, such an anomalous effect was observed within a distance of at most 4 cm from the sources [63], thus confirming the spatial threshold obtained from the analysis of the Cologne and Florence experiments (see Chap. 12). We can consider such an effect as the consequence of *an "hidden" (Lorentzian) interference*.

The purpose of the second experiment was to corroborate the results of the previous one [64, 65]. The experimental set-up was essentially the same (for instance, the dimensions of the apparatus, and the relevant quantities, like photon wavelength and aperture diameter, were identical to those of the first experiment). The main difference with respect to the equipment of the first experiment was in a right-to-left inversion along the bigger side of the box, and in the three used detectors, which were not photodiodes but phototransistors (of the type with a convergent lens). In this way, it was possible to study how the phenomenon changes under a spatial parity inversion and for a different type of detector. We want to point out that in this second experiment the time procedure to sample the signals on the detectors was different from that used in the first experiment. We indeed realized that the sampling time procedure was apparently crucial in order to observe the anomalous interference effect.

The results of this second experiment confirmed those of the first one. The value of the difference measured on detector A was $(0.008 \pm 0.003)\,\mu$V, which is consistent, within the error, with the difference $\Delta A \simeq 2.3\,\mu$V measured in the first experiment, *provided that the unlike efficiencies of the phototransistors with respect to those of the photodiodes are taken into account* [64].[2] The consistency between the results of the first two experiments shows apparently that the effect is not affected by the parity of the equipment and by the type of detector used (at least for photodiodes and phototransistors). Let us notice that one was compelled to use two different sampling time procedures for the two different types of detecting devices in order to make the effect evident. It turned out that there was apparently a sort of unavoidable bond between detector and sampling-time procedure, to be taken into account in order to reveal the effect.

The third experiment was planned and carried out in order to shed some light on this issue and to obtain a further evidence of the searched effect [66]. In order to test the apparent bond between detectors and sampling time procedures, the experiment was carried out by means of the box with photodiodes but using the sampling-time procedure adopted with phototransistors. The results of this third experiment were consistent with those of the two previous ones. By this statement we mean that the average value of the differences on detector A in the two lighting situations of the sources was below the threshold energy for the breakdown of LLI for the electromagnetic interaction, as required by the theory. In particular the maxima of $|\Delta A|$ accumulate around the value of $2.3\,\mu$V (see Fig. 13.2), in agreement with the results of the other two experiments.[3] One can conclude that the sampling time procedure, which permitted the effect to be evidenced on phototransistors, could reveal it on photodiodes as well.

---

[2]One can define the relative geometrical efficiency $\eta_g$ of the phototransistor (with respect to the photodiode) as the ratio of their respective sensitive areas, and their relative time efficiency $\eta_t$ as the ratio of their respective detection times. Then, one can define the relative total efficiency $\eta_T$ of the phototransistor with respect to the photodiode as the product $\eta_T = \eta_g \eta_t$ . From the values of $\eta_g$ and $\eta_t$ in this case, one gets $\eta_T = 0.0015$ [64].

Therefore, it was reasonable to foresee that the value of the expected phenomenon in the second experiment to be given by the product of the total relative efficiency times the value measured in the first experiment, i.e., $\eta_T[(2.3 \pm 0.5)\mu\text{V}] = (0.004 \pm 0.001)\,\mu V$, in agreement with the experimental result.

[3]Let us note that the photodiodes used as detectors in the first and third experiment were integrated to a transimpedance amplifier, transducing the photocurrent signal into a voltage signal. Such a voltage, measured by means of a multimeter, does not depend therefore on the value of the circuit resistances of the voltage measuring system.

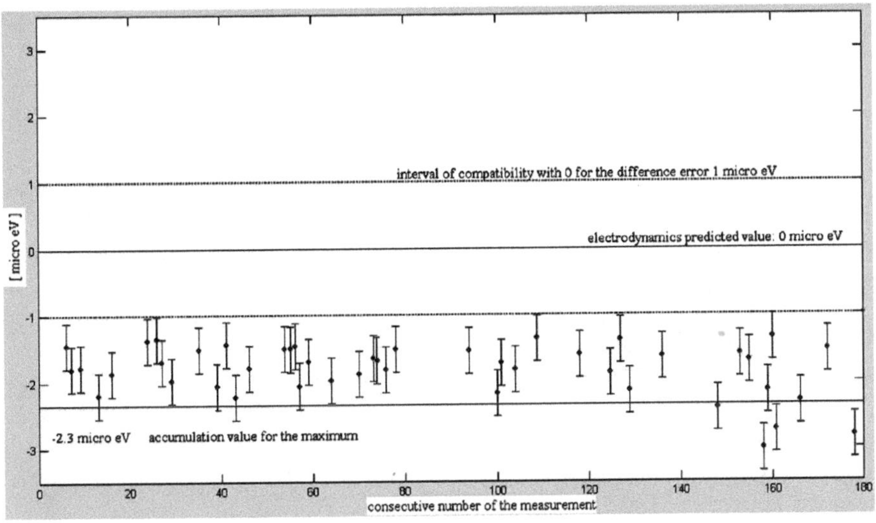


FIGURE 13.2. Value of the differences $\Delta A$ of signal sampled on detector A for the two lighting states of the sources $S_1$ on, $S_2$ off, and $S_1$ on, $S_2$ on (third experiment). The differences are clearly incompatible with zero

Therefore, it is possible to state that, although there is not such a tight bond between detector and time procedure, the latter plays a very important role in giving evidence to the effect. More explicitly, one could say that the phenomenon one tried to detect and to study possesses very complex features, which make it hard to be grasped both in literal and figurative sense. We already know that the anomalous Lorentzian interference manifests itself only under precise conditions, namely below an energy threshold and within some spatial threshold as well. In this sense, it is endowed with a peculiar structure both in energy and in space. The global view of these three experiments teaches us that there exists also some sort of threshold for the sampling time interval. Because of this time structure, the effect looks quite different depending on the time procedure adopted to sample the signals on the detectors, as it is apparent from the two responses we got from the first and the third experiments. In order to evidence this anomalous photon behavior, which is the consequence of a very complex physical phenomenon, i.e., the breakdown of LLI, one has to adapt the physical inquiry to it and be aware of the existence of these thresholds in energy, space and time.

We want to add that the third experiment was repeated several times over a whole period of four months in order to collect a fairly large amount of samples and hence have a significant statistical reproducibility of the results. Thanks to this large quantity of data, it was possible to study the distribution of the differences of signals on detector $A$, which is shown in Fig. 13.2. For clarity' sake, we reported only the differences $\Delta A$ outside the interval $[-1, 1]$, which is the interval of compatibility with zero of the values of $\Delta A$.

The circumstance that the majority of the differences $\Delta A$ is negative (namely $A(S_1 on\ S_2 off) < A(S_1 on\ S_2 on)$) might wrongly induce to deem that, when source $S_2$ was turned on, the signal detected on A increased. One might then be incorrectly tempted to account for this by stating that some photons of $S_2$ passed through aperture $F_2$ (see Fig. 13.1). Conversely, if one takes into account the mode of operation of the photodiodes chosen as detectors,[4] it becomes immediately apparent that the above inequality means exactly the opposite situation. Namely, when $S_2$ was turned on, detector A recorded a lower signal and hence received less photons, although there was a larger number of photons in the box because both sources were on.[5] On the other hand, it is impossible to account for this reduction of the signal on A when $S_2$ got turned on and $S_1$ was already on as a destructive interference between photons from the two sources, because the LEDs are incoherent sources of light.

## 13.2 Crossing Photon Beam Experiments

The results of the double-slit experiments suggest that similar anomalous effects can be observed also in different experimental situations involving photon systems, like e.g., in interference experiments. Further evidence for the anomalous photon system behavior (and for the related anomalous photon-photon cross section) was observed indeed in orthogonal crossing photon beams.

These interference experiments were carried out after our first one, one with microwaves emitted by horn antennas (see Fig. 13.2), at IFAC-CNR (Ranfagni and coworkers) [67–69], and the other with infrared $CO_2$ laser beams (Fig. 13.3), at INOA-CNR (Meucci and coworkers) [69]. Let us summarize the results obtained.

---

[4]In order to understand this point, let us give some brief details about the mode operation of the type of photodiode (OPT301 Burr Brown) used in the third experiment. First of all, we have to say that its pins were connected to the input pins of a trans-impedance operational amplifier which was integrated along with the photodiode on the same chip. The photodiode was not inversely polarized and the dark current was always greater than the photocurrent. As is well known, the two currents flow in opposite directions, and the total current flowing in the photodiode is given by their subtraction. When the total current increases, the op-amp output voltage increases too. However, a rise of the total current (and hence a rise of the output voltage) means a decrease of the photocurrent (the dark current cannot change) and this means a drop of the number of photons received by the photodiode. Thus, when both sources were on, the increase of the output voltage means that the photodiode A was receiving less photons.

[5]Needless to say, the stability of power supplies was constantly checked.

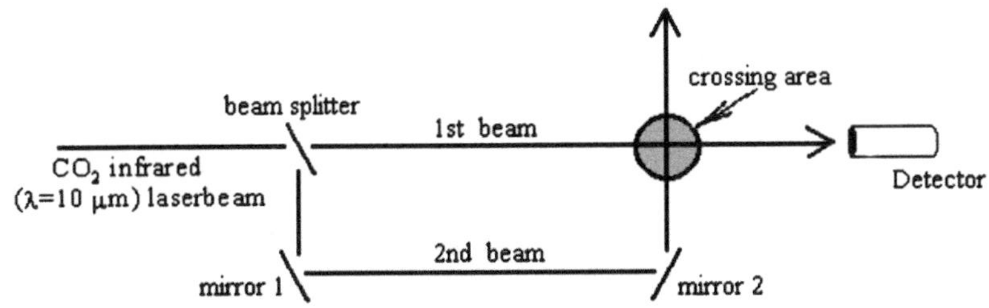


FIGURE 13.3. Schematic view of the crossed-beam experiment in the infrared range, exploiting a $CO_2$ laser emitting at 10.6 $\mu m$ on the fundamental $TEM_{00}$ Gaussian mode. The laser beam is split in two orthogonal beams (beam 1 and beam 2) by means of a beam splitter. By using two flat mirrors the two beams are directed to the crossing area within the near field of the Gaussian mode, estimated at 1.5 m from the out-coupler mirror of the laser cavity. Beam 2 is periodically interrupted by means of a chopper whose frequency is the reference frequency in a lock-in amplifier connected to the detector

The main result of the IFAC experiment consists in an unexpected transfer of modulation from one beam to the other, which cannot be accounted for by a simple interference effect. This confirms the presence of an anomalous behavior in photon systems, in the microwave range too.

In the optical experiment carried out at INOA-CNR [69], the wavelength of the used infrared laser beams was 10,600 nm, namely one order of magnitude higher than the wavelength of the sources (LEDs) used in our experiments (850 nm). Let us also remark that the energy of the photons of the three double-slit experiments was $10^4$ times higher than that of the photons in the Cologne and Florence experiments [32, 33, 35], and 10 times higher than that of the INOA CNR experiment [69].

The optimum alignment which can be achieved with lasers and the laser beam confinement make this optical set-up especially suitable for investigating the anomalous behavior of the photon systems. This allowed one to perform a statistical test on the averaged results [64, 69]. The signal statistics provided a significant variation in the mean values obtained with or without beam crossing. Hence the chance to have two identical statistics was rejected with a sufficient level of confidence. Moreover, it was estimated [64] that the actual shift of the crossed beam signal with respect to the single beam signal is $(2.08 \pm 0.13)\,\mu$V. This value agrees excellently with that obtained in our first experiment $\Delta A \simeq 2.3\,\mu$V. Notice that the laser experiment shows that the observed phenomenon does depend neither on the infrared wavelength, nor on the coherence properties of the light.

Although further checks are needed, one can conclude that the crossing photon-beam experiments do preliminarily support the evidence for an anomalous interference effect under the space and energy constraints obtained by the DSR formalism.

## 13.3 The Shadow of Light: Hollow Wave, LLI Breakdown and Violation of Electrodynamics

We want now to provide an interpretation, and discuss the implications, of the observed anomalous photon behavior.

Needless to say, the results obtained in different photon systems in different experiments are consistent with LLI breakdown. The signature of violation of LLI is provided by the marked threshold behavior the phenomenon exhibited. In fact, the anomalous effect was observed within a distance of at most 4 cm from the sources (1 cm in the second experiment), and the measured signal difference on detector A ranged from $\Delta A \simeq 2.3\,\mu$V (first and third experiment) to $\Delta A \simeq 0.008\,\mu$V (second experiment) [63–66]. These values are consistent with the space and energy threshold behavior for the electromagnetic breakdown of LLI, obtained in the framework of DSR (see Chapt. 4 and 12).

Moreover, in our opinion, the results of the photon experiments described earlier cannot be explained in the framework of the Copenhagen interpretation of quantum wave [65], or in its implementation in terms of path integrals in Feynman's approach.

Indeed, let us consider the difference $\Delta A$ in the signal measured by detector A according to whether only $S_1$ is turned on or both sources are on, and recall the role played by the three detectors in our experiments (see Sect. 13.1). On one side, detector C measures – and hence destroys – the superposition of states belonging to the photons emitted by $S_2$ (thus manifesting their corpuscle nature); on the other hand, detector B is always underneath the dark voltage threshold, thus ensuring no transit of photons through aperture $F_2$. Therefore in no way – according to the Copenhagen interpretation – photons from $S_2$ can interact with those from $S_1$, thus accounting for the signal difference on detector A.

On the contrary, such a result can be understood by interpreting – following Einstein, de Broglie and Bohm [70–72] – the quantum wave as a pilot (or hollow) wave.

In such a framework, pilot waves can interact with quantum objects (as assumed by de Broglie and Andrade y Silva [73]). Then, the region outside aperture $F_2$ is optically forbidden to the photons emitted by the source $S_2$, *but not to the hollow waves associated to them.* Thence, the photons emitted by the source $S_1$ can interact with the hollow waves of photons from the source $S_2$, which have gone through the aperture $F_2$. Consequently, the change $\Delta A$ in the A signal – in absence of any change in the response signal of detectors B and C – finds a natural explanation, in the Einstein–de Broglie–Bohm interpretation of quantum wave, in terms of the interaction of the $S_1$ photons (and their hollow waves) with the hollow waves (of $S_2$ photons) passed through $F_2$.

The role played by the aperture $F_2$ is fundamental, since, although hollow waves can penetrate in optically forbidden regions, nonetheless the mass distribution and density are expected to affect their propagation. Hence, they can pass only through space regions with a lower mass density.

Since, according to DSR, the breakdown of LLI is connected to a deformation of the Minkowski metrics, it is possible to put forward the hypothesis [63] that *the hollow wave (at least for photons) is nothing but a deformation of space–time geometry, intimately bound to the quantum entity ("shadow of light").*

This can be depicted as follows. Most of the energy of the photon is concentrated in a tiny extent; the remaining part is employed to deform the space–time surrounding it and, hence, it is stored in this deformation. It is just the deformations ("shadows") of the photons from $S_2$ that expand, go through $F_2$ and interact with the shadows of the photons emitted from $S_1$.

Therefore, in this view, the difference of signal measured by the detector A in all the double-slit experiments can be interpreted *as the energy absorbed by the space–time deformation itself*, which cannot be detected by the central detector B.[6] In other words, the experimental device, used in these experiments, "weighed" the energy corresponding to the space–time deformation by the measured difference on the first detector.

If the interpretation we have given here is correct, *the double-slit experiments do provide for the first time, among the others, direct evidence for the Einstein–de Broglie–Bohm waves and yield a measurement of the energy associated to them.*

The hypothesis of the hollow wave as space–time deformation is able to explain also the anomalous behavior observed in crossed photon-beam experiments (see Sect. 13.2). In fact, the shadow of the photon spreads beyond the border of the space and time sizes corresponding to the photon wavelength and period, respectively. This changes the photon-photon cross section (strongly depressed both in classical and in quantum electrodynamics),[7] and gives rise to the anomalous effects observed in the photon–photon interactions in crossing beams.

The earlier interpretation is of course incompatible with standard electrodynamics (either classical or quantum). This is also easily seen by the ensuing violation of LLI, on account of the strict connection between Lorentz invariance and electrodynamics (as is well known, the standard Lorentz group is the covariance group of Maxwell equations). We want now to show that a more detailed analysis of the measurements of the third experiment

[6] One might think to detect such an "energy of deformation field" (corresponding to the hollow waves of photons) by a detector operating by the gravitational interaction, rather than the electromagnetic one. However, this would still be impossible, because the deformation value lies within the energy interval for a flat (Minkowski) gravitational space–time, according to DSR (see Sect. 4.1).

[7] In fact it goes as $\alpha^4$ (with $\alpha$ being the fine structure constant).

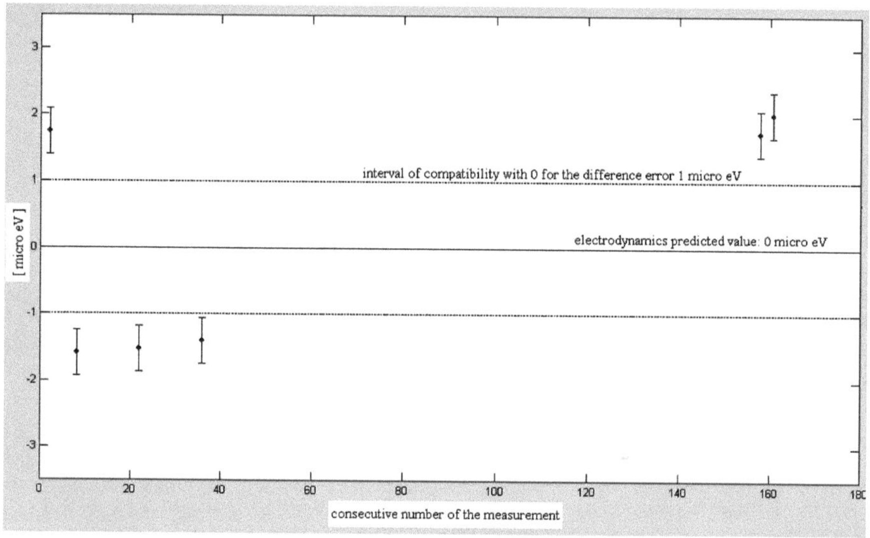


FIGURE 13.4. Values of the differences $\Delta A'$ of signal sampled on detector A with both sources on (third experiment). The differences are clearly compatible with zero

are just in favor of the anomalous (Lorentzian) interference observed as signature of a possible violation of electrodynamics.

This is easy to realize, by noting that the distribution of the results of the third experiment (reported in Fig. 13.2) is unmistakably different from that expected from the theoretical predictions of both quantum and classical electrodynamics.

With reference to Sect. 13.1, we recall that Fig. 13.2 shows the signal differences measured on A in correspondence to the two different states of lighting of the sources, $\Delta A = A(\mathrm{S_1on\ S_2off}) - A(\mathrm{S_1on\ S_2on})$. For comparison, we report in Fig. 13.4 the differences of the two values sampled on A in the same lighting condition of the sources, i.e., with both sources turned on: $\Delta A' = A(\mathrm{S_1on\ S_2on}) - A(\mathrm{S_1on\ S_2on})$. Again, for clarity' sake, we show only the differences outside the interval $[-1, 1]$. There is no surprise in observing that the differences are almost evenly distributed around zero, since the subtracted values belong to the same population. However, by the very design of the experimental box, according to either classical or quantum electrodynamics detector A was not to be affected by the state of lighting of the source $S_2$. Hence, one would expect that the mean value of these differences was zero and that the differences $\Delta A$ were uniformly distributed around it. In other words, one would expect to find roughly the same number of positive and negative differences, and therefore that Figs.13.2 and 13.4 displayed two compatible distributions of differences evenly scattered across zero. On the contrary, *the differences in Fig. 13.2 are not uniformly distributed around zero but are markedly shifted downward* (as compared to

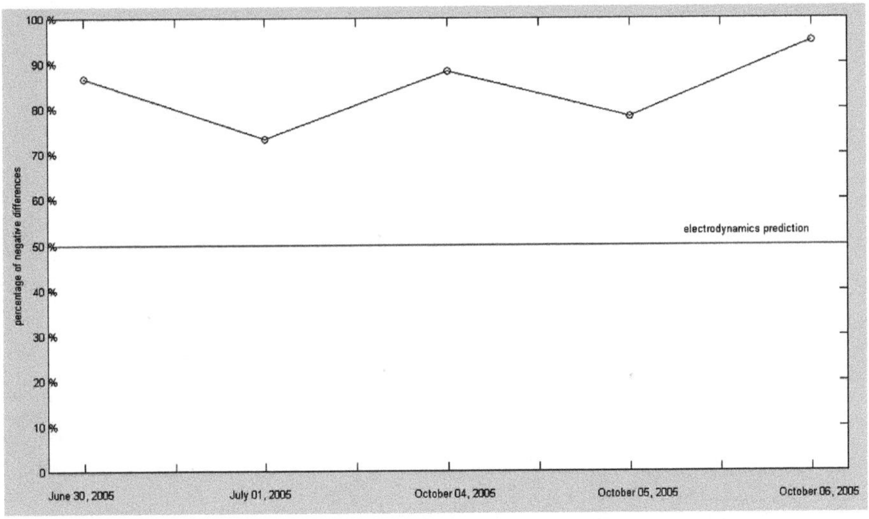


FIGURE 13.5. Oscillations of the percentage of differences $\Delta A$ in the five measurement sessions of the third experiment

those in Fig. 13.4), and hence the number of negative differences is larger than the positive ones.

One can go into this point in more depth by considering the oscillations of the percentage of negative differences $\Delta A$ (Fig. 13.5). The five points in such a figure represent the five percentages of negative differences attained in the five different sessions of the third experiment. It is quite evident that they do oscillate around a mean value as expected, but this mean value is approximately 85% and not 50% as predicted by electrodynamics. Then, it follows that the downward displacement of the differences in going from Fig. 13.4 to Fig. 13.2 is not a mere chance, but is a systematic result obtained every time the experiment was performed. Let us notice that each of the five sessions reported in Fig. 13.5 has actually to be counted as if it were four sessions, due to the particular procedure adopted to sample the signal on detector A [66]. Then one has 20 sessions of the experiment in which the percentage of negative differences is always much greater than 50%.

In order to further enforce the evidence for the difference of the two physical situations corresponding to Figs.13.2 and 13.4, we carried out a statistical analysis of the results found in the two cases (only the differences outside the interval $[-1, 1]$ have been considered), by taking also into account the instrumental drift. The Gaussian curves obtained are shown in Fig. 13.6. The dashed curve refers to the signal differences $\Delta A = A(\mathrm{S_1on\ S_2off}) - A(\mathrm{S_1on\ S_2on})$, whereas the solid one to $\Delta A' = A(\mathrm{S_1on\ S_2on}) - A(\mathrm{S_1on\ S_2on})$. The two curves differ by 2.5 $\sigma$, clearly showing that

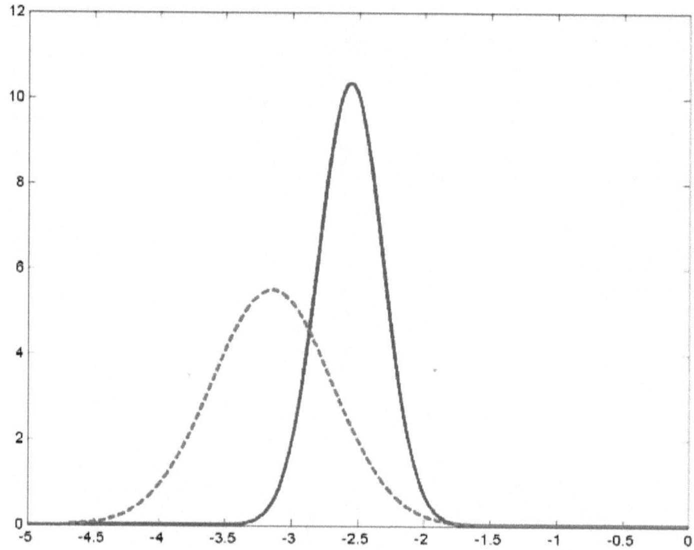


FIGURE 13.6. Gaussian curves (normal frequency vs. signal difference in $\mu V$) for the signal differences $\Delta A$ and $\Delta A'$ on detector A for the two cases of source $S_2$ off and on (*dashed* and *solid curve*, respectively). The instrumental drift has been taken into account. It is $\overline{\Delta A} = -3.15$ ($\sigma$ =0.45); $\overline{\Delta A'} = -2.56$ ($\sigma$ =0.24)

the two cases are statistically distinct, the latter one representing a mere fluctuation (unlike the former).

We can therefore conclude that the results obtained on the anomalous behavior of photon systems brings to light a more complex physics of the electromagnetic interaction, which again calls for giving up the local Lorentz invariance in order to be accounted for. They are apparently at variance with both standard quantum mechanics (in the Copenhagen interpretation) and usual (classical and quantum) electrodynamics.

The interpretation in terms of DSR is quite straightforward. Under the energy threshold $E_{0,\text{e.m.}}$=4.5 μeV, the metric of the electromagnetic interaction is no longer Minkowskian. The corresponding space–time is deformed. Such a space–time deformation shows up as the hollow wave accompanying the photon, and is able to affect the motion of other photons. This is the origin of the anomalous interference observed. It was noted at the beginning of this section that the difference of signal measured by the detector A in all the double-slit experiments can be regarded as the energy spent to deforme space–time. In space regions where the external electromagnetic field is present (regions of "standard" photon behavior), we can associate such energy to the difference $\Delta\mathcal{E}$, (3.124), between the energy density corresponding to the external e.m. field $F_{\mu\nu}$ and that of the deformed one $\widetilde{F}_{\mu\nu}$ given by (3.119).

But it is known from the experimental results that the anomalous interference effects observed can be explained in terms of the shadow of light, namely in terms of the hollow waves present in space regions where no external e.m. field occurs. How to account for this anomalous photon behavior within DSR? The answer is provided by the internal structure of the deformed Minkowski space discussed in Sect. 9.4. In fact, we have seen that the structure of the deformed Minkowski space $\widetilde{M}$ as Generalized Lagrange Space implies the presence of two internal e.m. fields, the horizontal field $\mathcal{F}_{\mu\nu}$ and the vertical one, $f_{\mu\nu}$. Whereas $\mathcal{F}_{\mu\nu}$ is strictly related to the presence of the external electromagnetic field $F_{\mu\nu}$, vanishing if $F_{\mu\nu} = 0$, the vertical field $f_{\mu\nu}$ is geometrical in nature, depending only on the deformed metric tensor $g_{\mathrm{DSR},\mu\nu}(E)$ of $\mathrm{GL}^4 = \widetilde{M}$ and on $E$. Therefore, it is present also in space–time regions where no external electromagnetic field occurs. In our opinion, the arising of the internal electromagnetic fields associated to the deformed metric of $\widetilde{M}$ as Generalized Lagrange space is at the very physical, *dynamic* interpretation of the experimental results on the anomalous photon behavior. Namely, *the dynamic effects of the hollow wave of the photon, associated to the deformation of space–time* – which manifest themselves in the photon behavior contradicting both classical and quantum electrodynamics –, *arise from the presence of the internal v-electromagnetic field* $f_{\mu\nu}$ (in turn strictly connected to the geometrical structure of $\widetilde{M}$).

Moreover, as is well known, in relativistic theories, the vacuum is nothing but Minkowski geometry. An LLI breaking connected to a deformation of the Minkowski space is therefore associated to a lack of Lorentz invariance of the vacuum. Then, the view by Kostelecky [55] that the breakdown of LLI is related to the lack of Lorentz symmetry of the vacuum accords with our results in the framework of DSR, provided that the quantum vacuum is replaced by the geometric vacuum. Notice also that in the Kostelecky formalism it is impossible to recover local Lorentz invariance. On the contrary, DSR recovers it in a generalized sense, in the form of deformed Lorentz invariance (see Sects. 3.3.5, 3.3.7). Let us also recall (as we shall see in Part IV) that, as already said, DSR admits a natural immersion in a 5D-space, and that the vacuum solutions of the Einstein equations in such a space reproduce the phenomenological metrics discussed in Sect. 4.1. In this connection, it was proved [74] that waves and particles admit a common geometrical interpretation as isometries of a 5D space. One can therefore hazard the view that local Lorentz invariance, apparently violated, is actually recovered in the 5D version of DSR *as an exact symmetry*, intimately related to the propagation of quantum waves in the 4D space time.

# 14
# The Coil Experiment

In Chap. 13, we discussed some experiments which provide evidence for a deformation of space–time of electromagnetic nature. In the present one, we shall describe an experimental test of Lorentz invariance whose results require recourse, in order to be explained, not only to an electromagnetic, but also to a gravitational deformation of the space–time geometry.

## 14.1 Experimental Setup and Results

At the end of the twentieth century we proposed, together with Bartocci [75], a new electromagnetic experiment aimed at testing LLI and able of providing direct evidence for its breakdown. The results obtained in a first, preliminary experimental run carried out in June 1998 – essentially aimed at providing new upper limits on the LLI breakdown parameter by an entirely new class of electromagnetic experiments – admitted as the most natural interpretation the fact that local Lorentz invariance is in fact broken [6, 75]. The experiment was just repeated in 1999 in a different place (100 Km far from the previous one), with a completely new and improved apparatus, and confirmed the positive evidence of the first one [6, 76].

The new proposed test is based on the possibility of detecting a nonzero Lorentz force between the magnetic field **B** generated by a stationary current $I$ circulating in a closed loop $\gamma$, and a charge $q$, on the assumption that both $q$ and $\gamma$ are at rest in the same inertial reference frame. Such a force is zero, according to the standard (relativistic) electrodynamics. The

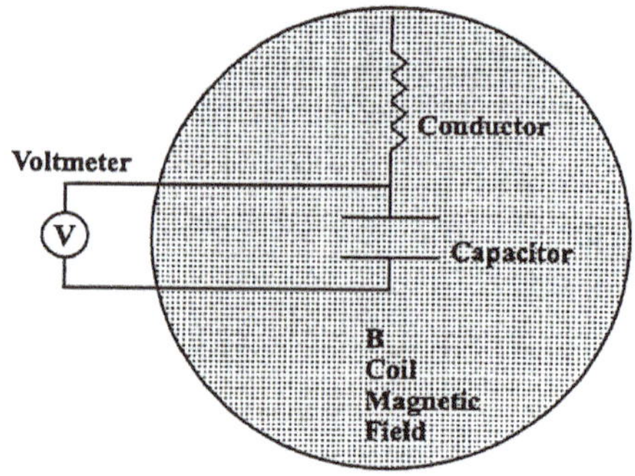


FIGURE 14.1. Schematic view of the experimental setup in the coil experiment

theoretical estimate of the effect related to the violation of the standard electrodynamic law can be found in [6].

The experimental setup was devised in order not only to put new upper limits on the breakdown of LLI, but also to test possible anisotropic effects in such limits. Measurements of the voltage "V" across the capacitor were carried out for the system lying in the different coordinate planes $(x, y)$, $(x, z)$, $(y, z)$, and at different values of the orientation angle $\alpha$ of the circuit in the plane considered (spaced by $\pi/4$). The orientation of the coil $\gamma$ and the verse of the current $I$ were chosen so that, when $\gamma$ lay on $(x, y)$, its magnetic field $\mathbf{B}$ was directed as $z$; when $\gamma$ was on $(y, z)$, $\mathbf{B}$ was directed as $x$; for $\gamma$ on $(x, z)$, $\mathbf{B}$ was directed as $\mathbf{B}_\mathrm{T}$ (Earth magnetic field).

The experimental device used is schematically depicted in Fig. 14.1. It consisted of a Helmholtz coil $\gamma$ and a copper conductor $R$ placed inside it on a plane orthogonal to the $\gamma$ axis. The conductor $R$ was connected in series to a capacitor $C$, and a voltmeter was connected in parallel to the capacitor, so as to measure the voltage due to a possible gradient of charge across $R$. The conductor could change its orientation in the coil plane. Moreover, the whole system of the RC circuit and the coil could turn to make its plane coincide with one of the coordinate planes. The center of the geometrical coordinate system coincided with the center of the coil. The coordinate system was chosen as follows: the $(x, y)$-plane tangent to the Earth surface, with the $y$-axis directed as the (local) Earth's magnetic field $\mathbf{B}_\mathrm{T}$; the $z$-axis directed as the outgoing normal to the Earth's surface, and the $x$-axis directed so that the coordinate system was left-handed. The conductor orientation in the plain coil was parameterized in terms of an angle $\alpha$ (ranging from 0 to $2\pi$). The rotation of $\alpha$ was chosen clockwise in the plain coil with respect to an observer oriented along the coordinate axis orthogonal to the coil plane. The first orientation of $R$ corresponding to the angle $\alpha = 0$ was along the negative direction of the $z$-axis in the case of the two vertical canonical planes and along the negative direction of the $y$-axis in the case of the horizontal plane. A steady-state current $I$ circulating in the coil produced a constant magnetic field $\mathbf{B}$ in which the

RC circuit was embedded. The circuit and the coil were mutually at rest in the laboratory frame.

In the first experiment, the measurements performed with the system lying on the planes $(x, y)$ and $(y, z)$ gave values of $V$ compatible with the instrument zero. Indeed, in such cases the statistical tests of correlation showed that each of the points outside the zero-voltage band is uncorrelated with the preceding and the subsequent point either, and the whole set of points was shown to be uncorrelated ($R^2 < 30\%$). Let us stress that each point was the average of five measurements, taken at the same angle. As to the measurements in the plane $(x, z)$, it was shown instead that *the four points outside the zero band were statistically correlated* ($R^2 > 80\%$), and so they represented a valid candidate for a nonzero signal.

A polynomial interpolating curve for these points is shown in Fig. 14.2. Such an interpolating procedure was essentially aimed at finding the angle $\alpha_{\max}$ corresponding to the maximum value of $V$, $V_{\text{xzmax}} = (3.6 \pm 1.0) \times 10^{-5}$ volt. The value found was $\alpha_{\max} = 3.757$ rad. The knowledge of $\alpha_{\max}$ was needed in order to determine the value of the anisotropic LLI violation parameter in our case [6, 76].

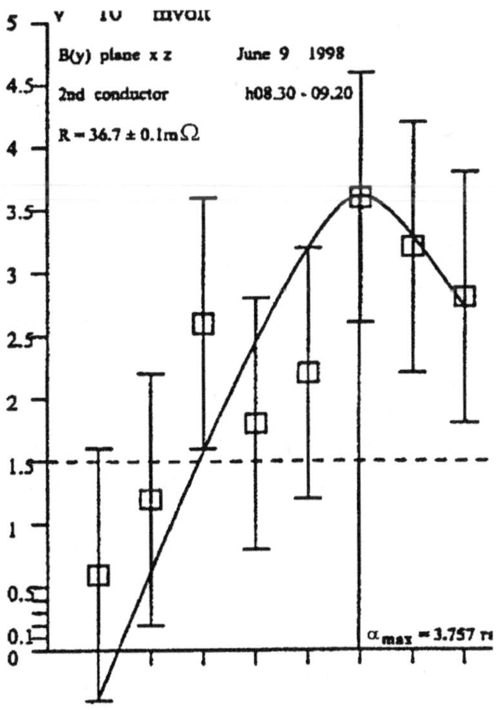


FIGURE 14.2. Curve interpolating the data obtained with the apparatus in the $(x, z)$ plane, showing the angle of maximum signal $\alpha_{\max} = 3.757\,rad$

In the second experiment, a signal candidate was analogously found in the plane $(x, z)$. For $B = B_1 = (5.14 \pm 0.01)\,\mathrm{mT}$, the average peak value was in excellent agreement with the result of the former experiment: $V'^{xz}_{\mathrm{max}} = (3.54 \pm 0.01) \times 10^{-5}$ volt. The signal was again highly anisotropic, and its behavior with $\alpha$ is the same as depicted in Fig. 14.2. However, possible signal candidates were now also found in the planes $(x, y)$ and $(y, z)$ (this was also a consequence of the higher sensitivity of the multimeter, improved by two orders of magnitude). In those planes, there was no dependence on $\alpha$, and therefore no spatial anisotropy. On the contrary, a time anisotropy was found in the $(x, y)$ plane, since the measurements taken a.m. gave values within the instrument zero. The average level values found for $B = B_1$ were $V'^{xy} = (3.07 \pm 0.01) \times 10^{-5}$ volt and $V'^{yz} = (2.66 \pm 0.01) \times 10^{-5}$ volt. The measurements taken with the halved value of the coil magnetic field, $B = B_2 = (2.58 \pm 0.01)mT$, gave similar results, with voltage values $V'^{xz}_{\mathrm{max}} = (4.18 \pm 0.01) \times 10^{-5}$ volt, $V'^{xy} = (3.44 \pm 0.01) \times 10^{-5}$ volt and $V'^{yz} = (3.06 \pm 0.01) \times 10^{-5}$ volt. Not only these values were not halved with respect to those obtained for $B = B_1$ (as expected in the case of a linear relation between $V$ and $I$, like that derived via the Lorentz force), but, surprisingly enough, they came out slightly higher! Moreover, a check was made by reversing the coil current. No change in the sign of $V$ occurred. This allows one to conclude that the observed effect *is independent of the magnitude and direction of the current.*

## 14.2 LLI Breakdown Parameter

First of all, let us stress that effects analogous to that observed in the coil experiment have been foreseen [77–84]. For instance, it was shown that a nonzero electric field is expected to exist outside wires and/or closed loops carrying a constant current, whereas a nonnull Lorentz force between a charge and a coil both at rest in the same reference frame was predicted by the classically interpreted Maxwell theory. Moreover, some claims of evidence for such anomalous electromagnetic phenomena are present in literature, although they are controversial [77–84]. However, all such (both theoretical and experimental) effects do depend on the magnitude and/or the verse of the current, and are fully isotropic, and therefore they have nothing to do with ours.

Among the possible interpretations, the effect can be thought to arise due to a kinematical decoupling of the magnetic field **B** from the coil that generates it. As a consequence, the coil and the conductor are at rest in the same frame (the laboratory frame), whereas the field **B** is at rest with respect to an absolute reference frame $\Sigma_0$ (see Sect. 3.3.7). In the framework of this interpretation, it is possible to give an estimate of the Earth's speed $v$ with respect to such an absolute frame. One obtains:

$$v = (5.906 \pm\ 0.001) \times 10^{-2} \mathrm{m\,s}^{-1}. \tag{14.1}$$

It is now easy to see why it is impossible to detect such an effect by means of an experiment of the Michelson–Morley-type. As is well known, the displacement $\Delta n$ of the interference fringe in an MM experiment is given by

$$\Delta n = \frac{\ell_1 + \ell_2}{\lambda}\left(\frac{v_{\mathrm{R}}}{c}\right)^2, \tag{14.2}$$

where $\ell_1$, $\ell_2$ are the lengths of the arms of the interferometer, $\lambda$ is the light wavelength, and $v_{\mathrm{R}} \simeq 3\ \times 10^8\,\mathrm{m\,s}^{-1}$ is the velocity of Earth's revolution. In the original MM experiment, it is $\ell_1 + \ell_2 = 22\ m$, $\lambda = 5.5\ \times 10^{-7}\ m$, $\Delta n = 0.4$. In our case, we have to replace $v_{\mathrm{R}}$ by the Earth's speed $v$ with respect to the absolute reference frame $\Sigma_0$, whose value, according to our experimental findings (and the interpretation we proposed), is given by the above estimate, $v \simeq 0.06\ m/s$. Then, by using the same parameters of the original MM experiment, one gets

$$\Delta n \simeq 0.2 \times 10^{-11} \tag{14.3}$$

a fringe displacement completely unobservable even by modern tools.

We want to stress that the estimated degree of breakdown of LLI ensuing from our experiments is in agreement with the existing limits [56]. A detailed discussion of this point is given in [76]. Here, we confine ourselves to summarizing the main results.

We recall that two different kinds of LLI violation parameters $\delta$ exist: Isotropic (essentially obtained by means of experiments based on the propagation of e.m. waves, e.g., of the Michelson–Morley type), and anisotropic parameters (obtained via experiments of the Hughes–Drever type [56], which test the isotropy of the nuclear levels). The smallest upper limit obtained in the former case is $\delta < 10^{-8}$, whereas the upper limits on the anisotropic parameter range from $\delta < 10^{-18}$ of the HD experiment to $\delta < 10^{-27}$ of the Washington experiment [56]. In either case, one has to consider, for the evaluation of $\delta$ according to (11.1), the effective LLI breakdown speed $v$. In our framework, the speed $v$ is the Earth's speed with respect to the absolute frame $\Sigma_0$ given by (14.1). Then, it is possible to show [76] that the isotropic LLI parameter corresponding to our effect has the value $\delta \simeq 4\ \times 10^{-10}$, which is lower by two orders of magnitude than the upper limit for the isotropic case. In the anisotropic case, the parameter $\delta$ is in the range $2\ \times 10^{-29} < \delta < 6\ \times 10^{-20}$, and therefore compatible with the anisotropic upper limits.

We want to stress that, in general, in the usual analysis of the LLI violating parameters, one looks for an LLI breakdown speed $v$ which is in a sense external to the interaction ruling the physical system under measurement.

Indeed, typical candidates for $v$ are the Earth revolution speed around the Sun ($v = 30\ Km/s$) or the drift speed of the solar system in the Galaxy ($v = 300\ Km/s$). In the present case, one is assuming that the LLI breaking speed is actually *internal to the system.* It parametrizes the amount of LLI breakdown inherent to (and characteristic of) the interaction ruling the process. This different point of view is typical of the DSR formalism, and is reflected in the derivation we made of the anisotropic parameters (see [76]). Indeed, in order to calculate $\delta$ in the anisotropic case, one exploits the values of the maximal causal speeds derived from the analysis of the strong interaction based on BE correlation (the only interaction at present described by a spatially anisotropic metric: see Sect. 4.1.3)

The present experimental status of the LLI parameters, in the light of our results, is summarized in Fig. 14.3.

In conclusion, in two experiments, carried out in different places, with different experimental apparatuses, there was observed an effect of a voltage induced across a conductor by a stationary magnetic field that could be

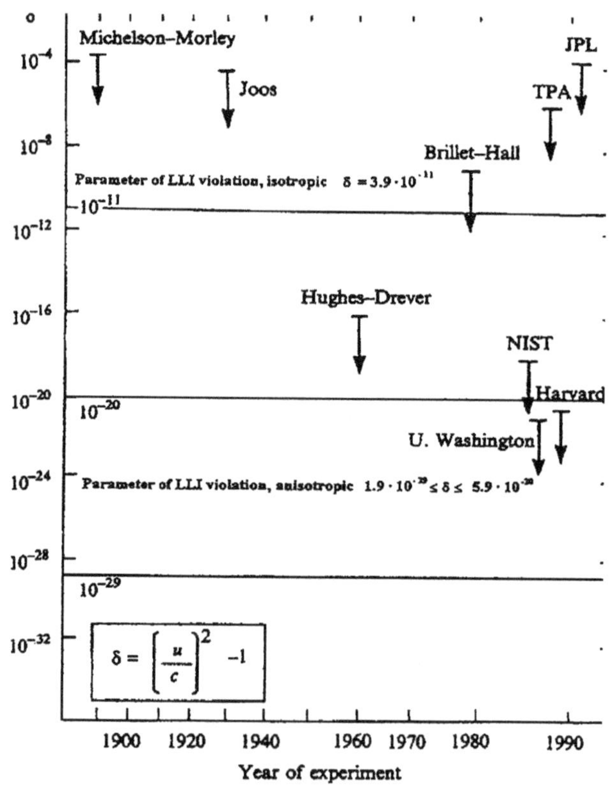


FIGURE 14.3. The present experimental situation of the limits on the LLI breakdown parameter d (adapted from Will, [49], p.322). The three horizontal straight lines are the limits obtained in the coil experiments. See the text

interpreted as a violation of local Lorentz invariance. Moreover, its parametrization in terms of an effective speed yields values of the LLI breakdown parameters consistent with the existing upper limits.

## 14.3 Interpretation in Terms of DSR

Let us now attempt to give an interpretation of the earlier results in terms of DSR.

To take account of the spatial anisotropy of the observed effect, we define the quantities

$$D = V_{am}^{xz} - V_{am}^{xy}, \tag{14.4}$$

$$\Delta =< V^{xz} > -V_{am}^{xy}. \tag{14.5}$$

The averages $< D >$, $< \Delta >$ are

$$\begin{aligned} < D > &= (21.1 \pm 0.9)\,\mu V; \\ < \Delta > &= (20.7 \pm 1.0)\,\mu \mathrm{V}. \end{aligned} \tag{14.6}$$

A possible (although preliminary) interpretation of these results requires recourse to an LLI breakdown not only electromagnetic, but also gravitational. Both experiments show a (peak) signal in the $(x, z)$-plane. Such an effect could, at a first sight, be attributed to a gravitational breakdown of the local Lorentz invariance. However, the results we obtained in the second experiment, which show a (level) signal in the plane $(x, y)$, suggest that the effect we observed is both gravitational and electromagnetic. If so, then we can conclude that the difference between the measured voltages in the two planes (namely, the quantities $< D >$, $< \Delta >$ provide a measure of the amount of the gravitational contribution to the measured voltage versus the electromagnetic one.

This is indeed supported by the agreement of the values (14.6) with the threshold energy for the gravitational metric $E_{0,\mathrm{grav}}$ (see (4.19)).

On the other hand, let us evaluate the differences in the voltages measured in the three planes. According to the second experiment with $B = B_1$, the transitions of the LLI breakdown value of the voltage:

1. From the space anisotropy in the plane $(x, z)$ $< V^{xz} >= (36.2 \pm 0.7)\,\mu V$ to the time anisotropy in the plane $(x, y)$ $(V_{pm}^{xy} = (30.7 \pm 1.5)\,\mu\mathrm{V})$
2. From the time anisotropy in the plane $(x, y)$ to the space–time isotropy in the plane $(y, z)$ $(< V^{yz} >= (26.5 \pm 1.8)\,\mu\mathrm{V})$

do occur for steps $< \Delta >= 4.85\,\mu V$:

$$\begin{aligned} &< V^{xz} > -V_{pm}^{xy} = 5.5 \mu V; \\ &V_{pm}^{xy} - < V^{yz} >= 4.1 \mu V. \end{aligned} \tag{14.7}$$

This does just agree with the value (4.4) of the threshold energy for the electromagnetic metric $E_{0.\mathrm{e.m.}}$ (see Sect. 4.1).

The earlier discussion shows that the observed space anisotropies of the effect are probably related to the behavior in energy of the phenomenological metrics describing, in DSR, the two interactions involved (electromagnetic and gravitational). In fact, the gravitational metric is over-Minkowskian, and reaches the limit of Minkowskian metric for decreasing values of $E$ (with $E > E_{0,\mathrm{grav}}$) (see (4.17)–(4.18)), whereas the electromagnetic metric is sub-Minkowskian and thus attains the Minkowskian limit for increasing values of the energy ($E < E_{0,\mathrm{e.m.}}$) (see (4.2)–(4.3)). Therefore the two metrics become Minkowskian for energy variations of opposite sign.

On the contrary, there is presently no explanation of the observed time anisotropy. We stress that just taking the time anisotropy into account permits to highlight the role of electromagnetic interaction in the breakdown of LLI and to get the related contributions, (14.7). However, it might be argued that some connection exists with the time structure observed in the measurement sampling in the double-slit experiments (see Chap. 13). Let us in fact recall that the plane $(x, y)$ (in which the time anisotropy was observed) is the same of the layout of the interference experiment (see Fig. 13.1). We shall come back to this point in Part V.

We can therefore conclude that, even if the first experiment left open the possibility of a mere gravitational explanation of the observed effect, the results of the second one are strongly in favor also of an inescapable electromagnetic contribution to the breakdown of LLI.

Moreover, the analysis of the experiments discussed in Chaps. 12–14, all based on the electromagnetic interaction, confirms that the electromagnetic breaking of the local Lorentz invariance occurs indeed at very low energies, in agreement with the sub-Minkowskian behavior of the electromagnetic metric (cf. Sect. 4.1.1).

# 15
# The Speed of Gravity

## 15.1 How Fast Is Gravity?

It is well known that gravitational interactions between bodies in all dynamic systems are always taken to happen instantaneously. In fact, a finite gravity speed would give rise to the aberration phenomenon, whereby the gravitational acceleration vector would be directed toward the retarded position of the source, not toward the instantaneous one. It is experimentally ascertained that gravitational effects do not present aberration (unlike the electromagnetic ones). This is why astronomers can calculate orbits using instantaneous forces. However, this instantaneous propagation of gravitational effects is indeed an instantaneous action-at-a-distance with all its unphysical consequences and problems. This has been a major concern for many physicists who have therefore carried out experiments in order to put a lower limit to the speed of gravity. As a matter of fact, the problem of the speed of transmission of gravitational effects is an old one. It can be traced back to Laplace, who, in his monumental work "Mecanique Celeste" in 1825, estimated $u_{\mathrm{grav}} \geq 10^8 c$ by analyzing the motion of the Moon with respect to the gravitational pull exerted on it by the Earth and the Sun.

Let us clarify what we mean by "speed of gravitational effects" . As is well known, General Relativity predicts the existence of gravitational waves, i.e., weak disturbances of the space–time metric[1] (obeying a Helmholtz wave

[1] In fact, gravitational radiation is a fifth order effect in $v/c$ .

equation) which propagate at the speed of light. Indirect confirmation of this fact was provided by Taylor and Hulse in 1974 by their analysis of binary pulsars (see the Nobel Lectures [85, 86]). Gravitational radiation does admit retarded-potential solutions of electromagnetic type. It therefore describes propagation of the perturbations of a *static* (or near static) gravitational potential field.

On the contrary, how much is the propagation speed *of the gravitational force*? By this we mean the speed at which *variations of the gravitational force* do travel. It answers the question of how much time a target body will take to respond to the acceleration of a source mass. Such a time is obviously zero, in Newtonian mechanics. Borrowing a beautiful analogy from [87], let us consider a buoy floating on sea surface. The buoy is connected by a chain to an anchor holding it in place. If the anchor is moved, the chain causes the buoy to move too. In turn, the buoy motion sets off water waves. Translated in gravitational language, the anchor is the source mass, the chain is the gravitational force, the buoy the target mass. The water waves caused by the buoy motion (induced by the anchor motion) travel at the sound speed in water, and are the analogous of gravitational waves: there is no connection between their speed and the speed of transmission of the force field from the anchor to the buoy by the chain. Variations of the gravitational force (namely, variations of the whole space–time geometry) originate from acceleration of the source mass; gravitational waves (i.e., small ripples of the space–time geometry) originate from acceleration of the target body.

Let us stress that, in the framework of unified theories of fundamental interactions (including gravity) based on multidimensional spaces (see [18] and Chap. 18), the gravity speed $u_{\text{grav}}$ is *different* both from the speed of light (namely, the speed of propagation of the electromagnetic signals, $u_{\text{e.m.}}$) and from the speed parameter $c$ entering the Lorentz transformation (the relation between them being $u_{\text{e.m.}} = c/\mathcal{N}$, with $\mathcal{N}$ being the vacuum refractive index: see Sect. 3.4.3).

The experiments testing the speed of gravity are essentially of three types: Solar system, astrophysical and laboratory experiments. The Laplace estimate was just of the first type. The same arguments by Laplace were applied, at an astrophysical level, by Van Flandern [88] to binary pulsars. Thus he got a lower limit of $2 \cdot 10^{10}\, c$ for the speed of gravity.[2] Due to the extremely high value of the speed of gravity, all attempts to measure it directly by means of tabletop setups could not produce any reasonable value (let us, however, quote the experiment performed by Walker and

---

[2] For reader's convenience, we recall that the equation for the gravitational speed used by Van Flandern is $v_{\text{g}} = \left[\frac{12\pi^2}{p\dot{p}}\left(\frac{a}{c}\right)\right] c$, where $p$ is the period of the orbit of the binary pulsar, $\dot{p}$ is its time variation, and $a$ is the major semiaxis of the orbit. The data refer to the binary pulsars PSR1913+16 and PSR1534+12.

Dual, which apparently confirms the superluminal propagation of gravitational effects [89]). The most recent attempt to evaluate the gravity speed was based on the analysis of astronomical interferometry data from measurement of the deflection of light from a quasar by planet Jupiter [90]. The conclusion was that the speed of gravity is between 0.8 and 1.2 times the speed of light, in agreement with the standard prediction of General Relativity. However, such a claim has been criticized by several physicists, on the grounds that the results of the measurements have been misinterpreted (see e.g., [91,92]).

What are the predictions of DSR for the speed of gravitational force? According to the general discussion of Sects. 3.2 and 3.3.5, the maximal causal speed for a given interaction is a function of the energy-dependent coefficients of the metric describing the interaction considered, and is different for each space direction for an anisotropic space (see (3.69)). We have seen in Sect. 4.1 that, for gravity, only the time coefficient $b_0(E)$ can be derived from the experimental data on clock rates. The gravitational metric is therefore given by (4.17) and (4.18). According to the possible guess on the spatial metric coefficients $b_k(E)$, one gets different results for the gravity maximum speed $u_{\text{grav}}$ [6]. In particular, by assuming a gravitational metric analogous to the strong one (see (4.17a), (4.18a)), and two of the space coefficients $b_k(E) = 1$ $(k = 1, 2)$, one finds both $u_{\text{grav}} = c$ and $u_{\text{grav}} \neq c$. In the latter case, the expression of the (energy-dependent) gravitational speed is

$$u_{\text{grav}}(E) = \left(1 + \frac{E}{E_{0,\text{grav}}}\right) c \tag{15.1}$$

with $E_{0,\text{grav}} \simeq 20.2\ \mu\text{eV}$ is the threshold energy for the gravitational interaction (see (4.19)).

An estimate of the lower limit of the gravitational speed $u_{\text{grav}}$ (15.1) can be given by considering, for the energy $E$, the rest energy associated to the gravitational object of minimal mass constituting the matter, i.e., the electron ($m_e c^2 \simeq 0.5\,\text{MeV}$). Replacing such a value in (15.1) yields [6]

$$u_{\text{grav}} \geq \left(1 + \frac{m_e c^2}{E_{0,\text{grav}}}\right) c = 2.5 \cdot 10^{10} c, \tag{15.2}$$

in astonishing agreement with the astronomical estimate by Van Flandern [88].

In spite of the difficulty in carrying out tabletop experiments, at the end of the twentieth century we designed and performed a laboratory experiment aimed at estimating a lower limit for the speed of gravity. It is essentially a Cavendish-like experiment. However – since the single delay of propagation due to the finite speed of gravity would be too short to be measured with a reasonable accuracy – we exploited the stratagem of prolonging the measurement for a sufficiently long amount of time in order to let these delays to accumulate. This produced a total delay easy to be measured, from which it was possible to estimate a lower limit to the speed of gravity.

## 15.2 Cavendish-like Experiment

### *15.2.1 Experimental Setup*

The experiment was carried out at L'Aquila in 1999–2000. The experimental setup used is shown in Fig. 15.1. It consisted of an asymmetric rotor, i.e., an unbalanced one, with only one sphere of mass $M = 1.5\,\mathrm{kg}$, which revolved in front of a Cavendish torsion balance contained in a suitable box in order to prevent disturbances from the surroundings (air displacements

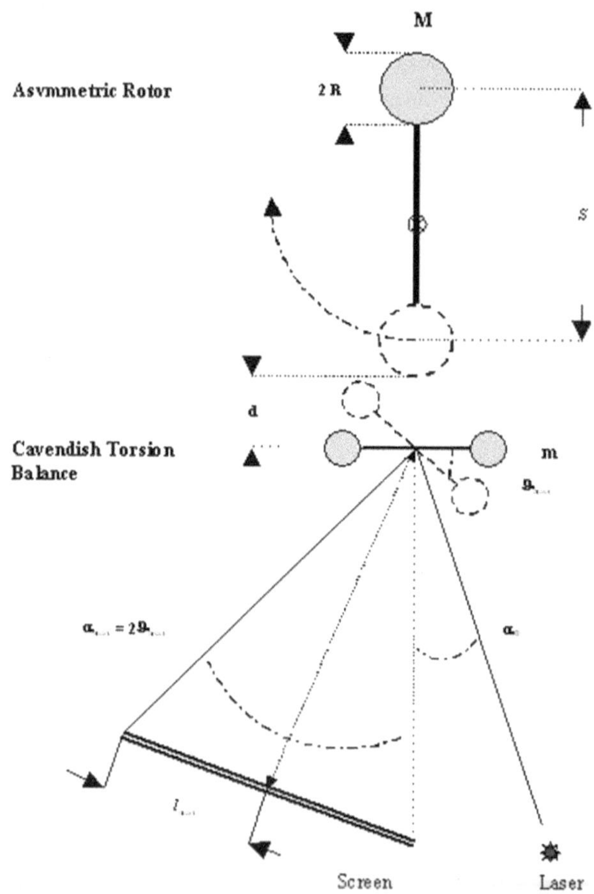


FIGURE 15.1. Experimental setup employed in the Cavendish-like experiment aimed at testing the speed of gravity

and vibrations). The spheres at the end of the torsion balance crossbar had a mass $m = 0.012\,\mathrm{Kg}$. At the beginning of the experiment the distance between the revolving sphere and the center of the crossbar was $d = 2\,\mathrm{cm}$, as indicated in the figure. The maximum excursion of the center of the revolving sphere from its closest position to the torsion balance crossbar was $S = 32.7\,\mathrm{cm}$. In order to measure the torsion of the balance, we conveyed a laser light onto a mirror attached to the polyethylene wire to which the crossbar was suspended, collected the reflected spot of light on a graduated screen and measured its excursion $l$. The distance of the screen from the center of the balance crossbar was $D = 213\,\mathrm{cm}$ (see Fig. 15.1). At the beginning, when the rotor occupied its closest position to the balance, the torsion of the wire was null as well as the excursion of the laser spot on the graduated screen. This position corresponded to an initial angle $\alpha_0 = 0.175\,\mathrm{rad}$. The maximum excursion of the laser spot (obtained in the measuring time of 200 min) was $l_{\max} = l(t = 200\,\mathrm{min}) = 3.6\,\mathrm{cm}$. The excursion of the spot on the graduated screen was sampled every $T_C = 1200$ s (20 min).

The purpose of the experiment was to estimate the delay, due to the (finite) speed of gravity, of the response of the Cavendish torsion balance gravitationally excited by a revolving weight, i.e., the asymmetric rotor.

The measurements were carried on for a long interval of time in order to accumulate this delay which otherwise would be at maximum 1 ns, because of the small dimensions of the apparatus (refer to Fig. 15.1).

The Cavendish torsion balance used in the experiment could oscillate both like a compound pendulum, with free-oscillation period $T_P = 1.1\,\mathrm{s}$, and like a torsion pendulum, with free-oscillation period $T_T = 160\,\mathrm{s}$. This means that the two types of oscillation were in phase roughly every 160 s, i.e., about 4 times every 10 min and about 8 times every 20 min. This circumstance suggested to choose a revolution frequency $\nu = 0.25\,\mathrm{Hz}$ (corresponding to a 4 s period) for the asymmetric rotor put in front of the of the torsion balance (see Fig. 15.1). The rotor exerted a gravitational pull on either sphere of the balance, which was thus forced to oscillate with frequency $\nu$.

It was verified that the torsion balance was indeed affected by this periodic forcing, by checking that a 4-second oscillation was superimposed on its free oscillations.

To this aim, we used a rectangular array of photo-resistances which was lain along and amid the graduated screen where the excursion of the laser spot, corresponding to the torsion of the balance, was measured. The photo-resistance width was so chosen that the laser spot was entirely contained within it as the laser swept the whole length of the photo-resistance. The electric signal collected at the photo-resistance was visualized by an oscillograph. As the laser spot got closer to the mid point of the photo-resistance, the trace on the oscillograph screen was a mounting ramp because of the increase of the received light. The 4-second oscillation was clearly visible as a periodic ripple superimposed on this ramp. This means that the length of the photo-resistance was long enough to make out the 4-second oscillation.

One could think that the delay, due to the finite speed of gravity, can be determined from the phase difference between the response signal of the torsion balance (whose period is 4 s) to the rotor gravitational pull and the 4-second signal corresponding to the rotation of the rotor. Unfortunately this method has two main problems. The first is the impossibility to extract a sufficiently clean and clear response signal from the complex oscillations of the balance, and hence to measure a significant phase difference. The second problem lies with the features of the delay. Namely, even if a phase difference would be measurable and the delay could be determined by this phase difference, it would still be impossible to know if the delay remains constant with the measuring time or if it has a rule of variability with it.

Thus we preferred to wait for the single delays to accumulate and then estimate the whole delay from the experimental data. Making such a choice meant to make some hypotheses, as to the way the single delays accumulate, in order to be able to estimate the speed of gravity from the whole delay. Obviously, at the end of the analysis of the experimental data, it was necessary to verify a posteriori the correctness of these hypotheses by comparing our estimate of the speed of gravity with the theoretical and experimental estimates known in literature.

### *15.2.2 Measurement Analysis and Results*

In general a kinematical estimate of the speed of gravitational effects would be given by

$$u_{\text{grav}} = \frac{s}{t_r}, \tag{15.3}$$

where $s$ is the distance between the sphere of the rotor, exerting the gravitational pull, and the axis of the balance, and $t_r$ is the time delay due to the finite gravity speed.

However, we deem that relation (15.3), being purely kinematical, is not completely exact, because it does not take into account the peculiar features of the experimental apparatus. In order to do this, one has to introduce suitable variation coefficients (or deformation coefficients) both for the distance $s$ in space and for the time delay $t_r$.

Since the oscillation period of the torsion pendulum is an entire multiple of the time taken by the rotor to complete one revolution, we inferred that within $T_C = 1,200\,\text{s}$ (interval of time between two samplings of the excursion) there are $T_C/T_T = 7.5$ gravitational pulls in phase out of a total number of pulls equal to $T_C \cdot \nu = 300$. Hence there are 7.5 pulls in phase and 292.5 pulls out of phase. This is to say that within 5 times 20 min (1,200 s), i.e., 100 min, there are $A_{lin} = 7.5 \cdot 5 = 37.5$ gravitational pulls in phase for which the effect of each delay is added linearly to the others. On the other hand, 300 − 7.5 = 292.5 times within 1,200 s (20 min), the torque exerted by the rotor is not in phase with the two types of oscillation of the balance and hence the effects of the delays cannot be added linearly.

Thus, we made the working hypothesis to add up the effects of all the delays within every sampling interval, and therefore to consider the whole final effect as the result of successive amplifications of the phenomenon of propagation delay.

One is guaranteed that this procedure is independent of the sampling interval by the fact that for different sampling intervals, being equal the rotation frequency of the rotor and the geometrical disposition of the whole apparatus, the measurement results of the torsion of the balance would be different.

Besides, we hypothesized that the effect due to the delay induces an amplification with gain $A$ of the oscillation.

Let $a = 292.5$ be the amplification coefficient and $k = 5$ the sampling coefficient, i.e., the number of times this amplification must be applied. Then, one has a nonlinear amplification gain $A = a^k = (292.5)^5$ over the first 100 min out of the whole measuring time of 200 min. In short, we can say that the phenomenon is described as the consequence of $k$ stages of amplification $a$, which yield a final amplification equal to $A = a^k$.

Thus, over the first 100 min the accumulated effect due to the delay is the sum of two contributions: the linear one, embodied by the amplification gain $A_{\text{lin}}$, and the nonlinear one given by the amplification gain $A$. Then the total amplification coefficient is $A_{\text{tot}} = A + A_{\text{lin}} \simeq A$ (since $A_{\text{lin}} \ll A$). We want to stress here that the definitions of linear and nonlinear amplification have to be understood as being related to the in-phase gravitational pulls and the out-of-phase ones, respectively.

On the other hand, if we take into account the entire measuring time of 200 min, the sampling coefficient $k'$ is twice as big as $k$ ($k' = 5 \cdot 2$), corresponding to an amplification coefficient $A' = a^{k'} = (292.5)^{5\cdot 2} = A^2$. Moreover, the response of the apparatus shows a phase inversion in moving from the first 100 min to the second 100 min (see Fig. 15.2). In order to take this decreasing effect into account, it is necessary to introduce, over 200 min, a coefficient with negative exponent. The total amplification gain is therefore given by

$$A'_{\text{tot}} = A' \cdot A^{-1} = A^2 \cdot A^{-1} = A. \tag{15.4}$$

The experimental results for the (maximum) excursions of the laser spot on the screen, $l$, as function of time are shown in Fig.15.2. The same figure reports also the graph $p$ of the least-squares fit polynomial, which approximates the experimental data with a correlation coefficient $R^2 = 0.99$, and two linear fits of the data: $r_1$ (continuous line), the least-squares line fitting the data with the minimum slope (this line goes through two main points: the point corresponding to the first measurement increased by its experimental error and the point corresponding to the last measurement decreased by its experimental error); $r_2$ (dashed line), the least-squares line

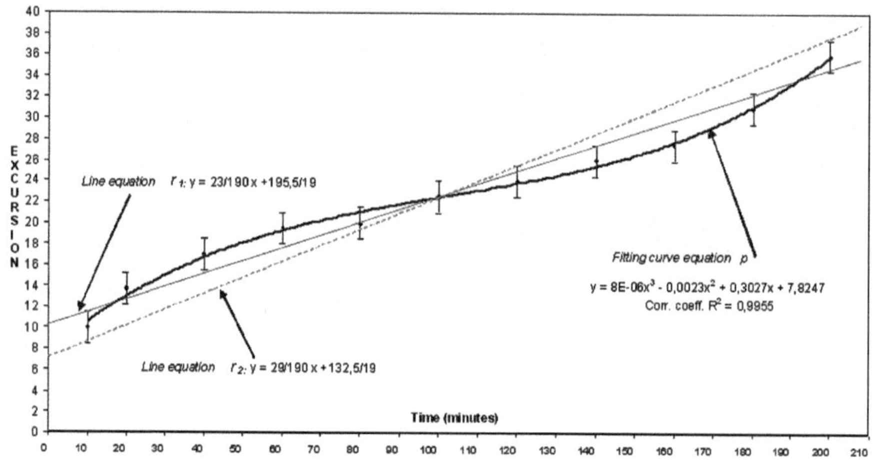


FIGURE 15.2. Laser spot excursions $l$ as function of time

fitting the data with the maximum slope (this line goes through two main points: the point corresponding to the first measurement decreased by its experimental error and the point corresponding to the last measurement increased by its experimental error).

The equations of the polynomial curve and of the lines are:

$$\begin{aligned} p &: \; l = 8 \cdot 10^{-6} t^3 - 0.0023 t^2 + 0.3027 t + 7.8247; \\ r_1 &: \; l = t\frac{23}{190} + \frac{195.5}{19}; \\ r_2 &: \; l = t\frac{29}{190} + \frac{132.5}{19}, \end{aligned} \qquad (15.5)$$

where the independent variable time ($t$) is expressed in "min", and the dependent variable excursion ($l$) in "mm".

With respect to these two lines, the experimental data show a periodic pseudosinusoidal wave shape. Thus the intersections of these two lines with the polynomial best fit allow us to estimate the value of the delay $\tau$ and of the corresponding error from the experimental data.

Indeed, the delay $\tau$ is given by the sum of the time interval between the two intersections of the lines and the polynomial curve at the beginning of the pseudosinusoidal wave shape and the time interval between the two intersections of the lines and the polynomial curve at the end of the pseudosinusoidal wave shape. Moreover, the time difference between the intersections of the two lines and the fitting curve near the flex point corresponds to the error $\Delta\tau$. In fact, passed the polynomial curve through the intersecting point of the two lines, there would be no error and furthermore the delay at the beginning and the one at the end of the pseudosinusoidal shape of the data would be identical.

It is quite evident that the intersections of the two lines with the polynomial best-fit do not depend on the experimental errors, since for different errors – measurement accuracy and sampling procedure being equal – we would get in general different data.

Thus, we have at our disposal the time delay and its error, i.e., $\tau \pm \Delta\tau$. This means that we can provide four kinematical estimates of the values of the speed of the gravitational effect, two of them related to the Einstein metric tensor for a weak gravitational field and the other two related to the deformed space–time. Of course, both couples correspond to the two possible values of the delay, i.e., $\tau_{\text{min}} = \tau - \Delta\tau$ and $\tau_{\text{max}} = \tau + \Delta\tau$.

The time coordinates of the intersections of the two lines $r_1$, $r_2$ with the curve $p$ are the following (see Fig. 15.3):

$$\begin{aligned} r_1 p \quad &: \quad t_{11} = 17.09 \text{ min}, \; t_{12} = 100.87 \text{ min}, \; t_{13} = 189.52 \text{ min}; \\ r_2 p \quad &: \quad t_{21} = -5.24 \text{ min}, t_{22} = 102.36 \text{ min}, t_{23} = 210.36 \text{ min}. \end{aligned} \tag{15.6}$$

The negative value of $t_{21}$ does not make physical sense and then we replace it by $t_{21} = 0$, which is the nearest value to have physical sense with regards to the measurements carried out. This is justified by the fact that, because of the initial transient effects (not reported in Fig. 15.2) in the experimental measuring apparatus, it is impossible to extrapolate to zero either the two lines or the polynomial curve.

With reference to Fig. 15.3, one gets the following values for the delay $\tau$ and its measurement error:

$$\begin{aligned} \tau &= (t_{11} - t_{21}) + (t_{23} - t_{13}) = 37.93 \text{ min} = 2276 \text{ s}; \\ \Delta\tau &= (t_{22} - t_{12}) = 1.49 \text{ min} = 89 \text{ s}. \end{aligned} \tag{15.7}$$

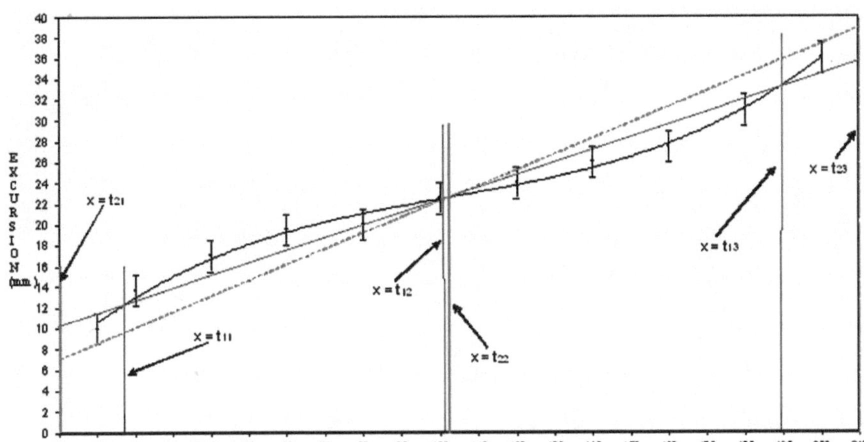


FIGURE 15.3. Intersections of the lines $r_1$, $r_2$ with the polynomial curve $p$

According to General Relativity, a local weak gravitational field affects the measure of time only (see (2.21)). Then, if one adheres to the GR view, the amplification effect has to be applied only to the measured value $\tau$ of the delay accumulated during the experiment. On the other hand, according to DSR, it is the whole 4D space–time to be in principle deformed, so the amplification would affect both time and space (in different ways of course).

Both in GR and in DSR, the relation between proper time and coordinate time is the same, $\tau = \sqrt{g_{00}}t$ (see (3.84)). Moreover, one has $t_r/\tau = 1/A$, so the delay $t_r$ in terms of the measured delay $\tau$ is given by the expression

$$t_r = \frac{\tau}{A}, \tag{15.8}$$

valid for both a flat space and a deformed one.

In terms of a flat space (i.e., with the Einstein metric tensor (2.21)), the measure of distances is not affected and hence $s = S$ (remember that $S$ is the maximum linear excursion of the center of the rotor exciting sphere from its closest position to the torsion balance crossbar: see Fig.15.1). On the other hand, according to DSR, the 3D space is deformed and therefore the distance $s$ is related to $S$ by

$$s = SA'_{\text{tot}} = SA^2 = 1.5 \times 10^{21}\text{Km} \tag{15.9}$$

(see (15.4) for $A'_{\text{tot}}$).

The expressions used to estimate the speed $u_{\text{grav}}$ of the gravitational effects by means of (15.3) in the two cases of flat space and deformed space–time are as follows:

– *Flat space*:

$$u_{\text{grav}} = \frac{SA}{\tau}; \tag{15.10}$$

– *Deformed space–time*:

$$u_{\text{grav}} = \frac{SA^2}{\tau} \tag{15.11}$$

We report in Tables 15.1 and 15.2 the minimum and maximum values of $u_{\text{grav}}$ corresponding to the estimated delays $\tau$ for the two theoretical hypotheses about the metric of space–time: Einstein metric (flat space) and deformed space–time, respectively. The speeds are expressed in units of the vacuum light speed $c$. From the values in Tables 15.1, 15.2 we obtain

TABLE 15.1. Values of $\tau$ and $u_{\text{grav}}$ for a flat space.

| | Flat Space | |
|---|---|---|
| $\tau$ (s) | $\tau_{\text{min}} = 2{,}187$ | $\tau_{\text{max}} = 2,365$ |
| $u_{\text{grav}}$ (units of $c$) | $u_{\text{grav max}} = 1.07$ | $u_{\text{grav min}} = 0.99$ |

TABLE 15.2. Values of $\tau$ and $u_{\mathrm{grav}}$ for a deformed space.

| | Deformed Space–Time | |
|---|---|---|
| $\tau$ (s) | $\tau_{\min} = 2{,}187$ | $\tau_{\max} = 2{,}365$ |
| $u_{\mathrm{grav}}$ (units of $c$) | $u_{\mathrm{grav\,max}} = 2.29 \times 10^{12}$ | $u_{\mathrm{grav\,min}} = 2.11 \times 10^{12}$ |

the following two different estimates of the speed of gravity $u_{\mathrm{grav}}$:

$$\text{Flat space} : u_{\mathrm{grav}} = (1.03 \pm 0.04)c; \qquad (15.12)$$

$$\text{Deformed space–time} \;: u_{\mathrm{grav}} = (2.20 \pm 0.09) \cdot 10^{12} c. \qquad (15.13)$$

## 15.3 Interpretation in Terms of DSR

We want now to discuss the compatibility of results (15.12), (15.13) with the theoretical predictions of DSR concerning the speed of gravitational effects and the variability of its value with energy. However, let us warn the reader about the physical difference between the experiment on clock rates (whose data have been utilized in order to derive the gravitational metric: see Sect. 4.1 and [6] for a detailed discussion) and the Cavendish-like experiment considered in Sect. 15.2. As a matter of fact, in the former case one worked in a *static* condition (the Earth gravitational field), i.e., in the language of DSR, inside an already (gravitationally) deformed space–time. On the contrary, in the latter case, we faced with a *nonstatic* (dynamic) situation, namely the settling of the gravitational deformation of space–time. Accordingly, the very meaning of the maximal causal speed $u_{\mathrm{grav}}$ for gravity must be changed, in the sense that, in this framework, it is better to be regarded as the speed of the gravitational action (product of energy by time) which brings about the deformation of space–time.

In spite of the earlier remarks, let us show that the experimental values (15.12), (15.13) of $u_{\mathrm{grav}}$ are in fact consistent with the DSR formalism. According to the discussion of Sect. 15.1, the value of the maximal causal speed for gravity obtained in DSR can be either $u_{\mathrm{grav}} = c$ or $u_{\mathrm{grav}} \gtrsim 2\cdot10^{10}c$. In particular, in the latter case the expression of $u_{\mathrm{grav}}$ is given by (15.1), with $E_{0,\mathrm{grav}} \simeq 20.2\ \mu\mathrm{eV} = 3.2 \cdot 10^{-24}$ J. Notice first of all that the value (15.13) $u_{\mathrm{grav}} \neq c$ is compatible with the DSR lower limit (15.2).

Let us apply (15.1) to the Cavendish-like experiment in order to get a theoretical estimate of $u_{\mathrm{grav}}$ based on the formalism of DSR. In this case, according to the very grounds of DSR (see Chap. 2), $E$ is the torsion energy of the Cavendish balance, measured in flat space conditions by the electric interaction. Accordingly, it is the consequence of the electric repulsion of the atoms of the wire to which the crossbar of the balance is suspended, and can be calculated from the torsion constant $K$.

The torsion energy $E_{tor}$ is defined as:

$$E_{tor} = \frac{1}{2} K \vartheta^2, \tag{15.14}$$

where $\vartheta$ is the torsion angle of the balance and, in our case, $K = 1.19 \cdot 10^{-7}\,\mathrm{N\,m\,rad^{-1}}$.

We use the maximum value $\vartheta_{max}$ of the torsion angle of the balance in order to estimate the maximum value of the measured energy. Since it is impossible to take into account all the energy losses due to the internal friction of the entire balance, the value of the energy estimated by $\vartheta_{max}$ is only a lower bound of the gravitational energy exchanged. Consequently, one can obtain only a lower limit of the speed of gravity $u_{grav}(E)$.

From the geometry of the apparatus (see Fig. 15.1) it is easy to get the following expression for $\vartheta_{max}$:

$$2\vartheta_{max} = \alpha_{max} = \operatorname{arctg}\left(\frac{l_{max}}{D}\right). \tag{15.15}$$

By putting the values of $l_{max}$ and $D$ (see Sect. 15.2.1) in (15.15), one finds $\vartheta_{max} = 0.01$ rad. Then, (15.14) yields for the torsion energy the value $E_{tor} = 4.28 \times 10^{-12}$ J, much greater than $E_{0,e.m.} = 4.5\,\mu\mathrm{eV} = 7.2 \times 10^{-25}$ J. Therefore, according to DSR, we are guaranteed that the measure (performed by means of the electromagnetic interaction) took place in Minkowskian conditions. Moreover, it is also $E_{tor} \gg E_{0,grav}$, namely the gravitational metric is over threshold and therefore in non-Minkowskian conditions, ensuring that gravity was indeed deforming space–time. We are thus entitled to use (15.1) in order to estimate $u_{grav}$.

Being $E_{tor}$ a lower limit of the energy of the gravitational effects, it provides only a lower limit of the speed of gravity $u_{grav}(E)$, which has to be verified by kinematical measurements of $u_{grav}$. By (15.1) such a value is:

$$u_{grav}(E_{tor}) = \left(1 + \frac{E_{tor}}{E_{0,grav}}\right) c = 1.5 \times 10^{12} c < u_{grav} = (2.20 \pm 0.09) \times 10^{12} c. \tag{15.16}$$

It is therefore possible to conclude that both measurement results for $u_{grav}$ obtained by the torsion balance experiment are in good agreement with the theoretical predictions of DSR, in particular as far as (15.1) is concerned.

DSR allows one to get a possible explanation of the huge value (15.9) obtained for $s$ in the case of deformed space, by hypothesizing that the deformation of space occurred in the form of a Riemann foliation. One can think that each pull of the rotor on the torsion balance produces a Riemann surface. The time $\Delta t$ of production of a surface is therefore given by

$$\Delta t = \frac{S}{u} = \begin{cases} \dfrac{\tau}{A}\,, & \text{flat space;} \\ \dfrac{\tau}{A^2}\,, & \text{deformed space,} \end{cases} \tag{15.17}$$

where $\tau \simeq 2.3 \times 10^3$ s is the cumulative delay time. So, the number $\mathcal{N}$ of Riemann surfaces produced in the total measuring time $T = 12 \times 10^3$ s is

$$\mathcal{N} = \frac{T}{\Delta t} = \begin{cases} \dfrac{T}{\tau} A \simeq 5.2 A \ , & \text{flat space;} \\ \dfrac{T}{\tau} A^2 \simeq 5.2 A^2 \ , & \text{deformed space.} \end{cases} \tag{15.18}$$

Needless to say, such a kind of space deformation has nothing to do with the simple deformations described by the phenomenological metrics of Sect. 4.1 (although let us recall that the experimental data on clock rates do not provide any information on the space metric coefficients). However, it must be again emphasized that – as already stressed before – in those cases the situation was a static one, whereas we faced, in this experiment, a truly dynamic behavior of a given interaction. In such cases, it is not enough – in order to account for the phenomenon *in fieri* – to consider energy as a parameter, but one has to change its nature into that of *a true coordinate*. As already said, this amounts to embed the deformed Minkowski space into a 5D Riemann space with energy as extra dimension (see Parts IV and V).

# 16
# Piezonuclear Reactions in Cavitated Water

## 16.1 Can Pressure Waves Trigger Nuclear Reactions?

At the end of the twentieth century and at the beginning of the twenty-first one, experiments of cavitating water and of explosions of foils in water have provided possible evidence for production of stable, unstable and artificial nuclides induced by ultrasounds and shock waves, i.e., for nuclear reactions catalyzed by pressure waves.

Let us recall that the process of cavitation [93, 94] consists in subjecting gaseous liquids to elastic pressure waves of suitable power and frequency (in particular to ultrasounds). The main physical phenomena occurring in a cavitated liquid (like e.g., sonoluminescence [95]) can be accounted for in terms of a hydrodynamic model based on the formation and the collapse of gas bubbles in the liquid [93, 94].

Three different experiments on cavitation carried out in the last years [96–98] provided evidence for an anomalous production of intermediate and high mass number (both stable, unstable and artificial) nuclides within a sample of water subjected to cavitation, in turn induced by ultrasounds with 20 kHz frequency. These results together seem to show that ultrasounds and cavitation are able to generate nuclear phenomena bringing to modifications of the nuclei involved in the process (in particular, sononuclear fusion). Let us also stress that the measurement of ionizing radiation performed in the first experiment yielded no signal out of the background.

Such findings (in particular those of the first experiment) are similar under many respects to those obtained by Russian teams at Kurchatov Institute and at Dubna JINR [99–102] in the experimental study of electric explosion of titanium foils in liquids. In a first experiment carried out in water, the Kurchatov group [99] observed change in concentrations of chemical elements and the absence of significant radioactivity. These results have been subsequently confirmed at Dubna [100]. Later, the experiments have been carried out in a solution of uranyl sulfate in distilled water, unambiguously showing [101] a distortion of the initial isotopic relationship of uranium and a violation of the secular equilibrium of $Th^{234}$. Moreover, the neutron flux was measured and found to be very low ($< 103$ neutron/electric explosion), so that the change in the uranium isotopic composition cannot be attributed to the induced fission. Due to the similarity of such results with the cavitation ones, it is more than likely that the two observed phenomena share a common origin. Namely, one might argue that the shock waves caused by the foil explosion act on the matter in a way similar to ultrasounds in cavitation. In other words, the results of the Russian teams, together with those obtained by cavitating water, support the evidence for nuclear reactions induced by high pressures (*piezonuclear reactions*).

A connection can also be envisaged with the experiment by Taleyarkhan *et al.* [103] on nuclear fusion induced by cavitation. In such an experiment, it was observed emission of neutrons in deuterated acetone subjected to cavitation. The neutron flux measured was compatible with d–d fusion during bubble collapse. This result was subsequently disclaimed by another Oak Ridge group [104], which measured a neutron flux three orders of magnitude smaller than that required for tritium production. Such a disproof has been rebutted by Taleyarkhan et al. [105]. Although therefore general agreement exists on the emission of neutrons in the phenomenon, the controversial point is whether or not the observed neutron flux is compatible with d–d fusion and consequent tritium production. Notice that, in the first cavitation experiment we carried out, proton number is conserved, whereas neutron number is apparently not. In our opinion, the Oak Ridge experiments have only shown that cavitation does affect nuclei, by inducing them to emit neutrons, but have not provided firm evidence for cavitation-generated nuclear reactions (in particular fusion). In our view, one could interpret the Oak Ridge experiments as a transmutation of nuclei induced by cavitation, in which the emission of neutrons, although not consistent with fusion of deuteron nuclei, could be due to other piezonuclear processes in bubble collapse. In fact, in no Oak Ridge experiment either a mass-spectrometer analysis of the liquid before and after cavitation was performed in order to match the detected neutron emission with possibly occurred nuclear reactions in the cavitated liquid. A further factor of confusion about the real validity and conclusions of the Oak Ridge experiments is due to the fact that the deuterated acetone was preliminarily irradiated with neutrons (before subjecting it to the ultrasounds), in order to generate deuterium gas bubbles. Then, it is not clear at all if the measured neutron flux was a consequence of the neutron irradiation or of the cavitation process (or even of both of them).

It follows from the earlier discussion that the emission of neutrons in the processes possibly involving piezonuclear reactions is a fundamental issue.

In this chapter, we shall give only a brief account of the three experiments which provided a possible evidence for transmutation of elements in water induced by cavitation (the reader is referred to [96–98] for a thorough examination of this subject). A model able to account for piezonuclear reactions will be also illustrated. We will instead discuss in detail in Chapter 17 some new recent cavitation experiments in which it was possible to detect emission of neutrons. The purpose is matching these new phenomena (the piezonuclear ones) with the predictions of DSR, in particular as far as the strong interaction (and its metric: see Sect. 4.1 and [6] is concerned.

## 16.2 Cavitating Water Experiments

### *16.2.1 First Experiment*

In the first experimental work (carried out at Perugia University in 1998), a sample of bidistilled and deionized water was cavitated by means of a new type of sonotrode ("cavitator") with a very long working time ($> 30$ min). The cavitation lasted without stopping for a total time of 210 min, at the constant power of 630 watt and the frequency of 20 kHz.

After cavitation, the cavitated water sample was analyzed, by confining the analysis to the stable chemical elements (from $Z = 1$ to $Z = 92$), and the results compared with those of the uncavitated water. Very surprisingly, relevant changes were found in the concentrations of the elements in the cavitated sample (despite the very low original concentrations). The analysis of the water both before and after cavitation was carried out by three different procedures, reaching the precision of one *p.p.b.* and with a SD on concentrations $\sigma = 10^{-5} \mu\,\mathrm{g\,L^{-1}}$, namely:

1. Mass atomic absorption (Inducted Coupled Plasma, ICP)
2. Cyclotron spectrometry (ICR)
3. Mass spectrometry (MS)

In order to asset the changes in the chemical elements, the variation factors have been accepted only for a value $\geq 2$ in the concentration ratios in water after and before cavitation. Evidence was found for 10 increasing and 19 decreasing elements. Let us notice, in particular, that the decrease concerned stable elements with low mass number.

One checked the possible contributions to such changes due to impurities, possibly arising from the titanium tip of the cavitator and the flint glass of the vessel, by three different methods:

1. Mass atomic absorption (ICP) on cavitated water sample
2. Electron microscopy on dusts of tip and vessel
3. X-ray microanalysis, carried out on both the dusts of tip and vessel and the dry residues of the two (cavitated and uncavitated) water samples

The results obtained excluded contributions due to impurities to the observed concentration changes [96].

During the cavitation, it was looked for possible emission of radiation by putting, on the external walls of the vessel, slabs of colloid LR115, sensitive to ionization energies in the range 100 KeV–4 MeV, which are typical e.g., of $\alpha$-particles. The results obtained were compatible with the flux intensity of the background radiation in the laboratory (210–150 Bqm$^{-2}$).

A basic point to be stressed is that *the number of protons between increasing and decreasing elements was conserved, whereas that of neutrons was not.*

Moreover, a huge increase in the concentration of uranium was found. This result caused to perform the second experiment.

### *16.2.2 Second Experiment*

The measurements of the first experiment were confined to the stable chemical elements. We therefore performed a second cavitation experiment by using a standard sonotrode, and analyzed the mass composition of cavitated water by a spectrometer in the mass region $210 < M < 271$.

A mass of bidistilled and deionized water of about 30 g was subjected to cavitation by a standard sonotrode. The cavitation was carried out in a different underground laboratory in a different place, in order to get rid of possible local background effects. The sonotrode worked at the constant power of about 300 watt and the frequency of 20 kHz. Four subsequent cavitation runs, of 10 min each, were carried out on the water mass, with a cooling interval between any two of them of 15 min.

Four mass measurements were carried out on samples from water cavitated one, two, three and four times. Therefore, the whole cavitation time ranged from 10 min for the first water sample to 40 min for the last water sample. Each measurement was performed immediately after each cavitation run. We confined ourselves to merely counting the different masses identified by the spectrometer. An error $\Delta M = 0.1$ was accepted in the mass value determination.

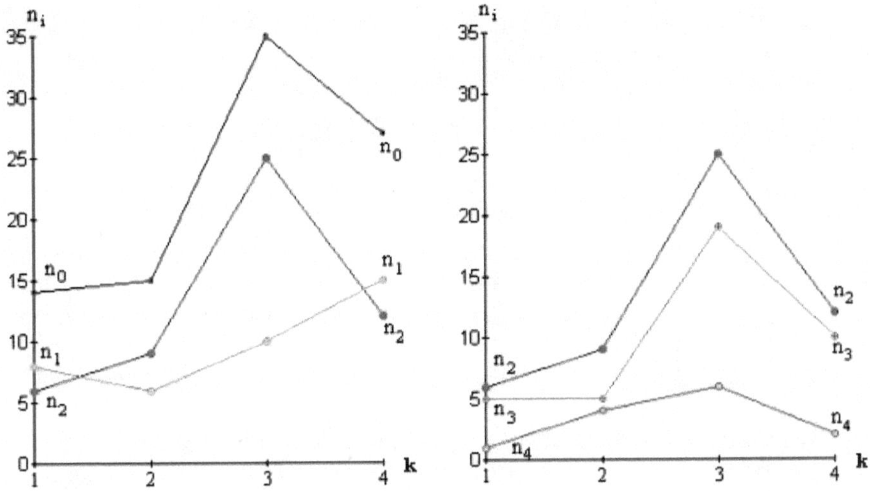


FIGURE 16.1. Numbers $n_i$ ($i = 0, \ldots, 4$) of nuclides vs. the number $k$ ($k = 1, \ldots, 4$) of cavitation runs. Left: Behavior of $n_0$, $n_1$, and $n_2$ Right: Behavior of $n_2$, $n_3$ and $n_4$

Let us label by $k$ and $i$ ($k = 1, \ldots, 4$; $i = 0, \ldots, 4$), respectively, the four measurement runs carried out and the mass intervals (the total interval 0: $210 < M < 271$, was split in the two subintervals 1: $210 < M < 238$, and 2: $238 < M < 271$; the subinterval 2 was in turn split in the subintervals 3: $238 < M < 264$, and 4: $264 < M < 271$). The numbers $n_{ik}$ are defined as:

$$n_{ik} \equiv N_{ik} - N_{Bi}, \tag{16.1}$$

where $N_{ik}$ is the number of nuclei measured in the $i$th mass interval in the $k$th water sample cavitated $k$ times, and $N_{Bi}$ is the number of nuclei measured in the $i$th mass interval in the background/blank measurement (which of course includes the masses of the calibration compound).

In Fig. 16.1 the numbers of nuclides $n_i$ ($i = 0, \ldots, 4$) are plotted as functions of the cavitation runs $k$. On the left it is seen that the excess in the number of masses identified in run 3, in the water sample cavitated 3 times, has to be ascribed to masses in region 2, $238 < M < 271$. In such a region 2, the contribution to the excess in the number of masses appears to be due to masses in the transuranic region, $238 < M < 264$, as shown on the right in Fig. 16.1.

One might argue that the excess in the number of masses could be due to the formation of macromolecules by processes like e.g., dimerization. However, such a possibility is unable to explain why the excess is just concentrated in a definite mass interval, i.e., in the transuranic region. On the other hand, possible contributions from cosmic-rays induced phenomena (like spallation) even at altitudes less than 200 m a.s.l.– as it is indeed the case of the experiment sites – could give rise to an excess of light nuclei, but cannot explain an increment of heavy nuclei (even less the presence of nuclei heavier than uranium). Moreover, the actual presence of transuranic elements is supported by the huge increase of uranium found in the first experiment, which might just be explained in terms of the formation and subsequent decay of transuranic elements. The results of the first experiment do agree with the results of the second one, showing the rearrangement of elements, with the decrease in the concentrations of light-medium elements, and the increase in those of the heavy ones. Although a definite conclusion can only be drawn by carrying out a further experiment in the whole mass range, including both stable and transuranic elements, it is reasonable to assume that the excess of masses in the transuranic region are compensated by a simultaneous deficit of masses in the stable element region.

Such a picture is supported by the different behavior of the $n_i$'s ($i = 1, 2, 3, 4$). The net increase of $n_1$ is due to the fact that in the mass range $210 < M < 238$ there are mainly stable nuclei, and therefore their number increases following the production. On the contrary, in the range $Z = 92/114$ and $238 < M < 270$, there are only unstable radionuclides (either experimentally known or theoretically predicted). Thus, their number first increases, then decreases (after the third cavitation) as soon as the decay rate overcomes the rate of production.

### *16.2.3 Third Experiment*

We have seen that the first two experiments provided evidence for a change in concentrations of stable and transuranic chemical elements in cavitated water [96]. The third experiment was aimed at looking for the production of the so-called rare earth elements, and was performed at the University of Rome "La Sapienza" [97]. To cavitate water, the same sonotrode device of the second experiment was employed, but with a reduced power setup (still able to induce cavitation), in order to account for the different contributions to the phenomenon due to both the cavitation and the ultrasounds at different powers. The sonotrode was made of a column of piezoelectrics directly connected to a steel tip, shaped like a truncated cone. The sonotrode tip was plunged into an open vessel at atmospheric pressure, filled with water at room temperature. The sonotrode was cooled by air at room temperature (20°C) and had a working frequency $\nu = 20\,\text{kHz}$, and transmitted power $P = 100\,\text{watt}$. The continuous operation folding time of this sonotrode was 15 min, followed by a cooling period (15 min) of the column of piezoelectrics. During the cooling period the sonotrode was off.

For the analysis of the water, we exploited an ICP mass spectrometer, with temperature of the ionization chamber $T > 9000°C$.

The water vessel was a Pyrex beaker, previously washed by using a sulphochromic mixture. It was filled by 300 cc of deionized and bidistilled water, whose resistivity was of $0.1\mu$S. A Teflon tube was used to transfer the water from the Pyrex beaker, where the cavitation occurred, to the ICP mass ionization chamber.

It was chosen to examine, through ICP, two mass intervals, from 90 to 150 amu and from 200 to 255 amu, since they include also the rare earth elements one was looking for.

For every mass value in the two intervals, subdivided into steps of 0.01 amu, the results of the ICP mass count were analyzed. The count series coincided with the measure series; the latter were obtained using the count data acquisition program PQ Vision supplied by Thermo Elemental. The aim of this analysis was to highlight the count variations (decrease or increase compared with the previous count).

For each mass the upper limit for the count background was evaluated, including both blank and noise. One took into consideration the masses whose upper count was higher than this upper limit. A given mass was taken into account only if its count was at least twice greater than the upper limit.

This criterion was applied to the counts obtained, for the same mass values, from measures performed with scanning times of both 10 and 150 s.

No mass whose counts satisfied the described criterion was identified for scanning times of 150 s, whereas only one mass was found for the scanning time of 10 s, namely $M = (137.93 \pm 0.01)$ amu.

We performed differential measurements with a time interval of 300 s, corresponding to the cavitation interval time. Then one got the differential counts, from which we derived the integral counts plotted in Fig. 16.2 as function of the cavitation interval time. In this way the data have been ensured from instrumental pile-up effects.

The identification of the observed peak with a given radionuclide required the determination of the lifetime. This could be done by analyzing Fig. 16.2, and interpreting it as two subsequent cycles of production and decay of the observed element [98]. Thus it was possible to evaluate its halftime $T_{1/2} = (12 \pm 1)$ s within 1.5 $\sigma$.

From the tables of nuclides [108] we got as possible candidate the isotope of europium $Eu_{63}^{138}$ [98].

During the first experiment the concentration of $Eu_{63}$ was not changed. It is known that the abundance on Earth of stable europium is less than 1.06 *ppm* (the natural $Eu$ is a mixture of two isotopes, $Eu_{63}^{151}$ with a percentage abundance of 47.77% and $Eu_{63}^{153}$ with a percentage abundance of 52.23%). The candidate identified during the third experiment does not exist in nature; it is an artificial radionuclide (discovered only in 1995–1997 [108]) that can be produced at the present time in nuclear reactors and by synchrotrons.

There are two ways whereby $Eu_{63}^{138}$ can be produced: by nuclear fission or by nuclear fusion. The former process requires less energy. However, from the results of the first two experiments [96,97], the quantity of heavy nuclei which can produce $Eu^{138}$ by nuclear fission is very much smaller

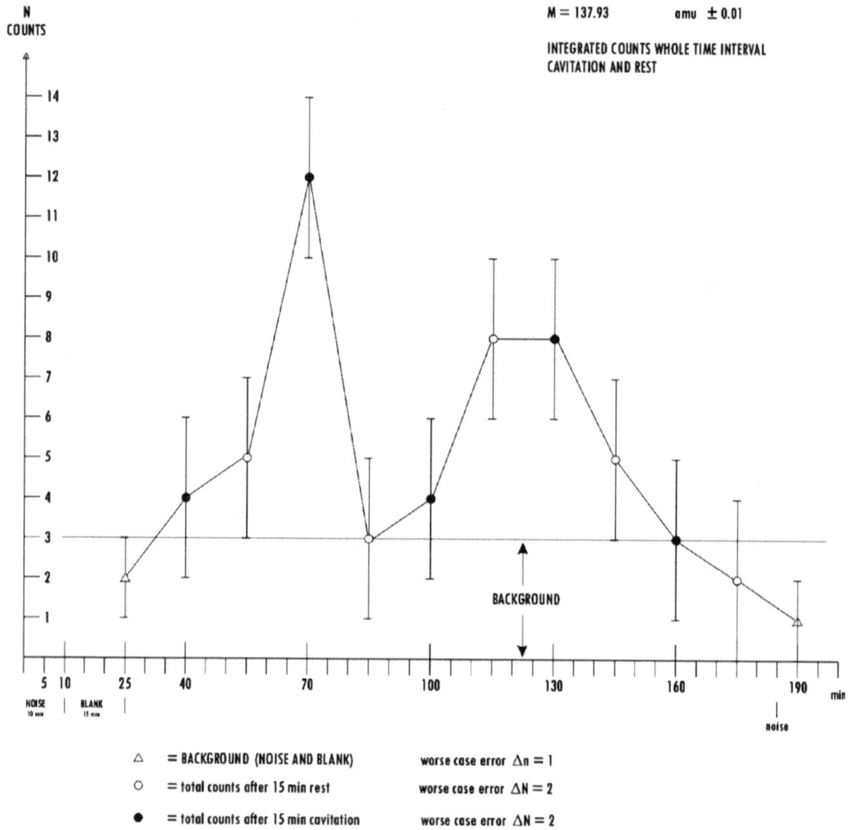


FIGURE 16.2. Integrated counts for $M = 137.93$ $amu$ in the third experiment of cavitating water

(by two-three orders of magnitude) than that of the intermediate nuclei which can produce it by fusion (on account also of the impurities of the tip). As a matter of fact, a rough estimate (just based on the detected abundance of heavy nuclei in the second experiment) yields a probability of $10^{-6} - 10^{-8}$ for the production of $Eu^{138}$ by fission. Moreover, nuclear fusion is the only possible explanation of the changes in concentration of stable elements, induced by cavitation, observed in the first experiment.

## 16.3 Phenomenological Model of Piezonuclear Reactions

### *16.3.1 Classical Cavitation Model*

The fundamental and intriguing question posed by the experiments of cavitating water is therefore: How can the pressure waves generated by

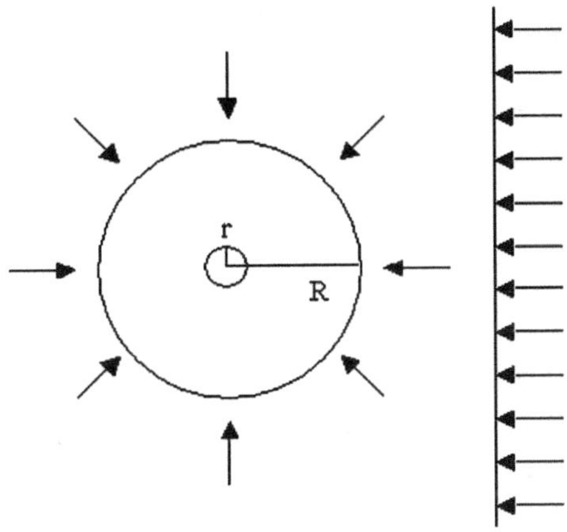


FIGURE 16.3. Conversion of a plane pressure wave into a spherical pressure wave on the bubble surface during cavitation process

cavitation trigger nuclear fusion? The answer comes possibly from the well-known fact that cavitation allows one to achieve an extreme concentration of energy for unit time in the collapsing bubble [106,107]. Indeed there exist speculations on the possibility that cavitation (in particular, sonoluminescence) might be a viable approach to inertial-confinement fusion (provided that the temperature attained in the process is substantially higher than that predicted by a simple thermal model, namely $10^4$°K) [106,107]. Let us illustrate a classical model of piezonuclear reactions, based on the earlier features of the cavitation process [109].

In order to explain how cavitation can produce the energies needed to induce nuclear fusion, let us take into consideration the physics underlying the cavitation process. It consists, as is well known, in the implosive collapse of a gas bubble within a liquid under suitable pressure conditions. In our case (water sample) the speed of sound is about $v \simeq 10^3\,\mathrm{m\,s^{-1}}$, which – on account of the used frequency $\nu \simeq 10^4\,\mathrm{Hz}$ – corresponds to a wavelength $\lambda \simeq 10^{-1}\,\mathrm{m}$. In order to get the gas bubble to implode, the plane pressure wave of the ultrasounds must be converted into a symmetric spherical shock wave on the bubble surface (see Fig. 16.3).

Therefore, the condition to be satisfied for an implosive collapse is that the wavelength must be much greater than the bubble size. Taking as example a spherical bubble with radius $R$, this means $\lambda \gg R$. Since the size of a bubble subjected to cavitation is of the order of magnitude $R = 10^{-6}\,\mathrm{m}$ (see [94]), the collapsing condition is respected in all the three experiments.

We now remark that – contrary to what previously believed – the only atoms influenced by the shock wave producing the cavitation are the ones lying on the surface of the bubble itself. These atoms are trapped by the surface tension of the bubble (generated by the combined electrostatic repulsion of the liquid and of the bubble) in a double-layer film at the border liquid-bubble. Namely, all other atoms inside the bubble volume escape to the outside during the collapse, due to the fact that the inner pressure, corresponding to the saturated vapor pressure of water, is far lower than the external pressure. The trapping of atoms at the bubble surface is expected to be more effective for metals. Indeed, it is known that usually metals and metal cations do not enter the cavitation bubble, but stay at the interface between bubble and bulk solution. At room temperature (300°K) – as it was the case of all experiments – the saturated vapor pressure of water is 0.02 bar, whereas we estimated that the pressure induced by the implosive shock wave, for a transmitted power of 100 watt is of the order of $10^9$ bar. This circumstance entails that there is no limit to the spatial size attainable by the collapse. It is therefore reasonable to hypothesize that the lower limit of this size can be identified with the nuclear size. As a matter of fact, there is in literature no experimental information on the value of the minimum size attained by a collapsing cavitation bubble. Therefore, we can suppose that at the end of collapse the bubble dimensions become near to the nuclear dimensions, commonly about $10^{-15}$ m (Fermi radius).

As stated before, if the wavelength $\lambda$ of the plane pressure wave satisfies the condition $\lambda \gg R$ , where $R$ is the bubble radius, the plane pressure wave becomes a spherical shock wave symmetrically acting on the bubble. Let $P$ be the power of the plane pressure wave, and $R$, $r$ the radius of the bubble before and after collapse, respectively. Then, the power density on the bubble before and after collapse are $D_P = P/(4\pi R^2)$ and $D'_P = P/(4\pi r^2)$. If the initial energy flux (i.e., the energy for unit time and unit surface) is conserved (due to the continuity equation and to energy conservation), we have

$$P = SD_P = (\pi r^2)D'_P \Longrightarrow D'_P = D_P(S/\pi r^2) = fD_P \qquad (16.2)$$

where $f = (S/\pi r^2)$ is the amplification factor.

As a consequence, the power density on the bubble surface, generated by the power of the plane pressure wave, produces after implosion a power density increased by the factor $f$ on the reduced bubble surface. Such a factor, caused by the implosion, ranges from $f_A \sim 10^8$ (when the bubble collapses to the atomic size, $r_A = 10^{-8}$ cm) to $f_N \sim 10^{18}$ (collapse to nuclear radius, $r_N = 10^{-13}$ cm). In the three experiments we used powers ranging from 100 to 630 watt, i.e., from $6 \times 10^{20}$ to $4 \times 10^{21}$ eV s$^{-1}$. Thus the final power density $D_N$ for collapse to the nuclear radius is $D_N \simeq 2\times 10^{46}$ eV (s cm$^{-2}$), corresponding to an equivalent temperature (for a low-density plasma) of at least $10^{20}$°K. We think that this power density allows a heavy-ion-fusion-like process between two parent nuclei to generate a son nucleus. When the bubble collapses and the bubble surface shrinks to the nuclear dimensions, trapped atoms do move together with the surface and come closer and closer. The collapsing surface of the bubble acts therefore as an "inertial accelerator" of neutral atoms and the squeezing of the surface to the nuclear dimension produces on the bubble surface the energy required to activate the fusion.[1] At the final stage of this process, atomic electrons are stripped away, and a kind of heavy-ion cold fusion (in the sense of Oganessian [110]) can occur.

### *16.3.2 Application to Europium*

Let us apply the model discussed earlier to the fusion of europium, in order to explain the results of the third experiment. Possible candidates as parent nuclei for $Eu_{63}^{138}$ are $Zr_{40}^{90}$ and $V_{23}^{51}$. A feasible reaction scheme could be:

$$Zr_{40}^{90} + V_{23}^{51} \longrightarrow Eu_{63}^{138} + 3n. \tag{16.3}$$

Where the nuclides $Zr$ and $V$ could come from? A possible answer to this question can be found in the impurities lying on the surface of the sonotrode tip. The latter was shaped through mechanical tools (lathes) made by alloys of iron, vanadium and zirconium, introduced to harden the tools themselves. During the manufacturing, small numbers of atoms should remain trapped inside the iron lattice of the sonotrode tip. Impurity atoms are more loosely bond to the iron lattice than the iron atoms, so the ultrasonic vibrations of the tip can remove them from the lattice. By the way, the possibility of the neutron excess (see (16.3)) could be explained by the observations already made by other research groups working on cavitation-induced nuclear reactions [105] (see Sect. 16.1).

The Coulomb barrier against fusion for $Zr$ ($Z_1 = 40$, $A_1 = 90$) and $V$ ($Z_2 = 23$, $A_2 = 51$) can be evaluated by the formula

$$E_{coul} = \frac{Z_1 Z_2}{A_1^{1/3} + A_2^{1/3}} \,\mathrm{MeV} = 112\,\mathrm{MeV} \tag{16.4}$$

or also by

$$E_{coul} = \frac{Z_1 Z_2}{d} \,\mathrm{MeV} = 140\,\mathrm{MeV}, \tag{16.5}$$

with $d = r_1 + r_2 + 2r_0$ (where $r_1 = 4.5 \times 10^{-13}$ cm and $r_2 = 3.7 \times 10^{-13}$ cm are the nuclear radii of $Zr$ and $V$, respectively, and $r_0 = 0.5 \times 10^{-13}$ cm is the characteristic Bohr–Wheeler nuclear length).

From the power density $D_{PN}$ on the bubble surface after collapsing from a radius of $10^{-4}$ cm to the nuclear radius $r_N$ estimated earlier, it is possible to evaluate the energy and the energy per nucleon needed to bring about the formation of europium 138 from vanadium and zirconium according to reaction (16.3).

---

[1]The fact that the collapsing bubble surface is responsible for the nuclear reaction ignition is confirmed by the evaluation of the number of interacting atoms, endowed with the velocity required to overcome the internuclear Coulomb barrier, present on the bubble surface for a given overpressure. Indeed, it can be shown that this number is incompatible with the number of atoms inside a (even rarefied) bubble (see [97]).

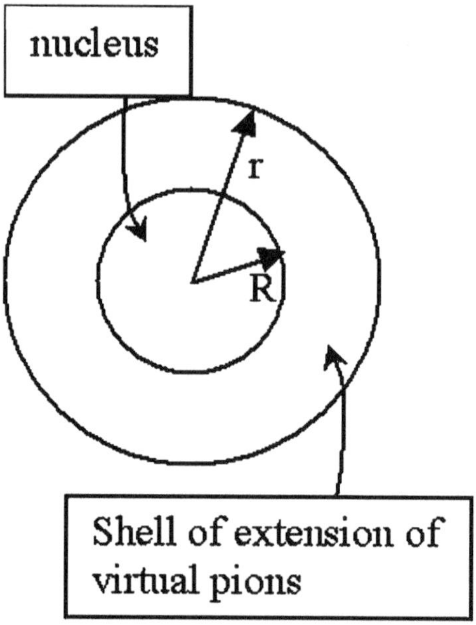


FIGURE 16.4. Nuclear radius and shell extension of the virtual pions around the nucleus

With reference to Fig. 16.4, let $r_N$ and $R_N$ be, respectively, the radius of the nucleus and the radius of the shell of extension of the virtual pions. One has

$$\begin{aligned} r_N &= 0.5 \times 10^{-13} \sqrt[3]{A}\text{cm}; \\ R_N &\simeq R_F = 0.5 \times 10^{-13}\text{cm}. \end{aligned} \tag{16.6}$$

Then, the effective nuclear radius $\overline{R}$ is given by

$$\begin{aligned} \overline{R} &= \frac{r_N + R_N}{2} = \frac{1}{2}\left(\frac{1}{2}\sqrt[3]{A} + \frac{1}{2}\right) 10^{-13}\text{cm} \\ &= 0.25 \times \left(\sqrt[3]{A} + 1\right) 10^{-13}\text{cm} \simeq r_N. \end{aligned} \tag{16.7}$$

The effective nuclear surface $S$ is therefore $S = 4\pi r_N^2 = A^{\frac{2}{3}}\pi 10^{-26}\,\text{cm}^2$. For vanadium and zirconium one gets, respectively:

$$\begin{aligned} &\text{For vanadium:} \quad S_{\text{V}} = 4.3 \times 10^{-25}\text{cm}^2; \\ &\text{For zirconium:} \quad S_{\text{Z}} = 6.3 \times 10^{-25}\text{cm}^2. \end{aligned} \tag{16.8}$$

The mean value is $\overline{S} \simeq 5.3 \times 10^{-25}\text{cm}^2$.

In order to find the energy achieved in this case, the collapse time of the bubble must be estimated. For an employed frequency $\nu = 20\,\text{kHz}$, the period is $T = 1/\nu = 5 \times 10^{-5}\text{s}$. The collapse time is obviously comprised between $T$ (upper limit) and the time $t_s = R/v_s$ taken by the pressure wave to travel the bubble radius (at the sound speed): $t_s < t_c < T$. In deionized and bidistilled water, the sound speed is $v_s \simeq 1.4 \times 10^3\,\mathrm{m\,s^{-1}}$. This corresponds to the lower time limit for collapse $t_s = 7 \times 10^{-10}\,\mathrm{s}$. The energy transmitted by the bubble implosion to the $V - Zr$ system is therefore

$$E_{c\,\min} = \overline{S} D_{PN} t_s < E_c < E_{c\,\max} = \overline{S} D_{PN} T \qquad (16.9)$$

One gets

$$\begin{aligned} E_{c\,\min} &= 3.3 \times 10^6\,\mathrm{MeV}; \\ E_{c\,\max} &= 2.5 \times 10^{11}\,\mathrm{MeV}. \end{aligned} \qquad (16.10)$$

The total number of nucleons is $N = 90 + 51 = 141$. Thus the energy for nucleon $\varepsilon = E/N$ ranges in the interval

$$\varepsilon_{\min} \simeq 2.3 \times 10^4\,\mathrm{MeV} < \varepsilon < \varepsilon_{\max} \simeq 1.8 \times 10^9\,\mathrm{MeV}. \qquad (16.11)$$

Let us notice that $\varepsilon_{\min}$ is much higher than the maximum estimated Coulomb barrier for the europium formation. Moreover, the range of values of $\varepsilon$ corresponds to the energies of the direct nuclear reactions induced by photons and to the production of pairs $e^+e^-$ due to photon conversion in the nuclear electric field, many orders of magnitude higher than the energy required to overcome the Coulomb barrier.

Then, one can conclude that the mechanism proposed for piezonuclear fusion is able indeed to account for europium formation generated by the cavitation process.

The simple phenomenological model discussed earlier may of course be also responsible of the results we got in the first two experiments. In this case, the atoms of some elements (the decreasing ones of the first experiment) may have been subjected to the inertial-fusion-like process due to bubble collapse to generate new elements (the increasing ones and/or the transuranic elements observed in the second experiment).

Moreover, this mechanism avoids the need for the introduction of a nuclear shape deformation, invoked sometimes to increase the nuclear tunneling probability. We remind, to this regard, that $Zr$ and $V$ are typically spherically shaped nuclei, at least in their fundamental state, so that a shape deformation would be very difficult to justify.

On the other hand, an Oganessian-like low temperature heavy ion nuclear fusion [110] could be possible even by the following mechanism. Let us suppose that there be some kind of oversaturated vapor of the $\pi$-meson (boson) gas, caused by the final power density, all around the parent nuclei. The $\pi$-mesons are emitted, but they are absorbed more slowly – as if they would condensate (something like a Bose–Einstein condensation) –, typically after a time which, on the basis of the previous values of the energy-space–time density, we estimate as $10^{-16}\,\mathrm{s}$ (to be compared with the typical nuclear time of $10^{-23}$ s). As a consequence, the probability of nuclear interactions is enhanced and there is an anomalous increase of the nucleus–nucleus cross section.

### *16.3.3 Limits of the Classical Model*

In spite of its effectiveness in accounting for the europium production, the classical model illustrated earlier suffers from some drawbacks we are going to discuss [109].

First of all, it must be noticed that not all of the power from the sonotrode is concentrated in one bubble of nuclear dimension. Also, much of the power put in is dissipated in other processes (in fact, energy is subtracted e.g., by sonoluminescence and by endothermic sonochemical reactions. This could be taken into account by an approximated energy efficiency factor, useful for nonsonochemical reactions; the latter are to be identified with those reactions that involve parent nuclei to generate composite nuclei with a greater mass. Moreover, a crucial parameter is just represented by the final size reached by the implosed bubble (which critically determines the energy available for the fusion process).

Another intrinsic limit of this model is due to the very phenomenology of nuclear fusion of heavy nuclei. At energies below the barrier, the fusion probability is low and rises exponentially with energy. At energies just above the barrier, the probability for parent nuclei to fuse and form a composite nucleus does increase further but not indefinitely. As a matter of fact, at energies of about 10 MeV per nucleon, the cross section for forming a composite system is low and diminishes at a rate of $1/E_{\mathrm{cm}}$ (where $E_{\mathrm{cm}}$ is the center-of-mass energy of the relative motion of the two nuclei). Such a decrease occurs because the composite system (for instance, $V + Zr$), formed at such high relative energies, is a highly excited nucleus and has large angular momentum. This causes the composite system to fission instantaneously (with a time frame much shorter than $10^{-22}$ s). For nuclear collisions at the energies we estimated in Sect. 16.3.2, the fusion does not occur at all.

### *16.3.4 Deformed Space–Time of Strong Interaction*

A possible answer to both these questions may come from the formalism of DSR. In particular, we have to take into account the hadronic deformation of space–time described by the strong metric, (4.10)–(4.14).

Let us recall that the energy $E_c$ corresponding attained for collapse of a cavitating bubble to the nuclear radius ranges in the interval between $E_{\mathrm{cmin}} \sim 10^6$ MeV and $E_{cmax} \sim 10^{11}$ MeV. Then, $E_{\mathrm{cmin}}$ is about one order of magnitude higher than the threshold energy of the DSR strong interaction, $E_{0,\mathrm{strong}} \simeq 4\times10^5$ MeV. This entails that *the hadronic interaction between nuclei occurs in the non-Minkowskian part of the strong interaction metric, i.e., the (hadronic) space–time geometry is deformed in the final collapse region.* As already noted, in the (usual) flat Minkowskian metric the interacting nuclei, although overcoming the Coulomb barrier, produce, after fusion, a nucleus in an highly excited state, and therefore with high probability of spontaneous fission. On the contrary, in the deformed space–time produced by the over-threshold hadronic condition, the excess energy after fusion goes in deforming the space–time region, thus leaving the son nucleus (produced by the fusion of the two parent nuclei) in a low-excited (or even unexcited) state.

Therefore, *the stability of the nuclei produced by cavitation is due* (according to DSR) *to the deformation of space–time in the collapse region ensuing from the non-Minkowskian behavior of the strong interaction in the range* $E_c > E_{0,\text{strong}}$. This explains why no emission of radiation was observed in the first cavitation experiment.

### *16.3.5 Threshold Energy for Piezonuclear Reactions*

Let us show that DSR is also able to predict the cavitation power needed to produce piezonuclear reactions in a stable way. This is a consequence of the law of time deformation in an hadronic field, (4.16), we rewrite here for reader's convenience:

$$\frac{dt_{hadr.}}{dt_{\text{e.m.}}} = \frac{E_{0,\text{strong}}}{E}. \tag{16.12}$$

A way to read (16.12) is as follows. It can be regarded as an action-reaction relation, i.e., as the equality between two energy speeds: An electromagnetic speed $W_{\text{e.m.}}$ of supplying energy to the atoms by the electromagnetic interaction (action) and an hadronic speed $W_{\text{strong}}$ of response by the strong interaction of nuclei (reaction)

$$W_{\text{strong}} = \frac{E_{0,\text{strong}}}{dt_{hadr.}} = \frac{E}{dt_{\text{e.m.}}} = W_{\text{e.m.}}. \tag{16.13}$$

In order to attain the threshold of LLI breakdown for strong nuclear interaction, during the time taken by a generic cavitating bubble to collapse, for a given electric energy $E$, an energy speed $W_{\text{e.m.}}$ must be supplied such to equate the nuclear one.

Let $dt_{\text{hadr.}}$ the nuclear reaction time given by

$$dt_{\text{hadr.}} = \gamma_{\text{strong}} \Delta t \tag{16.14}$$

where $\gamma_{\text{strong}}$is the deformed strong relativistic factor and $\Delta t = h/m_\pi$ is the Yukawa time (nuclear year).

An estimate of $dt_{\text{hadr.}}$ at the energy threshold $E_{0,\text{strong}}$ can be gotten by means of the relation $\gamma_{\text{strong}} = E_{0,\text{strong}}/m_\pi$ (on account of the well-known fact that $\gamma = E/m$ is the relativistic factor of time dilation in Minkowskian conditions for $E \leq E_{0,\text{strong}}$, and by recalling that the process occurs approaching $E_{0,\text{strong}}$ from below). Replacing such an expression of $\gamma_{\text{strong}}$ in (16.13) yields

$$dt_{\text{hadr.}} = \frac{h}{m_\pi^2} E_{0,\text{strong}} \tag{16.15}$$

($h = 4.136 \times 10^{-15}$ eV s; $m_\pi = (m_\pi^\pm + m_\pi^0)/2 = 1.373\times108$ eV).

For the energy of the electric action we have

$$E = dt_{\text{e.m.}} E_{0,\text{strong}} \frac{m_\pi^2}{E_{0,\text{strong}} h} = dt_{\text{e.m.}} W_{\text{strong}}. \tag{16.16}$$

Since $E = dt_{\text{e.m.}}(m_\pi^2/h)$, $W_{\text{strong}}$ reads

$$W_{\text{strong}} = m_\pi^2/h = 4.8 \times 10^{30}\text{eV s}^{-1} = 7.6 \times 10^{11}\text{ W}. \tag{16.17}$$

Let us assume for $dt_{\text{e.m.}}$ the time taken by a microbubble of radius $R$ to collapse to the nuclear size with $r \sim 10^{-13}$ cm (due to the electric repulsion of the water atoms subjected to the ultrasonic pressure wave). The collapse can occur at the velocity of sound in distilled water, $v = v_s = 1.4 \times 10^3\, m\, s^{-1}$, or at the velocity of the shock wave, $v = v_u = 4v_s$. Because the ultrasound wavelength is much greater than the microbubble diameter, it is $dt_{\text{e.m.}} = R/v$ in either case.

Therefore we have, for the threshold energy $E_{thres}$:

$$E_{thres} = \frac{Rm_\pi^2}{vh}. \tag{16.18}$$

The values of $E_{thres}$ deduced from (16.18) range from $5\times10^2$ J to $2\times10^3$ J for the collapse speed $v_s$ (with radius $R$ of the collapsing microbubbles varying from 1 $\mu$ to 4 $\mu$) and from $10^2$ J to $2\times10^3$ J for $v_u$ (with 1 $\mu < R < 8\ \mu$).

In order to produce stable piezonuclear reactions, and therefore a stable emission of nuclear radiation, it is necessary to supply constantly an energy $E \geq E_{\text{thres}}$ to the system of distilled water and solute. Such a condition permits to trigger piezonuclear reactions in presence of broken local Lorentz invariance.

By using a cavitator absorbing 2,000 watt and able to provide a stable supply from 100 J up to a maximum of 2,000 J, it is possible to investigate the collapse of bubbles with size ranging from 1 $\mu$ to 8 $\mu$, by taking either $v_s$ or $v_u = 4v_s$ as collapsing speed.

However, the previous experiments were carried out with energies of the order of one hundred joules and provided evidence for the occurring of nuclear reactions. This is in favor of the velocity of the shock wave as collapsing velocity, and therefore of the model in which the ultrasound pressure plane wave generates a symmetrical, spherical shock wave around the bubble.

Nothing can be said about the total mass of water and compound to be cavitated, or about the amplitude of the ultrasonic wave. Both of them are phenomenological parameters, to be determined empirically. Needless to say, a higher supplied energy corresponds to a greater amplitude and a lower energy is available for cavitation and bubble collapse if a greater mass is subjected to ultrasounds.

As we have seen, the existence of the threshold $E_{\text{thres}}$ is a direct consequence of the existence of of the energy threshold $E_{0,\text{strong}}$ for the hadronic interaction. This circumstance allows one to discriminate between signals coming from the nuclear reactions produced by cavitation.

In fact, if the nuclear radiation emitted by the nuclei after interaction are neutrons, the son nuclei are left in an excited rotational state. Therefore, due to angular momentum conservation, they decay to a lower state by emitting $\gamma$ radiation. Then, from a classical viewpoint, such a disexcitation process is accompanied by both neutron and photon radiation. However, if the interaction among nuclei occurs in non-Minkowskian conditions (for $E > E_{0,\text{strong}}$), the excess energy is partly absorbed by the hadronic space–time deformation, so that *there is no emission of $\gamma$ radiation.*

The two facts of the energy threshold overcoming, $E > E_{0,\text{strong}}$, and of the neutron emission in absence of $\gamma$ radiation do provide the complete signature of piezonuclear reactions produced by the cavitating collapse of gas bubbles of water in non-Minkowskian conditions.

By measuring the radiation produced by the piezonuclear reactions generated by cavitation at or over threshold it is possible, in principle, to determine the radiative calorimetry of the produced process.

In conclusion, let us remark that the hypothesis of power conservation – which is at the very basis of the model of piezonuclear processes – made in Sect. 16.3.1, can be drawn from a different interpretation of the hadronic law of time deformation (16.11). This latter can be indeed regarded as an equality of energies per unit time, i.e., an equality of powers, and hence interpreted as a relation of power conservation.

In turn, this new interpretation of (16.11) in terms of power conservation acquires a deeper and more natural meaning within the frame of the pentadimensional geometrical representation of interactions by a 5D metric in which the extra dimension is the energy (see Parts IV and V). In this framework, such a relation means moving along the fifth dimension at a constant speed. In other words, *the hadronic law of time deformation can be regarded as a principle of inertia regarding the energy.*

# 17
# Piezonuclear Reactions in Cavitated Solutions

We have seen in Sect. 16.2.1 that the analysis of the change in concentration of stable chemical elements, observed in the first cavitation experiment, leads one to conclude that in such processes the number of protons is conserved, whereas that of neutrons does not [96, 97]. Moreover, according to the model of piezonuclear reactions discussed in Sect. 16.3, a possible signature of non-Minkowskian conditions in such processes is provided by neutron emission without a concomitant photon radiation. This is why we carried out new experiments, explicitly aimed at detecting and analyzing the radiation possibly emitted during cavitation.

## 17.1 The Thorium Experiment

### *17.1.1 Experimental Setup and Results*

We recall that a Russian team observed a violation of the thorium-234 secular equilibrium induced by electric foil explosion [101, 102]. In order to check the possible effects of cavitation on thorium decay, an experiment different from the previous ones was carried out. Instead of water, a solution of thorium in water was subjected to cavitation. Precisely, we prepared 12 identical solutions of $Th^{228}$ in deionized bidistilled water, with concentration ranging from 0.01 to 0.03 ppb (part per billion). $Th^{228}$ is an unstable element whose half life is $t_{1/2} = 1.9$ years $= 9.99\times10^5$ min. It decays by emitting 6 $\alpha$ and 3 $\beta^-$. The minimum emission energy of the alpha particles is 5.3 MeV, which is nearly equal to the energy of the $\alpha$'s emitted by radon

222. This alikeness allowed us to use the detector CR39, a polycarbonate whose energy calibration is just designed to detect those emitted by radon 222.

Eight solutions out of the twelve were divided into two groups of four, and each of them was cavitated for $t_c = 90$ min at a frequency of 20 kHz and a power of 100 watt. The remaining four were not cavitated, and regarded as reference solutions.

The 12 detectors CR39 corresponding to the 12 solutions were examined and the traces impressed on them by the alpha particles counted.

The results obtained are depicted in Fig. 17.1. Precisely, the first column shows the four detectors CR39 used with the four noncavitated solutions taken as reference, whereas in the second and third columns one sees the eight detectors used with the eight cavitated solutions. The circles in the

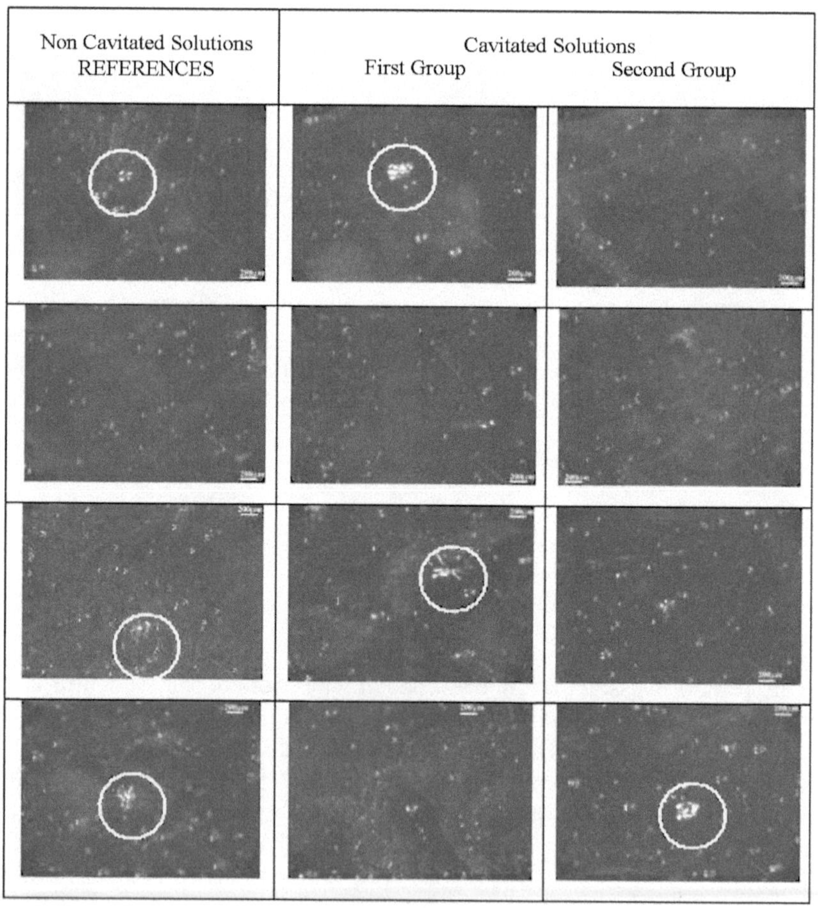


FIGURE 17.1. Traces left by $\alpha$-particles emitted from thorium decay on detectors CR39 (*circles*) for noncavitated (left) and cavitated (right) solutions

TABLE 17.1. Content of $Th^{228}$ in noncavitated (reference) solutions

| Mass-spectrometer analysis | cps | ppb |
|---|---|---|
| Sample 1 | 287 | 0.020 |
| Sample 3 | 167 | 0.012 |
| Sample 4 | 363 | 0.026 |
| Mean values | 272.33 | 0.019 |

TABLE 17.2. Content of $Th^{228}$ in cavitated solutions

| Mass-spectrometer analysis | cps | ppb |
|---|---|---|
| Sample 1 (first group) | 231 | 0.016 |
| Sample 3 (first group) | 57 | 0.004 |
| Sample 4 (second group) | 79 | 0.006 |
| Mean values | 122.33 | 0.009 |
| Ratio of mean values noncavitated/cavitated | 2.2 | 2.1 |

figure highlight the traces left by the particles $\alpha$ produced by thorium decay. On the four CR39 used with the four reference solutions one counted 3 traces of alpha particles in all. The same number of traces was counted on the eight CR39 used with the eight cavitated solutions.

Then, the ratio of the number of traces and the number of solutions is 3/4 for the reference solutions and 3/8 for the cavitated ones. Thus, there is evidence of a reduction of the number of traces of alpha particles from thorium decay in the cavitated solutions with respect to those in the noncavitated ones. In particular, it is evident that the reduction in the number of traces from the former to the latter is by a factor 2.

The content of $Th^{228}$ in those solutions providing evidence of alpha particles from thorium decay was analyzed by a mass spectrometer. The amount of $Th^{228}$ (both in ppb and in counts per second, cps) found in the three cavitated solutions (whose CR39 showed the traces of the alpha particles emitted by thorium) is half of that in the three reference solutions corresponding to an $\alpha$-emission. The results are reported in Tables 17.1 and 17.2, and allow one to conclude that *the process of cavitation reduced the content of* $Th^{228}$ *in the solutions.*

### *17.1.2 Hadro-Leptonic Thorium Decay in DSR*

The spontaneous decay of $Th^{228}$ through the weak interaction halves it in a time $t_{1/2}$= 1.9 years = $9.99\times10^5$ min. The ratio between the half life of thorium, $t_{1/2}$ , and the time interval of cavitation, $t_c$ = 90 min, is $t_{1/2}/t_c = 10^4$. This means that cavitation brought about the transformation of $Th^{228}$ at a rate $10^4$ times faster than the natural leptonic decay would do. On the other hand, the experiments of the Russian team [101, 102] provided no evidence of spontaneous fission. One is therefore led to deem that we are facing a phenomenon of accelerated thorium decay, and that the transformation of thorium into some other nuclide, induced by cavitation, is rather due to strong interaction, in particular to its non-Minkowskian part.

By bearing this in mind, it is possible to interpret $t_{1/2}/t_c = 10^4$ as the ratio between the decaying time of $Th^{228}$ via the leptonic interaction (leptonic time $t_{\rm lep}$), and the transformation time of $Th^{228}$ via the hadronic interaction (hadronic time $t_{\rm had}$). Namely, one has:

$$\frac{t_{1/2}}{t_c} = 10^4 = \frac{t_{\rm lep}}{t_{\rm had}}. \tag{17.1}$$

Let us recall that the time coefficients of both metrics of electromagnetic and weak interactions, $b_{0,\rm e.m.}$ and $b_{0,\rm weak}$, are equal to each other, energy independent, and always equal to 1 (see Sect. 4.1). Either metrics is therefore always Minkowskian in time. Moreover, the space coefficients of both metrics have the same energy behavior (cf. (4.3), (4.8)). Thank to this circumstance, it will be always true, for the intervals of time $dt_{\rm e.m.}$ and $dt_{\rm weak}$, that $dt_{\rm e.m.}=dt_{\rm weak}$. Hence we can write:

$$\frac{t_{\rm weak}}{t_{\rm had}} = \frac{t_{\rm e.m.}}{t_{\rm had}} = 10^4. \tag{17.2}$$

On account of the hadronic law of time deformation (16.12), the same relation of proportionality holds between the threshold hadronic energy and the hadronic time and between the electromagnetic energy and the electromagnetic time. In the present case, (16.12) can be rewritten in terms of the time intervals $t_{\rm weak} = t_{\rm e.m.}$ and $t_{\rm had}$ – whose ratio, experimentally determined, is given by (17.2) – as:

$$\frac{t_{\rm had}}{t_{\rm weak}} = \frac{E_{0,\rm had}}{E_{\rm e.m.}}. \tag{17.3}$$

From the earlier relation it is possible to estimate the unknown variable $E_{\rm e.m.}$, i.e., the energy transferred by the electrical (Minkowskian) interaction to the nuclei of thorium (which get transformed into other nuclides by the strong interaction). One gets

$$E_{\rm e.m.} = E_{0,\rm had}\frac{t_{\rm e.m.}}{t_{\rm had}} = 367.5\,\text{GeV} \times 10^4 = 3.675 \times 10^{15}\,\text{eV}. \tag{17.4}$$

This value of energy is compatible with the maximum energy for nucleon $\varepsilon_{\rm max}$ estimated for the cavitation experiment which provided evidence of the production of the europium isotope 138 (see (16.11)). Let us recall that such an estimate was done by considering power conservation, according to a continuity equation applied to the bubble collapse due to cavitation.

## 17.2 Evidence for Neutron Emission in Non-Minkowskian Conditions

Both the signature of the occurring of piezonuclear reactions in non-Minkowskian conditions, i.e., the production of neutrons without concomitant emission of $\gamma$ radiation, and the energy required to induce them according to the prediction of DSR, have been checked in experiments on cavitation of water solutions, carried out at CNR National Laboratories (Rome 1 Area) and Italian Army technical facilities in 2004–2006. We report in the following the details of these experiments. Two separate investigations have been carried out, so the section is divided into two parts.

### *17.2.1 First Investigation*

Experimental Setup

The sonotrode employed was endowed with a compressed air cooling system. The ultrasound parameters were the same for all the experiments, i.e., the oscillation amplitude of the piezoelectric ceramics was kept to 50% (the sonotrode worked at a frequency of 20 kHz , with an amplitude of 30 $\mu$m at tip,and hence the power transmitted into the solutions was constant and equal) and the water was always deionized and bidistilled. The first investigation was made up of five different cavitation experiments. We subjected to cavitation not only pure water, but also solutions of four different salts in $H_2O$ (with a concentration of 1 ppm), namely:

1. 250 ml of $H_2O$
2. 250 ml of $H_2O$ solution of iron chloride $Fe(Cl)_3$
3. 250 ml of $H_2O$ solution of aluminium chloride $Al(Cl)_3$
4. 250 ml of $H_2O$ solution of lithium chloride LiCl
5. 500 ml of $H_2O$ solution of iron nitrate $Fe(NO_3)_3$

The first four cavitations lasted 90 min while the fifth one did 120 min.

The immersion depth of the sonotrode in the solution was 6 cm. It was suitably studied with reference to the cavitation chamber, in order to maximize the concentration of energy utilized for cavitation (by taking advantage also of the pressure waves reflected by the bottom of the cavitation chamber toward the sonotrode tip, and reducing the energy dispersion in the piezoconvective motions of the cavitated liquid).

Measurements of ionizing ($\alpha$, $\beta$, and $\gamma$) radiation background were carried out, along with measurements of neutron radiation background. The latter were conducted during every cavitation of the five solutions

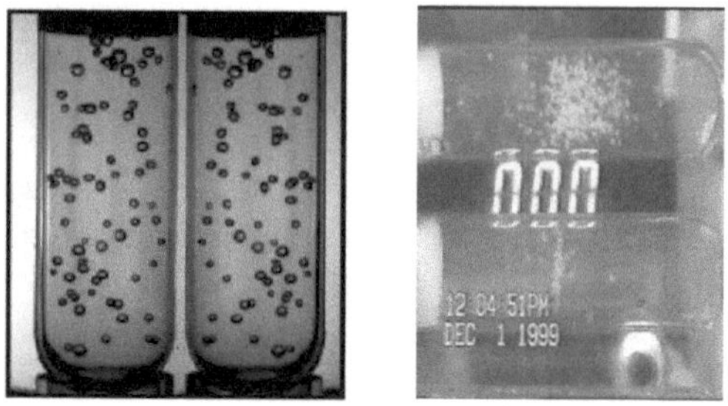


FIGURE 17.2. Morphology and distribution of bubbles produced in a Defender by the passage of neutrons (left); heavy ions (right)

above with the same type of detectors in a room suitably separated from that in which cavitation was taking place.

The detector of ionizing radiation was a Geiger counter with a mica window transparent to $\alpha$, $\beta$, and $\gamma$ radiation. It was provided with two aluminium filters 1 and 3 mm thick, to screen $\alpha$ radiation and $\alpha$ and $\beta$, respectively. During every cavitation, further measurements of $\alpha$ radiation were carried out by CR39 polycarbonate detectors, sensitive to ionizing radiation in the energy range 40 keV–4 MeV. A CR39 plate was positioned underneath each cavitation chamber in order to detect the radiation possibly emitted during the process. Moreover, in the first experiment of cavitation of mere water, a CR39 detector was put inside the bottle, in order to get a response without the filter of the glass of the bottle.

To detect neutrons we used integrating, passive dosimeters called Defender and produced by BTI.[1] They consist of minute droplets of a superheated liquid dispersed throughout an elastic polymer gel. When neutrons strike these droplets, they form small gas bubbles that remain fixed in the polymer. The number of bubbles is directly related to the amount and the energy of neutrons, so the obtained bubble pattern provides an immediate visual record of the dose. Due to their very operation way, the Defenders are sensitive to ionizing radiation too. However, the morphology and the distribution of the bubbles are quite different, as shown by Fig. 17.2 (supplied by the manufacturer, BTI). In the picture on the left one sees the effect produced by neutrons in two Defenders. The bubbles are big and spread out the whole volume. Conversely, the two detectors on the right show a different type and a different distribution of bubbles (generated by heavy ions), gathered in a cluster and much smaller than those produced by neutrons. The detectors Defender are sensitive to neutron energies in

[1] The matriculation number of the Defenders employed was composed by 6 figures. However, since the first three were the same (100) for all, in the following they will be identified by using only the last three digits.

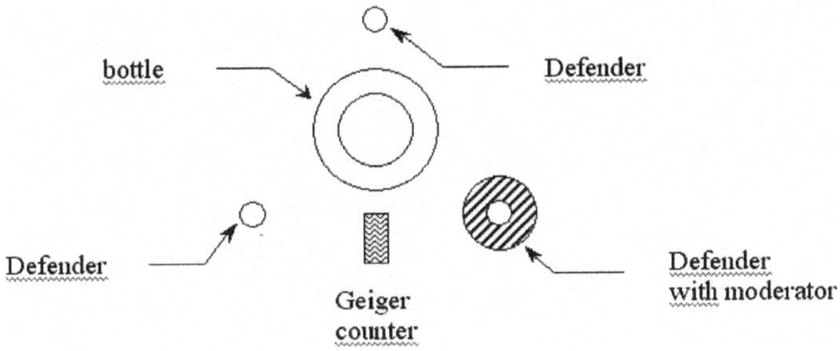


FIGURE 17.3. Layout of the experimental setup used to cavitate solutions and measure the emitted radiation

the interval 10 keV–15 MeV. The choice of using passive neutron detectors, instead of electronic ones, was made in order to avoid the well-known problems encountered with electronic devices (like background electronic noise, dead times, pileups).

The schematic layout of the experimental equipment is shown in Fig. 17.3. The cavitation chamber (bottle) was in the center and the sonotrode has to be imagined perpendicular to the plane of the figure, just over the bottle and lined up with it. For each cavitation experiment, three detectors Defender (of cylindrical shape) were used. They were placed vertically, coaxially to the sonotrode, arranged as shown in Fig. 17.3. One of the Defenders was screened by immersing it in a cylinder of carbon (moderator) 3 cm thick. The Geiger counter was pointed towards the area inside the bottle where cavitation took place. A second equal arrangement of the three Defenders and the bottle containing the uncavitated solution (blank), placed in a different room, was used to measure the neutron radiation background at the same time when cavitation was taking place.

## Radiation Measurements

The radiations $\alpha\ \beta\ \gamma$, $\beta\ \gamma$ and $\gamma$, measured *in all the cavitation runs*, turned out to be compatible with the background radiation.[2]

In particular, all the polycarbonate detectors CR39, i.e., both those used for cavitation experiments and those used in background measurement, recorded a radiation compatible with a normal background level. Therefore, either no ionizing radiation is produced during cavitation (at least the type of radiation with energy within the range of CR39's sensitivity, namely in the interval 40 keV–4 MeV) or the ionizing radiation able to affect CR39 could not escape the bottle or, if it got out, its energy was outside the energy range of CR39 detectors.

[2]This agrees with the results on the absence of radiation emission in the first cavitation experiment (see Sect. 16.2.1).

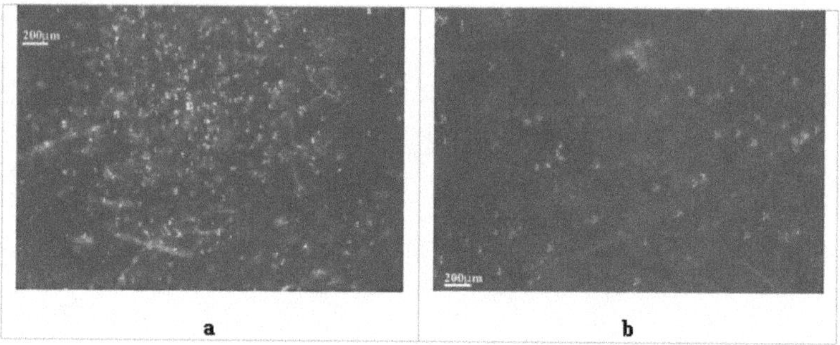


FIGURE 17.4. Showing the magnified central parts of the CR39 plate immersed in water (**a**) and of that underneath the bottle (**b**)

Due to their calibration, the CR39 are sensitive not only to $\alpha$-particles, but to heavy ions and protons too, provided their ionizing energy is within the detector energy range. Unfortunately all these particles can be easily slowed down by the water in which they are emitted, and then almost completely screened by the thick glass of the bottle in which cavitation took place. On the contrary, gamma rays can easily pass through the water and the glass of the bottle, but CR39 are not sensitive to them. However, as we stated before, gamma ray measurements carried out by a Geiger counter yielded doses of this radiation absolutely compatible with gamma background doses.

We remind that, in the first cavitation run with mere water, a CR39 was immersed directly into the water and hence was not screened by the glass of the bottle. We report in Fig. 17.4 two pictures showing the magnified central parts of the CR39 plate immersed in water and of that underneath the bottle.

It is impressive the difference in the amount of traces between the detector immersed in water on the left and that outside the bottle on the right. This fact induces one to deem that a great deal of radiation compatible with the CR39 detecting features was emitted inside the bottle and remained trapped in it. As just said, this radiation cannot consist of gamma rays, as they would have been detected by the Geiger counter outside the bottle, and moreover CR39 is not affected by them. Thus, one might suppose that the tracks on the CR39 plate were impressed by alpha particles or heavy ions (fission or fusion fragments) or by protons and electrons emitted by beta decays of those neutrons which had not succeeded to escape outside the bottle.

As to the neutron radiation, the measurements carried out in the experiments with $H_2O$, aluminium chloride and lithium chloride were compatible with the background level. On the contrary, in the second and the fifth experiment, with iron chloride and iron nitrate, respectively, the measured neutron radiation was *incompatible* with the neutron background level; moreover, in the last 30 min of cavitation the measured dose was significantly higher than the background.

Let us comment these results. The negative outcomes of the three experiments with water, aluminium chloride and lithium chloride allow one to conclude that the neutrons emitted during the cavitation of iron chloride cannot be related to the presence of $H_2O$ and Cl in this experiment. As to the positive outcomes of the fifth experiment with iron nitrate, although it cannot be excluded that the neutron emission be related to the presence of nitrogen, we can certainly state that this emission took place when iron was part of the cavitated solution.

Moreover, the absence of ionizing radiation $\alpha$, $\beta$, and $\gamma$ above the background level in all the experiments – even in those two in which we got the evidence of neutron emission – means that neutrons were produced without the usual consequent emission of gamma radiation.

One can therefore state that *only the presence of iron in the cavitated solution gave rise to neutron emission (and therefore to nuclear processes induced by cavitation)*, but without the accompanying emission of $\gamma$ radiation. According to the discussion of Sect. 16.3, *such a pattern of radiation emission agrees with the features (and is the signature) of the occurrence of piezonuclear reactions in presence of non-Minkowskian strong interactions.*

It is now clear, the reasons whereby we chose to cavitate solutions with salts of lithium, aluminium, and iron. Actually, this choice was just aimed at checking the role of non-Minkowskian strong interaction in piezonuclear reactions. Fe is the nucleus with the highest value of the bond energy per nucleon. In Minkowskian conditions, it is nuclei with a lower bond energy per nucleon, like Li and Al, which would be expected to give rise to nuclear signals, once subjected to stresses able to affect and make unstable their structure. This is exactly the contrary of what was observed: iron, neither lithium nor aluminium, is effective in producing nuclear signals. This is easily understood by the considerations carried out in Sect. 16.3.4. It is just iron which, with increasing energy in non-Mikowskian conditions, can overcome the hadronic threshold energy $E_{0,\mathbf{strong}}$ before all other nuclei, in particular the light nuclei preceding it (namely those with mass number $A < 56$). The purpose of repeating the experiment with two chemical solutions containing two different iron salts, $Fe(Cl)_3$ and $Fe(NO_3)_3$, was to exclude spurious effects and to assert beyond any doubt the role of iron in generating nuclear signals, thus stating the validity of the non-Minkowskian model of piezonuclear reactions.

### Evidence for Neutrons

Due to their physical relevance, let us discuss in detail the two experiments that provided evidence for neutron emission, namely those performed with solutions of $Fe(Cl)_3$ and $Fe(NO_3)_3$.

FIGURE 17.5. Defenders used to detect neutron emission induced by cavitation for solution 1 (300 ml of $H_2O$ with 300 μl of iron chloride). Left: Defenders 171 and 176 in presence of cavitation (case ii); right: Defenders 168 and 179 in absence of cavitation (blank). The morphology and the distribution of the bubbles in the former case corresponds to neutron detection (compare with Fig. 17.2)

*Iron Chloride Solution*

In the cavitation run with $Fe(Cl)_3$ the ultrasonic power transmitted into the solution was about 100 W by using a power set up of 1,000 W.

The typical aspect of the Defender detectors obtained is shown in Fig. 17.5, which refers to cavitated and uncavitated (blank) $Fe(Cl)_3$ solution (2)). The Defender 171 and 176 at left correspond to the cavitated solution, whereas the Defender 168 and 179 at right to the blank (namely to uncavitated solution). Notice the difference in bubble number and size between the two detector pairs. In fact, the bubbles of Defender 171 and 176 are about three times higher in number and larger in size than those of 168 and 179.[3] Moreover, a comparison with Fig. 17.2 shows that *the morphology and the distribution of the bubbles of the former pair correspond to detection of neutrons emitted as consequence of cavitation.*

---

[3]The larger number of bubbles of Defender 171 with respect to the 176 (as seen in Fig. 17.5) is a consequence of the different response of the two kinds of detectors, in agreement with the calibration provided by the manufacturer. Actually the measurement after calibration showed that both Defenders detected the same neutron dose. This testifies both their correct working and the reality of the observed phenomenon of neutron emission.

There is one more interesting difference between the two detector pairs. The pair on the right contains some bubbles as well – due to the detector thermodynamic background – but they are distributed almost uniformly in the whole volume of the two detectors. Conversely, the distribution of the bubbles in the two detectors on the left is not uniform at all, as the bubbles are concentrated in the lower half of the active volume. This is easily seen to be due to the very geometry of the apparatus. Indeed, the lower half of the active volume, where the bubbles gather, was contained in the semi-space which extended from the tip of the sonotrode downwards. In this region, and especially between the tip of the sonotrode and the bottom of the bottle, the process of cavitation was more vigorous and hence the emission of neutrons more likely.

The evidence for neutron emission is enforced by comparing the content of bubbles in Defender 172 – screened by inserting it in the cylinder with the moderator (carbon) – and in Defenders 171 and 176. The former displays a lower number of bubbles, and hence a lower dose of neutrons (comparable with the dose in all Defenders used in the background case). Thus, we can conclude that the moderator, by which Defender 172 was screened, reduced the energy of the neutrons that struck it to a value below the lower energy threshold of the detector (10 keV). This fact constitutes a further proof of the neutron nature of the detected signals, and allows one to get a rough estimate of the lower energy value of the neutrons emitted during cavitation. These neutrons were emitted in the water inside the bottle and hence slowed down by hydrogen before escaping the bottle. Since the Defender sensitivity is in the energy range 10 keV–15 MeV, the neutrons emitted by cavitation had to be at least epithermal with an energy of 15 keV. This allowed them to come out of the bottle with an energy content still above 10 keV, which is the minimal energy threshold for neutrons to be revealed by the two unscreened defenders (171 and 176). However, they could not be detected by Defender 172, since they had still to pass a layer of moderator which surrounded it and lowered the neutron energy below 10 keV. This consideration does not mean that neutrons emitted during could be only epithermal or fast. They could be thermal neutrons as well, which, however, were condemned to be trapped within the bottle and eventually decayed.

Analogous considerations hold for the neutron background measurement. The content and the spreading of bubbles in Defender 175, screened by the moderator, came out comparable with the content of bubbles of the other two Defenders (168 and 179) and of Defender 172 as well.

A more quantitative behavior of the phenomenon is illustrated by Fig. 17.6, which reports the neutron doses in nano-Sievert as function of the cavitation time. The horizontal line represents the measured neutron background level (chosen pessimistically among all the performed background measurements). The cavitation process lasted 90 min and one checked the number of bubbles in the detectors before starting cavitation and then

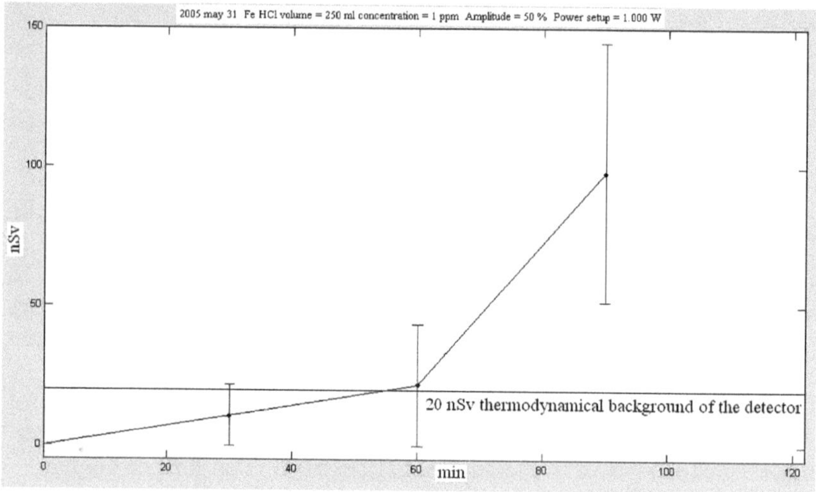


FIGURE 17.6. Neutron dose (nSv) vs. the cavitation time for $Fe(Cl)_3$ solution. The horizontal line represents the background level

every 30 min. It is seen from Fig. 17.6 that until approximately 60 min of cavitation the neutron dose remained well below the background level. After 60 min, the curve derivative increases and the dose of neutrons grows bigger over the background value.

The behavior of the neutron dose plotted in Fig. 17.6 entails the necessity to exceed a threshold in energy in order to start the emission of neutrons. Taken for granted that one transmits into the solution a power higher than the required threshold, the emission of neutrons begins only after a certain time interval. Clearly, amount of minutes is synonym of amount of energy.

In summary, the experiment with iron chloride provided the following main evidences in favor of neutron emission by cavitation. First of all, the morphology and distribution of the bubbles obtained from cavitation in the detectors do match those stated by the Defender manufacturer to be due to neutrons. Moreover, there occurs what one might call a "tropism" in the response of the detectors, which is localized exactly in the part of the active volume next to the area where cavitation is more intense. Finally, the neutron emission shows a threshold behavior in time (and therefore in energy).

*Iron Nitrate Solution*

Let us now consider the cavitation run carried out with the solution (5) of $Fe(NO_3)_3$, in order to show that the evidences collected are absolutely consistent with those found in the experiment with iron chloride. For this run, a power setup of 2,000 W was applied in the last 30 min of cavitation.

We stress that in this case the whole active volume of the detector (and not only half of it) laid inside the semi-space that extended from

FIGURE 17.7. Showing both the unscreened Defenders used in the cavitation run with the Iron Nitrate solution. The bubbles are spread all over the detecting volume

the plane containing the sonotrode tip downwards. Then, if the bubble tropism (which was inferred from the bubble distribution in the detectors in the previous experiment) is a correct deduction, one would have to get bubbles in the whole active volume of the detectors. This is exactly the case, as seen from Fig. 17.7 of the unscreened Defenders used in the cavitation run.

The measurement of the neutron background was carried out and the detector response was compatible with that obtained in the previous experiment.

The behavior of the neutron as function of cavitation time is shown in Fig. 17.8. We recall that in this case the cavitation process lasted 120 min. During the initial 90 min the bubbles in the detectors have been counted every 30 min, while in the last 30 min of cavitation the bubble number was checked every 10 min. One sees that the dose curve stays well below the background level during all the first 90 min. Thereafter, the curve begins to rise and the dose increases fairly over the background one in the final 30 min. We therefore find again the energy-threshold behavior noted with the iron chloride. As expected, the amount of minutes (and therefore the energy) needed to trigger neutron emission depends on the quantity of cavitated solution (60 min for 250 ml in the $Fe(Cl)_3$ case and 90 min for 500 ml of the $Fe(NO_3)_3$ solution).

Therefore, the cavitation run carried out with iron nitrate confirms the evidences for neutron emission found with iron chloride. However, it

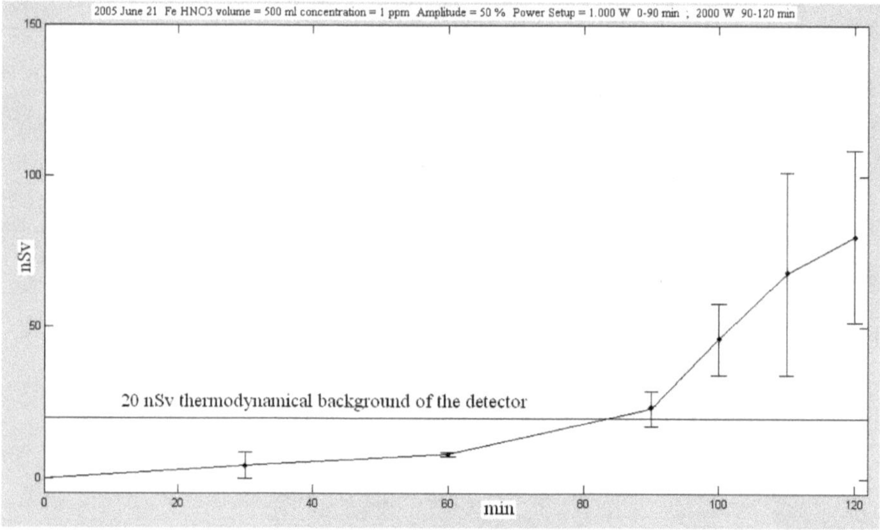


FIGURE 17.8. Neutron dose (nSv) vs. the cavitation time for $Fe(NO_3)_3$ solution. The horizontal line represents the background level

remains to dispel any possible doubt about the fact that the bubbles observed in the detectors were indeed produced by neutrons. By having excluded ionizing radiation, the only other physical cause might be the ultrasounds employed to induce cavitation. But we remind the reader that all five experiments were carried out with the same geometrical arrangement of the detectors with respect to the source of ultrasounds, and that the ultrasonic power was always the same for all of them. Therefore, would ultrasounds instead of neutrons be the cause of the bubbles, one would have had to observe the same bubble pattern in all cavitation runs. On the contrary, only two experiments out of five gave evidences of neutron emission (i.e., of a number of bubbles greater than the number corresponding to the neutron background level). In the other three experiments, the bubbles in the detectors near cavitation were absolutely compatible with those in the detectors far from cavitation. This fact demonstrates that ultrasounds cannot be considered the cause of the bubbles observed in the Defenders.

### *17.2.2 Second Investigation*

Experimental Setup

The second investigation was carried out in order to get more evidences about neutron emission with a different experimental setup, different detectors, and in a different geographical area with a soil of completely diverse geological origin.

Since the first investigation confirmed the hypothesis of the basic role of iron in producing piezonuclear reactions in non-Minkowskian conditions (according to the predictions of DSR based on the behavior of the strong metric with energy), the second one was devoted to a systematic study of such an evidence, by using solutions with only iron nitrate. Then, six cavitation runs (each lasting 90 min) were carried out on the same quantity (250 ml) of deionized and bidistilled water and of a solution of $Fe(NO_3)_3$ with different concentration, subjected to ultrasounds of different power, namely:

(1') $H_2O$ (oscillation amplitude 50%)

(2') $H_2O$ (oscillation amplitude 70%)

(3') $H_2O$ solution of $Fe(NO_3)_3$ (concentration 1 ppm oscillation amplitude 50%)

(4') $H_2O$ solution of $Fe(NO_3)_3$ (concentration 10 ppm; oscillation amplitude 50%)

(5') $H_2O$ solution of $Fe(NO_3)_3$ (concentration 1 ppm; oscillation amplitude 70%)

(6') $H_2O$ solution of $Fe(NO_3)_3$ (concentration 10 ppm; oscillation amplitude 70%)

Therefore, the cavitated solutions could have three possible concentrations, 0, 1, and 10 ppm. Moreover, the oscillation amplitude and hence the transmitted ultrasonic power assumed two different values, 50% and 70%, corresponding to about 100 and 130 W, respectively. The energy delivered to the solution within the whole cavitation time was 0.54 and 0.70 MJ in the two cases.

During each cavitation we carried out ionizing radiation measurements by two Geiger counters and neutron radiation measurements by five new neutron detectors of the Defender type XL, but with higher sensitivity (by one order of magnitude). Background neutron measurements were also conducted, but – unlike the previous investigation – they were accomplished only at the beginning of the whole set of cavitations.

Two pictures of the experimental apparatus used in the six cavitation runs are shown in Fig. 17.9. The bottle in which cavitation took place is visible in the middle of both pictures and the sonotrode, the vertical tapered metal stick, is aligned with and inserted in it. The green pipe surrounding the sonotrode was part of the compressed air cooling system and conveyed the cooling air onto the sonotrode surface. The three horizontal greyish cylinders with a black cylindrical endcap are the neutron detectors. Two of them were positioned next to the bottle at a height with respect to the tip of the sonotrode, in order to be struck by horizontally emitted neutrons. They were not screened from neutrons by anything but a

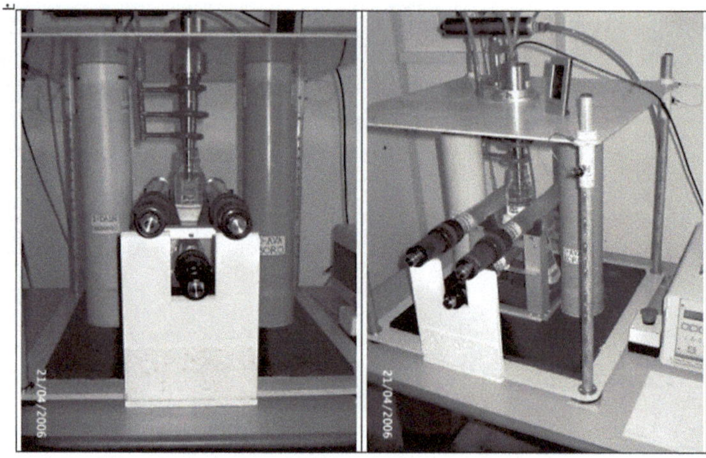


FIGURE 17.9. Experimental apparatus used in the second investigation. The cavitation chamber is visible in the middle of both pictures and the sonotrode, the vertical tapered metal stick, is aligned with and inserted in it. The green pipe surrounding the sonotrode conveyed the cooling air onto the sonotrode surface. The three horizontal greyish cylinders with a black cylindrical endcap are the neutron detectors. The two orange and creamy vertical cylinders contained the two screened Defenders

1.5 cm thick layer of water of the bottle. The third detector was placed underneath the bottle in order to collect the vertically emitted neutrons (a vertex detector). Like the other two, it was not screened from neutrons by anything but a 7 cm thick layer of water of the bottle. The two orange and creamy vertical cylinders contained one neutron detector each, of the same type of the three horizontal ones. The Defenders in the orange and creamy cylinders were surrounded, and hence screened, by boron powder (thermal neutron absorber) and by carbon powder (neutron moderator), respectively. Eventually two orange Geiger counters, pointed towards the cavitation area, can be spotted underneath the bottle. One of them was used to measure gamma rays only, the other one was used to measure alpha, beta and gamma radiation.

A difference with respect to the previous investigation to be stressed is that the immersion depth of the sonotrode in the solution was 1 cm.

## Neutron Measurements

In all of the six experiments evidence of neutron emission was got in the unscreened Defenders, namely *the phenomenon was perfectly reproducible.* The screened Defenders (both by boron and carbon) always detected a reduced neutron dose, comparable with the background one, as in the first

FIGURE 17.10. Example of the bubble pattern obtained in the second investigation

investigation. Analogously, the ionizing radiation measured by the Geiger counters was always compatible with the background level.

Thus, the results of the second investigation too provided evidence for the emission of anomalous nuclear radiation, since neutrons were not accompanied by gamma rays. Hence the piezonuclear reactions producing such a radiation were consequence of non-Minkowskian strong interactions, according to the discussion of Sect. 16.3.5.

An example of the neutron bubble pattern obtained in the six experiments is shown in Fig. 17.10. This figure refers to one of those two detectors which were laid horizontally beside the bottle.

The bubbles are distributed in the whole volume but they are mostly concentrated in the central part of the Defender. The part in which there are more bubbles was just the closest to the cavitation area. From this evidence and from all the rest of the pictures of bubbles in the detectors, it is possible to infer that the neutron emission from the cavitation area has roughly the shape of a semi-sphere.

As for the first series of cavitations, we plotted the measured doses (in nano-Sievert), for all the six experiments, as function of the cavitation time. In this case, the number of bubbles was counted every 10 min. Each curve corresponds to one concentration of the $Fe(NO_3)_3$ solution, from 0 to 10 ppm, and one oscillation amplitude (and therefore ultrasonic power), 50% (100 W) or 71% (130 W). The six graphs are reported in Fig. 17.11. They are displaced in a Cartesian coordinate system with concentration on the $y$-axis and amplitude (power) on the $x$-axis. The horizontal black line represents the neutron background level of 3.5 nSv, which is due to the thermodynamical behavior of the detector.

The examination of the six graphs of Fig. 17.15 shows immediately that the second series of cavitations supports the conclusions already drawn from the experiments with iron solutions in the first series. One sees indeed that in all experiments there exists a certain threshold of energy which has to be exceeded in order to start the emission of neutrons, provided that the power transferred to the solution is already higher than the threshold power. According to the discussion of Sect. 16.3.5, this amounts to say that the velocity in the coordinate energy is greater than the threshold velocity.

Moreover, Fig. 17.11 disproves further the possible criticism about a possible generation of the bubbles by ultrasounds rather than by neutrons.

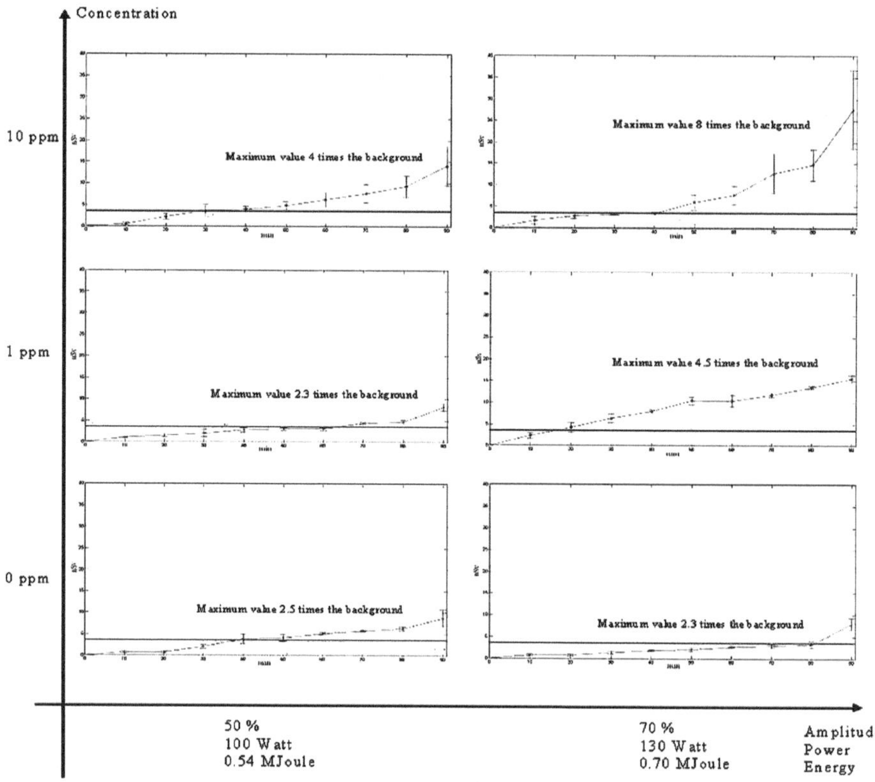


FIGURE 17.11. The six graphs (one for each cavitation of the second series) showing the neutron dose (in nSv) as a function of time in minutes (time interval 10 min). Each curve corresponds to one value of concentration and one of the amplitude. The horizontal line in all graphs corresponds to the thermodynamical background of 3.5 nSv. The graphs are displaced in a Cartesian plane with concentration (in ppm) on the $y$-axis and amplitude (power) on the $x$-axis

Indeed, by looking at the compound graph and reading it along its columns, i.e., keeping the amplitude (power) fixed, it is seen that the curves are different, while the ultrasonic power is always the same. Conversely, had ultrasounds been the real cause of the bubbles, one should have had equal effects.

A difference between the two series of cavitations is instead provided by the different amount of neutron doses in the two cases. Actually (as it can be seen by comparing Fig. 17.11 with Figs. 17.4 and 17.8) the neutron doses yielded by cavitation in the experiments of the first investigation are nearly one order of magnitude higher than the neutron doses yielded by cavitation in the second one. This fact is even more puzzling since one would expect higher doses in those experiments carried out with a greater amplitude and hence higher ultrasonic power. A possible explanation can be found in the different immersion depth of the sonotrode in the solution. Indeed, the immersion depth in the first investigation, about 6 cm, was just chosen in order to get the cavitation process more effective. However, this circumstance needs some further investigations in new experiments in which neutron emission could be studied as a function of the immersion depth of the sonotrode, by investigating both how the sonotrode tip wears out (in connection with its oscillation mode, first or second Bessel harmonic) and how the piezoconvective motions of the cavitated liquid go on.

Let us also remark that in the second investigation one got evidence for neutron emission also in cavitating pure water, unlike the case of the first one. This is obviously due to the higher sensitivity of the detectors employed in the second investigation. Such a result agrees with the indirect evidence for neutron emission obtained in the first experiment of water cavitation, in which the changes in concentration of the stable elements occurred with a variation in neutron number (see Sect. 16.2.1).

A fundamental point we want to stress is that piezonuclear reactions are completely incomprehensible from a standard, quanto-mechanical point of view. Atomic sizes are of the order of $10^{-8}$ cm, whereas the nuclear ones are $10^{-13}$ cm, five orders of magnitude smaller. On the contrary, the corresponding energies are $\backsim 10$ and $10^6$ eV, respectively. This means that the interactions of atoms and nuclei are inconceivable in standard quantum mechanics. We have discussed in Sects. 16.3.4 and 16.3.5 possible DSR models whereby atoms can indeed affect nuclei. The production of neutrons by cavitation is therefore another example (besides the shadow of light discussed in Chap. 13) of a phenomenon contradicting quantum mechanics.

In conclusion of the review of the cavitation experiments, let us notice that the cavitation process (like the case of the Cavendish experiment for the measurement of the gravity speed discussed in Chap. 15) is a *dynamic* one, in the sense that energy changes during the process. Indeed, the macroscopic energy (namely, that supplied to the liquid by the sonotrode) is fixed, but in the single microscopic processes of cavitation the energy transferred to the bubbles can vary depending on the bubble radius (see Sect. 16.3.5, in particular (16.18)). Such a kind of physical dynamic phenomenon is better dealt with in the framework of DR5 than in DSR. This point will be tackled again in Part V.

# Part IV

# DEFORMED SPACE–TIME IN FIVE DIMENSIONS: GEOMETRY

# 18
# Multidimensional Space–Time

Einstein's relativity theories contain, among the others, two basic precepts. First, they stated beyond any reasonable doubt that our Universe is (at least) (3+1)-dimensional. Moreover, Einstein taught us that introducing extra dimensions (time, in this case) can provide a better description of (and even simplify) the laws of nature (electromagnetism for Special Relativity and gravity for General Relativity).

This latter teaching by Einstein was followed by Kaluza [111] and Klein [112], who added a fifth dimension to the four space–time ones in order to unify electromagnetism and gravitation (let us recall however that Nordström [113] was the first to realize that in a 5D space–time the field equations do split naturally into Einstein's and Maxwell's equations). Although unsuccessful, the Kaluza–Klein (KK) theory constituted the first attempt to unification of fundamental interactions within a multidimensional space–time. The KK scheme, in which the coefficient of the fifth coordinate is constant, was later generalized by Jordan [114] and Thiry [115], who considered it to be a general function of the space–time coordinates. The 5D KK formalism has been later extended to higher dimensions, also in the hope of achieving unification of all interactions, including weak and strong forces [116–118]. Six dimensions were e.g., considered by Ingraham [117] and Podolanski [118]. However, a true revival of multidimensional theories starting from 1970 was due to the advent of string theory [119] and supersymmetry [120]. As is well known, their combination, superstring theory [121], provides a framework for gravity quantization. Modern generalizations [122,123] of the Kaluza–Klein scheme require a minimum number of 11 dimensions in order to accommodate the

Standard Model of electroweak and strong interactions; let us recall that 11 is also the maximum number of dimensions required by supergravity theories [124]. For an exhaustive review of Kaluza–Klein theories we refer the reader to [125].

A basic problem in any multidimensional scheme is the hidden nature of the extra dimensions, namely to explain why the Universe looks 4D. A possible solution (first proposed by Klein) is assuming that each extra dimension is *compactified*, namely it is curled up in a circle, which from a mathematical standpoint is a compact set, whence the name (*cylindricity condition*).[1] In such a view, therefore, space–time is endowed with a cylindrical geometrical structure. The radius $R$ of the circle is taken to be so small (roughly of the order of the Planck length) as to make the extra dimension unobservable at distances exceeding the compactification scale.

In the last decade of the past century the hypothesis of noncompactified extra dimensions began to be taken into serious consideration. Five-dimensional theories of the Kaluza–Klein type, but with no cylindricity condition on the extra dimension, were built up e.g., by Wesson (the so-called "Space–Time–Mass" (STM) theory, in which the fifth dimension is the rest mass [123]) and by Fukui ("Space–Time–Mass–(Electric) Charge" (STMC) theory with charge as extra dimension [126]). In such a kind of theories, the (noncompactified) extra dimension is a physical quantity, but not a spatial one in its strict sense. More recently, instead, the idea of a true hyperspace, with large *space* dimensions, was put forward (starting from the 1998 pioneering ADD model [127]). The basic assumption is that gravity is the only force aware of the extra space dimensions. This leads to view the physical world as a multidimensional space ("bulk") in which the ordinary space is represented by a 3D surface ("3D brane"). Electromagnetic, weak, and strong forces are trapped within the brane, and do not "feel" the extra space dimensions. On the contrary, gravitons escape the brane and spread out the whole hyperspace. In this picture, the reason the gravitational force appears to be so weak is because it is diluted by the extra dimensions. In the ADD scheme (and similar), the size $L$ of the extra dimensions is related to their number. Only one extra dimension would have a size greater than the solar system, and therefore would have been already discovered. For two extra dimensions, it is possible to show that $L \sim 0.2\,\mathrm{mm}$, whereas for three it is $L \sim 1\,\mathrm{nm}$. In this framework, therefore, the extra dimensions are still finite (although not microscopic as in the KK theories).

The ADD model has been also modified in order to allow for infinite (in the sense of unlimited) extra dimensions. In the "warped-geometric" model [128], it is still assumed that the ordinary space is confined to a

---

[1] Compactification of the extra dimensions can be achieved also by assuming compact spaces different from (and more complex than) a circle.

3D brane embedded in a bulk. Only the gravitons can escape the brane along the extra dimension, but they feel the gravitational field of the brane and therefore do not venture out of it to long distances. This amounts to say that the geometry of the extra dimension is warped (the 5D metric of the hyperspace contains an exponentially decreasing "warp factor," namely the probability of finding a graviton decays outside the brane along the extra dimension). An analogous result can be obtained by hypothesizing the existence of two branes, put a distance $L$ apart along the extra dimension, one trapping gravity and the other not. If the two branes have opposite tension, the geometry of space between the branes is warped too. Models with both an infinite, warped extra dimension and a finite compact one have been also considered [25].

Needless to say, it is impossible to get direct evidence of extra dimensions. However, one can probe them in an indirect way, because the existence of extra dimensions has several observational consequences, both in astrophysics and cosmology and in particle physics, depending on the model considered. In compactified theories, new excited states appear within the extra dimensions (*Kaluza–Klein towers*), with energies $E_n = nhc/R$. They affect the carriers of the electromagnetic, weak, and strong forces by turning them into a family of increasingly massive clones of the original particle, thus magnifying the strengths of the nongravitational forces. Such effects can therefore be detected in particle accelerators. A research carried at CERN's Large Electron–Positron (LEP) Collider provided no evidence of such extradimensional influences at up to an energy of 4 TeV. This result puts a limit of $0.5 \times 10^{-19}$ m to the size $L$ of the extra dimensions [129]. The ADD model foresees deviations from the Newton's inverse-square law of gravity for objects closer together than the size of the extra dimension. Such a stronger gravitational attraction ($\sim r^{-4}$) could be observed in tabletop experiments (of the Cavendish type ). In general, in models with a 3D brane, gravitons leaving the brane into, or entering it from, the extra dimensions could provide signatures for the hyperspace in accelerator experiments. In the former case, one has to look for missing energy in a collision process, due to the disappearing of the gravitons into the extra dimensions. In the latter, since gravitons can decay into pairs of photons, electrons, or muons, detecting an excess of these particles at specific energy and mass levels would indirectly provide evidence for the existence of dimensions beyond our own.

It must also be noted that other fields besides the gravitational one are expected to be present in the bulk. Bulk gauge fields are associated with "new forces," the strength of which is predicted to be roughly a million times stronger than gravity. These stronger forces can manifest themselves in different ways, and could be detectable also in tabletop experiments. For example, they can simulate antigravity effects on submillimeter distance scales, since gauge forces between like-charged objects are naturally repulsive.

The hypothesis of a multidimensional space–time allows one not only to unify interactions and quantize gravity, but to solve or at least to address from a more basic viewpoint a number of fundamental problems still open in particle physics and astrophysics. Let us mention for instance the weakness of the gravitational force, the abundance of matter over antimatter, the extraordinarily large number of elementary particles, the nature of dark matter and the smallness of the cosmological constant.

In the following, we shall see that DSR admits a natural extension to a further extra dimension, in the sense that the deformed Minkowski space–time $\widetilde{M}$ can be embedded in a 5D Riemannian space.

# 19
# Embedding Deformed Minkowski Space in a 5D Riemann Space

## 19.1 From LLI Breakdown to Energy as Fifth Dimension

Both the analysis of the physical processes considered in deriving the phenomenological energy-dependent metrics for the four fundamental interactions, and the experiments discussed in Part III, seem to provide evidence (indirect and direct, respectively) for a breakdown of LLI invariance (at least in its usual, special-relativistic sense). But it is well known that, in general, the breakdown of a symmetry is the signature of the need for a *wider, exact* symmetry. In the case of the breaking of a space–time symmetry – as the Lorentz one – this is often related to the possible occurrence of higher-dimensional schemes. It will be shown that this is indeed the case, and that *energy does in fact represent an extra dimension.*

In the description of interactions by energy-dependent metrics, we saw that energy plays in fact a *dual* role. On one side, as more and more stressed, it constitutes a dynamic variable. On the other hand, it represents a parameter characteristic of the phenomenon considered (and therefore, for a given process, it cannot be changed at will). In other words, when describing a given process, the deformed geometry of space–time (in the interaction region where the process is occurring) is "frozen" at the situation described by those values of the metric coefficients $\left\{b_{\mu}^{2}(E)\right\}_{\mu=0,1,2,3}$ corresponding to the energy value of the process considered. Namely, a fixed value of $E$ determines the space–time structure of the interaction region at that given energy. In this respect, therefore, the energy of the process has to be considered as *a geometrical quantity* intimately related to the very geometrical structure of the physical world. In other words, from a geometrical point of view, all goes on as if were actually working on "slices" (sections) of a 5D space, in which the extra dimension is just represented by the energy. Then, the 4D, deformed, energy-dependent space–time is just a manifestation (or a "shadow," to use the famous word of Minkowski) of a larger space with energy as fifth dimension.

The simplest way to take account of (and to make explicit) the double role of energy in DSR is assuming that $E$ represents an extra metric dimension – on the same footing of space and time – and therefore embedding the 4D deformed Minkowski space $\widetilde{M}(E)$ of DSR in a 5D (Riemann) space $\Re_5$. This leads to build up a "Kaluza–Klein-like" scheme, with energy as fifth dimension, we shall refer to in the following as *5D Deformed Relativity* (DR5) [6,130,131].

Let us recall that the use of momentum components as metric variables on the same foot of the space–time ones can be traced back to Ingraham [117]. On the contrary, it was just shown by Lee that time (namely, a space–time coordinate) can be used as a (discrete) dynamic variable [132]. Moreover, many authors (starting from Dirac [133]) treated mass as a dynamic variable in the context of scale-invariant theories of gravity [134,135]. Such a point of view has been advocated also in the framework of modern Kaluza–Klein theories by the already quoted "Space–Time–Mass" (STM) theory [123].

It is worth stressing that, apart from the previous considerations, we already ran across some clues of a possible 5D structure underlying DSR. One such an indication is provided e.g., by generalized energy-momentum dispersion law (3.100), which – as already stressed – is typical of multidimensional theories. Another one is provided by the form of the phenomenological metric of strong interaction (see Sect. 4.1.3), in particular expressions (4.12), (4.13) of the space coefficients $b_{2,\text{strong}} = \sqrt{2}/5$ and $b_{3,\text{strong}} = 2/5$. Indeed, the 5 at the denominators are reminiscent of the same factor entering the relation between the Ricci tensor and the scalar curvature in a 5D Riemann space, $R_{AB} = (R/5)g_{AB}$, with $g_{AB}$ being the 5D metric tensor.[1] Another clue is the interpretation of the

[1] In fact, let us consider the vacuum Einstein equations with a cosmological constant $\Lambda$ in a $N$-dimensional Riemann space:

$$R_{AB} - \frac{1}{2}Rg_{AB} = \Lambda g_{AB}.$$

By contracting on $A, B$ and using the well-known property $g^{DA}g_{AB} = \delta^D_B$, one gets

$$R = \frac{2n}{2-n}\Lambda.$$

Then

$$R_{AB} = \left(\frac{1}{2}R + \Lambda\right) g_{AB} = \frac{R}{n} g_{AB}$$

(M. Mamone Capria, private communication).

hadronic law of time deformation, (17.3), as a relation of power conservation, $W = const.$ (needed to explain the mechanism of piezonuclear reactions: see Sect. 16.3.5). As already stressed, in a 5D optics this means moving along the extra dimension energy at constant speed (namely it amounts to a principle of inertia for energy). Another possible experimental inkling of the fifth dimension can be found in the double-slit-like experiments (Chap. 13). Indeed, we have seen that, in order to put the anomalous interference effect into evidence, it is necessary to employ a suitable time sampling of the measurements. On account of the fact that the phenomenon has a threshold behavior both in space and in energy, we can state it to occur in a well-defined space–time–energy, 5D region. Moreover, the crucial dependence on the time sampling can be interpreted as follows. As is well known, a way to realize one lives on a curved manifold is by means of the geodesic deviation. For instance, on Earth surface, moving from Equator along two meridians shows that the meridian separation decreases, thus implying Earth surface is curved (Wheeler's "parable of the two travelers"). However, the travelers are able to discern the decrease of their relative separation only if they move an appreciable distance (compared to the Earth radius of curvature). Otherwise, no separation is seen and they remain convinced that Earth is flat. In our opinion, the anomalous interference effect is not only related to the deformation of space–time (and therefore to the breakdown of LLI), but also to the Gaussian curvature of the 5D space–time–energy manifold $\Re_5$. Selecting the suitable time sampling amounts therefore to choose the time magnitude scale necessary to detect the curvature of the 5D region in which the anomalous effect shows up.

## 19.2 The 5D Space–Time–Energy Manifold $\Re_5$

On the basis of the arguments of Sect. 19.1, we assume therefore that physical phenomena do occur in a world which is actually described by a 5D space–time–energy manifold $\Re_5$ endowed with the energy-dependent metric:[2]

$$g_{AB,\mathrm{DR5}}(E) \equiv \mathrm{diag}(b_0^2(E), -b_1^2(E), -b_2^2(E), -b_3^2(E), f(E)) \overset{\mathrm{ESC\ off}}{=}$$

$$= \delta_{AB}\left(b_0^2(E)\delta_{A0} - b_1^2(E)\delta_{A1} - b_2^2(E)\delta_{A2} - b_3^2(E)\delta_{A3} + f(E)\delta_{A5}\right). \tag{19.1}$$

[2]In the following, capital Latin indices take values in the range $\{0, 1, 2, 3, 5\}$, with index 5 labeling the fifth dimension. We choose to label by 5 the extra coordinate, instead of using 4, in order to avoid confusion with the notation often adopted for the (imaginary) time coordinate in a (formally) Euclidean Minkowski space.

It follows from (19.1) that $E$, which is an independent *nonmetric* variable in DSR, becomes a *metric* coordinate in $\Re_5$. Then, whereas $g_{\mu\nu,\text{DSR}}(E)$ (given by (2.17)) is a deformed, Minkowskian metric tensor, $g_{AB,\text{DR5}}(E)$ is a genuine Riemannian metric tensor.

Therefore, the infinitesimal interval of $\Re_5$ is given by

$$\begin{aligned} ds^2_{\text{DR5}}(E) &\equiv dS^2(E) \equiv g_{AB,\text{DR5}}(E)dx^A dx^B \\ &= b_0^2(E)\left(dx^0\right)^2 - b_1^2(E)\left(dx^1\right)^2 - b_2^2(E)\left(dx^2\right)^2 - b_3^2(E)\left(dx^3\right)^2 + f(E)\left(dx^5\right)^2 \\ &= b_0^2(E)c^2\left(dt\right)^2 - b_1^2(E)\left(dx^1\right)^2 - b_2^2(E)\left(dx^2\right)^2 - b_3^2(E)\left(dx^3\right)^2 + f(E)l_0^2\left(dE\right)^2, \end{aligned} \tag{19.2}$$

where we have put

$$x^5 \equiv l_0 E \ , \ l_0 > 0. \tag{19.3}$$

The constant $l_0$ provides the dimensional conversion energy→ length, and it has therefore the dimensions of the inverse of a force. On physical grounds, it is expected to be a fundamental constant of DR5, so it is worth trying to guess a possible identification of $l_0$. Let us recall that in Sect. 4.2 we already came across a quantity built up by fundamental constants with dimensions of a force: the Kostro constant or Planck force $K = c^4/G$ (see (4.27), which can be interpreted as the greatest possible force in Nature [40]). Then, it is natural to assume

$$l_0 = \frac{1}{K} = \frac{G}{c^4} = \frac{1}{8\pi}\kappa, \tag{19.4}$$

where $\kappa$ is the gravitational coupling constant of the usual, four dimensional Einstein equations $G_{\mu\nu} = \kappa T_{\mu\nu}$ (with $G_{\mu\nu} = R_{\mu\nu} - \frac{1}{2}g_{\mu\nu}R$ and $T_{\mu\nu}$ being the Einstein curvature tensor and the energy–momentum tensor, respectively). Therefore, identifying $l_0$ with the inverse of the Kostro constant has as consequence that it coincides with the gravitational constant $\kappa$ apart from the numerical factor $8\pi$ (which however is essentially due to the choice of the unit system). As is well known, in General Relativity $\kappa$ determines the effectiveness of the energy density of the source in deforming space–time and can be interpreted as the force per unit area required to give space–time a unit curvature.[3] If the identification (19.4) is correct, then $l_0$, and consequently $\kappa$, plays an analogous role in DR5, namely it is related in an essential way to the curvature of $\Re_5$ – which in turn reflects itself in the deformation of the 4D space–time $\widetilde{M}$ – *whatever the interaction involved.* In the framework of DR5, therefore, the gravitational constant rises, from mere coupling constant for the gravity only, to the role of *universal constant of deformation, valid for all interactions.*

Since the space–time metric coefficients are dimensionless, it can be assumed that they are functions of the ratio $E/E_0$, where $E_0$ is an energy

[3]Remember that curvature has dimensions $l^{-2}$.

scale characteristic of the interaction (and the process) considered (for instance, the energy threshold in the phenomenological metrics of Sect. 4.1). The coefficients $\{b_\mu^2(E)\}$ of the metric of $\widetilde{M}(E)$ can be therefore expressed as

$$\left\{b_\mu\left(\frac{E}{E_0}\right)\right\} \equiv \left\{b_\mu\left(\frac{x^5}{x_0^5}\right)\right\} = \{b_\mu(x^5)\} \quad \forall\mu = 0, 1, 2, 3, \tag{19.5}$$

where we put

$$x_0^5 \equiv l_0 E_0 \ . \tag{19.6}$$

As to the fifth metric coefficient, one assumes that it too is a function of the energy only: $f = f(E) \equiv f(x^5)$ (although, in principle, nothing prevents from assuming that, in general, $f$ may depend also on space–time coordinates $\{x^\mu\}$, $f = f(\{x^\mu\}, x^5)$). Unlike the other metric coefficients, it may be $f(E) \lessgtr 0$. Therefore, a priori, the energy dimension may have either a time-like or a space-like signature in $\Re_5$, depending on $\mathrm{sgn}\,(f(E)) = \pm 1$. In the following, it will be sometimes convenient assuming $f(E) \in R_0^+$ and explicitly introducing the double sign in front of the fifth coefficient.

In terms of $x^5$, the (*covariant*) metric tensor can be written as

$$\begin{aligned} g_{AB,\mathrm{DR5}}(x^5) &= \mathrm{diag}(b_0^2(x^5), -b_1^2(x^5), -b_2^2(x^5), -b_3^2(x^5), \pm f(x^5)) \\ &\overset{ESC\ off}{=} \delta_{AB}\left[b_0^2(x^5)\delta_{A0} - b_1^2(x^5)\delta_{A1} - b_2^2(x^5)\delta_{A2} - b_3^2(x^5)\delta_{A3} \pm f(x^5)\delta_{A5}\right]. \end{aligned} \tag{19.7}$$

On account of the relation

$$g^{AB}_{\mathrm{DR5}}(x^5) g_{BC,\mathrm{DR5}}(x^5) = \delta^A_{\ C} \ , \tag{19.8}$$

the *contravariant* metric tensor reads

$$\begin{aligned} g^{AB}_{\mathrm{DR5}}(x^5) &= \mathrm{diag}(b_0^{-2}(x^5), -b_1^{-2}(x^5), -b_2^{-2}(x^5), \ \ b_3^{-2}(x^5), \pm\left(f(x^5)\right)^{-1}) \\ &\overset{ESC\ off}{=} \delta_{AB}\left[\begin{array}{c} b_0^{-2}(x^5)\delta_{A0} - b_1^{-2}(x^5)\delta_{A1} - b_2^{-2}(x^5)\delta_{A2} - b_3^{-2}(x^5)\delta_{A3} \\ \pm\left(f(x^5)\right)^{-1}\delta_{A5} \end{array}\right]. \end{aligned} \tag{19.9}$$

The space $\Re_5$ has the following "slicing property"

$$\Re_5|_{dx^5=0 \Leftrightarrow x^5=\overline{x^5}} = \widetilde{M}(\overline{x^5}) = \left\{\widetilde{M}(x^5)\right\}_{x^5=\overline{x^5}} \tag{19.10}$$

(where $\overline{x^5}$ is a fixed value of the fifth coordinate) or, at the level of the metric tensor:

$$\begin{aligned} &g_{AB,\mathrm{DR5}}(x^5)\big|_{dx^5=0 \Leftrightarrow x^5=\overline{x^5} \in R_0^+} \\ &= \mathrm{diag}\left(b_0^2(\overline{x^5}), -b_1^2(\overline{x^5}), -b_2^2(\overline{x^5}), -b_3^2(\overline{x^5}), \pm f(\overline{x^5})\right) = g_{AB,\mathrm{DSR}}(\overline{x^5}). \end{aligned} \tag{19.11}$$

We recall that in general, in the framework of 5D Kaluza–Klein (KK) theories, the fifth dimension must be necessarily space-like, since, in order to avoid the occurrence of causal (loop) anomalies, the number of time-like dimensions cannot be greater than one. But it is worth to stress that the present theory is not a Kaluza–Klein one. In "true" KK theories, due to the lack of observability of the extra dimensions, it is necessary to impose to them the cylindricity condition. This is not required in the framework of DR5, since the fifth dimension (energy) is a physically observable quantity (think to the Minkowski space of standard SR: There is no need to hide the fourth dimension, since time is an observable quantity). Actually, in DR5 not only the cylindricity condition is not implemented, but it is even reversed. In fact, the metric tensor $g_{AB,\mathrm{DR5}}(x^5)$ depends only on the fifth coordinate $x^5$. Therefore, one does not assume the compactification of the extra coordinate (one of the main methods of implementing the cylindricity condition in modern hyperdimensional KK theories, as discussed in Chap. 18), which remains therefore extended (i.e., with infinite compactification radius). The problem of the possible occurrence of causal anomalies in presence of more time-like dimensions is then left open in the "pseudo-Kaluza–Klein" context of DR5. This is reflected in the uncertainty in the sign of the energy metric coefficient $f(x^5)$. In particular, it cannot be excluded a priori that the signature of $x^5$ can change. This occurs whenever the function $f(x^5)$ does vanish for some energy values. As a consequence, in correspondence to the energy values which are zeros of $f(x^5)$, the metric $g_{AB,\mathrm{DR5}}(x^5)$ is *degenerate*.

DR5 belongs therefore to the class of noncompactified KK theories . Moreover, it has some connection with Wesson's STM theory [123]. Both in the DR5 formalism and in the STM theory (at least in its more recent developments) it is assumed that all metric coefficients do in general depend on the fifth coordinate. Such a feature distinguishes either models from true Kaluza–Klein theories. However, DR5 differs from the STM model – as well as from similar ones, like e.g., the Fukui STMC [126] – at least in the following main respects:

(1) Its physical motivations are based on the phenomenological analysis of Part I and on the experimental results of Part III, and therefore are not merely speculative.

(2) The fact of assuming *energy* (which is a true variable), and not rest mass (which instead is an invariant), as fifth dimension.[4]

(3) The *local* (and not *global*) nature of the 5D space $\Re_5$, whereby the energy-dependent deformation of the 4D space–time is assumed to provide a geometrical description of the interactions.

---

[4] In this respect, therefore, the DR5 formalism rensembles more the one due to Ingraham [117].

We want to stress that, in embedding the deformed Minkowski space $\widetilde{M}(x^5)$ in $\Re_5$, *energy does lose its character of dynamic parameter* (the role it plays in DSR), *by taking instead that of a true metrical coordinate,* on the same footing of the space–time ones. This has a number of basic implications. The first one is of geometrical nature, and is just the passage from a (flat) pseudoeuclidean metric to a genuine (curved) Riemannian one. The others consequences pertain to both symmetries and dynamics, as we shall see in this Part and in the next one. In such a change of role of energy, with the consequent passage from $\widetilde{M}(x^5)$ to $\Re_5$, some of the geometrical and dynamic features of DSR are lost, whereas others are still present and new properties appear. Among the former, we recall the basic one – valid at the slicing level $x^5 = const.$ ($\mathrm{d}x^5 = 0$) – related to the Generalized Lagrange Space structure of $\widetilde{M}(x^5)$, which implies *the natural arising of gauge fields,* intimately related to the inner geometry of the deformed Minkowski space (see Part II). Let us also stress that, in the framework of $\Re_5$, the dependence of the metric coefficients on a true metric coordinate make them fully analogous to the gauge functions of non-abelian gauge theories, thus implementing DR5 as a metric gauge theory (in the sense specified in Sect. 4.4).

## 19.3 Phenomenological 5D Metrics of Fundamental Interactions

Let us now consider the 4D metrics of the deformed Minkowski spaces $\widetilde{M}(x^5)$ for the four fundamental interactions (electromagnetic, weak, strong, and gravitational) (see Sect. 4.1). In passing from the deformed, special-relativistic 4D framework of DSR to the general-relativistic 5D one of DR5 – geometrically corresponding to the embedding of the deformed 4D Minkowski spaces $\left\{\widetilde{M}(x^5)\right\}_{x^5 \in R_0^+}$ (where $x^5$ is a parameter) in the 5D Riemann space $\Re_5$ (where $x^5$ is a metric coordinate), in general the phenomenological metrics (4.2)–(4.3), (4.7)–(4.8), (4.10)–(4.13), and (4.17)–(4.18) take the following 5D form ($f(x^5) \in R_0^+ \ \forall x^5 \in R_0^+$):

$$
\begin{aligned}
&g_{AB,\mathrm{DR5,e.m.}}(x^5) \\
&= \mathrm{diag}\left(1, -\left\{1 + \widehat{\Theta}(x^5_{0,\mathrm{e.m.}} - x^5)\left[\left(\frac{x^5}{x^5_{0,\mathrm{e.m.}}}\right)^{1/3} - 1\right]\right\},\right. \\
&\quad -\left\{1 + \widehat{\Theta}(x^5_{0,\mathrm{e.m.}} - x^5)\left[\left(\frac{x^5}{x^5_{0,\mathrm{e.m.}}}\right)^{1/3} - 1\right]\right\}, \\
&\left. -\left\{1 + \widehat{\Theta}(x^5_{0,\mathrm{e.m.}} - x^5)\left[\left(\frac{x^5}{x^5_{0,\mathrm{e.m.}}}\right)^{1/3} - 1\right]\right\}, \pm f(x^5)\right); \qquad (19.12)
\end{aligned}
$$

$$
\begin{aligned}
&g_{AB,\mathrm{DR5,weak}}(x^5) \\
&= \mathrm{diag}\left(1, -\left\{1+\widehat{\Theta}(x^5_{0,\mathrm{weak}}-x^5)\left[\left(\frac{x^5}{x^5_{0,\mathrm{weak}}}\right)^{1/3}-1\right]\right\},\right. \\
&\quad -\left\{1+\widehat{\Theta}(x^5_{0,\mathrm{weak}}-x^5)\left[\left(\frac{x^5}{x^5_{0,\mathrm{weak}}}\right)^{1/3}-1\right]\right\}, \\
&\left. -\left\{1+\widehat{\Theta}(x^5_{0,\mathrm{weak}}-x^5)\left[\left(\frac{x^5}{x^5_{0,\mathrm{weak}}}\right)^{1/3}-1\right]\right\}, \pm f(x^5)\right);
\end{aligned}
\tag{19.13}
$$

$$
\begin{aligned}
&g_{AB,\mathrm{DR5,strong}}(x^5) \\
&= \mathrm{diag}\left(1+\widehat{\Theta}(x^5-x^5_{0,\mathrm{strong}})\left[\left(\frac{x^5}{x^5_{0,\mathrm{strong}}}\right)^{2}-1\right], -\left(\frac{\sqrt{2}}{5}\right)^2,\right. \\
&\left. -\left(\frac{2}{5}\right)^2, -\left\{1+\widehat{\Theta}(x^5-x^5_{0,\mathrm{strong}})\left[\left(\frac{x^5}{x^5_{0,\mathrm{strong}}}\right)^{2}-1\right]\right\}, \pm f(x^5)\right);
\end{aligned}
\tag{19.14}
$$

$$
\begin{aligned}
&g_{AB,\mathrm{DR5,grav.}}(x^5) \\
&= \mathrm{diag}\left(1+\widehat{\Theta}(x^5-x^5_{0,\mathrm{grav.}})\left[\frac{1}{4}\left(1+\frac{x^5}{x^5_{0,\mathrm{grav.}}}\right)^{2}-1\right], -b^2_{1,\mathrm{grav.}}(x^5),\right. \\
&\left. -b^2_{2,\mathrm{grav.}}(x^5), -\left\{1+\widehat{\Theta}(x^5-x^5_{0,\mathrm{grav.}})\left[\frac{1}{4}\left(1+\frac{x^5}{x^5_{0,\mathrm{grav.}}}\right)^{2}-1\right]\right\}, \pm f(x^5)\right).
\end{aligned}
\tag{19.15}
$$

As we are going to show, all the earlier metrics – derived on a mere phenomenological basis, from the experimental data on some physical phenomena ruled by the four fundamental interactions, at least as far as their space–time part is concerned – can be recovered as solutions of the vacuum Einstein equations in the 5D space $\Re_5$, natural covering of the deformed Minkowski space $\widetilde{M}(x^5)$.

# 20
# Einstein's Field Equations in $\Re_5$ and Their Solutions

## 20.1 Riemannian Structure of $\Re_5$

We have seen that, unlike $\widetilde{M}(x^5)$, which is a flat pseudoeuclidean space, $\Re_5$ is a genuine Riemann one. Its affine structure is determined by the 5D affine connection $\Gamma^A_{BC}(x^5)$ (which rules the parallel transport of vectors in $\Re_5$), defined by

$$\Gamma^A_{BC}(x^5) \equiv \frac{\partial x^A}{\partial \xi^D} \frac{\partial^2 \xi^D}{\partial x^B \partial x^C}, \tag{20.1}$$

where $\left\{\xi^A\right\}$, $\left\{x^A\right\}$ are the coordinates in a locally inertial (Lorentzian) frame and in a generic frame, respectively. Let us recall that $\Gamma^A_{BC}$ *is not* a true tensor, since it vanishes in a locally inertial frame (namely, in absence of a gravitational field).

Due to the compatibility between affine geometry and metric geometry in Riemann spaces (characterized by the vanishing of the covariant derivative of the metric, and therefore torsion-free), it is possible to express the connection components in terms of the metric tensor as

$$\Gamma^I_{AB}(x^5) = \frac{1}{2} g^{IK}_{\mathrm{DR5}} (\partial_B g_{KA,\mathrm{DR5}} + \partial_A g_{KB,\mathrm{DR5}} - \partial_K g_{AB,\mathrm{DR5}}) = \left\{ \begin{array}{c} I \\ AB \end{array} \right\}, \tag{20.2}$$

where the quantities $\left\{ \begin{array}{c} I \\ AB \end{array} \right\}$ are the second-kind Christoffel symbols.

Then, it is easy to check that the only nonzero components of $\Gamma^A_{BC}(x^5)$ are (the prime denotes derivation with respect to $x^5$)[1]

$$\begin{cases} \Gamma^0_{05} = \Gamma^0_{50} = \dfrac{b'_0}{2b_0}\,; & \Gamma^1_{15} = \Gamma^1_{51} = \dfrac{b'_1}{2b_1}; \\ \Gamma^2_{25} = \Gamma^2_{52} = \dfrac{b'_2}{2b_2}\,; & \Gamma^3_{35} = \Gamma^3_{53} = \dfrac{b'_3}{2b_3}; \\ \Gamma^5_{00} = -\dfrac{b_0 b'_0}{2f}\,; \quad \Gamma^5_{11} = \dfrac{b_1 b'_1}{2f}\,; & \Gamma^5_{22} = \dfrac{b_2 b'_2}{2f}; \\ \Gamma^5_{33} = \dfrac{b_3 b'_3}{2f}\,; \quad \Gamma^5_{55} = \dfrac{f'}{2f}\,. \end{cases} \tag{20.3}$$

The Riemann–Christoffel (curvature) tensor in $\Re_5$ is given by

$$R^A_{BCD}(x^5) = \partial_C \Gamma^A_{BD} - \partial_D \Gamma^A_{BC} + \Gamma^A_{KC}\Gamma^K_{BD} - \Gamma^A_{KD}\Gamma^K_{BC}. \tag{20.4}$$

Let us give, for readers' convenience, the only nonzero components of the (covariant) Riemann–Christoffel tensor $R_{ABCD}(x^5)$:

$$\begin{aligned} R_{0101} &= \frac{b'_0 b'_1}{4f};\ R_{0202} = \frac{b'_0 b'_2}{4f};\ R_{0303} = \frac{b'_0 b'_3}{4f}; \\ R_{0505} &= \frac{b'_0 f' b_0 + (b'_0)^2 - 2b''_0 b_0 f}{4b_0 f}; \\ R_{1212} &= -\frac{b'_1 b'_2}{4f};\ R_{1313} = -\frac{b'_1 b'_3}{4f};\ R_{1515} = \frac{b'_1 f' b + (b'_1)^2 - 2b''_1 b_1 f}{4b_1 f}; \end{aligned} \tag{20.5}$$

$$\begin{aligned} R_{2323} &= -\frac{b'_2 b'_3}{4f};\ R_{2525} = \frac{b'_2 f' b_2 + (b'_2)^2 - 2b''_2 b_2 f}{4cf}; \\ R_{3535} &= \frac{b'_3 f' b_3 + (b'_3)^2 - 2b''_3 b_3 f}{4b_3 f}. \end{aligned}$$

By contraction of $R^A_{BCD}$ on two and four indices, respectively, we get as usual the 5D Ricci tensor $R_{AB}(x^5)$, given explicitly by

$$R_{AB}(x^5) = \partial_I \Gamma^I_{AB} - \partial_B \Gamma^I_{AI} + \Gamma^I_{AB}\Gamma^K_{IK} - \Gamma^K_{AI}\Gamma^I_{BK}, \tag{20.6}$$

and the scalar curvature $R(x^5) = R^A_A(x^5)$.

---

[1]Henceforth, in order to simplify the notation, we adopt units such that $c =$ (velocity of light)$= 1$ and $\ell_0 = 1$.

The nonvanishing components of the Ricci tensor $R_{AB}(x^5)$ read thence as follows:

$$R_{00} = -\frac{1}{2}\frac{b_0''}{f} - \frac{b_0'}{4f}\left(-\frac{b_0'}{b_0} + \frac{b_1'}{b} + \frac{b_2'}{b_2} - \frac{f'}{f}\right); \tag{20.7}$$

$$R_{11} = \frac{1}{2}\frac{b_1''}{f} + \frac{b_1'}{4f}\left(\frac{b_0'}{b_0} - \frac{b_1'}{b_1} + \frac{b_2'}{b_2} + \frac{b_3'}{b_3} - \frac{f'}{f}\right); \tag{20.8}$$

$$R_{22} = \frac{1}{2}\frac{b_2''}{f} + \frac{b_2'}{4f}\left(\frac{b_0'}{b_0} + \frac{b_1'}{b_1} - \frac{b_2'}{b_2} + \frac{b_3'}{b_3} - \frac{f'}{f}\right); \tag{20.9}$$

$$R_{33} = \frac{1}{2}\frac{b_3''}{f} + \frac{b_3'}{4f}\left(\frac{b_0'}{b_0} + \frac{b_1'}{b_1} + \frac{b_2'}{b_2} - \frac{b_3'}{b_3} - \frac{f'}{f}\right); \tag{20.10}$$

$$\begin{aligned} R_{44} = &-\frac{1}{2}\left(\frac{b_0'}{b_0} + \frac{b_1'}{b_1} + \frac{b_2'}{b_2} + \frac{b_3'}{b_3}\right)' + \frac{f'}{4f}\left(\frac{b_0'}{b_0} + \frac{b_1'}{b_1} + \frac{b_2'}{b_2} + \frac{b_3'}{b_3}\right) \\ &-\frac{1}{4}\left[\left(\frac{b_0'}{b_0}\right)^2 + \left(\frac{b_1'}{b_1}\right)^2 + \left(\frac{b_2'}{b_2}\right)^2 + \left(\frac{b_3'}{b_3}\right)^2\right]. \end{aligned} \tag{20.11}$$

The scalar curvature in five dimensions, $R(x^5)$, is finally given by the lengthy expression:

$$\begin{aligned} R(x^5) - &\frac{b_1 b_1' f(b_2' b_3 b_0 + b_3' b_2 b_0 + b_0' b_2 b_3) + b_2 b_3\left[2b_1'' b_1 f - b_0(b_1')^2 f - b_0 b_1' f' b_1\right]}{4b_1^2 f^2 b_2 b_3 b_0} \\ + &\frac{b_2' b_2 f(b_1' b_3 b_0 + b_3' b_0 b_1 + b_0' b_3 b_1) + b_0 b_1 b_3\left[2b_2'' b_2 f - (b_2')^2 f - b_2' f' b_2\right]}{4b_2^2 f^2 b_3 b_0 b_1} \\ + &\frac{b_3' b_3 f(b_1' b_0 b_2 + b_2' b_0 b_1 + b_0' b_2 b_1) + b_0 b_1 b_2\left[2b_3'' b_3 f - (b_3')^2 f - b_3' f' b_3\right]}{4b_3^2 f^2 b_2 b_0 b_1} \\ + &\frac{b_0' b_0 f\,(b_1' b_2 b_3 + b_2' b_3 b_1 + b_3' b_2 b_1) + b_1 b_2 b_3\left[2b_0'' b_0 f - (b_0')^2 f - b_0' f' b_0\right]}{4b_0^2 f^2 b_2 b_3 b_1} \\ + &\frac{1}{4f^2 b_1^2 b_2^2}\left\{b_2^2\left[2b_1'' b_1 f - (b_1')^2 f - b_1' f' b_1\right] + b_1^2\left[2b_2'' b_2 f - (b_2')^2 f - b_2' f' b_2\right]\right\} \\ + &\frac{1}{4f^2 b_0^2 b_3^2}\left\{b_0^2\left[2b_3'' b_3 f - (b_3')^2 f - b_3' f' b_3\right] + b_3^2\left[2b_0'' b_0 f - (b_0')^2 f - b_0' f' b_0\right]\right\}. \end{aligned} \tag{20.12}$$

## 20.2 Vacuum Einstein's Equations

From the knowledge of the Riemann–Christoffel tensor and of its contractions it is possible to derive the Einstein equations in the space $\Re_5$ by exploiting the Hamilton principle. The 5D Hilbert–Einstein action in $\Re_5$ reads

$$S_{\mathrm{DR5}} = -\frac{1}{16\pi\widetilde{G}}\int \mathrm{d}^5x\sqrt{\pm\widetilde{g}(x^5)}R(x^5) - \Lambda_{(5)}\int \mathrm{d}^5x\sqrt{\pm\widetilde{g}(x^5)}, \qquad (20.13)$$

where $\widetilde{g}(x^5) = \det g_{\mathrm{DR5}}(x^5)$, $\widetilde{G}$ is the 5D "gravitational" constant, and $\Lambda_{(5)}$ is the "cosmological" constant in $\Re_5$. The form of the second term of the action (20.13) clearly shows that $\Lambda_{(5)}$ is assumed to be a genuine constant, although it might also, in principle, depend on both the fifth coordinate (namely, on the energy $E$) and the space–time coordinates $x$: $\Lambda_{(5)} = \Lambda_{(5)}(x, x^5)$. The double sign in the square root accords to that in front of the fifth metric coefficient $f(x^5)$. Among the problems concerning $S_{\mathrm{DR5}}$, let us quote its physical meaning (as well as that of $\widetilde{G}$) and the meaning of those energy values $\overline{x^5}$ such that $S_{\mathrm{DR5}}(\overline{x^5}) = 0$ (due to a possible degeneracy of the metric).

Then, a straightforward use of the variational methods yields the (vacuum) Einstein equations in $\Re_5$ in the form

$$R_{AB}(x^5) - \frac{1}{2}g_{AB,\mathrm{DR5}}(x^5)R(x^5) = \Lambda_{(5)}g_{AB,\mathrm{DR5}}(x^5). \qquad (20.14)$$

We want here to consider some special cases of the 5D Einstein equations (20.14) which – according to the discussion of Part I– seem to have a special physical relevance. They are (i) the case of spatial isotropy; and (ii) the case in which all the metric coefficients are pure powers in the energy (*Power Ansatz*).

### 20.2.1 Case (i): Space Isotropy

For a spatial isotropic deformation, it is $b_1(E) = b_2(E) = b_3(E) = b(E)$, so that the metric reduces simply to

$$g_{\mathrm{DR5}}(E) = \mathrm{diag}(b_0^2(E), -b^2(E), -b^2(E), -b^2(E), f(E)). \qquad (20.15)$$

The independent Einstein equations obviously reduce to the following three ones (for simplicity of notation, we omit the explicit functional dependence of all quantities on $E$):

$$\begin{cases} 3(-2b''f + b'f') = 4\Lambda_{(5)}bf^2; \\ f\left[b_0^2(b')^2 - 2b_0b_0'bb' - 4b_0^2bb'' - 2b_0b_0''b^2 + b^2(b_0')^2\right] \\ \qquad +b_0bf'(2b_0b' + b_0'b) = 4\Lambda_{(5)}b_0^2b^2f^2; \\ 3b'(b_0b)' = -4\Lambda_{(5)}b_0b^2f. \end{cases} \qquad (20.16)$$

### 20.2.2 Case (ii): Power Ansatz

We have seen in Sect. 19.2 that the space–time metric coefficients can be considered functions of the ratio $E/E_0$ (see (19.5)). Therefore, for the metric $g_{\mathrm{DR5}}$ written in the form (19.1), we can put, following also the hints from phenomenology (see Sect. 4.1):

$$\begin{cases} b_0^2(E) &= (E/E_0)^{q_0}\ ; \\ b_1^2(E) &= (E/E_0)^{q_1}\ ; \\ b_2^2(E) &= (E/E_0)^{q_2}\ ; \\ b_3^2(E) &= (E/E_0)^{q_3}\ , \end{cases}$$
$$q_\mu \in R\ \forall \mu = 0,1,2,3. \tag{20.17}$$

In the following we shall refer to the form (20.17) as the "Power Ansatz". For the dimensional parameter $f(E)$ we assume here simply (in order to simplify solution of the Einstein equations)

$$f(E) = E^r \tag{20.18}$$

($r \in R$), being understood that the characteristic parameter $E_0$ is possibly contained in $\ell_0$. Of course, the Einstein equations (20.14) reduce now to the following algebraic equations in the five exponents $q_0$, $q_1$, $q_2$, $q_3$, $r$:

$$\begin{cases} (2+r)(q_3+q_1+q_2) - q_1^2 - q_2^2 - q_3^2 - q_1q_2 - q_1q_3 - q_2q_3 = 4\Lambda_{(5)}E^{r+2}\ ; \\ (2+r)(q_3+q_0+q_2) - q_2^2 - q_3^2 - q_0^2 - q_2q_3 - q_2q_0 - q_3q_0 = 4\Lambda_{(5)}E^{r+2}\ ; \\ (2+r)(q_3+q_0+q_1) - q_1^2 - q_3^2 - q_0^2 - q_1q_3 - q_1q_0 - q_3q_0 = 4\Lambda_{(5)}E^{r+2}\ ; \\ (2+r)(q_0+q_1+q_2) - q_1^2 - q_2^2 - q_0^2 - q_1q_2 - q_1q_0 - q_2q_0 = 4\Lambda_{(5)}E^{r+2}\ ; \\ q_1q_2 + q_1q_3 + q_1q_0 + q_2q_3 + q_2q_0 + q_3q_0 = -4\Lambda_{(5)}E^{r+2}\ . \end{cases} \tag{20.19}$$

Of course, for consistency one would have to impose the compatibility condition that $\Lambda_{(5)}$, too, is a power of the energy, and precisely one should assume $\Lambda_{(5)}(E) \backsim E^{-(r+2)}$. The energy dependence of $\Lambda_{(5)}$ is however in contrast with the action (20.13). Needless to say, the vacuum prescription $\Lambda_{(5)} = 0$ is compatible with this hypothesis.

In the following, we shall also use for the 5D metric in the Power Ansatz the form

$$g_{AB,\mathrm{DR5power}}(\widetilde{\mathbf{q}}, x^5) =$$
$$= \mathrm{diag}\left(\left(\frac{x^5}{x_0^5}\right)^{q_0}, -\left(\frac{x^5}{x_0^5}\right)^{q_1}, -\left(\frac{x^5}{x_0^5}\right)^{q_2}, -\left(\frac{x^5}{x_0^5}\right)^{q_3}, \pm\left(\frac{x^5}{x_0^5}\right)^{r}\right) \tag{20.20}$$

($q_0, q_1, q_2, q_3, r \in Q$, $A, B = 0,1,2,3,5$), in which the double sign of the energy coefficient has been made clear and the (fake) five-vector $\widetilde{\mathbf{q}} \equiv (q_0, q_1, q_2, q_3, r)$ introduced.[2]

### 20.2.3 Phenomenological Metrics in the Power Ansatz

We have seen in Sect. 19.3 that embedding the DSR phenomenological metrics for the four interactions in $\Re_5$ leads to expressions (19.12)–(19.15). In the context of the Power Ansatz, and by making their piecewise structure explicit, they can be written in the form

$$g_{AB,\mathrm{DR5,e.m.,weak}}(x^5) = \begin{cases} \operatorname{diag}\left( \begin{array}{c} 1, -\left(\dfrac{x^5}{x^5_{0,e.m.,\mathrm{weak}}}\right)^{1/3}, -\left(\dfrac{x^5}{x^5_{0,e.m.,\mathrm{weak}}}\right)^{1/3}, \\ -\left(\dfrac{x^5}{x^5_{0,e.m.,\mathrm{weak}}}\right)^{1/3}, \pm\left(\dfrac{x^5}{x^5_{0,e.m.,\mathrm{weak}}}\right)^{r} \end{array} \right), \\ \qquad 0 < x^5 < x^5_{0,e.m.,\mathrm{weak}}; \\ \operatorname{diag}\left(1,-1,-1,-1,\pm\left(\dfrac{x^5}{x^5_{0,e.m.,\mathrm{weak}}}\right)^{r}\right), \\ \qquad x^5 \geq x^5_{0,e.m.,\mathrm{weak}}; \end{cases} \tag{20.21}$$

$$g_{AB,\mathrm{DR5,strong}}(x^5) = \begin{cases} \operatorname{diag}\left( \begin{array}{c} \left(\dfrac{x^5}{x^5_{0,\mathrm{strong}}}\right)^{2}, -\left(\dfrac{\sqrt{2}}{5}\right)^{2}, -\left(\dfrac{2}{5}\right)^{2}, \\ -\left(\dfrac{x^5}{x^5_{0,\mathrm{strong}}}\right)^{2}, \pm\left(\dfrac{x^5}{x^5_{0,\mathrm{strong}}}\right)^{r} \end{array} \right), \\ \qquad x^5 > x^5_{0,\mathrm{strong}}; \\ \operatorname{diag}\left(1,-\left(\dfrac{\sqrt{2}}{5}\right)^{2},-\left(\dfrac{2}{5}\right)^{2},-1,\pm\left(\dfrac{x^5}{x^5_{0,\mathrm{strong}}}\right)^{r}\right), \\ \qquad 0 < x^5 \leq x^5_{0,\mathrm{strong}}; \end{cases} \tag{20.22}$$

[2] In the following, we shall use the tilded-bold notation $\widetilde{\mathbf{v}}$ for a (true or fake) vector in $\Re_5$, in order to distinguish it from a vector $\mathbf{v}$ in the usual 3Dspace.

$$g_{AB,\text{DR5,grav.}}(x^5)$$
$$= \begin{cases} \text{diag}\begin{pmatrix} \frac{1}{4}\left(1+\frac{x^5}{x^5_{0,\text{grav.}}}\right)^2, -b^2_{1,\text{grav.}}(x^5), -b^2_{2,\text{grav.}}(x^5), \\ \frac{1}{4}\left(1+\frac{x^5}{x^5_{0,\text{grav.}}}\right)^2, \pm\left(\frac{x^5}{x^5_{0,\text{grav.}}}\right)^r \end{pmatrix}, \\ \qquad x^5 > x^5_{0,\text{grav.}}; \\ \text{diag}\left(1, -b^2_{1,\text{grav.}}(x^5), -b^2_{2,\text{grav.}}(x^5), -1, \pm\left(\frac{x^5}{x^5_{0,\text{grav.}}}\right)^r\right), \\ \qquad 0 < x^5 \le x^5_{0,\text{grav.}}. \end{cases} \tag{20.23}$$

In the gravitational metric $g_{AB,\text{DR5,grav.}}(x^5)$ the expressions of the two space coefficients $b^2_{1,\text{grav.}}(x^5)$ and $b^2_{2,\text{grav.}}(x^5)$ have not been made explicit, due to their indeterminacy at experimental level.

It must be stressed that the embedding of the phenomenological 4D metrics, obtained in the framework of DSR, in the 5D space $\Re_5$, implies – in the context of the Power Ansatz – a dependence of the corresponding 5D metrics on the parameter $r$ (exponent of the metric coefficient of the fifth, energetic dimension). This reflects itself in the dynamics of $\Re_5$, as we shall see in Part V.

The phenomenological 5D metrics in the Power Ansatz are therefore characterized by the parameter sets

$$\widetilde{\mathbf{q}}_{\text{e.m./weak}} = \begin{cases} \left(0, \frac{1}{3}, \frac{1}{3}, \frac{1}{3}, r\right), & 0 < x^5 < x^5_{0,\text{e.m./weak}}; \\ (0,0,0,0,r), & x^5 \ge x^5_{0,e.m./\text{weak}}; \end{cases}; \tag{20.24}$$

$$\widetilde{\mathbf{q}}_{\text{strong}} = \begin{cases} (2,(0,0),2,r), & x^5 > x^5_{0,\text{strong}}; \\ (0,(0,0),0,r), & 0 < x^5 \le x^5_{0,\text{strong}}; \end{cases}; \tag{20.25}$$

$$\widetilde{\mathbf{q}}_{\text{grav.}} = \begin{cases} (\underline{2},?,?,\underline{2},r), & x^5 > x^5_{0,\text{grav.}}; \\ (0,?,?,,0,r), & 0 < x^5 \le x^5_{0,\text{grav.}} \end{cases}, \tag{20.26}$$

where the question marks "?" reflect the unknown nature of the two gravitational spatial coefficients.

Let us clarify the notation adopted for $\widetilde{\mathbf{q}}_{\text{strong}}$ and $\widetilde{\mathbf{q}}_{\text{grav.}}$. The zeros in brackets in $\widetilde{\mathbf{q}}_{\text{strong}}$ reflect the fact that such exponents do not refer to the metric tensor $g_{AB,\text{DR5power}}(x^5)$ (20.20), but to the more general tensor

$$g_{AB,\text{DR5power-conform}}(x^5) =$$
$$= \text{diag}\left(\vartheta_0\left(\frac{x^5}{x^5_0}\right)^{q_0}, -\vartheta_1\left(\frac{x^5}{x^5_0}\right)^{q_1}, -\vartheta_2\left(\frac{x^5}{x^5_0}\right)^{q_2}, -\vartheta_3\left(\frac{x^5}{x^5_0}\right)^{q_3}, \pm\vartheta_5\left(\frac{x^5}{x^5_0}\right)^r\right) \tag{20.27}$$

with $\widetilde{\boldsymbol{\vartheta}}= (\vartheta_A)$ being a constant five-vector. Equation (20.27) can be written in matrix form as

$$\mathbf{g}_{\text{DR5power-conform}}(x^5) = \mathbf{g}_{\text{DR5power}}(x^5)\widetilde{\boldsymbol{\vartheta}} \tag{20.28}$$

where $\widetilde{\boldsymbol{\vartheta}}$ is meant to be a column vector. The passage from the metric tensor $g_{AB,\text{DR5power}}(x^5)$ to $g_{AB,\text{DR5power-conform}}(x^5)$ is obtained by means of the tensor transformation law in $\Re_5$ (ESC on)

$$g_{AB,\text{DR5power-conform}}(x^5) = \frac{\partial x^K}{\partial x'^A}\frac{\partial x^L}{\partial x'^B} g_{KL,\text{DR5power}}(x^5) \tag{20.29}$$

induced by the following 5$D$ *anisotropic rescaling* of the coordinates of $\Re_5$:

$$\mathrm{d}x^A = \sqrt{\vartheta_A}\mathrm{d}x'_A \leftrightarrow x^A = \sqrt{\vartheta_A}x'_A. \tag{20.30}$$

Such a transformation allows one to get, in the Power Ansatz, metric coefficients *constant* (i.e., independent of the energy) *but different.* This is just the case of the two constant space coefficients $b_1^2(x^5)$, $b_2^2(x^5)$ in the strong metric. In this case, the vector $\widetilde{\boldsymbol{\vartheta}}$ explicitly reads

$$\widetilde{\boldsymbol{\vartheta}}_{\text{strong}} = \left(0, \left(\frac{\sqrt{2}}{5}\right)^2, \left(\frac{2}{5}\right)^2, 0, ?\right), \tag{20.31}$$

where the question mark "?" reflects again the unknown nature of $\pm f(x^5)$. As we shall see in Part V, the coordinate rescaling (20.30) does not modify the dynamics of DR5 (the geodesics in $\Re_5$ remain unchanged).

The underlined 2, $\underline{2}$, in $\widetilde{\mathbf{q}}_{\text{grav.}}$ are due to the fact that actually the functional form of the related metric coefficients is not $\left(\frac{x^5}{x^5_{0,\text{grav.}}}\right)^2$ but $\frac{1}{4}\left(1+\frac{x^5}{x^5_{0,\text{grav.}}}\right)^2$. Again, it is possible to recover the phenomenological 5D metric $g_{AB,\text{DR5,grav.}}(x^5)$ from the Power Ansatz form $g_{AB,\text{DR5power}}(x^5)$ by a rescaling and a translation of the energy. In fact, one has

$$x^5 \longrightarrow x^{5\prime} = x^5 - \overline{x_0^5} \Leftrightarrow \mathrm{d}x^{5\prime} = \mathrm{d}x^5. \tag{20.32}$$

Such a translation in energy is allowed because we are just working in the framework of DR5. Therefore

$$b_0^2(x^5) = b_0^2(x^{5\prime} + \overline{x_0^5}) = b^2_{0,\text{new}}\,(x^{5\prime}) = \left(\frac{x^{5\prime} + \overline{x_0^5}}{x_0^5}\right)^2 = \left(\frac{x^{5\prime}}{x_0^5} + \frac{\overline{x_0^5}}{x_0^5}\right)^2. \tag{20.33}$$

By rescaling the threshold energy (in a physically consistent way, because it amounts to a redefinition of the scale of measure of energy)

$$x_0^5 \longrightarrow x_0^{5\prime} = x_0^5 \left( \frac{\widetilde{x_0^5}}{x_0^5} \right), \tag{20.34}$$

one gets

$$b_{0,\text{new}}^2 \left( x^{5\prime} \right) = \left( \frac{x^{5\prime}}{x_0^{5\prime}} \frac{\widetilde{x_0^5}}{x_0^5} + \frac{\widetilde{x_0^5}}{x_0^{5\prime}} \right)^2 = \left( \frac{\widetilde{x_0^5}}{x_0^{5\prime}} \right)^2 \left( 1 + \frac{x^{5\prime}}{x_0^5} \right)^2. \tag{20.35}$$

This metric time coefficient is of the gravitational type, except for the factor $\left( \frac{\widetilde{x_0^5}}{x_0^{5\prime}} \right)^2$, which, however, can be got rid of by the following rescaling of the time coordinate:

$$x^0 \longrightarrow x'^0 = \frac{\widetilde{x_0^5}}{x_0^{5\prime}} x^0. \tag{20.36}$$

This is a conformal transformation corresponding to a redefinition of the scale of measure of time. Notice that the above rescaling procedure of energy and time does not account for the factor 1/4 in front of $\left( 1 + \frac{x^5}{x_{0,\text{grav.}}^5} \right)^2$. This can be dealt with by the method followed for $\widetilde{\mathbf{q}}_{\text{strong}}$, namely by considering the generalized metric $g_{AB,\text{DR5power-conform}}(x^5)$, where now the vector $\widetilde{\boldsymbol{\vartheta}}$ is given by $\widetilde{\boldsymbol{\vartheta}} = \left( \frac{1}{4}, ?, ?, \frac{1}{4}, ? \right)$ (as before, the question marks reflect the unknown nature of the related metric coefficients).

Notice that both in (20.21)–(20.23) and (20.24)–(20.26) it was *assumed* that

$$q_{\mu,\text{int.}}(x_{0,\text{int.}}^5) = 0, \quad \mu = 0, 1, 2, 3 \text{ , int.} = \text{e.m., weak, strong, grav.,} \tag{20.37}$$

for simplicity reasons, since – as already stressed in Chap. 4 – nothing can be said on the behavior of the metrics at the energy thresholds.

Let us introduce the *left and right specifications* $\widehat{\Theta}_{\text{L}}(x)$, $\widehat{\Theta}_{\text{R}}(x)$ *of the Heaviside theta function*, defined, respectively, by

$$\widehat{\Theta}_{\text{L}}(x) \equiv \begin{cases} 1, x > 0, \\ 0, x \leq 0, \end{cases} \tag{20.38}$$

$$\widehat{\Theta}_{\text{R}}(x) \equiv \begin{cases} 1, x \geq 0, \\ 0, x < 0, \end{cases} \tag{20.39}$$

and satisfying *the complementarity relation*

$$1 - \widehat{\Theta}_{\text{R}}(x) = \widehat{\Theta}_{\text{L}}(x). \tag{20.40}$$

Then, the exponent sets (20.24)–(20.26) can be written in compact form as

$$\widetilde{\mathbf{q}}_{\text{e.m./weak}} = \left(0, \frac{1}{3}\widehat{\Theta}_{\text{L}}\left(x^5_{0,\text{e.m./weak}} - x^5\right),\right.$$
$$\left.\frac{1}{3}\widehat{\Theta}_{\text{L}}\left(x^5_{0,\text{e.m./weak}} - x^5\right), \frac{1}{3}\widehat{\Theta}_{\text{L}}\left(x^5_{0,\text{e.m./weak}} - x^5\right), r\right), \quad (20.41)$$

$$\widetilde{\mathbf{q}}_{\text{strong}} = \left(2\widehat{\Theta}_{\text{L}}\left(x^5 - x^5_{0,\text{strong}}\right), (0,0), 2\widehat{\Theta}_{\text{L}}\left(x^5 - x^5_{0,\text{strong}}\right), r\right), \quad (20.42)$$

$$\widetilde{\mathbf{q}}_{\text{grav.}} = \left(\widetilde{2\widehat{\Theta}_{\text{L}}}\left(x^5 - x^5_{0,\text{grav.}}\right), ?, ?, \widetilde{2\widehat{\Theta}_{\text{L}}}\left(x^5 - x^5_{0,\text{grav.}}\right), r\right), \quad (20.43)$$

where the tilde and the question marks have the same meaning as earlier.

## 20.3 Solving Einstein's Equations

Solving Einstein's equationsin the 5D, deformed space $\Re_5$ in the general case is quite an impossible task. On the contrary, one can show that, in the two special cases considered above, some classes of solutions can be found for (20.16) and (20.19) (corresponding, respectively, to spatial isotropy and metric coefficients which are powers of the energy), at least for $\Lambda_{(5)} = 0$. Notice that assuming a vanishing cosmological constant has the physical motivation (at least as far as gravitation is concerned and one is not interested into quantum effects) that $\Lambda_{(5)}$ is related to the vacuum energy; experimental evidence shows that, at least in our 4D space, $\Lambda \simeq 3 \cdot 10^{-52}\,\text{m}^{-2}$.

We recall moreover that (20.14) imply[3] $R(E) = -\frac{10}{3}\Lambda_{(5)}$. Since $\Lambda_{(5)} = 0$ (and consequently $R(E) = 0$) the spaces we will find are obviously Ricci flat. However, they differ, in general, from a 5D flat space, as it can be easily checked by showing explicitly that some components of the Riemann curvature tensor do not vanish.

In the former case (spatial isotropy), by putting $\Lambda_{(5)} = 0$, the system of differential equations (20.16) takes the form

$$\begin{cases} -2b''f + b'f' = 0\ ; \\ f\left[b_0^2(b')^2 - 2b_0b_0'bb' - 4b_0^2bb'' - 2b_0b_0''b^2 + b^2(b_0')^2\right] \\ \qquad +b_0bf'(2b_0b' + b_0'b) = 0\ ; \\ b'(b_0b)' = 0\ . \end{cases} \quad (20.44)$$

If $b_0 = \text{const.}$, then the third equation (20.44) entails $b' = 0$; it is then easy to see that the remaining equations are identically satisfied. Hence system (20.44) admits only the solution $b = const.$, $f(E)$ undetermined,

[3]See the footnote of Sect. 19.1.

which can be shown to correspond (modulo rescaling) to a flat 5D space. This entails, as one should suspect, that a 5D Minkowski space can be a solution of our system.

If $b_0$ is not a constant, then the third equation implies either (1) $b' = 0, (b_0 b)' \neq 0$ or (2) $b' \neq 0, (b_0 b)' = 0$. Let us consider these two cases.

(1) In this case $b =$ const. and the system (20.44) admits solutions with $b_0(E)$ arbitrary and $f(E)$ determined by the only remaining nontrivial equation, namely:

$$f[(b_0')^2 - 2b_0 b_0''] = -b_0 b_0' f'. \tag{20.45}$$

Putting

$$A(E) = \frac{2b_0 b_0'' - (b_0')^2}{b_0 b_0'} = \frac{f'}{f} \tag{20.46}$$

we get therefore

$$f(E) = k e^{\int^E A(\xi) d\xi}, \tag{20.47}$$

where $k$ is an integration constant. We remark that, if $f(E) =$ const., (20.45) becomes

$$(b_0')^2 - 2b_0 b_0'' = 0. \tag{20.48}$$

It is easy to see that (20.48) admits the only solution

$$b_0^2(E) = \left(1 + \frac{E}{E_0}\right)^2 \tag{20.49}$$

with $E_0$ constant. Therefore, this shows that the gravitational metric (19.15) corresponds to $f =$ const., in the case of spatial isotropy.

(2) In this second case, it is not difficult to get the following class of solutions:

$$\begin{aligned} f(E) &= k\,[b'(E)]^2\ ; \\ b_0(E) &= b(E)^{-1}\ , \end{aligned} \tag{20.50}$$

where $k$ is a constant (which fixes the sign of $f$) and $b(E)$ is an arbitrary function of $E$.

Let us now discuss the case of the metric coefficients which are pure powers of the energy. For $\Lambda_{(5)} = 0$ (20.19) admit 12 possible classes of solutions, which can be classified according to the values of the 5D set $\widetilde{\mathbf{q}} \equiv (q_0, q_1, q_2, q_3, r)$ built up from the energy exponents of the metric coefficients (see (20.17), (20.20)). Explicitly one has:

Class (I) $\widetilde{\mathbf{q}}_{\mathrm{I}} = \left(q_2, -q_2\left(\frac{2q_3 + q_2}{2q_2 + q_3}\right), q_2, q_3, \frac{q_3^2 - 2q_3 + 2q_2 q_3 - 4q_2 + 3q_2^2}{2q_2 + q_3}\right).$

Class (II) $\widetilde{\mathbf{q}}_{\mathrm{II}} = (0, q_1, 0, 0, q_1 - 2).$

Class (III) $\widetilde{\mathbf{q}}_{\mathrm{III}} = (q_2, -q_2, q_2, q_2, -2(1-q_2))$.

Class (IV) $\widetilde{\mathbf{q}}_{\mathrm{IV}} = (0, 0, 0, q_3, q_3 - 2)$.

Class (V) $\widetilde{\mathbf{q}}_{\mathrm{V}} = (-q_3, -q_3, -q_3, q_3, -(1+q_3))$.

Class (VI) $\widetilde{\mathbf{q}}_{\mathrm{VI}} = (q_0, 0, 0, 0, q_0 - 2)$.

Class (VII) $\widetilde{\mathbf{q}}_{\mathrm{VII}} = (q_0, -q_0, -q_0, -q_0, -2 - q_0)$.

Class (VIII) $\widetilde{\mathbf{q}}_{\mathrm{VIII}} = (0, 0, 0, 0, r)$.

Class (IX) $\widetilde{\mathbf{q}}_{\mathrm{IX}} = (0, 0, q_2, 0, -2 + q_2)$.

Class (X) $\widetilde{\mathbf{q}}_{\mathrm{X}} = \left(q_0, -\dfrac{q_3 q_0 + q_2 q_3 + q_2 q_0}{q_2 + q_3 + q_0}, q_2, q_3, r_{\mathrm{X}}\right)$ with

$$r_{\mathrm{X}} = \frac{q_3^2 + q_3 q_0 - 2q_3 + q_2 q_3 - 2q_2 + q_2 q_0 + q_2^2 - 2q_0 + q_0^2}{q_2 + q_3 + q_0}.$$

Class (XI) $\widetilde{\mathbf{q}}_{\mathrm{XI}} = \left(q_0, -\dfrac{q_2(2q_0 + q_2)}{2q_2 + q_0}, q_2, q_2, \dfrac{3q_2^2 - 4q_2 + 2q_2 q_0 - 2q_0 + q_0^2}{2q_2 + q_0}\right)$.

Class (XII) $\widetilde{\mathbf{q}}_{\mathrm{XII}} = \left(q_0, q_2, q_2, -\dfrac{q_2(2q_0 + q_2)}{2q_2 + q_0}, r_{\mathrm{XII}}\right)$ with

$$r_{\mathrm{XII}} = \frac{q_3^2 + q_3 q_0 - 2q_3 + q_2 q_3 - 2q_2 + q_2 q_0 + q_2^2 - 2q_0 + q_0^2}{q_2 + q_3 + q_0}.$$

In the following section, we shall discuss the physical relevance of the earlier solutions.

## 20.4 Discussion of Solutions

As we said in the earlier section, in the case of spatial isotropy the analytical solution of (20.45) for $f$ = const., yields immediately the gravitational metric (19.15).

On the other hand, the 12 classes of solutions found in the Power Ansatz allow one to recover, as special cases, *all* the phenomenological metrics discussed in Sect. 4.1. Let us write explicitly the interval in $\Re_5$ in such a case:

$$\mathrm{d}S^2 = \left(\frac{E}{E_0}\right)^{q_0} \mathrm{d}t^2 - \left(\frac{E}{E_0}\right)^{q_1} \mathrm{d}x^2 - \left(\frac{E}{E_0}\right)^{q_2} \mathrm{d}y^2 - \left(\frac{E}{E_0}\right)^{q_3} \mathrm{d}z^2 + E^r \mathrm{d}E^2. \tag{20.51}$$

Then, it is easily seen that the Minkowski metric is recovered from *all* classes of solutions. Solution (VIII) corresponds directly to a Minkowskian space–time, with the exponent $r$ of the fifth coefficient undetermined. In the other cases, we have to put: $q_1 = 0$ for class (II); $q_2 = 0$ for classes (III) and (IX); $q_3 = 0$ for (IV) and (V); $q_0 = 0$ for (VI) and (VII) (for all the previous solutions, it is $r = -2$); $q_2 = q_3 = 0$ for class (I); $q_2 = q_3 = q_0 = 0$ for class (X); $q_2 = q_0 = 0$ for class (XI); $q_2 = q_0 = 0$ for class (XII). The latter four solutions have $r = 0$ , and therefore correspond to a 5D Minkowskian flat space.

If we set $q_1 = 1/3$ in class (II), $q_3 = 1/3$ in class (IV), or $q_2 = 1/3$ in class (IX) (corresponding in all three cases to the value $r = 5/3$ for the exponent of the fifth metric coefficient), we get a metric of the "electroweak type" (see (19.12),(19.13)), i.e., with unit time coefficient and one space coefficient behaving as $(E/E_0)^{1/3}$, *but spatially anisotropic*, since two of the space metric coefficients are constant and Minkowskian (precisely, the $y$, $z$ coefficients for class (II); the $x$, $y$ coefficients for class (IV); and the $x$, $z$ ones for class (IX)). Notice that such an anisotropy does not disagree with the phenomenological results; indeed, in the analysis of the experimental data one was forced to assume spatial isotropy in the electromagnetic and in the weak cases, simply because of the lack of experimental information on two of the space dimensions.

Putting $q_0 = 1$ in class (VI), we find a metric which is spatially Minkowskian, with a time coefficient linear in $E$, i.e., a (gravitational) metric of the Einstein type (2.21).

Class (I) allows one to get as a special case a metric of the strong type (see (19.14)). This is achieved by setting $q_2 = 2$, whence $q_1 = -4(q_3+1)/(q_3+4)$; $r = (q_3^2 + 2q_3 + 4)/(q_3 + 4)$. Moreover, for $q_3 = 0$, it is $q_1 = -1$; $r = 1$. In other words, one finds a solution corresponding to $b_0(E) = b(E) = (E/E_0)$ and spatially anisotropic, i.e., a metric of the type (20.22).

Finally, the three classes (X)–(XII) admit as special case the gravitational metric (20.23), which is recovered by putting $q_0 = 2$ and $q_1 = q_2 = q_3 = 0$ (whence also $r = 0$) and by a rescaling and a translation of the energy (see Sect. 20.2.3).

In conclusion, we can state that *the formalism of DR5 permits to recover, as solutions of the vacuum Einstein equations, all the phenomenological energy-dependent metrics of the electromagnetic, weak, strong and gravitational type* (and also the gravitational one of the Einstein kind, (2.21)).

Let us mention that the functional form of the metric parameter $b_0(E)$ in the gravitational case (see (20.23)) suggests to introduce a modified proper time function $\tau(t, E)$ by setting:

$$\tau = \left(1 + \frac{E}{E_0}\right) t. \tag{20.52}$$

With this position, the gravitational interval takes the form:

$$
\begin{aligned}
dS^2 &\equiv d\tau^2 - b(E)[dx^2 + dy^2 + dz^2] \\
&+ \left[f(E) + \frac{\tau^2}{(E+E_0)^2}\right] dE^2 - 2\frac{\tau}{(E+E_0)} d\tau dE,
\end{aligned}
\tag{20.53}
$$

which shows a 5D "Gaussian behavior" (with lapse function equal to one).

## 20.5 DR5 and Warped Geometry

We have seen in Chap. 18 that in some multidimensional space–time theories the geometry of the extra (spatial) dimension(s) is warped, in order to avoid recourse to compactification.

If $\zeta$ is the extra coordinate, a typical 5D warped interval reads:

$$
dS^2 = a(\zeta) g_{\mu\nu} dx^\mu dx^\nu - d\zeta^2, \tag{20.54}
$$

where $g_{\mu\nu}$ is the Minkowski metric and $a(\zeta)$ the warp factor, given by

$$
a(\zeta) = e^{-k|\zeta|}. \tag{20.55}
$$

The decay constant $k > 0$ is proportional to $\sigma$, the energy density (per unit three-volume) of the brane. It is assumed of course that the brane is located at $\zeta = 0$. As a consequence, the metric induced on the brane is Minkowskian.

It can be shown that metric (20.54) is a solution of the 5D Einstein equations with a cosmological constant proportional to the square of the energy density of the brane: $\Lambda_{(5)} \sim \sigma^2$.

A generalization of the above warped interval is

$$
\begin{aligned}
dS^2 &= a(\zeta) c^2 dt^2 - b(\zeta) d\mathbf{x}^2 - d\zeta^2; \\
a(0) &= b(0) = 1,
\end{aligned}
\tag{20.56}
$$

where the warp factors of time and 3D space, $a(\zeta)$ and $b(\zeta)$, are different. This metric accounts for Lorentz-violating effects, provided the wave function of the particles spreads in the fifth dimension (this may be the case for gravitons).

We want to remark that intervals (20.54), (20.56) are just special cases of the DR5 interval (19.2). The by now expert reader recognizes that either metric is, for the space–time part, a spatially isotropic metric of the type (20.15), and with $f(E) = -1$.

Of course, the warped metrics (20.54), (20.56) and the DR5 metrics have a profound physical difference. In the former ones, $\zeta$ is an hidden space dimension: the (exponentially decaying) warp factors are just introduced in order to make this extra dimension unobservable. In the DR5 framework, the fifth dimension is the energy, and therefore an observable physical quantity. This entails, among the others, that the metric coefficients may have any functional form. Moreover – as repeatedly stressed – DSR and DR5 metrics are assumed to provide a *local* description of the physical processes ruled by one of the fundamental interactions; on the contrary, warped metrics do describe the physical world at a *large, global* scale.

However, the similar structure of the two types of metrics has as consequence that the mathematical study of the formal properties of the space $\Re_5$ of DR5 can be of some utility for warped geometry, too. Results obtained e.g., for DR5 isometries may hold in some cases for the warped models (or be adapted with suitable changes). This provides a further reason to exploring the mathematical features of DR5.

# 21
# Killing Equations in the Space $\Re_5$

In the present and in the following chapters, we shall deal with the problem of the isometries of the space $\Re_5$ of DR5. This will allow us to determine the symmetry properties of DR5, by getting also preliminary information on the infinitesimal structure of the related algebras. Let us recall that the metric homomorphisms of $\Re_5$ are strictly connected to the invariance under what we called the Metric Gaugement Process of DSR (see Sect. 4.4).

## 21.1 General Case

Let us discuss the Killing symmetries of the space $\Re_5$ [136].

In $\Re_5$, the Lie derivative $\mathcal{L}$ of a rank-2 covariant tensor field $T_{AB}$ along the five-vector field $\widetilde{\boldsymbol{\xi}} = \{\xi_A(x, x^5) \equiv \xi_A(x^B)\}$ is given as usual by

$$\underset{\widetilde{\boldsymbol{\xi}}}{\mathcal{L}} T_{AB} = T^C_A \xi_{C;B} + T^C_B \xi_{C;A} + T_{AB;C}\xi^C \tag{21.1}$$

with "$;A$" denoting covariant derivative with respect to $x^A$. If the tensor coincides with the metric tensor $g_{AB}$ (whose covariant derivative vanishes), its Lie derivative becomes

$$\underset{\widetilde{\boldsymbol{\xi}}}{\mathcal{L}} g_{AB} = \xi_{A;B} + \xi_{B;A} = \xi_{[A;B]}, \tag{21.2}$$

where the bracket [..] means symmetrization with respect to the enclosed indices (see (5.17)).

Then, a five-vector $\underset{\sim}{\boldsymbol{\xi}}$ is a Killing vector if the Lie derivative of the metric tensor with respect to $\boldsymbol{\xi}$ vanishes, i.e.,

$$\underset{\underset{\sim}{\boldsymbol{\xi}}}{\mathcal{L}} g_{AB} = 0 \Leftrightarrow \xi_{[A;B]} = 0 \Leftrightarrow \xi_{A;B} + \xi_{B;A} = 0 \tag{21.3}$$

are the Killing equations in $\Re_5$. Since the Lie derivative is nothing but the generalization of directional derivative, this means that the Killing vectors correspond to isometric directions, namely one recovers the property expressed by (5.11), (5.16). The integrability conditions of (21.3) are given by

$$\xi_{A;BC} = \xi_{C;[BA]} = R^D{}_{CBA}\xi_D \Leftrightarrow \underset{\underset{\sim}{\boldsymbol{\xi}}}{\mathcal{L}} \Gamma^A_{BC} = 0, \tag{21.4}$$

where $R_{ABCD}$ and $\Gamma^A_{BC}$ are the 5D Riemann–Christoffel tensor and affine connection (20.4), (20.1), respectively. In turn, (21.4) are integrable under the conditions

$$\underset{\underset{\sim}{\boldsymbol{\xi}}}{\mathcal{L}} R_{ABCD} = 0. \tag{21.5}$$

For metric (19.7), from the Christoffel symbols $\Gamma^A_{BC}$ of the metric $g_{AB,\text{DR5}}(x^5)$, (21.3) take the form of the following system of 15 coupled, partial derivative differential equations in $\Re_5$ for the Killing vector $\xi_A(x^B)$:

$$f(x^5)\xi_{0,0}(x^A) \pm b_0(x^5)b_0'(x^5)\xi_5(x^A) = 0; \tag{21.6}$$

$$\left.\begin{array}{l} \xi_{0,1}(x^A) + \xi_{1,0}(x^A) = 0 \\ \xi_{0,2}(x^A) + \xi_{2,0}(x^A) = 0 \\ \xi_{0,3}(x^A) + \xi_{3,0}(x^A) = 0 \end{array}\right\} \text{type I conditions;} \tag{21.7}$$

$$b_0(x^5)(\xi_{0,5}(x^A) + \xi_{5,0}(x^A)) - 2b_0'(x^5)\xi_0(x^A) = 0 \;\} \text{ type II condition;} \tag{21.8}$$

$$f(x^5)\xi_{1,1}(x^A) \mp b_1(x^5)b_1'(x^5)\xi_5(x^A) = 0; \tag{21.9}$$

$$\left.\begin{array}{l} \xi_{1,2}(x^A) + \xi_{2,1}(x^A) = 0 \\ \xi_{1,3}(x^A) + \xi_{3,1}(x^A) = 0 \end{array}\right\} \text{type I conditions;} \tag{21.10}$$

$$b_1(x^5)(\xi_{1,5}(x^A) + \xi_{5,1}(x^A)) - 2b_1'(x^5)\xi_1(x^A) = 0 \;\} \text{ type II condition;} \tag{21.11}$$

$$f(x^5)\xi_{2,2}(x^A) \mp b_2(x^5)b_2'(x^5)\xi_5(x^A) = 0; \tag{21.12}$$

$$\xi_{2,3}(x^A) + \xi_{3,2}(x^A) = 0 \ \} \text{ type I condition;} \tag{21.13}$$

$$b_2(x^5)(\xi_{2,5}(x^A) + \xi_{5,2}(x^A)) - 2b_2'(x^5)\xi_2(x^A) = 0 \ \} \text{ type II condition;} \tag{21.14}$$

$$f(x^5)\xi_{3,3}(x^A) \mp b_3(x^5)b_3'(x^5)\xi_5(x^A) = 0; \tag{21.15}$$

$$b_3(x^5)(\xi_{3,5}(x^A) + \xi_{5,3}(x^A)) - 2b_3'(x^5)\xi_3(x^A) = 0 \ \} \text{ type II condition;} \tag{21.16}$$

$$2f(x^5)\xi_{5,5}(x^A) - f'(x^5)\xi_5(x^A) = 0, \tag{21.17}$$

where now "$, A$" denotes ordinary derivative with respect to $x^A$.

Equations (21.6)–(21.17) can be divided in "fundamental" equations and "constraint" equations (of type I and II). The earlier system is in general overdetermined, i.e., its solutions will contain numerical coefficients satisfying a given algebraic system. Explicitly solving it yields

$$\xi_\mu(x^A) = F_\mu(x^{A \neq \mu})$$
$$\pm(-\delta_{\mu 0} + \delta_{\mu 1} + \delta_{\mu 2} + \delta_{\mu 3})b_\mu(x^5)b_\mu'(x^5)(f(x^5))^{-1/2} \int \mathrm{d}x^\mu F_5(x); \tag{21.18}$$

$$\xi_5(x^A) = (f(x^5))^{1/2} F_5(x). \tag{21.19}$$

The five unknown functions $F_A(x^{B \neq A})$ are restricted by the two following types of conditions:

(I) Type I (Cardinality 4, $\mu \neq \nu \neq \rho \neq \sigma$):

$$\pm A_\mu(x^5)G_{,\nu\rho\sigma}(x) + B_\mu(x^5)G_{,\mu\mu\nu\rho\sigma}(x)$$
$$+ b_\mu(x^5)F_{\mu,5}(x^{A \neq \mu}) - 2b_\mu'(x^5)F_\mu(x^{A \neq \mu}) = 0. \tag{21.20}$$

(II) Type II (Cardinality 6, symm. in $\mu, \nu$, $\mu \neq \nu \neq \rho \neq \sigma$):

$$F_{\mu,\nu}(x^{A \neq \mu}) + F_{\nu,\mu}(x^{A \neq \nu}) +$$
$$\pm(-\delta_{\mu 0} + \delta_{\mu 1} + \delta_{\mu 2} + \delta_{\mu 3})b_\mu(x^5)b_\mu'(x^5)(f(x^5))^{-1/2}G_{,\nu\nu\rho\sigma}(x) +$$
$$\pm(-\delta_{\nu 0} + \delta_{\nu 1} + \delta_{\nu 2} + \delta_{\nu 3})b_\nu(x^5)b_\nu'(x^5)(f(x^5))^{-1/2}G_{,\mu\mu\rho\sigma}(x) = 0, \tag{21.21}$$

where we introduced the fake four-vectors[1] (ESC off)

$$A_\mu(x^5) \equiv (-\delta_{\mu 0} + \delta_{\mu 1} + \delta_{\mu 2} + \delta_{\mu 3}) b_\mu(x^5)(f(x^5))^{-1/2}$$
$$\times \left[ -\left(b'_\mu(x^5)\right)^2 + b_\mu(x^5) b''_\mu(x^5) - \frac{1}{2} b_\mu(x^5) b'_\mu(x^5) f'(x^5)(f(x^5))^{-1} \right]; \tag{21.22}$$

$$B_\mu(x^5) \equiv b_\mu(x^5)(f(x^5))^{1/2}; \tag{21.23}$$

and defined the function

$$G(x) \equiv \int \mathrm{d}^4 x F_5(x). \tag{21.24}$$

## 21.2 The Hypothesis $\Upsilon$ of Functional Independence

In order to get the explicit forms of the functions $F_A(x^{B \neq A})$ (and therefore of the Killing vector (21.18), (21.19)), it is necessary to analyze conditions I and make suitable simplifying hypotheses.

To this aim, consider the following equation in the (suitably regular) functions $\alpha_1(x^5)$, $\alpha_2(x^5)$ and $\beta_1(x^\mu)$, $\beta_2(x^\mu)$:

$$\alpha_1(x^5)\beta_1(x^\mu) + \alpha_2(x^5)\beta_2(x^\mu) = 0. \tag{21.25}$$

If $\alpha_1(x^5) \neq 0$, $\alpha_2(x^5) \neq 0$, the solutions of (21.25) are given by the following two cases:

(1) $\exists \gamma \in R_0 : \alpha_1(x^5) = \gamma\alpha_2(x^5)$ $(\forall x^5 \in R_0)$ (*functional linear dependence* between $\alpha_1(x^5)$ and $\alpha_2(x^5)$). Then $\beta_2(x^\mu) = -\gamma\beta_1(x^\mu)$ $(\forall x^\mu \in R$, $\mu = 0, 1, 2, 3)$.

(2) $\nexists \gamma \in R_0 : \alpha_1(x^5) = \gamma\alpha_2(x^5)$ $(\forall x^5 \in R_0)$ (*functional linear independence* between $\alpha_1(x^5)$ and $\alpha_2(x^5)$). Then $\beta_2(x^\mu) = 0 = \beta_1(x^\mu)$ $(\forall x^\mu \in R$, $\mu = 0, 1, 2, 3)$.

Let us now consider type I conditions for $\mu = 0$. By taking their derivative with respect to $x^0$ one gets:

$$\partial_0 \left( (I)|_{\mu=0} \right) :$$
$$\pm A_0(x^5) G_{,0123}(x) + B_0(x^5) G_{,000123}(x) = 0$$
$$\Leftrightarrow \pm A_0(x^5) F_5(x) + B_0(x^5) F_{5,00}(x) = 0. \tag{21.26}$$

---

[1] Indeed, it is in general (ESC off) $(i = 1, 2, 3)$

$$b_\mu(x^5) = g_{\mu\mu,\mathrm{DR5}}(x^5)\left(\delta_{\mu 0} - \delta_{\mu i}\right),$$
$$f_\mu(x^5) = \pm g_{55,\mathrm{DR5}}(x^5),$$

which clearly show the nonvector nature of $A_\mu(x^5)$ and $B_\mu(x^5)$ in the 4D subspaces of $\Re_5$.

If $A_0(x^5) \neq 0$ and $B_0(x^5) \neq 0$, we have the following two possibilities:

(1) $\exists c_0 \in R_0 : \pm A_0(x^5) = c_0 B_0(x^5)$ $(\forall x^5 \in R_0^+)$. From (21.26) one gets $(\forall x^0, x^1, x^2, x^3 \in R)$:

$$G_{,000123}(x) = \mp c_0 G_{,0123}(x) \Leftrightarrow F_{5,00}(x) = \mp c_0 F_5(x); \qquad (21.27)$$

(2) $\nexists c_0 \in R_0 : \pm A_0(x^5) = c_0 B_0(x^5)$ $(\forall x^5 \in R_0^+)$. It follows from (21.26) $(\forall x^0, x^1, x^2, x^3 \in R)$:

$$G_{,000123}(x) = 0 = G_{,0123}(x) \Leftrightarrow F_{5,00}(x) = 0 = F_5(x). \qquad (21.28)$$

In general, let us take the derivative with respect to $x^\mu$ of type I conditions (ESC off):

$$\begin{aligned} &\partial_\mu I) : \\ &\pm A_\mu(x^5) G_{,\mu\nu\rho\sigma}(x) + B_\mu(x^5) G_{,\mu\mu\mu\nu\rho\sigma}(x) = 0 \\ &\Leftrightarrow \pm A_\mu(x^5) F_5(x) + B_\mu(x^5) F_{5,\mu\mu}(x) = 0. \end{aligned} \qquad (21.29)$$

If $G(x)$ satisfies the Schwarz lemma at any order, since $\mu \neq \nu \neq \rho \neq \sigma$, one gets

$$G_{,\mu\nu\rho\sigma}(x) = G_{,0123}(x)(= F_5(x)), \qquad (21.30)$$

namely the function $F_5(x)$ is present in $\partial_\mu I)$ $\forall \mu = 0, 1, 2, 3$. It is therefore *sufficient* to assume that there exists at least a special index

$$\overline{\mu} \in \{0,1,2,3\} : \begin{cases} \nexists c_{\overline{\mu}} \in R_0 : \pm A_{\overline{\mu}}(x^5) = c_{\overline{\mu}} B_{\overline{\mu}}(x^5) (\forall x^5 \in R_0^+) \\ A_{\overline{\mu}}(x^5) \neq 0, B_{\overline{\mu}}(x^5) \neq 0 \end{cases} \qquad (21.31)$$

in order that $(\forall x^0, x^1, x^2, x^3 \in R)$

$$(F_5(x) =) G_{,0123}(x) = 0 = G_{,\overline{\mu}\overline{\mu}\overline{\mu}123}(x) (= F_{5,\overline{\mu}\overline{\mu}}(x)) \underset{\not\Leftarrow}{\overset{\Rightarrow}{\text{(in gen.)}}}$$

$$\underset{\not\Leftarrow}{\overset{\Rightarrow}{\text{(in gen.)}}} G_{,\mu\mu\mu123}(x) (= F_{5,\mu\mu}(x)) = 0, \ \forall \mu = 0, 1, 2, 3. \qquad (21.32)$$

In the following the existence hypothesis:

$$\exists \text{ (at least one) } \overline{\mu} \in \{0,1,2,3\} : \\ \begin{cases} \nexists c_{\overline{\mu}} \in R_0 : \pm A_{\overline{\mu}}(x^5) = c_{\overline{\mu}} B_{\overline{\mu}}(x^5) (\forall x^5 \in R_0^+) \\ A_{\overline{\mu}}(x^5) \neq 0, B_{\overline{\mu}}(x^5) \neq 0 \end{cases} \qquad (21.33)$$

will be called *"$\Upsilon$-hypothesis" of functional independence.*

The earlier reasoning can be therefore summarized as

$$\underbrace{\text{Hp.}\Upsilon : \begin{cases} \exists \text{ (at least one) } \overline{\mu} \in \{0,1,2,3\} : \\ \nexists c_{\overline{\mu}} \in R_0 : \pm A_{\overline{\mu}}(x^5) = c_{\overline{\mu}} B_{\overline{\mu}}(x^5)(\forall x^5 \in R_0^+); \\ A_{\overline{\mu}}(x^5) \neq 0, B_{\overline{\mu}}(x^5) \neq 0, \end{cases}}$$

$$\Downarrow$$

$$\underbrace{(F_5(x) =) G_{,0123}(x) = 0, \forall x^0, x^1, x^2, x^3 \in R,}$$

$$\Downarrow$$

$$(F_{5,\mu\mu}(x) =) G_{,\mu\mu\mu 123}(x) = 0 \ \forall x^0, x^1, x^2, x^3 \in R \ \forall \mu = 0,1,2,3. \quad (21.34)$$

## 21.3 Solving Killing Equations in $\Re_5$ in the $\Upsilon$-Hypothesis

Then, by assuming the hypothesis $\Upsilon$ of functional independence to hold, and replacing (21.34) in (21.18) and (21.19), one gets for the covariant Killing five-vector $\xi_A(x, x^5)$:

$$\left.\begin{array}{r} \xi_\mu(x^A) = F_\mu(x^{A \neq \mu}), \ \forall \mu = 0,1,2,3, \\ \xi_5(x^A) = 0 \end{array}\right\} \Rightarrow \xi_A(x^B) = \left(F_\mu(x^{A \neq \mu}), 0\right). \quad (21.35)$$

The conditions to be satisfied now by the 4 unknown functions $F_\mu(x^{A \neq \mu})$ ($\forall \mu, \nu = 0,1,2,3$) are obtained by substituting (21.34) in (21.20) and (21.21) and read

(I) Type I (Cardinality 4):

$$b_\mu(x^5) F_{\mu,5}(x^{A \neq \mu}) - 2b'_\mu(x^5) F_\mu(x^{A \neq \mu}) = 0; \quad (21.36)$$

(II) Type II (Cardinality 6, symmetry in $\mu, \nu$, $\mu \neq \nu$):

$$F_{\mu,\nu}(x^{A \neq \mu}) + F_{\nu,\mu}(x^{A \neq \nu}) = 0.$$

Solving the equations of type I yields

$$F_\mu(x^{A \neq \mu}) = b_\mu^2(x^5) \widetilde{F_\mu}(x^{\nu \neq \mu}), \forall \mu = 0,1,2,3 \quad (21.37)$$

and eventually ($\forall \mu, \nu = 0,1,2,3$, $\mu \neq \nu$)

$$\left.\begin{array}{r} F_{\mu,\nu}(x^{A \neq \mu}) + F_{\nu,\mu}(x^{A \neq \nu}) = 0 \\ F_\mu(x^{A \neq \mu}) = b_\mu^2(x^5) \widetilde{F_\mu}(x^{\nu \neq \mu}) \end{array}\right\} \underset{\Leftarrow}{\overset{\Rightarrow}{\text{(in gen.)}}}$$

$$\underset{\Leftarrow}{\overset{\Rightarrow}{\text{(in gen.)}}} \left\{ b_\mu^2(x^5) \frac{\partial \widetilde{F_\mu}(x^{\rho \neq \mu})}{\partial x^\nu} + b_\nu^2(x^5) \frac{\partial \widetilde{F_\nu}(x^{\rho \neq \nu})}{\partial x^\mu} = 0. \right. \quad (21.38)$$

Summarizing, we can state that, in the hypothesis $\Upsilon$ of functional independence, the covariant Killing five-vector $\xi_A(x, x^5)$ has the form (ESC off)

$$\xi_A(x^B) = \left(b_\mu^2(x^5)\widetilde{F_\mu}(x^{\nu\neq\mu}), 0\right), \tag{21.39}$$

where the four unknown real functions of three real variables $\left\{\widetilde{F_\mu}(x^{\rho\neq\mu})\right\}$ are solutions of the following system of six (due to the symmetry in $\mu$ and $\nu$) nonlinear, partial derivative equations:

$$b_\mu^2(x^5)\frac{\partial\widetilde{F_\mu}(x^{\rho\neq\mu})}{\partial x^\nu} + b_\nu^2(x^5)\frac{\partial\widetilde{F_\nu}(x^{\rho\neq\nu})}{\partial x^\mu} = 0, \quad \mu, \nu = 0, 1, 2, 3, \ \mu \neq \nu, \tag{21.40}$$

which is in general overdetermined, i.e., its explicit solutions will depend on numerical coefficients obeying a given system.

Solving system (21.40) is quite easy (although cumbersome: for details, see [136]). The final solution yields the following expressions for the components of the *contravariant* Killing five-vector $\xi^A(x, x^5)$ satisfying the 15 Killing equations (21.6)–(21.17) in the hypothesis $\Upsilon$ of functional independence (21.34):

$$\begin{aligned} \xi^0(x^1, x^2, x^3) &= \widetilde{F_0}(x^1, x^2, x^3) \\ &= d_8x^1x^2x^3 + d_7x^1x^2 + d_6x^1x^3 + d_4x^2x^3 \\ &\quad (d_5 + a_2)x^1 + d_3x^2 + d_2x^3 + (a_1 + d_1 + K_0); \end{aligned} \tag{21.41}$$

$$\begin{aligned} \xi^1(x^0, x^2, x^3) &= -\widetilde{F_1}(x^0, x^2, x^3) \\ &= -h_2x^0x^2x^3 - h_1x^0x^2 - h_8x^0x^3 - h_4x^2x^3 \\ &\quad - (h_7 + e_2)\, x^0 - h_3x^2 - h_6x^3 - (K_1 + h_5 + e_1)\, ; \end{aligned} \tag{21.42}$$

$$\begin{aligned} \xi^2(x^0, x^1, x^3) &= -\widetilde{F_2}(x^0, x^1, x^3) \\ &= -l_2x^0x^1x^3 - l_1x^0x^1 - l_6x^0x^3 - l_4x^1x^3 \\ &\quad - (l_5 + e_4)\, x^0 - l_3x^1 - l_8x^3 - (l_7 + K_2 + e_3); \end{aligned} \tag{21.43}$$

$$\xi^3(x^0,x^1,x^2) = -\widetilde{F_3}(x^0,x^1,x^2)$$

$$= -m_8x^0x^1x^2 - m_7x^0x^1 - m_6x^0x^2 - m_4x^1x^2$$

$$-(m_5+g_2)\,x^0 - m_3x^1 - m_2x^2 - (m_1+g_1+c); \tag{21.44}$$

$$\xi^5 = 0 \neq \xi^5(x,x^5). \tag{21.45}$$

where (some of) the real parameters $d_i$ , $h_i$, $l_i$ , $m_i$ ($i = 1,2,\ldots,8$), $e_k$ ($k = 1,2,3,4$), $g_l$, $a_l$ ($l = 1,2$) satisfy the following algebraic system of six constraints:

$$\begin{cases}
(01) & \begin{cases} b_0^2(x^5)\left[d_8x^2x^3 + d_7x^2 + d_6x^3 + (d_5+a_2)\right] \\ +b_1^2(x^5)\left[h_2x^2x^3 + h_1x^2 + h_8x^3 + (h_7+e_2)\right] = 0; \end{cases} \\
(02) & \begin{cases} b_0^2(x^5)\left(d_8x^1x^3 + d_7x^1 + d_4x^3 + d_3\right) \\ +b_2^2(x^5)\left[l_2x^1x^3 + l_1x^1 + l_6x^3 + (l_5+e_4)\right] = 0; \end{cases} \\
(03) & \begin{cases} b_0^2(x^5)\left(d_8x^1x^2 + d_6x^1 + d_4x^2 + d_2\right) \\ +b_3^2(x^5)\left[m_8x^1x^2 + m_7x^1 + m_6x^2 + (m_5+g_2)\right] = 0; \end{cases} \\
(12) & \begin{cases} b_1^2(x^5)\left(h_2x^0x^3 + h_1x^0 + h_4x^3 + h_3\right) \\ +b_2^2(x^5)\left(l_2x^0x^3 + l_1x^0 + l_4x^3 + l_3\right) = 0; \end{cases} \\
(13) & \begin{cases} b_1^2(x^5)\left(h_2x^0x^2 + h_8x^0 + h_4x^2 + h_6\right) \\ +b_3^2(x^5)\left(m_8x^0x^2 + m_7x^0 + m_4x^2 + m_3\right) = 0; \end{cases} \\
(23) & \begin{cases} b_2^2(x^5)\left(l_2x^0x^1 + l_6x^0 + l_4x^1 + l_8\right) \\ +b_3^2(x^5)\left(m_8x^0x^1 + m_6x^0 + m_4x^1 + m_2\right) = 0. \end{cases}
\end{cases} \tag{21.46}$$

## 21.4 Power Ansatz and Reductivity of the Hypothesis $\Upsilon$

We want now to investigate if and when the *simplifying* $\Upsilon$-hypothesis (21.34) – we exploited in order to solve the Killing equations in $\Re_5$ – is *reductive*. To this aim, one needs to consider explicit forms of the 5D Riemannian metric $g_{AB,\text{DR5}}(x^5)$. As we have seen in Chap. 20, the "Power Ansatz" allows one to recover all the phenomenological metrics derived for the four fundamental interactions. So it is worth considering such a case, corresponding to a 5D metric of the form (20.20) with $\widetilde{\mathbf{q}}\equiv(q_0, q_1, q_2, q_3, r) = (q_\mu, r)$.

In the Power Ansatz, the (fake) four-vectors $A_\mu(x^5)$ and $B_\mu(x^5)$ ((21.22), (21.23)) take the following explicit forms:

$$\begin{aligned} A_{\mu,power}(x^5) &= \frac{1}{(x_0^5)^2}(\delta_{\mu 0} - \delta_{\mu 1} - \delta_{\mu 2} - \delta_{\mu 3})\frac{q_\mu}{2}\left(1 + \frac{r}{2}\right)\left(\frac{x^5}{x_0^5}\right)^{\frac{3}{2}q_\mu - \frac{1}{2}r - 2} \\ &= A_{\mu,power}(\widetilde{\mathbf{q}}; x^5); \end{aligned} \tag{21.47}$$

$$B_{\mu,power}(x^5) = \left(\frac{x^5}{x_0^5}\right)^{(1/2)q_\mu + (1/2)r} = B_{\mu,power}(\widetilde{\mathbf{q}}; x^5). \tag{21.48}$$

Therefore

$$\frac{\pm A_{\mu,power}(\widetilde{\mathbf{q}}; x^5)}{B_{\mu,power}(\widetilde{\mathbf{q}}; x^5)} = \pm\frac{1}{(x_0^5)^2}(\delta_{\mu 0} - \delta_{\mu 1} - \delta_{\mu 2} - \delta_{\mu 3})\frac{q_\mu}{2}\left(1 + \frac{r}{2}\right)\left(\frac{x^5}{x_0^5}\right)^{q_\mu - r - 2}. \tag{21.49}$$

Since $x^5 \in R_0^+$, one gets, respectively

$$A_{\mu,power}(\widetilde{\mathbf{q}}; x^5) \neq 0 \Leftrightarrow \frac{q_\mu}{2}\left(1 + \frac{r}{2}\right) \neq 0$$

$$\Leftrightarrow \begin{cases} q_\mu \neq 0 \\ 1 + \frac{r}{2} \neq 0 \Leftrightarrow 2 + r \neq 0 \end{cases}; \tag{21.50}$$

$$B_{\mu,power}(\widetilde{\mathbf{q}}; x^5) \neq 0, \forall q_\mu, r \in Q. \tag{21.51}$$

Then

$$\frac{\pm A_{\mu,power}(\widetilde{\mathbf{q}}; x^5)}{B_{\mu,power}(\widetilde{\mathbf{q}}; x^5)} = c_{(\mu; q_\mu, r)} \in R_{(0)}, \forall x^5 \in R_0^+ \Leftrightarrow q_\mu - r - 2 = 0. \tag{21.52}$$

It follows that, if $A_{\mu,power}(\widetilde{\mathbf{q}}; x^5) \neq 0$ and $B_{\mu,power}(\widetilde{\mathbf{q}}; x^5) \neq 0$, assuming the Power Ansatz form for $g_{AB,\mathrm{DR5}}(x^5)$ amounts to express the hypothesis $\Upsilon$ of functional independence (21.34) as

$$\exists \text{ (at least one) } \overline{\mu} \in \{0, 1, 2, 3\} : \left\{ \begin{array}{c} q_{\overline{\mu}} - (r + 2) \neq 0 \\ \begin{cases} q_{\overline{\mu}} \neq 0 \\ r + 2 \neq 0 \end{cases} \end{array} \right\}$$

$$\Leftrightarrow q_{\overline{\mu}} \neq 0, r + 2 \neq 0, q_{\overline{\mu}} \neq r + 2. \tag{21.53}$$

In other words, in the framework of the "Power Ansatz" for the metric tensor *the reductive nature of the $\Upsilon$-hypothesis depends on the value of the rational parameters* $q_0, q_1, q_2, q_3$ *and* $r$, exponents of the components of $g_{AB,\mathrm{DR5power}}(x^5)$.

A similar result holds true if one assumes for the $\Re_5$ metric the generalized form $g_{AB,\text{DR5power-conform}}(x^5)$ (20.27), obtained by the anisotropic rescaling (20.30). We have, in this case, for $A_\mu(x^5)$ and $B_\mu(x^5)$:

$$A_{\mu,\text{power-conform}}(x^5)$$
$$= \frac{1}{(x_0^5)^2} \frac{(\vartheta_\mu)^{3/2}}{(\vartheta_5)^{\frac{1}{2}}} (\delta_{\mu 0} - \delta_{\mu 1} - \delta_{\mu 2} - \delta_{\mu 3}) \frac{q_\mu}{2} \left(1 + \frac{r}{2}\right) \left(\frac{x^5}{x_0^5}\right)^{(3/2)q_\mu - (1/2)r - 2}$$
$$= \frac{(\vartheta_\mu)^{3/2}}{(\vartheta_5)^{1/2}} A_{\mu,power}(x^5); \tag{21.54}$$

$$B_{\mu,\text{power-conform}}(x^5)$$
$$= (\vartheta_\mu \vartheta_5)^{1/2} \left(\frac{x^5}{x_0^5}\right)^{(1/2)q_\mu + (1/2)r} = (\vartheta_\mu \vartheta_5)^{1/2} B_{\mu,power}(x^5) \tag{21.55}$$

whence

$$\frac{\pm A_{\mu,power}(\widetilde{\mathbf{q}}; x^5)}{B_{\mu,power}(\widetilde{\mathbf{q}}; x^5)}$$
$$= \pm \frac{1}{(x_0^5)^2} \frac{\vartheta_\mu}{\vartheta_5} (\delta_{\mu 0} - \delta_{\mu 1} - \delta_{\mu 2} - \delta_{\mu 3}) \frac{q_\mu}{2} \left(1 + \frac{r}{2}\right) \left(\frac{x^5}{x_0^5}\right)^{q_\mu - r - 2} \tag{21.56}$$

and

$$\frac{\pm A_{\mu,power}(\widetilde{\mathbf{q}}; x^5)}{B_{\mu,power}(\widetilde{\mathbf{q}}; x^5)} = c_{(\mu; q_\mu, r)} \in R_{(0)} \quad \forall x^5 \in R_0^+ \Leftrightarrow q_\mu - r - 2 = 0. \tag{21.57}$$

Therefore conditions (21.50)–(21.52), obtained in the power case, hold unchanged, together with expression (21.53) of the $\Upsilon$-hypothesis. Thus, we can conclude that, independently of a possible anisotropic, conformal rescaling of the coordinates of the type (20.30), the reductive nature of the $\Upsilon$-hypothesis depends only on the value of the parameters $q_0, q_1, q_2, q_3$, and $r$.

The discussion of the possible reductivity of the $\Upsilon$-hypothesis for all the 12 classes of solutions of the 5D Einstein equations in vacuum derived in Sect. 4.1 (labeled by the 5D set $\widetilde{\mathbf{q}} \equiv (q_0, q_1, q_2, q_3, r)$) allows one to state that in five general cases such hypothesis of functional independence is reductive indeed. The Killing equations can be explicitly solved in such cases. We refer the reader to Appendix A for these general cases, and go to discuss the special cases of the 5D phenomenological power metrics describing the four fundamental interactions (see Sect. 19.3).

# 22

# Killing Symmetries for the 5D Metrics of Fundamental Interactions

We want now to investigate the possible reductivity of the hypothesis $\Upsilon$ of functional independence (21.34) for the 5D metrics (19.12)–(19.15), and solve the related Killing equations. Due to the piecewise structure of these phenomenological metrics, we shall distinguish the two energy ranges $x^5 \geqslant x_0^5$ (above threshold, case (a) and $0 < x^5 < x_0^5$ (below threshold, case (b) for sub-Minkowskian metrics (electromagnetic and weak), and $0 < x^5 \leq x_0^5$ (below threshold, case (a') and $x^5 > x_0^5$ (above threshold, case (b') for over-Minkowskian metrics (strong and gravitational). Needless to say, cases (a, a') correspond to the Minkowskian behavior of the related metrics, whereas (b, b') refer to the non-Minkowskian one.

## 22.1 Electromagnetic and Weak Interactions

### 22.1.1 Validity of the $\Upsilon$-Hypothesis

**(Case a) (Minkowskian conditions).** In the energy range $x^5 \geqslant x_0^5$ the 5D metrics (19.12), (19.13) read:

$$g_{AB,\mathrm{DR5}}(x^5) = \mathrm{diag}\left(1, -1, -1, -1, \pm f(x^5)\right). \tag{22.1}$$

This metric is a special case of

$$g_{AB,\mathrm{DR5}}(x^5) = \mathrm{diag}\left(a, -b, -c, -d, \pm f(x^5)\right) \tag{22.2}$$

$(a, b, c, d, f(x^5) \in R_0^+)$. From definitions (21.22) and (21.23) one gets:

$$\left.\begin{aligned} A_\mu(x^5) &= 0 \\ B_\mu(x^5) &= \left(f(x^5)\right)^{\frac{1}{2}} \end{aligned}\right\} \quad \forall \mu = 0, 1, 2, 3. \tag{22.3}$$

Therefore the hypothesis $\Upsilon$ of functional independence (21.34) is not satisfied $\forall \mu \in \{0, 1, 2, 3\}$. The 15 Killing equations corresponding to the e.m. and weak metrics (19.12), (19.13) are:

$$\left\{\begin{array}{c} f(x^5)\xi_{0,0}(x^A) = 0; \\ \xi_{0,1}(x^A) + \xi_{1,0}(x^A) = 0; \\ \xi_{0,2}(x^A) + \xi_{2,0}(x^A) = 0; \\ \xi_{0,3}(x^A) + \xi_{3,0}(x^A) = 0; \\ \xi_{0,5}(x^A) + \xi_{5,0}(x^A) = 0; \\ f(x^5)\xi_{1,1}(x^A) = 0; \\ \xi_{1,2}(x^A) + \xi_{2,1}(x^A) = 0; \\ \xi_{1,3}(x^A) + \xi_{3,1}(x^A) = 0; \\ \xi_{1,5}(x^A) + \xi_{5,1}(x^A) = 0; \\ f(x^5)\xi_{2,2}(x^A) = 0; \\ \xi_{2,3}(x^A) + \xi_{3,2}(x^A) = 0; \\ \xi_{2,5}(x^A) + \xi_{5,2}(x^A) = 0; \\ f(x^5)\xi_{3,3}(x^A) = 0; \\ \xi_{3,5}(x^A) + \xi_{5,3}(x^A) = 0; \\ 2f(x^5)\xi_{5,5}(x^A) - f'(x^5)\xi_5(x^A) = 0. \end{array}\right. \tag{22.4}$$

Solving this system requires some cumbersome algebra but is trivial. The result for the contravariant Killing vector is

$$\xi^0(x^1, x^2, x^3, x^5) = T^0 - B^1 x^1 - B^2 x^2 - B^3 x^3 + \Xi^0 F(x^5); \tag{22.5}$$

$$\xi^1(x^0, x^2, x^3, x^5) = T^1 - B^1 x^0 + \Theta^3 x^2 - \Theta^2 x^3 - \Xi^1 F(x^5); \tag{22.6}$$

$$\xi^2(x^0, x^1, x^3, x^5) = T^2 - B^2 x^0 - \Theta^3 x^1 + \Theta^1 x^3 - \Xi^2 F(x^5); \tag{22.7}$$

$$\xi^3(x^0, x^1, x^2, x^5) = T^3 - B^3 x^0 + \Theta^2 x^1 - \Theta^1 x^2 - \Xi^3 F(x^5); \tag{22.8}$$

$$\xi^5(x, x^5) = \pm\left(f(x^5)\right)^{-\frac{1}{2}}[T^5 - \Xi^0 x^0 - \Xi^1 x^1 - \Xi^2 x^2 - \Xi^3 x^3]. \tag{22.9}$$

Here, we have put

$$F(x^5) = \int \mathrm{d}x^5 \left(f(x^5)\right)^{1/2} \tag{22.10}$$

and omitted an unessential integration constant in (22.10) (it would only amount to a redefinition of $T^\mu$, $\mu = 0, 1, 2, 3$ in (22.5)–(22.9)). By inspection of such equations, it is easy to get the following physical interpretation of the real parameters entering into the expression of $\xi^A(x, x^5)$:

$$\underset{\text{space–space angles of true rotations}}{\Theta^1, \Theta^2, \Theta^3 \in R} \quad \overset{\text{periodicity } T=2\pi}{\longrightarrow} \quad \Theta^1, \Theta^2, \Theta^3 \in [0, 2\pi);$$

$$\underset{\text{space–time angles of pseudorotations}}{\overset{\text{boost rapidity}}{B^1, B^2, B^3 \in R}} \quad ;$$

$$\underset{\text{space–time translation parameters}}{T^0, T^1, T^2, T^3 \in R} \quad . \tag{22.11}$$

As to the parameters $\Xi^\mu$ ($\mu = 0, 1, 2, 3$) and $T^5$, their physical meaning (if any) depends on the signature of the fifth dimension.

The metric (22.2) can be dealt with along the same lines with only minor changes. In particular, the contravariant Killing vector is obtained from (22.5)–(22.9) by a suitable rescaling of the parameters in the space–time components, namely

$$\xi^0(x^1, x^2, x^3, x^5) = \frac{1}{a}\left[T^0 - B^1x^1 - B^2x^2 - B^3x^3 + \Xi^0 F(x^5)\right]; \tag{22.5'}$$

$$\xi^1(x^0, x^2, x^3, x^5) = \frac{1}{b}\left[T^1 - B^1x^0 + \Theta^3x^2 - \Theta^2x^3 - \Xi^1 F(x^5)\right]; \tag{22.6'}$$

$$\xi^2(x^0, x^1, x^3, x^5) = \frac{1}{c}\left[T^2 - B^2x^0 - \Theta^3x^1 + \Theta^1x^3 - \Xi^2 F(x^5)\right]; \tag{22.7'}$$

$$\xi^3(x^0, x^1, x^2, x^5) = \frac{1}{d}\left[T^3 - B^3x^0 + \Theta^2x^1 - \Theta^1x^2 - \Xi^3 F(x^5)\right]; \tag{22.8'}$$

$$\xi^5(x, x^5) = \pm\left(f(x^5)\right)^{-\frac{1}{2}}\left[T^5 - \Xi^0x^0 - \Xi^1x^1 - \Xi^2x^2 - \Xi^3x^3\right]. \tag{22.9'}$$

**(Case b)(Non-Minkowskian conditions)** In this energy range the form of the metrics (19.12), (19.13) is

$$g_{AB,\mathrm{DR5}}(x^5) = \mathrm{diag}\left(1, -\left(\frac{x^5}{x_0^5}\right)^{1/3}, -\left(\frac{x^5}{x_0^5}\right)^{1/3}, -\left(\frac{x^5}{x_0^5}\right)^{1/3}, \pm f(x^5)\right). \tag{22.12}$$

We have, from the definitions (21.22) and (21.23) of the "vectors" $A_\mu(x^5)$ and $B_\mu(x^5)$:

$$A_0(x^5) = 0;$$

$$A_i(x^5) = -\left(\frac{x^5}{x_0^5}\right)^{1/6} \frac{1}{6\sqrt{f(x^5)}\,(x^5)^{\frac{2}{3}}\,(x_0^5)^{\frac{1}{3}}}\left[\frac{1}{x^5} + \frac{1}{2}\frac{f'(x^5)}{f(x^5)}\right], \quad \forall i = 1, 2, 3; \tag{22.13}$$

$$B_0(x^5) = 0;$$
$$B_i(x^5) = \sqrt{f(x^5)} \left(\frac{x^5}{x_0^5}\right)^{1/6} \quad \forall i = 1,2,3. \tag{22.14}$$

Therefore

$$\frac{\pm A_i(x^5)}{B_i(x^5)} = \mp \frac{1}{6f(x^5)\,(x^5)^{\frac{2}{3}}\,(x_0^5)^{1/3}} \left[\frac{1}{x^5} + \frac{1}{2}\frac{f'(x^5)}{f(x^5)}\right] \quad \forall i = 1,2,3. \tag{22.15}$$

Then, the $\Upsilon$ hypothesis (21.34) is not satisfied for $\mu = 0$ but it does for $\mu = i = 1,2,3$ under the following condition:

$$\frac{1}{x^5} + \frac{1}{2}\frac{f'(x^5)}{f(x^5)} \neq cf(x^5)\,(x^5)^{2/3}, \quad c \in R. \tag{22.16}$$

Therefore, on the basis of the results of Sect. 21.4, it is easy to get that the contravariant components of the 5D Killing vector $\xi^A(x, x^5)$ for metric (22.12) in the range $0 < x^5 < x_0^5$ are given by (21.41)–(21.45), where (some of) the real parameters satisfy the following system (namely system (21.46) for metric (22.12)):

$$\begin{aligned}
&(01)\ \begin{cases} \left[d_8 x^2 x^3 + d_7 x^2 + d_6 x^3 + (d_5 + a_2)\right] \\ + \left(\frac{x^5}{x_0^5}\right)^{1/3} \left[h_2 x^2 x^3 + h_1 x^2 + h_8 x^3 + (h_7 + e_2)\right] = 0; \end{cases} \\
&(02)\ \begin{cases} \left(d_8 x^1 x^3 + d_7 x^1 + d_4 x^3 + d_3\right) \\ + \left(\frac{x^5}{x_0^5}\right)^{1/3} \left[l_2 x^1 x^3 + l_1 x^1 + l_6 x^3 + (l_5 + e_4)\right] = 0; \end{cases} \\
&(03)\ \begin{cases} \left(d_8 x^1 x^2 + d_6 x^1 + d_4 x^2 + d_2\right) \\ + \left(\frac{x^5}{x_0^5}\right)^{1/3} \left[m_8 x^1 x^2 + m_7 x^1 + m_6 x^2 + (m_5 + g_2)\right] = 0; \end{cases} \\
&(12)\ \begin{cases} \left(\frac{x^5}{x_0^5}\right)^{1/3} \left(h_2 x^0 x^3 + h_1 x^0 + h_4 x^3 + h_3\right) \\ + \left(\frac{x^5}{x_0^5}\right)^{1/3} \left(l_2 x^0 x^3 + l_1 x^0 + l_4 x^3 + l_3\right) = 0; \end{cases} \\
&(13)\ \begin{cases} \left(\frac{x^5}{x_0^5}\right)^{1/3} \left(h_2 x^0 x^2 + h_8 x^0 + h_4 x^2 + h_6\right) \\ + \left(\frac{x^5}{x_0^5}\right)^{1/3} \left(m_8 x^0 x^2 + m_7 x^0 + m_4 x^2 + m_3\right) = 0; \end{cases} \\
&(23)\ \begin{cases} \left(\frac{x^5}{x_0^5}\right)^{1/3} \left(l_2 x^0 x^1 + l_6 x^0 + l_4 x^1 + l_8\right) \\ + \left(\frac{x^5}{x_0^5}\right)^{1/3} \left(m_8 x^0 x^1 + m_6 x^0 + m_4 x^1 + m_2\right) = 0. \end{cases}
\end{aligned} \tag{22.17}$$

Solving system (22.17) yields:

$$\begin{array}{l} d_2 = d_3 = d_4 = d_6 = d_7 = d_8 = 0; \\ m_4 = m_6 = m_7 = m_8 = 0; \quad m_5 = -g_2; \\ h_1 = h_2 = h_4 = h_8 = 0; \quad h_3 = -l_3; \quad h_6 = -m_3; \quad h_7 = -e_2; \quad h_8 = -m_7; \\ l_1 = l_2 = l_4 = l_6 = 0; \quad l_5 = -e_4; \quad l_8 = -m_2; \\ a_2 = -d_5. \end{array} \tag{22.18}$$

Then, one gets the following expression for $\xi^A(x, x^5)$:

$$\xi^0 = \widetilde{F_0} = (a_1 + d_1 + K_0); \tag{22.19}$$

$$\xi^1(x^2, x^3) = -\widetilde{F_1}(x^2, x^3) = l_3 x^2 + m_3 x^3 - (K_1 + h_5 + e_1)\,; \tag{22.20}$$

$$\xi^2(x^1, x^3) = -\widetilde{F_2}(x^1, x^3) = -l_3 x^1 + m_2 x^3 - (l_7 + K_2 + e_3); \tag{22.21}$$

$$\xi^3(x^0, x^1, x^2) = -\widetilde{F_3}(x^0, x^1, x^2) = -m_3 x^1 - m_2 x^2 - (m_1 + g_1 + c); \tag{22.22}$$

$$\xi^5 = 0. \tag{22.23}$$

### *22.1.2 Killing Isometries for Electromagnetic and Weak Metrics*

The 5D contravariant Killing vector $\xi^A(x, x^5)$ for the whole range of energies, (22.5)–(22.9) and (22.19)–(22.23), can be cast in a compact form by using the distribution $\widehat{\Theta}_{\mathrm{R}}(x^5 - x^5_0)$ (right specification of the Heaviside distribution: see (20.39)), by redenominating ($\forall i = 1, 2, 3$)

$$\begin{array}{l} B^i \equiv \zeta^i; \\ \Theta^i \equiv \theta^i; \\ \Xi^0 \equiv \zeta^5 \end{array} \tag{22.24}$$

and putting

$$\begin{array}{l} (a_1 + d_1 + K_0) = T^0; \\ -(K_1 + h_5 + e_1) = T^1; \\ -(l_7 + K_2 + e_3) = T^2; \\ -(m_1 + g_1 + c) = T^3; \\ l_3 = \Theta^3; \\ m_3 = -\Theta^2; \\ m_2 = \Theta^1. \end{array} \tag{22.25}$$

Then, one gets

$$\begin{aligned} &\xi^0(x^1, x^2, x^3, x^5) \\ &= \widehat{\Theta}_{\mathrm{R}}(x^5 - x^5_0)\left[-\zeta^1 x^1 - \zeta^2 x^2 - \zeta^3 x^3 + \zeta^5 F(x^5)\right] + T^0; \end{aligned} \tag{22.26}$$

$$\xi^1(x^0, x^2, x^3, x^5)$$
$$= \widehat{\Theta}_{\mathrm{R}}(x^5 - x_0^5)\left[-\zeta^1 x^0 - \Xi^1 F(x^5)\right] + \theta^3 x^2 - \theta^2 x^3 + T^1; \tag{22.27}$$

$$\xi^2(x^0, x^1, x^3, x^5)$$
$$= \widehat{\Theta}_{\mathrm{R}}(x^5 - x_0^5)\left[-\zeta^2 x^0 - \Xi^2 F(x^5)\right] - \theta^3 x^1 + \theta^1 x^3 + T^2; \tag{22.28}$$

$$\xi^3(x^0, x^1, x^2, x^5)$$
$$= \widehat{\Theta}_{\mathrm{R}}(x^5 - x_0^5)\left[-\zeta^3 x^0 - \Xi^3 F(x^5)\right] + \theta^2 x^1 - \theta^1 x^2 + T^3; \tag{22.29}$$

$$\xi^5(x, x^5)$$
$$= \widehat{\Theta}_{\mathrm{R}}(x^5 - x_0^5)\left\{\mp \left(f(x^5)\right)^{-\frac{1}{2}}\left[\zeta^5 x^0 + \Xi^1 x^1 + \Xi^2 x^2 + \Xi^3 x^3 - T^5\right]\right\}, \tag{22.30}$$

valid for both ranges $0 < x^5 < x_0^5$ and $x^5 \geqslant x_0^5$.

By considering slices of $\Re_5$ at $dx^5 = 0$, one gets:

$$x^5 = \overline{x^5} \in R_0^+ \left\{ \begin{array}{l} \Leftrightarrow dx^5 = 0 \underset{\substack{(\text{in gen.}) \\ \not\Leftarrow}}{\Rightarrow} F(x^5) = \int dx^5 \left(f(x^5)\right)^{\frac{1}{2}} = 0; \\ \qquad\qquad\quad \underset{\substack{(\text{in gen.}) \\ \not\Leftarrow}}{\Rightarrow} \xi^5(x, x^5) = 0. \end{array} \right. \tag{22.31}$$

Therefore, it easily follows from the expression of the Killing vector (22.26)–(22.30) that, in the energy range $x^5 \geqslant x_0^5$, the 5D Killing group of such constant-energy sections is the standard Poincaré group $P(1,3)$:

$$P(1,3)_{\mathrm{STD.}} = \mathrm{SO}(1,3)_{\mathrm{STD.}} \otimes_s \mathrm{Tr.}(1,3)_{\mathrm{STD.}} \tag{22.32}$$

(as it must be), whereas, in the energy range $0 < x^5 < x_0^5$, the Killing group is given by

$$\mathrm{SO}(3)_{\mathrm{STD.}} \otimes_s \mathrm{Tr.}(1,3)_{\mathrm{STD.}}. \tag{22.33}$$

### *22.1.3 Solution of Killing Equations below Threshold with Violated $\Upsilon$-Hypothesis*

In case (b), the hypothesis $\Upsilon$ of functional independence (21.34) does not hold for any value of $\mu$ if the metric coefficient $f(x^5)$ satisfies the following equation

$$\frac{1}{2}\frac{f'(x^5)}{f(x^5)} - cf(x^5)\left(x^5\right)^{2/3} + \frac{1}{x^5} = 0, \quad c \in R. \tag{22.34}$$

Such ordinary differential equation (ODE) belongs to the homogeneous class of type G and to the special rational subclass of Bernoulli's ordinary differential equations (it becomes separable for $c = 0$). The only solution of (22.34) is

$$f(x^5) = \frac{1}{6c(x^5)^{5/3} + \gamma(x^5)^2}, \quad c, \gamma \in R. \tag{22.35}$$

Since $f(x^5)$ must be dimensionless, it is convenient to make this feature explicit by introducing the characteristic parameter $x_0^5 \in R_0^+$ (which, as by now familiar, is the threshold energy of the interaction considered) so that $f(x^5) \equiv f\left(\frac{x^5}{x_0^5}\right)$. Equation (22.34) can be therefore rewritten as

$$f(x^5) \equiv f\left(\frac{x^5}{x_0^5}\right) = \frac{1}{6c\left(\frac{x^5}{x_0^5}\right)^{\frac{5}{3}} + \gamma\left(\frac{x^5}{x_0^5}\right)^2}, c, \gamma \in R, \tag{22.36}$$

where of course a rescaling of constants $c$ and $\gamma$ occurred. Moreover, because in general $f(x^5)$ has to be strictly positive $\forall x^5 \in R_0^+$, $c$ and $\gamma$ must necessarily satisfy the condition:

$$\begin{aligned} &c, \gamma \in R : 6c + \gamma\left(\frac{x^5}{x_0^5}\right)^{1/3} > 0 \ \forall x^5 \in R_0^+ \Leftrightarrow \\ &\Leftrightarrow c, \gamma \in R^+ \text{ (not both zero)}. \end{aligned} \tag{22.37}$$

Therefore, *imposing the complete violation of the $\Upsilon$-hypothesis of functional independence allows one to determine the functional form of the fifth metric coefficient.* This result will be seen to hold also for the strong and the gravitational interaction above threshold (see Sects. 22.2.3 and 22.3.2). On account of the possible dynamical role of the coefficient $f(x^5)$ in the 5D description of processes occurring in the standard 4D space–time (cf. Chap. 26), it can be ventured that the $\Upsilon$-hypothesis is perhaps something more than a mere mathematical simplifying assumption.

Then, we get the following expression for the 5D metric describing e.m. and weak interactions in the energy range $0 < x^5 < x_0^5$ if the $\Upsilon$-hypothesis is not satisfied by any value of $\mu$:

$$\begin{aligned} g_{AB,\mathrm{DR5}}(x^5) = \mathrm{diag}\Bigg(&1, -\left(\frac{x^5}{x_0^5}\right)^{1/3}, -\left(\frac{x^5}{x_0^5}\right)^{1/3}, -\left(\frac{x^5}{x_0^5}\right)^{1/3}, \\ &\pm\left(6c\left(\frac{x^5}{x_0^5}\right)^{5/3} + \gamma\left(\frac{x^5}{x_0^5}\right)^2\right)^{-1}\Bigg). \end{aligned} \tag{22.38}$$

The solution of the Killing equations is cumbersome in this case, too. After some tedious and lengthy algebra, one gets the following expression for the contravariant Killing five-vector $\xi^A(x, x^5)$ corresponding to the e.m. and weak metric (22.38):

$$\xi^0 = c_0; \tag{22.39}$$

$$\xi^1(x^2, x^3) = -\left(a_2 x^2 + a_3 x^3 + a_4\right)\left(x_0^5\right)^{1/3}; \tag{22.40}$$

$$\xi^2(x^1, x^3) = \left(a_2 x^1 b_1 x^3 - b_6\right)\left(x_0^5\right)^{1/3}; \tag{22.41}$$

$$\xi^3(x^1, x^2) = \left(a_3 x^1 - b_1 x^2 - b_2\right)\left(x_0^5\right)^{1/3}; \tag{22.42}$$

$$\xi^5 = 0, \tag{22.43}$$

where the dimensions of the real transformation parameters are (on account of the fact that $\xi$ has the dimension of a length)

$$[a_2] = [a_3] = [b_1] = l^{-1/3}, \quad [a_4] = [b_2] = [b_6] = l^{2/3}, \quad [c_0] = l. \tag{22.44}$$

The 5D Killing group of isometries is therefore

$$\mathrm{SO}(3)_{\mathrm{STD}.(E_3)} \otimes_s \mathrm{Tr}.(1,3)_{\mathrm{STD}}. \tag{22.45}$$

with $E_3$ being the 3D manifold with metric $g_{ij} = -\left(\frac{x^5}{x_0^5}\right)^{1/3} \operatorname{diag}(1,1,1)$.

## 22.2 Strong Interaction

### 22.2.1 *Validity of the $\Upsilon$-Hypothesis*

**(Case a') (Minkowskian conditions).** In the energy range $0 < x^5 \leqslant x_0^5$ the metric (20.22) for strong interaction reads:

$$g_{AB,\mathrm{DR5}}(x^5) = \operatorname{diag}\left(1, -\frac{2}{25}, -\frac{4}{25}, -1, \pm f(x^5)\right). \tag{22.46}$$

From (21.22) and (21.23) one finds, for the fake vectors $A_\mu(x^5)$, $B_\mu(x^5)$ in this case:

$$\begin{aligned} A_\mu(x^5) &= 0 \forall \mu = 0,1,2,3; \\ B_0(x^5) &= B_3(x^5) = \frac{5}{\sqrt{2}} B_1(x^5) = \frac{5}{2} B_2(x^5) = \left(f(x^5)\right)^{1/2}. \end{aligned} \tag{22.47}$$

Therefore the $\Upsilon$-hypothesis (21.34) is not satisfied by any value of $\mu \in \{0,1,2,3\}$. The 15 Killing equations corresponding to metric (22.46) are given by (22.4), i.e., coincide with those relevant to the 5D e.m. and weak metrics in the range $x^5 \geqslant x_0^5$. Since the contravariant metric tensor is

$$g^{AB}_{\mathrm{DR5}}(x^5) = \mathrm{diag}\left(1, -\frac{25}{2}, -\frac{25}{4}, -1, \pm\left(f(x^5)\right)^{-1}\right), \tag{22.48}$$

the components of the contravariant Killing five-vector $\xi^A(x,x^5) \overset{\mathrm{ESC\ on}}{=} g^{AB}_{\mathrm{DR5}}(x^5)\xi_B(x,x^5)$ are given by

$$\xi^0(x^1,x^2,x^3,x^5) = -B^1x^1 - B^2x^2 - B^3x^3 + \Xi^0 F(x^5) + T^0; \tag{22.49}$$

$$\xi^1(x^0,x^2,x^3,x^5) = \frac{25}{2}\left[-B^1x^0 + \Theta^3x^2 - \Theta^2x^3 - \Xi^1F(x^5) + T^1\right]; \tag{22.50}$$

$$\xi^2(x^0,x^1,x^3,x^5) = \frac{25}{4}\left[-B^2x^0 - \Theta^3x^1 + \Theta^1x^3 - \Xi^2F(x^5) + T^2\right]; \tag{22.51}$$

$$\xi^3(x^0,x^1,x^2,x^5) = -B^3x^0 + \Theta^2x^1 - \Theta^1x^2 - \Xi^3F(x^5) + T^3; \tag{22.52}$$

$$\xi^5(x,x^5) = \mp\left(f(x^5)\right)^{-\frac{1}{2}}\left[\Xi^0x^0 + \Xi^1x^1 + \Xi^2x^2 + \Xi^3x^3 - T^5\right], \tag{22.53}$$

in the same notation of (22.5)–(22.10).

**(Case b') (Non-Minkowskian conditions).** In the energy range $x^5 > x_0^5$ the 5D strong metric takes the form:

$$g_{AB,\mathrm{DR5}}(x^5) = \mathrm{diag}\left(\left(\frac{x^5}{x_0^5}\right)^2, -\frac{2}{25}, -\frac{4}{25}, -\left(\frac{x^5}{x_0^5}\right)^2, \pm f(x^5)\right). \tag{22.54}$$

From (21.22) and (21.23) one gets:

$$\begin{aligned} A_0(x^5) &= -A_3(x^5) = \frac{(x^5)^2}{(x_0^5)^3}\left(f(x^5)\right)^{-\frac{1}{2}}\left(\frac{1}{x^5} + \frac{1}{2}\frac{f'(x^5)}{f(x^5)}\right); \\ A_1(x^5) &= A_2(x^5) = 0; \end{aligned} \tag{22.55}$$

$$\begin{aligned} B_0(x^5) &= B_3(x^5) = \frac{x^5}{x_0^5}\left(f(x^5)\right)^{\frac{1}{2}}; \\ B_1(x^5) &= \frac{1}{\sqrt{2}}B_2(x^5) = \frac{\sqrt{2}}{5}\left(f(x^5)\right)^{1/2}; \end{aligned} \tag{22.56}$$

$$\frac{\pm A_0(x^5)}{B_0(x^5)} = \frac{\mp A_3(x^5)}{B_3(x^5)} = \pm\frac{x^5}{f(x^5)\,(x_0^5)^2}\left(\frac{1}{x^5} + \frac{1}{2}\frac{f'(x^5)}{f(x^5)}\right). \tag{22.57}$$

Therefore, on account of the strict positiveness of $f(x^5)$, the hypothesis $\Upsilon$ of functional independence is satisfied by $\mu = 0, 3$ under the following constraints:

$$\left.\begin{array}{l} \dfrac{1}{x^5} + \dfrac{1}{2}\dfrac{f'(x^5)}{f(x^5)} \neq 0; \\ \dfrac{x^5}{f(x^5)}\left(\dfrac{1}{x^5} + \dfrac{1}{2}\dfrac{f'(x^5)}{f(x^5)}\right) \neq c \Leftrightarrow \dfrac{1}{x^5} + \dfrac{1}{2}\dfrac{f'(x^5)}{f(x^5)} \neq c\dfrac{f(x^5)}{x^5}, c \in R_0 \end{array}\right\} \Leftrightarrow \frac{1}{x^5} + \frac{1}{2}\frac{f'(x^5)}{f(x^5)} \neq c\frac{f(x^5)}{x^5}, c \in R. \tag{22.58}$$

The case is analogous to the case (b) of the e.m. and weak metrics. Thus, the components of the contravariant Killing five-vector $\xi^A(x, x^5)$ for the phenomenological strong metric in the range $x^5 > x_0^5$ are given by (21.41)–(21.45), where (some of) the real parameters are constrained to obey the following system (see (22.17)):

$$\begin{array}{ll}
(01) & \left\{\begin{array}{l} \left(\dfrac{x^5}{x_0^5}\right)^2 \left[d_8 x^2 x^3 + d_7 x^2 + d_6 x^3 + (d_5 + a_2)\right] \\ + \dfrac{2}{25}\left[h_2 x^2 x^3 + h_1 x^2 + h_8 x^3 + (h_7 + e_2)\right] = 0; \end{array}\right. \\
(02) & \left\{\begin{array}{l} \left(\dfrac{x^5}{x_0^5}\right)^2 \left(d_8 x^1 x^3 + d_7 x^1 + d_4 x^3 + d_3\right) \\ + \dfrac{4}{25}\left[l_2 x^1 x^3 + l_1 x^1 + l_6 x^3 + (l_5 + e_4)\right] = 0; \end{array}\right. \\
(03) & \left\{\begin{array}{l} \left(\dfrac{x^5}{x_0^5}\right)^2 \left(d_8 x^1 x^2 + d_6 x^1 + d_4 x^2 + d_2\right) \\ + \left(\dfrac{x^5}{x_0^5}\right)^2 \left[m_8 x^1 x^2 + m_7 x^1 + m_6 x^2 + (m_5 + g_2)\right] = 0; \end{array}\right.
\end{array}$$

$$\begin{array}{ll}
(12) & \left\{\begin{array}{l} \dfrac{2}{25}\left(h_2 x^0 x^3 + h_1 x^0 + h_4 x^3 + h_3\right) \\ + \dfrac{4}{25}\left(l_2 x^0 x^3 + l_1 x^0 + l_4 x^3 + l_3\right) = 0; \end{array}\right. \\
(13) & \left\{\begin{array}{l} \dfrac{2}{25}\left(h_2 x^0 x^2 + h_8 x^0 + h_4 x^2 + h_6\right) \\ + \left(\dfrac{x^5}{x_0^5}\right)^2 \left(m_8 x^0 x^2 + m_7 x^0 + m_4 x^2 + m_3\right) = 0; \end{array}\right. \\
(23) & \left\{\begin{array}{l} \dfrac{4}{25}\left(l_2 x^0 x^1 + l_6 x^0 + l_4 x^1 + l_8\right) \\ + \left(\dfrac{x^5}{x_0^5}\right)^2 \left(m_8 x^0 x^1 + m_6 x^0 + m_4 x^1 + m_2\right) = 0. \end{array}\right.
\end{array} \tag{22.59}$$

The solutions of this system are given by:

$$\begin{aligned}
&d_3 = d_4 = d_6 = d_7 = d_8 = 0; \quad d_2 = -\left(m_5 + g_2\right); \\
&m_2 = m_3 = m_4 = m_6 = m_7 = m_8 = 0; \\
&h_1 = h_2 = h_4 = h_6 = h_8 = 0; \\
&l_1 = l_2 = l_4 = l_6 = l_8 = 0; \quad l_5 = -e_4; \\
&h_3 = -2l_3; \quad h_7 = -e_2; \quad a_2 = -d_5.
\end{aligned} \tag{22.60}$$

Replacing (22.60) into (21.41)–(21.45) yields the explicit form of the Killing five-vector $\xi^A(x, x^5)$:

$$\xi^0(x^3) = \widetilde{F_0}(x^3) = d_2 x^3 + (a_1 + d_1 + K_0); \tag{22.61}$$

$$\xi^1(x^2) = -\widetilde{F_1}(x^2) = 2l_3 x^2 - (K_1 + h_5 + e_1)\,; \tag{22.62}$$

$$\xi^2(x^1) = -\widetilde{F_2}(x^1) = -l_3 x^1 - (l_7 + K_2 + e_3); \tag{22.63}$$

$$\xi^3(x^0) = -\widetilde{F_3}(x^0) = d_2 x^0 - (m_1 + g_1 + c); \tag{22.64}$$

$$\xi^5 = 0. \tag{22.65}$$

### *22.2.2 Killing Isometries for Strong Metric*

As in the e.m. and weak case, it is possible to express the contravariant Killing vector of the phenomenological strong metric in a unique form, valid in the whole energy range. This is done by redenominating the parameters in (22.49)–(22.53) as follows:

$$\begin{cases} B^i \equiv \zeta^i; \\ \Theta^i \equiv \theta^i; \\ \Xi^0 \equiv \zeta^5 \end{cases} \tag{22.66}$$

($\forall i - 1, 2, 3$) and putting, in (22.61)–(22.65):

$$\begin{aligned}
&(a_1 + d_1 + K_0) = T^0; \\
&-\left(K_1 + h_5 + e_1\right) = \frac{25}{2} T^1; \\
&-(l_7 + K_2 + e_3) = \frac{25}{4} T^2; \\
&-(m_1 + g_1 + c) = T^3; \\
&l_3 = \frac{25}{4} \Theta^3 : \\
&d_2 = -B^3.
\end{aligned} \tag{22.67}$$

Then, exploiting the right specification $\widehat{\Theta}_{\mathrm{R}}(x_0^5 - x^5)$ of the step function, we get the following general form of the contravariant Killing five-vector $\xi^A(x, x^5)$ for the 5D phenomenological metric of the strong interaction:

$$\begin{aligned}
&\xi^0(x^1, x^2, x^3, x^5) \\
&= \widehat{\Theta}_{\mathrm{R}}(x_0^5 - x^5)\left[-\zeta^1 x^1 - \zeta^2 x^2 + \zeta^5 F(x^5)\right] - \zeta^3 x^3 + T^0;
\end{aligned} \tag{22.68}$$

$$\xi^1(x^0, x^2, x^3, x^5)$$
$$= \frac{25}{2}\widehat{\Theta}_{\mathrm{R}}(x_0^5 - x^5)\left[-\zeta^1 x^0 - \theta^2 x^3 - \Xi^1 F(x^5)\right] + \frac{25}{2}\theta^3 x^2 + \frac{25}{2}T^1; \quad (22.69)$$

$$\xi^2(x^0, x^1, x^3, x^5)$$
$$= \frac{25}{4}\widehat{\Theta}_{\mathrm{R}}(x_0^5 - x^5)\left[-\zeta^2 x^0 + \theta^1 x^3 - \Xi^2 F(x^5)\right] - \frac{25}{4}\theta^3 x^1 + \frac{25}{4}T^2; \quad (22.70)$$

$$\xi^3(x^0, x^1, x^2, x^5)$$
$$= \widehat{\Theta}_{\mathrm{R}}(x_0^5 - x^5)\left[\theta^2 x^1 - \theta^1 x^2 - \Xi^3 F(x^5)\right] - \zeta^3 x^0 + T^3; \quad (22.71)$$

$$\xi^5(x, x^5)$$
$$= \widehat{\Theta}_{\mathrm{R}}(x_0^5 - x^5)\left\{\mp\left(f(x^5)\right)^{-\frac{1}{2}}\left[\zeta^5 x^0 + \Xi^1 x^1 + \Xi^2 x^2 + \Xi^3 x^3 - T^5\right]\right\}. \quad (22.72)$$

By redefining

$$\frac{25}{2}\begin{cases} T^1 \\ \zeta^1 \\ \theta^2 \\ \Xi^1 \end{cases} \equiv \begin{cases} T^{1\prime} \\ \zeta^{1\prime} \\ \theta^{2\prime} \\ \Xi^{1\prime} \end{cases}$$
$$\frac{25}{4}\begin{cases} T^2 \\ \zeta^2 \\ \theta^1 \\ \theta^3 \\ \Xi^2 \end{cases} \equiv \begin{cases} T^{2\prime} \\ \zeta^{2\prime} \\ \theta^{1\prime} \\ \theta^{3\prime} \\ \Xi^{2\prime} \end{cases} \quad (22.73)$$

in (22.68)–(22.72) (and omitting the apices) one finds eventually:

$$\xi^0(x^1, x^2, x^3, x^5)$$
$$= \widehat{\Theta}_{\mathrm{R}}(x_0^5 - x^5)\left[-\frac{2}{25}\zeta^1 x^1 - \frac{4}{25}\zeta^2 x^2 + \zeta^5 F(x^5)\right] - \zeta^3 x^3 + T^0; \quad (22.74)$$

$$\xi^1(x^0, x^2, x^3, x^5)$$
$$= \widehat{\Theta}_{\mathrm{R}}(x_0^5 - x^5)\left[-\zeta^1 x^0 - \theta^2 x^3 - \Xi^1 F(x^5)\right] + 2\theta^3 x^2 + T^1; \quad (22.75)$$

$$\xi^2(x^0, x^1, x^3, x^5)$$
$$= \widehat{\Theta}_{\mathrm{R}}(x_0^5 - x^5)\left[-\zeta^2 x^0 + \theta^1 x^3 - \Xi^2 F(x^5)\right] - \theta^3 x^1 + T^2; \quad (22.76)$$

$$\xi^3(x^0, x^1, x^2, x^5)$$
$$= \widehat{\Theta}_{\mathrm{R}}(x_0^5 - x^5)\left[\frac{2}{25}\theta^2 x^1 - \frac{4}{25}\theta^1 x^2 - \Xi^3 F(x^5)\right] - \zeta^3 x^0 + T^3; \quad (22.77)$$

$$\xi^5(x, x^5)$$
$$= \widehat{\Theta}_{\mathrm{R}}(x_0^5 - x^5)\left\{\mp\left(f(x^5)\right)^{-\frac{1}{2}}\left[\zeta^5 x^0 + \frac{2}{25}\Xi^1 x^1 + \frac{4}{25}\Xi^2 x^2 + \Xi^3 x^3 - T^5\right]\right\}. \quad (22.78)$$

The 5D strong metric (22.46) in the energy range $0 < x^5 \leqslant x_0^5$ can be put in the form:

$$g_{AB,\mathrm{DR5}}(x^5) = \mathrm{diag}\left(g_{\mu\nu,\overline{M_4}}(x^5), \pm f(x^5)\right), \quad (22.79)$$

where $\overline{M_4}$ is a standard 4D Minkowskian manifold with the following coordinate rescaling (contraction):

$$\begin{aligned} x^1 &\longrightarrow \frac{\sqrt{2}}{5}x^1 \underset{\substack{(\text{in gen.})\\ \not\Leftarrow}}{\Rightarrow} \mathrm{d}x^1 \longrightarrow \frac{\sqrt{2}}{5}\mathrm{d}x^1; \\ x^2 &\longrightarrow \frac{2}{5}x^2 \underset{\substack{(\text{in gen.})\\ \not\Leftarrow}}{\Rightarrow} \mathrm{d}x^2 \longrightarrow \frac{2}{5}\mathrm{d}x^2. \end{aligned} \quad (22.80)$$

Considering slices at $\mathrm{d}x^5 = 0$ of $\Re_5$ entails $\xi^5(x, x^5) = 0$ (see (22.31)). Then, the explicit form (22.74)–(22.78) of the Killing vector entails that – as expected – the Killing group is the standard Poincaré group $P(1,3)$ (suitably rescaled):

$$[P(1,3)_{\mathrm{STD.}} = \mathrm{SO}(1,3)_{\mathrm{STD.}} \otimes_s \mathrm{Tr.}(1,3)_{\mathrm{STD.}}]|_{x^1 \longrightarrow \frac{\sqrt{2}}{5}x^1, x^2 \longrightarrow \frac{2}{5}x^2}. \quad (22.81)$$

For $x^5 > x_0^5$ the strong metric is given by (22.54). Therefore, it easily follows from (22.74)–(22.78) that the 5D Killing group of the constant-energy sections of $\Re_5$ is

$$\left(\mathrm{SO}(2)_{\mathrm{STD.},\Pi(x^1,x^2 \longrightarrow \sqrt{2}x^2)} \otimes B_{x^3,STD.}\right) \otimes_s \mathrm{Tr.}(1,3)_{\mathrm{STD.}}. \quad (22.82)$$

Here $\mathrm{SO}(2)_{\mathrm{STD.},\Pi(x^1,x^2 \longrightarrow \sqrt{2}x^2)} = \mathrm{SO}(2)_{\mathrm{STD.},\Pi(x^1 \longrightarrow \frac{\sqrt{2}}{5}x^1, x^2 \longrightarrow \frac{2}{5}x^2)}$ is the 1-parameter group (generated by the usual, special-relativistic generator $S^3_{\mathrm{SR}}|_{x^2 \longrightarrow \sqrt{2}x^2}$) of the 2D rotations in the plane $\Pi(x^1, x^2)$ (characterized by the coordinate contractions (22.80)), and $B_{x^3,STD.}$ is the usual one-parameter group (generated by the special-relativistic generator $K^3_{\mathrm{SR}}$) of the standard Lorentzian boosts along $\widehat{x^3}$. The direct and semidirect nature of the group products in (22.81) and (22.82) has the following explanation.

In general (independently of contractions and/or dilations of coordinates) the standard mixed Lorentz algebra is given by the commutation relations (ESC on):

$$[S^i_{\mathrm{SR}}, K^i_{\mathrm{SR}}] = \epsilon_{ijl} K^l_{\mathrm{SR}}, \forall i, j = 1, 2, 3, \tag{22.83}$$

where as usual $\epsilon_{ijl}$ is the Levi-Civita 3-tensor of rank 3 and $S^i_{\mathrm{SR}}$ and $K^i_{\mathrm{SR}}$ are the $i$th generator of (true) rotations and Lorentz boosts, respectively. It follows that:

$$[S^3_{\mathrm{SR}}, K^3_{\mathrm{SR}}]\big|_{x^2 \longrightarrow \sqrt{2}x^2} \left(= [S^3_{\mathrm{SR}}\big|_{x^2 \longrightarrow \sqrt{2}x^2}, K^3_{\mathrm{SR}}]\right) = 0 \tag{22.84}$$

what justifies the presence of the direct group product in (22.82). The semi-direct product of $\left(\mathrm{SO}(2)_{\mathrm{STD.},\Pi(x^1,x^2 \longrightarrow \sqrt{2}x^2)} \otimes B_{x^3,\mathrm{STD.}}\right)$ by $\mathrm{Tr.}(1,3)_{\mathrm{STD.}}$ is due instead to the fact that the standard mixed Poincaré algebra (independently of contractions and/or dilations of coordinates) is defined by the following commutation relations (ESC on) ($\forall i, j, k = 1, 2, 3$):

$$\begin{aligned}
&[K^i_{\mathrm{SR}}, \Upsilon^0_{\mathrm{SR}}] = -\Upsilon^i_{\mathrm{SR}};\\
&[K^i_{\mathrm{SR}}, \Upsilon^j_{\mathrm{SR}}] = -\delta^{ij} \Upsilon^0_{\mathrm{SR}};\\
&[S^i_{\mathrm{SR}}, \Upsilon^0_{\mathrm{SR}}] = 0;\\
&[S^i_{\mathrm{SR}}, \Upsilon^k_{\mathrm{SR}}] = \epsilon_{ikl} \Upsilon^l_{\mathrm{SR}},
\end{aligned} \tag{22.85}$$

where $\Upsilon^0_{\mathrm{SR}}, \Upsilon^1_{\mathrm{SR}}, \Upsilon^2_{\mathrm{SR}}$ and $\Upsilon^3_{\mathrm{SR}}$ are the generators of the standard space–time translations.

### *22.2.3 Solution of Strong Killing Equations above Threshold with Violated $\Upsilon$-Hypothesis*

In the energy range $x^5 > x^5_0$ , if condition (22.58) is not satisfied, the hypothesis $\Upsilon$ of functional independence (21.34)does not hold for any value of $\mu$. In this case the metric coefficient $f(x^5)$ obeys the equation

$$\frac{1}{2}\frac{f'(x^5)}{f(x^5)} - c\frac{f(x^5)}{x^5} + \frac{1}{x^5} = 0, c \in R. \tag{22.86}$$

Such ODE is separable $\forall c \in R$. By solving it, one gets the following form of the 5D metric of the strong interaction (for $x^5 > x^5_0$ and when the $\Upsilon$-hypothesis (21.34) is violated):

$$g_{AB,\mathrm{DR5}}(x^5)$$

$$= \mathrm{diag}\left(\left(\frac{x^5}{x^5_0}\right)^2, -\frac{2}{25}, -\frac{4}{25}, -\left(\frac{x^5}{x^5_0}\right)^2, \pm\frac{1}{\gamma\left(\frac{x^5}{x^5_0}\right)^2 + c}\right) \tag{22.87}$$

with

$$c, \gamma \in R : \gamma \left( \frac{x^5}{x_0^5} \right)^2 + c > 0, \forall x^5 \in R_0^+ \Leftrightarrow c, \gamma \in R^+ \text{(not both zero)}. \quad (22.88)$$

Solving the related Killing equations (after lengthy but elementary calculations) yields for the contravariant Killing five-vector $\xi^A(x, x^5)$ the following compact form (valid for $c, \gamma \in R^+$ but not vanishing simultaneously):

$$\xi^0(x^3; c, \gamma) = (1 - \delta_{c,0}) \left[ - \left(x_0^5\right)^2 \left( (1 - \delta_{\gamma,0}) \, d_3 x^3 + T_0 \right) \right]; \quad (22.89)$$

$$\xi^1(x^2; \gamma) = - (1 - \delta_{\gamma,0}) \frac{25}{2} d_2 x^2 - \frac{25}{2} T_1; \quad (22.90)$$

$$\xi^2(x^1; \gamma) = (1 - \delta_{\gamma,0}) \frac{25}{4} d_2 x^1 - \frac{25}{4} T_2; \quad (22.91)$$

$$\xi^3(x^0; c, \gamma) = - (1 - \delta_{c,0}) (1 - \delta_{\gamma,0}) \left(x_0^5\right)^2 \left( d_3 x^0 + T_3 \right); \quad (22.92)$$

$$\xi^5 \left(x^5; c, \gamma\right) = \pm \delta_{c,0} \frac{\gamma \alpha}{\left(x_0^5\right)^2} x^5, \quad (22.93)$$

where we identified $-\varepsilon = T_0$ in (22.89), highlighted the parametric dependence of $\xi^A$ on $c$ and $\gamma$, and introduced the Kronecker $\delta$.

The dimensions and ranges of the transformation parameters are

$$[\alpha] = l^2, \ [d_3] = l^{-2}, \ [T_0] = [T_3] = l^{-1}, \ [T_1] = [T_2] = l, \ [d_2] = l^0; \quad (22.94)$$

$$\alpha, d_2, d_3, T_0, T_1, T_2, T_3 \in R. \quad (22.95)$$

The 4D Killing group (i.e., the isometry group of the slices of $\Re_5$ at $dx^5 = 0$), too, can be written in the compact form

$$\left[ \text{Tr.}_{\widehat{x^1}, \widehat{x^2} \text{STD.}} \otimes (1 - \delta_{c,0}) \, \text{Tr.}_{\widehat{x^0} \text{STD.}} \otimes (1 - \delta_{c,0}) (1 - \delta_{\gamma,0}) \, \text{Tr.}_{\widehat{x^3} \text{STD.}} \right] \otimes_s$$
$$\otimes_s \left[ (1 - \delta_{\gamma,0}) \, \text{SO}(2)_{\text{STD.}(\Pi_2)} \otimes (1 - \delta_{c,0}) (1 - \delta_{\gamma,0}) \, B_{\text{STD.} \widehat{x^3}} \right]. \quad (22.96)$$

Here, $\Pi_2$ is the 2D manifold $\left(x^1, x^2\right)$ with metric rescaling $x^2 \longrightarrow \sqrt{2} x^2$ with respect to the Euclidean level, and $\text{SO}(2)_{\text{STD.}(\Pi_2)}$, $B_{\text{STD.} \widehat{x^3}}$ are the 1-parameter abelian groups generated by $S_{\text{SR}}^3 |_{x^2 \longrightarrow \sqrt{2} x^2}$ and $K_{\text{SR}}^3$, respectively. The semidirect nature of the group product is determined by the following commutation relations of the mixed, rototranslational, space–time Lorentz algebra (ESC on):

$$\left[ S_{\text{SR}}^i, \Upsilon_{\text{SR}}^j \right] = \epsilon_{ijl} \Upsilon_{\text{SR}}^l. \quad (22.97)$$

The direct product of $SO(2)_{STD.(\Pi_2)}$ and $B_{\text{STD.} \widehat{x^3}}$ and of the translation groups $\text{Tr.}_{\widehat{x^1},\widehat{x^2}\ \text{STD.}}$, $\text{Tr.}_{\widehat{x^0}\ \text{STD.}}$, $\text{Tr.}_{\widehat{x^3}\ \text{STD.}}$ is instead a consequence of the commutativity of the generators:

$$\left[S^3_{\text{SR}}, K^3_{\text{SR}}\right]\big|_{x^2 \longrightarrow \sqrt{2}x^2} = 0; \tag{22.98}$$

$$[\Upsilon^\mu_{\text{SR}}, \Upsilon^\nu_{\text{SR}}] = 0. \tag{22.99}$$

## 22.3 Gravitational Interaction

### *22.3.1 Validity of the $\Upsilon$-Hypothesis*

**(Case a') (Minkowskian conditions).** In the energy range $0 < x^5 \leqslant x^5_0$ the 5D metric for gravitational interaction (20.23) becomes:

$$g_{AB,\text{DR5}}(x^5) = \text{diag}\left(1, -b_1^2(x^5), -b_2^2(x^5), -1, \pm f(x^5)\right). \tag{22.100}$$

The "vectors" $A_\mu(x^5)$ and $B_\mu(x^5)$ (21.19) and (21.23) read therefore:

$$\begin{aligned} A_0(x^5) &= A_3(x^5) = 0; \\ B_0(x^5) &= B_3(x^5) = \left(f(x^5)\right)^{1/2}; \end{aligned} \tag{22.101}$$

$$\begin{gathered} A_i(x^5) = b_i(x^5)(f(x^5))^{-1/2} \\ \cdot\left[-\left(b_i'(x^5)\right)^2 + b_i(x^5)b_i''(x^5) - \frac{1}{2}b_i(x^5)b_i'(x^5)f'(x^5)(f(x^5))^{-1}\right], \\ i = 1, 2; \end{gathered} \tag{22.102}$$

$$B_i(x^5) = b_i(x^5)(f(x^5))^{1/2}, \quad i = 1, 2; \tag{22.103}$$

$$\begin{aligned} \frac{\pm A_i(x^5)}{B_i(x^5)} = &\pm (f(x^5))^{-1}\left[-\left(b_i'(x^5)\right)^2 + b_i(x^5)b_i''(x^5)\right. \\ &\left.-\frac{1}{2}b_i(x^5)b_i'(x^5)f'(x^5)(f(x^5))^{-1}\right], \\ &i = 1, 2. \end{aligned} \tag{22.104}$$

One has

$$\left.\begin{array}{l} A_i(x^5) \neq 0 \Leftrightarrow \\ \Leftrightarrow \left[ -\left(b_i'(x^5)\right)^2 + b_i(x^5) b_i''(x^5) - \frac{1}{2} b_i(x^5) b_i'(x^5) f'(x^5) (f(x^5))^{-1} \right] \neq 0; \\ B_i(x^5) \neq 0 \ \ \forall x^5 \in R_0^+ \text{(no condition)} ; \\ \dfrac{\pm A_i(x^5)}{B_i(x^5)} \neq c, c \in R_0 \\ \Leftrightarrow -\left(b_i'(x^5)\right)^2 + b_i(x^5) b_i''(x^5) - \frac{1}{2} b_i(x^5) b_i'(x^5) f'(x^5) (f(x^5))^{-1} \neq c f(x^5), \\ c \in R_0 \end{array}\right\}$$

$$\begin{aligned} &\Leftrightarrow -\left(b_i'(x^5)\right)^2 + b_i(x^5) b_i''(x^5) - \frac{1}{2} b_i(x^5) b_i'(x^5) f'(x^5) (f(x^5))^{-1} \\ &\qquad \neq c f(x^5), \quad c \in R, i = 1, 2. \end{aligned} \tag{22.105}$$

Therefore the validity for $\mu = 1, 2$ of the $\Upsilon$-hypothesis (21.34) (not satisfied for $\mu = 0, 3$) depends on the nature and the functional form of the metric coefficients $b_1^2(x^5)$ and $b_2^2(x^5)$. In general the 15 Killing equations corresponding to metric (22.100) are:

$$\left\{\begin{array}{c} f(x^5)\xi_{0,0}(x^A) = 0; \\ \xi_{0,1}(x^A) + \xi_{1,0}(x^A) = 0; \\ \xi_{0,2}(x^A) + \xi_{2,0}(x^A) = 0; \\ \xi_{0,3}(x^A) + \xi_{3,0}(x^A) = 0; \\ \xi_{0,5}(x^A) + \xi_{5,0}(x^A) = 0; \\ f(x^5)\xi_{1,1}(x^A) \mp b_1(x^5) b_1'(x^5) \xi_5(x^A) = 0; \\ \xi_{1,2}(x^A) + \xi_{2,1}(x^A) = 0; \\ \xi_{1,3}(x^A) + \xi_{3,1}(x^A) = 0; \\ b_1(x^5)(\xi_{1,5}(x^A) + \xi_{5,1}(x^A)) - 2b_1'(x^5)\xi_1(x^A) = 0; \\ f(x^5)\xi_{2,2}(x^A) \mp b_2(x^5) b_2'(x^5) \xi_5(x^A) = 0; \\ \xi_{2,3}(x^A) + \xi_{3,2}(x^A) = 0; \\ b_2(x^5)(\xi_{2,5}(x^A) + \xi_{5,2}(x^A)) - 2b_2'(x^5)\xi_2(x^A) = 0; \\ f(x^5)\xi_{3,3}(x^A) = 0; \\ \xi_{3,5}(x^A) + \xi_{5,3}(x^A) = 0; \\ 2f(x^5)\xi_{5,5}(x^A) - f'(x^5)\xi_5(x^A) = 0. \end{array}\right. \tag{22.106}$$

By making suitable assumptions on the functional form of the metric coefficients $b_i^2(x^5)$ $(i = 1, 2)$, it is possible in 11 cases (which include all those of physical and mathematical interest) to solve the relevant Killing equations for the gravitational interaction and get the related isometries (see Appendix B).

**(Case b') (Non-Minkowskian conditions).** In the energy range $x^5 > x_0^5$ the 5D gravitational metric (20.23) reads:

$$g_{AB,\mathrm{DR5}}(x^5)$$
$$= \mathrm{diag}\left(\frac{1}{4}\left(1+\frac{x^5}{x_0^5}\right)^2, -b_1^2(x^5), -b_2^2(x^5), -\frac{1}{4}\left(1+\frac{x^5}{x_0^5}\right)^2, \pm f(x^5)\right). \tag{22.107}$$

Equations (21.22) and (21.23) yield

$$\begin{aligned} A_0(x^5) &= -A_3(x^5) = \\ &= \frac{1}{8}\left(1+\frac{x^5}{x_0^5}\right)\frac{x^5}{(x_0^5)^2}\left(f(x^5)\right)^{-\frac{1}{2}}\left[\frac{1}{x^5}+\frac{1}{2}\left(1+\frac{x_0^5}{x^5}\right)\frac{f'(x^5)}{f(x^5)}\right]; \\ B_0(x^5) &= B_3(x^5) = \frac{1}{2}\left(1+\frac{x_0^5}{x^5}\right)\left(f(x^5)\right)^{1/2}; \end{aligned} \tag{22.108}$$

$$\frac{\pm A_0(x^5)}{B_0(x^5)} = \frac{\mp A_3(x^5)}{B_3(x^5)} = \pm\frac{1}{4}\frac{1}{f(x^5)}\frac{x^5}{(x_0^5)^2}\left[\frac{1}{x^5}+\frac{1}{2}\frac{f'(x^5)}{f(x^5)}+\frac{1}{2}\frac{x_0^5}{x^5}\frac{f'(x^5)}{f(x^5)}\right], \tag{22.109}$$

whereas (22.101)–(22.104) of the previous case still hold for $A_i(x^5)$ and $B_i(x^5)$ $(i = 1, 2)$. Then, since $f(x^5)$ is strictly positive and

$$x^5, x_0^5 \in R_0^+ \underset{\not\Leftarrow}{\overset{\Rightarrow}{\text{(in gen.)}}} \left(1+\frac{x_0^5}{x^5}\right) \in R_0^+, \tag{22.110}$$

the $\Upsilon$-hypothesis of functional independence for the gravitational metric over threshold is satisfied at least for $\mu = 0, 3$ under the following conditions:

$$\left.\begin{aligned} &\frac{1}{x^5}+\frac{1}{2}\frac{f'(x^5)}{f(x^5)}+\frac{1}{2}\frac{x_0^5}{x^5}\frac{f'(x^5)}{f(x^5)} \neq 0; \\ &\frac{x^5}{f(x^5)}\left[\frac{1}{x^5}+\frac{1}{2}\frac{f'(x^5)}{f(x^5)}+\frac{1}{2}\frac{x_0^5}{x^5}\frac{f'(x^5)}{f(x^5)}\right] \neq \mathsf{C} \Leftrightarrow \\ &\frac{1}{x^5}+\frac{1}{2}\frac{f'(x^5)}{f(x^5)}+\frac{1}{2}\frac{x_0^5}{x^5}\frac{f'(x^5)}{f(x^5)} \neq \mathsf{C}\frac{f(x^5)}{x^5}, \mathsf{C} \in R_0, \end{aligned}\right\}$$

$$\Leftrightarrow \frac{1}{x^5}+\frac{1}{2}\frac{f'(x^5)}{f(x^5)}+\frac{1}{2}\frac{x_0^5}{x^5}\frac{f'(x^5)}{f(x^5)} \neq \mathsf{C}\frac{f(x^5)}{x^5}, \mathsf{C} \in R. \tag{22.111}$$

Therefore, in the energy range $x^5 > x_0^5$, if the $\Upsilon$-hypothesis of functional independence for the gravitational metric is not satisfied for $\mu = 0, 3$, the metric coefficient $f(x^5)$ obeys the following equation:

$$\frac{1}{2}\frac{f'(x^5)}{f(x^5)} + \frac{1}{2}\frac{x_0^5}{x^5}\frac{f'(x^5)}{f(x^5)} - \mathtt{C}\frac{f(x^5)}{x^5} + \frac{1}{x^5} = 0, \mathtt{C} \in R, \tag{22.112}$$

which, since $x_0^5$, $x^5$, $f(x^5) \in R_0^+$, can be rewritten as

$$f'(x^5) + \frac{2}{x^5 + x_0^5} f(x^5) - \frac{2\mathtt{C}}{x^5 + x_0^5}\left(f(x^5)\right)^2 = 0, \mathtt{C} \in R. \tag{22.113}$$

Such ordinary differential equation belongs to the separable subclass of the Bernoulli type $\forall \mathtt{C} \in R$. Its only solution is

$$f(x^5) = \frac{1}{\gamma(x^5 + x_0^5)^2 + \mathtt{C}}, \mathtt{C}, \gamma \in R, \tag{22.114}$$

which can be expressed in dimensionless form as (by suitably rescaling the constants $\mathtt{C}, \gamma$):

$$\begin{cases} f(x^5) \equiv f\left(\dfrac{x^5}{x_0^5}\right) = \dfrac{1}{\gamma\left(1 + \dfrac{x^5}{x_0^5}\right)^2 + \mathtt{C}}, \\ c, \gamma \in R : \gamma\left(1 + \dfrac{x^5}{x_0^5}\right)^2 + \mathtt{C} > 0 \ \forall x^5 \in R_0^+ \Leftrightarrow \mathtt{C}, \gamma \in R^+ \text{ (not both zero)}. \end{cases} \tag{22.115}$$

The corresponding 5D gravitational metric is therefore

$$g_{AB,\mathrm{DR5}}(x^5) = \mathrm{diag}\left(\begin{array}{c} \frac{1}{4}\left(1 + \frac{x^5}{x_0^5}\right)^2, -b_1^2(x^5), -b_2^2(x^5), \\ -\frac{1}{4}\left(1 + \frac{x^5}{x_0^5}\right)^2, \pm\left(\gamma\left(1 + \frac{x^5}{x_0^5}\right)^2 + \mathtt{C}\right)^{-1} \end{array}\right) \tag{22.116}$$

Thus, in both energy ranges the $\Upsilon$-hypothesis is violated for $\mu = 0, 3$. Below threshold this is automatically ensured by the form of the gravitational metric (22.100), whereas *above threshold such a requirement determines the expression of the fifth metric coefficient* $f(x^5)$.

### 22.3.2 The 5D $\Upsilon$-Violating Metrics of Gravitation

We want now to discuss the 5D gravitational metrics which violate the $\Upsilon$-hypothesis of functional independence $\forall \mu = 0, 1, 2, 3$ in either energy range $0 < x^5 \leqslant x^5_{0,\mathrm{grav}}$ and $x^5 > x^5_{0,\mathrm{grav}}$.

It was shown in Sect. 22.3.1 that, when the $\Upsilon$-hypothesis is not satisfied by $\mu = 0, 3$, the gravitational metric is given by (22.100) and (22.116) below threshold and over threshold, respectively. If one imposes in addition that the $\Upsilon$-hypothesis is violated also by $\mu = 1$ and/or 2, such a requirement permits to get the expressions of the metric coefficients $b_1^2(x^5)$, $b_2^2(x^5)$ in terms of $f(x^5)$.

Indeed, in this case it follows from (22.105) that the metric coefficients $b_1^2(x^5)$, $b_2^2(x^5)$ and $f(x^5)$ satisfy the following ODE (ESC off)[1]

$$-\left(b_k'(x^5)\right)^2 + b_k(x^5)b_k''(x^5) - \frac{1}{2}b_k(x^5)b_k'(x^5)f'(x^5)(f(x^5))^{-1} - c_k f(x^5) = 0,$$
$$c_k \in R, \quad k = 1 \text{ and/or } 2, \tag{22.117}$$

whose solution for $f(x^5)$ in terms of $b_k(x^5)$ is:

$$f(x^5) = \frac{\left(b_k'(x^5)\right)^2}{d_k b_k^2(x^5) - c_k} \Leftrightarrow$$
$$\Leftrightarrow d_k b_k^2(x^5) f(x^5) - \left(b_k'(x^5)\right)^2 - c_k f(x^5) = 0,$$
$$k = 1 \text{ and/or } 2,\ d_k \in R^+, c_k \in R^- \text{(not both zero)}. \tag{22.118}$$

In dimensionless form for $f(x^5)$ and $b_k^2(x^5)$, we have

$$f(x^5) \equiv f\left(\frac{x^5}{x_0^5}\right) = \frac{\left(b_k'\left(\frac{x^5}{x_0^5}\right)\right)^2}{d_k b_k^2\left(\frac{x^5}{x_0^5}\right) - c_k}$$
$$\Leftrightarrow d_k b_k^2\left(\frac{x^5}{x_0^5}\right) f\left(\frac{x^5}{x_0^5}\right) - \left(b_k'\left(\frac{x^5}{x_0^5}\right)\right)^2 - c_k f\left(\frac{x^5}{x_0^5}\right) = 0,$$
$$k = 1 \text{ and/or } 2. \tag{22.119}$$

By assuming $f(x^5)$ known, one gets the following implicit solution of (22.117) for $b_k(x^5)$ ($d_k > 0$):

$$\alpha_k \pm \int^{x^5} dx^{5\prime}\sqrt{-f(x^{5\prime})} + \frac{1}{\sqrt{d_k}} \arctan\left(\frac{\sqrt{d_k}\, b_k(x^5)}{\sqrt{c_k - d_k b_k^2(x^5)}}\right) = 0,$$
$$k = 1 \text{ and/or } 2, \quad \alpha_k \in R, \tag{22.120}$$

under the constraint

$$c_k - d_k b_k^2\left(\frac{x^5}{x_0^5}\right) > 0, k = 1 \text{ and/or } 2\ \forall x^5 \in R_0^+ \Leftrightarrow d_k \in R^+, c_k \in R^-. \tag{22.121}$$

[1] In the following, the lower index "k" in the constants means that, in general, these depend on the metric coefficient $b_k$ considered.

Equation (22.120) can be solved for all possible pair of values $(d_k, c_k)$ (even in the limit case of $b_k$ constant). Precisely, one can distinguish the following three cases ($k = 1, 2$):

(I) $(d_k, c_k) \in R_0^+ \times R_0^-$:

$$b_k^2(x^5) = \frac{c_k}{d_k} \frac{\tanh^2\left[\sqrt{d_k}\left(\alpha_k \mp F(x^5)\right)\right]}{\left\{\tanh^2\left[\sqrt{d_k}\left(\alpha_k \mp F(x^5)\right)\right] - 1\right\}}$$
$$= -\frac{c_k}{d_k}\left\{\cosh\left[2\sqrt{d_k}\left(\alpha_k \mp F(x^5)\right)\right] - 1\right\}, \alpha_k \in R, \qquad (22.122)$$

where $F(x^5)$ is still given by (22.10).

(II) $(d_k, c_k) \in \{0\} \times R_0^-$:

$$b_k^2(x^5) = \left(\pm\sqrt{-c_k}F(x^5) + \delta_k\right)^2, \delta_k \in R. \qquad (22.123)$$

(III) $(d_k, c_k) \in R_0^+ \times \{0\}$:

$$b_k^2(x^5) = \kappa_k^2 \exp\left(2\sqrt{d_k}F(x^5)\right)$$
$$= \kappa_{k1}^2 \exp\left(2\sqrt{d_k}F(x^5)\right) + \kappa_{k2}^2 \exp\left(-2\sqrt{d_k}F(x^5)\right),$$
$$\kappa_k, \kappa_{ki} \in R_0, i = 1, 2. \qquad (22.124)$$

Let us note that all the previous results hold true in general for any $\mu \in \{0, 1, 2, 3\}$. They have been discussed by considering $\mu = k = 1$ and/or 2 in order to apply the results to the case of the DR5 metric of the gravitational interaction, characterized by the indeterminacy of the metric coefficients $b_k(x^5)$, $k = 1, 2$, and requiring therefore a treatment of the $\Upsilon$-violation for $\mu = k = 1$ and/or 2.

The earlier general formalism allows one to deal with the 5D metrics of DR5 for the gravitational interaction which violate $\Upsilon$ $\forall\mu = 0, 1, 2, 3$ in the energy ranges $0 < x^5 \leqslant x^5_{0,\mathrm{grav}}$ and $x^5 > x^5_{0,\mathrm{grav}}$.

In the first case the functional form of $f(x^5)$ is undetermined, since in general it must only satisfy the condition $f > 0$ $\forall x^5 \in R_0^+$. As to the space coefficients $b_k(x^5)$, $k = 1, 2$, one has nine possible cases, obtained by considering all the possible pairs $(\mathcal{I}_1, \mathcal{I}_2)$ $(\mathcal{I}_1, \mathcal{I}_2 = I,\ II,\ III)$ of the functional typologies for $b_k(x^5)$ corresponding to the pairs of values $(d_k, c_k)$ (see (22.122)–(22.124)). Since the two space coefficients are expressed in terms of the fifth metric coefficient, one gets "$f(x^5)$-dependent," i.e., in general "functionally parametrized" metrics.

In the energy range $x^5 > x^5_{0,\text{grav}}$, the fifth coefficient is determined by (22.112) with solution (22.115). The $\Upsilon$-violating gravitational metric has the form (22.116), and the space coefficients $b_1^2(x^5)$, $b_2^2(x^5)$ are still given by (22.122)–(22.124). However, now the function $F(x^5)$ can be explicitly evaluated. One has

$$
\begin{aligned}
F(x^5) &\equiv \int \mathrm{d}x^5 \left(f(x^5)\right)^{1/2} = + \int \mathrm{d}x^5 \sqrt{f(x^5)} \\
&= x_0^5 \int \frac{\mathrm{d}x^5}{\sqrt{\gamma\left(x^5\right)^2 + 2\gamma x_0^5 x^5 + (\mathtt{C}+\gamma)\left(x_0^5\right)^2}}
\end{aligned}
$$

$$
= \begin{cases}
\dfrac{x^5}{\sqrt{\mathtt{C}}} & \text{for } (\gamma, \mathtt{C}) \in \{0\} \times R_0^+ \quad \text{(case A);} \\[2ex]
\dfrac{x_0^5}{\sqrt{\gamma}} \ln\left(1 + \dfrac{x^5}{x_0^5}\right) & \text{for } (\gamma, \mathtt{C}) \in R_0^+ \times \{0\} \quad \text{(case B);} \\[2ex]
\dfrac{x_0^5}{\sqrt{\gamma}} \operatorname{arcsinh}\left[\sqrt{\dfrac{\gamma}{\mathtt{C}}}\left(1 + \dfrac{x^5}{x_0^5}\right)\right] & \text{for } (\gamma, \mathtt{C}) \in R_0^+ \times R_0^+ \text{ (case C).}
\end{cases}
\tag{22.125}
$$

Replacing the earlier expressions (22.125), corresponding to the three possible pairs $(\gamma, \mathtt{C})$, in (22.122)–(22.124), one gets all the possible forms of the coefficients $b_1^2(x^5)$ and $b_2^2(x^5)$ of the $\Upsilon$-violating gravitational metric above threshold. The functional typologies of the spatial coefficients can be labeled by the pair $(\mathcal{L}, \mathcal{I})$ (with $\mathcal{L} = A, B, C$ labeling the three cases of (22.125) and $\mathcal{I} = \mathrm{I}, \mathrm{II}, \mathrm{III}$ referring as before to the three expressions (22.122)–(22.124)). Then, one gets 27 possible forms for the 5D gravitational metrics violating the hypothesis $\Upsilon$ in the energy range. They can be labeled by $(\mathcal{L}_1\mathcal{I}_1, \mathcal{L}_2\mathcal{I}_2)$ $(\mathcal{L}_1, \mathcal{L}_2 = A, B, C,\ \mathcal{I}_1, \mathcal{I}_2 = \mathrm{I}, \mathrm{II}, \mathrm{III})$, in the notation exploited for the functional typology of the metric coefficients $b_k^2(x^5)$ for the indices 1 and 2, according to the earlier discussion. Their explicit form is easy to write down, and can be found in [136].

In correspondence to the different gravitational metrics, one gets 27 systems of 15 Killing equations, which would require an explicit solution (or at least not exploiting the$\Upsilon$-hypothesis), in order to find the corresponding isometries. However, solving these systems is far from being an easy task, even by using symbolic-algebraic manipulation programs.

A possible method of partial solution could be the Lie structural approach, based on the Lie symmetries obeyed by the system equations. Such a resolution could in principle be also applied to the general system (22.106), in which no assumption is made on the functional forms of the metric coefficients $b_\mu^2(x^5)$ $(\mu = 0, 1, 2, 3)$ and $f(x^5)$.

## 22.4 Infinitesimal-Algebraic Structure of Killing Symmetries in $\Re_5$

From the knowledge of the Killing vectors for the 5D metrics of the four fundamental interactions, we can now discuss the algebraic-infinitesimal structure of the related Killing isometries.

As is well known, the $M$ independent Killing vector fields of a differentiable, $N$-dimensional manifold $S_N$ do span a linear space $\mathcal{K}$. The maximum number of independent Killing vectors (i.e.,the maximum dimension of $\mathcal{K}$) is $N(N+1)/2 \geq M$. On the basis of the general discussion of Sect. 5.2, an infinitesimal transformation in $S_N$ can be written as (cf. (5.8)):

$$\begin{aligned} x'^A(x,\alpha) &= x^A + \delta x^A(x,\alpha) + \mathrm{O}(\alpha^2) \\ &= x^A + \xi^A(x,\alpha) + \mathrm{O}(\alpha^2), \quad A = 1,2,...,N, \end{aligned} \tag{22.126}$$

where $\{\alpha^{\mathcal{A}}\}$ $(\mathcal{A} = 1,2,\ldots,M)$ is the parametric $M$-vector and the contravariant Killing N-vector $\xi^A(x,\alpha)$ of the manifold reads (see (5.18)–(5.20)) (ESC on):

$$\xi^A(x,\alpha) = \xi^A_{\mathcal{A}}(x)\alpha^{\mathcal{A}}. \tag{22.127}$$

Quantities $\xi^A_{\mathcal{A}}(x)$ are the components of the linearly-independent Killing vectors of $\mathcal{K}$, and are given by

$$\xi^A_{\mathcal{A}}(x) = \left.\frac{\partial x'^A(x,\alpha)}{\partial \alpha^{\mathcal{A}}}\right|_{\substack{\alpha^{\mathcal{B}}=0 \\ \forall \mathcal{B}=1,\ldots,15}}. \tag{22.128}$$

The general form of an infinitesimal metric automorphism of $S_N$ is therefore

$$x'^A(x,\alpha) \overset{ESC\ \text{on}}{=} x^A + \xi^A_{\mathcal{A}}(x)\alpha^{\mathcal{A}} + O(\alpha^2), \quad A = 1,2,...,N, \tag{22.129}$$

By introducing the canonical vector basis $\{\partial_A \equiv \partial/\partial x^A\}$ in $S_N$, one has, for the Killing $N$-vector $\widetilde{\boldsymbol{\xi}}(x)$ (ESC on on $A$ and $\mathcal{A}$):[2]

$$\widetilde{\boldsymbol{\xi}}(x) = \xi^A_{\mathcal{A}}(x)\alpha^{\mathcal{A}}\partial_A = \widetilde{\boldsymbol{\xi}}_{\mathcal{A}}(x)\alpha^{\mathcal{A}}, \tag{22.130}$$

where (ESC on)[3]

$$\widetilde{\boldsymbol{\xi}}_{\mathcal{A}}(x) = \xi^A_{\mathcal{A}}(x)\partial_A. \tag{22.131}$$

[2] In this section, of course, the notation $\widetilde{\mathbf{v}}$ means a $N$-vector.

[3] Care must be exercised in distinguishing the two different vector spaces involved in (22.130)–(22.131). On one hand, $\xi^A(x)$ are the (contravariant) components of the $N$-dimensional Killing vector $\xi(x)$, belonging to the (tangent space of) the manifold $S_N$. On the other side, $\widetilde{\boldsymbol{\xi}}_{\mathcal{A}}(x)$ are the components of the $M$-dimensional vector belonging to the $M$-d Killing space. According to (22.131), each $\widetilde{\boldsymbol{\xi}}_{\mathcal{A}}(x)$ is in turn a vector in (the tangent space of) the manifold $S_N$.

The $M$ vectors $\widetilde{\boldsymbol{\xi}}_{\mathcal{A}}(x)$ are the infinitesimal generators of the algebra of the Killing symmetries of $S_N$. The product of this algebra is, as usual, the commutator

$$\left[\widetilde{\boldsymbol{\xi}}_{\mathcal{A}}(x), \widetilde{\boldsymbol{\xi}}_{\mathcal{B}}(x)\right], \quad \mathcal{A}, \mathcal{B} = 1, 2, \ldots, \mathrm{M}. \tag{22.132}$$

The Killing algebra is then specified by the set of commutation relations

$$\left[\widetilde{\boldsymbol{\xi}}_{\mathcal{A}}(x), \widetilde{\boldsymbol{\xi}}_{\mathcal{B}}(x)\right] = \mathcal{C}^{\mathcal{C}}_{\mathcal{AB}} \widetilde{\boldsymbol{\xi}}_{\mathcal{C}}(x), \tag{22.133}$$

where $\mathcal{C}^{\mathcal{C}}_{\mathcal{AB}} = -\mathcal{C}^{\mathcal{C}}_{\mathcal{BA}}$ are the $M(M-1)/2$ structure constants of the algebra.

In the present case of the space $\Re_5$, it is obviously $A = 0, 1, 2, 3, 5$. As to the dimension $M$ ($\leq 15$) of the Killing manifold, it depends on the explicit solution $\xi^A(x, x^5)$ of the 15 Killing equations (21.6)–(21.15), and therefore on the metric $g_{\mathrm{DR5,int.}}(x^5)$. In the following, we shall consider all possible cases of metrics of physical relevance.

### 22.4.1 Metric with Constant Space–Time Coefficients

Let us consider the 5D-metric (22.2) $g_{AB,\mathrm{DR5}}(x^5) = \mathrm{diag}\,(a, -b, -c, -d, \pm f(x^5))$, special cases of which are the electromagnetic and weak metrics above threshold ($x^5 \geq x^5_{0\mathrm{e.m.,weak}}$), (22.1), and the strong metric below threshold ($0 < x^5 \leq x^5_{0,\mathrm{strong}}$), (22.46).

The solution of the related Killing system for the contravariant Killing five-vector $\xi^A(x, \alpha)$ is given by (22.5')–(22.9'), we rewrite here for reader's convenience:

$$\begin{cases} \xi^0(x^1, x^2, x^3, x^5) = \dfrac{1}{a}\left[T^0 - B^1 x^1 - B^2 x^2 - B^3 x^3 + \varXi^0 F(x^5)\right]; \\ \xi^1(x^0, x^2, x^3, x^5) = \dfrac{1}{b}\left[T^1 - B^1 x^0 + \varTheta^3 x^2 - \varTheta^2 x^3 - \varXi^1 F(x^5)\right]; \\ \xi^2(x^0, x^1, x^3, x^5) = \dfrac{1}{c}\left[T^2 - B^2 x^0 - \varTheta^3 x^1 + \varTheta^1 x^3 - \varXi^2 F(x^5)\right]; \\ \xi^3(x^0, x^1, x^2, x^5) = \dfrac{1}{d}\left[T^3 - B^3 x^0 + \varTheta^2 x^1 - \varTheta^1 x^2 - \varXi^3 F(x^5)\right]; \\ \xi^5(x, x^5) = \pm\left(f(x^5)\right)^{-\frac{1}{2}}\left[T^5 - \varXi^0 x^0 - \varXi^1 x^1 - \varXi^2 x^2 - \varXi^3 x^3\right] \end{cases} \tag{22.134}$$

$\left(F(x^5) = \int \mathrm{d}x^5 \left(f(x^5)\right)^{1/2}\right)$.

Without loss of generality we can make the following identifications:

$$\begin{aligned} \alpha^1 &= \frac{1}{a}T^0; \quad \alpha^2 = \frac{1}{a}T^1; \quad \alpha^3 = \frac{1}{a}T^2; \quad \alpha^4 = \frac{1}{a}T^3; \\ \alpha^5 &= -B^1; \quad \alpha^6 = -B^2; \quad \alpha^7 = -B^3; \\ \alpha^8 &= \varTheta^3; \quad \alpha^9 = \varTheta^2; \quad \alpha^{10} = \varTheta^1; \\ \alpha^{11}_{(\pm)} &= \pm T^5; \quad \alpha^{12} = \varXi^0; \quad \alpha^{13} = \varXi^1; \quad \alpha^{14} = \varXi^2; \quad \alpha^{15} = \varXi^3. \end{aligned} \tag{22.135}$$

Equation (22.134) becomes therefore

$$\begin{cases} \xi^0(x^1,x^2,x^3,x^5) = \alpha^1 + \dfrac{1}{a}\left[\alpha^5 x^1 + \alpha^6 x^2 + \alpha^7 x^3 + \alpha^{12} F(x^5)\right]; \\ \xi^1(x^0,x^2,x^3,x^5) = \alpha^2 + \dfrac{1}{b}\left[\alpha^5 x^0 + \alpha^8 x^2 - \alpha^9 x^3 - \alpha^{13} F(x^5)\right]; \\ \xi^2(x^0,x^1,x^3,x^5) = \alpha^3 + \dfrac{1}{c}\left[\alpha^6 x^0 - \alpha^8 x^1 + \alpha^{10} x^3 - \alpha^{14} F(x^5)\right]; \\ \xi^3(x^0,x^1,x^2,x^5) = \alpha^4 + \dfrac{1}{d}\left[\alpha^7 x^0 + \alpha^9 x^1 - \alpha^{10} x^2 - -\alpha^{15} F(x^5)\right]; \\ \xi^5(x,x^5) = \left(f(x^5)\right)^{-\frac{1}{2}}\left(\alpha^{11} \mp \alpha^{12} x^0 \mp \alpha^{13} x^1 \mp \alpha^{14} x^2 \mp \alpha^{15} x^3\right). \end{cases} \tag{22.136}$$

Then, it follows from (22.128):

$$\xi^A_{\mathcal{A}}(x) = \begin{pmatrix} \xi^{\mu}_{\mathcal{A},\mathrm{SR}}(x) \\ \\ 0 \end{pmatrix}, \mathcal{A} = 1,2,3,4; \tag{22.137}$$

$$\xi^A_5(x^0,x^1) = \begin{pmatrix} \dfrac{x^1}{a} \\ \dfrac{x^0}{b} \\ 0 \\ 0 \\ 0 \end{pmatrix} = \delta^A_0 \frac{x^1}{a} + \delta^A_1 \frac{x^0}{b}; \tag{22.138}$$

$$\xi^A_6(x^0,x^2) = \begin{pmatrix} \dfrac{x^2}{a} \\ 0 \\ \dfrac{x^0}{c} \\ 0 \\ 0 \end{pmatrix} = \delta^A_0 \frac{x^2}{a} + \delta^A_2 \frac{x^0}{c}; \tag{22.139}$$

$$\xi^A_7(x^0,x^3) = \begin{pmatrix} \dfrac{x^3}{a} \\ 0 \\ 0 \\ \dfrac{x^0}{d} \\ 0 \end{pmatrix} = \delta^A_0 \frac{x^3}{a} + \delta^A_3 \frac{x^0}{d}; \tag{22.140}$$

$$\xi^A_8(x^1,x^2) = \begin{pmatrix} 0 \\ \dfrac{x^2}{b} \\ -\dfrac{x^1}{c} \\ 0 \\ 0 \end{pmatrix} = \delta^A_1 \frac{x^2}{b} - \delta^A_2 \frac{x^1}{c}; \tag{22.141}$$

$$\xi_9^A(x^1,x^3) = \begin{pmatrix} 0 \\ -\dfrac{x^3}{b} \\ 0 \\ \dfrac{x^1}{d} \\ 0 \end{pmatrix} = -\delta^A_{\ 1}\frac{x^3}{b} + \delta^A_{\ 3}\frac{x^1}{d}; \tag{22.142}$$

$$\xi_{10}^A(x^2,x^3) = \begin{pmatrix} 0 \\ \\ \dfrac{x^3}{c} \\ -\dfrac{x^2}{d} \\ 0 \end{pmatrix} = \delta^A_{\ 2}\frac{x^3}{c} - \delta^A_{\ 3}\frac{x^2}{d}; \tag{22.143}$$

$$\xi_{11}^A(x^5) = \begin{pmatrix} 0 \\ 0 \\ 0 \\ 0 \\ \mp\left(f(x^5)\right)^{-1/2} \end{pmatrix} = \mp\delta^A_{\ 5}\left(f(x^5)\right)^{-1/2}; \tag{22.144}$$

$$\xi_{12}^A(x^0,x^5) = \begin{pmatrix} \dfrac{F(x^5)}{a} \\ 0 \\ 0 \\ 0 \\ \mp\left(f(x^5)\right)^{-1/2} x^0 \end{pmatrix} = \delta^A_{\ 0}\frac{F(x^5)}{a} \mp \delta^A_{\ 5}\left(f(x^5)\right)^{-1/2} x^0; \tag{22.145}$$

$$\xi_{13}^A(x^1,x^5) = \begin{pmatrix} 0 \\ \dfrac{F(x^5)}{b} \\ 0 \\ 0 \\ \mp\left(f(x^5)\right)^{-1/2} x^1 \end{pmatrix} = \delta^A_{\ 1}\frac{F(x^5)}{b} \mp \delta^A_{\ 5}\left(f(x^5)\right)^{-1/2} x^1; \tag{22.146}$$

$$\xi_{14}^A(x^2,x^5) = \begin{pmatrix} 0 \\ 0 \\ -\dfrac{F(x^5)}{c} \\ 0 \\ \mp\left(f(x^5)\right)^{-1/2} x^2 \end{pmatrix} = -\delta^A_{\ 2}\frac{F(x^5)}{c} \mp \delta^A_{\ 5}\left(f(x^5)\right)^{-1/2} x^2; \tag{22.147}$$

$$\xi_{15}^{A}(x^3, x^5) = \begin{pmatrix} 0 \\ 0 \\ 0 \\ -\dfrac{F(x^5)}{d} \\ \mp \left(f(x^5)\right)^{-1/2} x^3 \end{pmatrix} = -\delta_3^A \frac{F(x^5)}{d} \mp \delta_5^A \left(f(x^5)\right)^{-1/2} x^3. \tag{22.148}$$

It is easy to see that the four Killing five-vectors $\xi_{\mathcal{A}}^{A}(x)$, $\mathcal{A} = 1,2,3,4$ (22.138) are the generators $\Upsilon^{\mu}$ ($\mu = 0,1,2,3$) of the standard translation group Tr.$(1,3)$,[4] whereas the six Killing five-vectors $\xi_{\mathcal{A}}^{A}(x)$, $\mathcal{A} = 5,\dots,10$ ((22.139)–(22.145)) are the generators of the deformed Lorentz group $\mathrm{SO}(1,3)_{\mathrm{DEF.}}$ (see Sect. 6.3). As is well known, the six generators of $\mathrm{SO}(1,3)_{\mathrm{DEF.}}$in the self-representation basis are given by the three-vectors $\mathbf{S}_{\mathrm{DSR}}$ and $\mathbf{K}_{\mathrm{DSR}}$ (associated to deformed space rotations and boosts, respectively: see Sect. 6.3.1).

The other five Killing five-vectors $\xi_{\mathcal{A}}^{A}(x)$, $\mathcal{A} = 11, ..., 15$, can be identified with the *new* generators of the Killing algebra as follows.[5] By its expression (22.144), the Killing vector $\xi_{11}^{A}(x^5)$ is of course the translation generator along $x^5$:

$$\xi_{11}^{A}(x^5) = \Upsilon^{5A}(x^5). \tag{22.149}$$

As to the other $\xi_{\mathcal{A}}^{A}(x)$ for $\mathcal{A} = 12,\dots,15$, it is easily seen from (22.145)–(22.148) that their interpretation as rotation or boost generators depends on the signature of the fifth coordinate $x^5$, namely on its time-like or space-like nature:

$$\xi_{12}^{A}(x^0, x^5) = \begin{cases} \text{“}+\text{”} : \Sigma^{1A}(x^0, x^5); \\ \text{“}-\text{”} : \Gamma^{1A}(x^0, x^5); \end{cases} \tag{22.150}$$

$$\xi_{13}^{A}(x^1, x^5) = \begin{cases} \text{“}+\text{”} : \Gamma^{1A}(x^1, x^5); \\ \text{“}-\text{”} : \Sigma^{1A}(x^0, x^5); \end{cases} \tag{22.151}$$

$$\xi_{14}^{A}(x^2, x^5) = \begin{cases} \text{“}+\text{”} : \Gamma^{2A}(x^2, x^5); \\ \text{“}-\text{”} : \Sigma^{2A}(x^2, x^5); \end{cases} \tag{22.152}$$

$$\xi_{15}^{A}(x^3, x^5) = \begin{cases} \text{“}+\text{”} : \Gamma^{3A}(x^3, x^5); \\ \text{“}-\text{”} : \Sigma^{3A}(x^3, x^5); \end{cases} \tag{22.153}$$

where $\Sigma^{iA}$ and $\Gamma^{iA}$ ($i = 1,2,3$) denote the new generators (with respect to the SR ones) corresponding, respectively, *to rotations and boosts involving the fifth coordinate.*

---

[4]This is due to the adopted choice of embodying the constants $a, b, c, d$ in the definitions of the translational parameters $\alpha^{\mu}$ (see (22.135)). It is easily seen that, on the contrary, the identification $\alpha^{\mu} = T^{\mu}$ leads to the generators of the deformed translation group Tr.$(1,3)_{\mathrm{DEF.}}$ (with the consequent changes in the commutator algebra).

[5]Needless to say, the identifications of the Killing vectors $\xi_{\mathcal{A}}^{A}(x)$ with the generators of the Killing symmetry algebra hold except for a sign here and in the subsequent cases.

By direct evaluation of the commutators (22.133),[6] it follows that the ensuing Killing algebra is made of two pieces: The deformed Poincaré algebra $P(1,3)_{\text{DEF.}} = \{\text{su}(2)_{\text{DEF.}} \otimes \text{su}(2)_{\text{DEF.}}\} \otimes_s \text{tr.}(1,3)_{\text{STD.}}$, generated by $\mathbf{S}_{DSR.}$, $\mathbf{K}_{\text{DSR}}$ and $\Upsilon^{\mu}$ ($\mu = 0,1,2,3$) (corresponding to $\mathcal{A} = 1,\dots,10$), expressed by (8.37)–(8.39), and the "mixed" algebra, involving also generators related to the energy dimension. This latter depends on the time-like or space-like nature of $x^5$(see (22.150)–(22.153)) and is specified by the following commutation relations ($\forall \mu = 0,1,2,3$ and $\forall i,j,k = 1,2,3$):[7]

(1) Timelike $x^5$:

$$
\text{``}+\text{''}\left\{
\begin{array}{c}
\left[\widetilde{\Upsilon}^5, \widetilde{\Upsilon}^{\mu}\right] = 0\ ; \\[2ex]
\left[\widetilde{\Upsilon}^0, \widetilde{\boldsymbol{\Sigma}}^1\right] = -\widetilde{\Upsilon}^5; \quad \left[\widetilde{\Upsilon}^0, \widetilde{\Gamma}^i\right] = 0; \\[2ex]
\left[\widetilde{\Upsilon}^i, \widetilde{\Gamma}^j\right] = -\widetilde{\Upsilon}^5 \delta^{ij}; \quad \left[\widetilde{\Upsilon}^i, \widetilde{\boldsymbol{\Sigma}}^1\right] = 0; \\[2ex]
\left[\widetilde{\mathbf{K}}^i, \widetilde{\Gamma}^j\right] = \dfrac{1}{g_{ii,\text{DR5}}} \widetilde{\boldsymbol{\Sigma}}^1 \delta^{ij}; \quad \left[\widetilde{\mathbf{K}}^i, \widetilde{\boldsymbol{\Sigma}}^1\right] = \dfrac{1}{a}\widetilde{\Gamma}^i = \dfrac{1}{g_{00,\text{DR5}}}\widetilde{\Gamma}^i; \\[2ex]
\left[\widetilde{\mathbf{K}}^i, \widetilde{\Upsilon}^5\right] = 0; \quad \left[\widetilde{\mathbf{S}}^i, \widetilde{\Upsilon}^5\right] = 0; \\[2ex]
\left[\widetilde{\mathbf{S}}^i, \widetilde{\boldsymbol{\Sigma}}^1\right] = 0; \quad \left[\widetilde{\mathbf{S}}^i, \widetilde{\Gamma}^j\right] \overset{ESC\ \text{on on k}}{=} \epsilon_{ijk} \dfrac{1}{g_{jj,\text{DR5}}} \widetilde{\Gamma}^k; \\[2ex]
\left[\widetilde{\Upsilon}^5, \widetilde{\boldsymbol{\Sigma}}^1\right] = \dfrac{1}{a}\widetilde{\Upsilon}^0 = \dfrac{1}{g_{00,\text{DR5}}}\widetilde{\Upsilon}^0; \quad \left[\widetilde{\Upsilon}^5, \widetilde{\Gamma}^i\right] = -\widetilde{\Upsilon}^i; \\[2ex]
\left[\widetilde{\boldsymbol{\Sigma}}^1, \widetilde{\Gamma}^i\right] = \widetilde{\mathbf{K}}^i; \quad \left[\widetilde{\Gamma}^i, \widetilde{\Gamma}^j\right] \overset{ESC\ on}{=} -\epsilon_{ijk}\widetilde{\mathbf{S}}^k.
\end{array}
\right.
\tag{22.154}
$$

[6]Remember that the products in the commutator (22.132) has to be meant as row×column products of matrices, so that

$$
\begin{aligned}
\left[\boldsymbol{\xi}_{\mathcal{A}}(x), \boldsymbol{\xi}_{\mathcal{B}}(x)\right] &= \left[\xi^B_{\mathcal{A}}(x)\partial_B, \xi^A_{\mathcal{B}}(x)\partial_A\right] \\
&= \left[\xi^B_{\mathcal{A}}(x)\xi^A_{\mathcal{B},B}(x) - \xi^B_{\mathcal{B}}(x)\xi^A_{\mathcal{A},B}(x)\right]\partial_A \rightarrow \\
\rightarrow \left[\boldsymbol{\xi}_{\mathcal{A}}(x), \boldsymbol{\xi}_{\mathcal{B}}(x)\right]^A &= \xi^B_{\mathcal{A}}(x)\xi^A_{\mathcal{B},B}(x) - \xi^B_{\mathcal{B}}(x)\xi^A_{\mathcal{A},B}(x).
\end{aligned}
$$

[7]In the following, for simplicity, we shall omit the DSR specification in the generator symbols.

It contains the following subalgebras:

$$\left.\begin{array}{r}\left[\widetilde{\mathbf{S}}^i,\widetilde{\mathbf{S}}^j\right] \overset{ESC\ \underline{on}\ \text{on k}}{=} -\epsilon_{ijk}\dfrac{1}{g_{kk,\mathrm{DR5}}}\widetilde{\mathbf{S}}^k;\\ \left[\widetilde{\mathbf{S}}^i,\widetilde{\Gamma}^j\right] \overset{ESC\ \underline{on}\ \text{on k}}{=} \epsilon_{ijk}\dfrac{1}{g_{jj,\mathrm{DR5}}}\widetilde{\Gamma}^k;\\ \left[\widetilde{\Gamma}^i,\widetilde{\Gamma}^j\right] \overset{ESC\ on}{=} -\epsilon_{ijk}\widetilde{\mathbf{S}}^k,\end{array}\right\} \mathrm{su}(2)_{\mathrm{DEF.}} \otimes su(2)_{\mathrm{DEF.}}; \tag{22.155}$$

$$\left.\begin{array}{r}\left[\widetilde{\Upsilon}^\nu,\widetilde{\Upsilon}^\mu\right] = 0\ ;\\ \left[\widetilde{\Upsilon}^5,\widetilde{\Upsilon}^\mu\right] = 0\ ,\end{array}\right\} \mathrm{tr.}(2,3)_{\mathrm{STD.}}. \tag{22.156}$$

(2) Space-like $x^5$:

$$\text{“}-\text{”}\left\{\begin{array}{c}\left[\widetilde{\Upsilon}^5,\widetilde{\Upsilon}^\mu\right] = 0\ ;\\ \left[\widetilde{\Upsilon}^0,\widetilde{\Gamma}^1\right] = \widetilde{\Upsilon}^5;\quad \left[\widetilde{\Upsilon}^0,\widetilde{\mathbf{\Sigma}}^i\right] = 0;\\ \left[\widetilde{\Upsilon}^i,\widetilde{\mathbf{\Sigma}}^j\right] = \widetilde{\Upsilon}^5\delta^{ij};\quad \left[\widetilde{\Upsilon}^i,\widetilde{\Gamma}^1\right] = 0;\\ \left[\widetilde{\mathbf{K}}^i,\widetilde{\mathbf{\Sigma}}^j\right] = -\dfrac{1}{g_{ii,\mathrm{DR5}}}\widetilde{\Gamma}^1\delta^{ij};\quad \left[\widetilde{\mathbf{K}}^i,\widetilde{\Gamma}^1\right] = \dfrac{1}{a}\widetilde{\mathbf{\Sigma}}^i = \dfrac{1}{g_{00,\mathrm{DR5}}}\widetilde{\mathbf{\Sigma}}^i;\\ \left[\widetilde{\mathbf{K}}^i,\widetilde{\Upsilon}^5\right] = 0;\quad \left[\widetilde{\mathbf{S}}^i,\widetilde{\Upsilon}^5\right] = 0;\\ \left[\widetilde{\mathbf{S}}^i,\widetilde{\Gamma}^1\right] = 0;\quad \left[\widetilde{\mathbf{S}}^i,\widetilde{\mathbf{\Sigma}}^j\right] \overset{ESC\ \underline{on}\ \text{on k}}{=} -\epsilon_{ijk}\dfrac{1}{g_{jj,\mathrm{DR5}}}\widetilde{\mathbf{\Sigma}}^k;\\ \left[\widetilde{\Upsilon}^5,\widetilde{\Gamma}^1\right] = \dfrac{1}{g_{00,\mathrm{DR5}}}\widetilde{\Upsilon}^0;\quad \left[\widetilde{\Upsilon}^5,\widetilde{\mathbf{\Sigma}}^i\right] = -\dfrac{1}{g_{ii,\mathrm{DR5}}}\widetilde{\Upsilon}^i;\\ \left[\widetilde{\Gamma}^1,\widetilde{\mathbf{\Sigma}}^i\right] = -\widetilde{\mathbf{K}}^i;\quad \left[\widetilde{\mathbf{\Sigma}}^i,\widetilde{\mathbf{\Sigma}}^j\right] \overset{ESC\ on}{=} -\epsilon_{ijk}\widetilde{\mathbf{S}}^k\end{array}\right. \tag{22.157}$$

with the subalgebras

$$\left.\begin{array}{r}\left[\widetilde{\mathbf{S}}^i,\widetilde{\mathbf{S}}^j\right] \overset{\mathrm{ESC}\ \underline{\mathrm{on}}\ \mathrm{on}\ k}{=} -\epsilon_{ijk}\dfrac{1}{g_{kk,\mathrm{DR5}}}\widetilde{\mathbf{S}}^k;\\ \left[\widetilde{\mathbf{S}}^i,\widetilde{\mathbf{\Sigma}}^j\right] \overset{\mathrm{ESC}\ \underline{\mathrm{on}}\ \mathrm{on}\ k}{=} -\epsilon_{ijk}\dfrac{1}{g_{jj,\mathrm{DR5}}}\widetilde{\mathbf{\Sigma}}^k;\\ \left[\widetilde{\mathbf{\Sigma}}^i,\widetilde{\mathbf{\Sigma}}^j\right] \overset{\mathrm{ESC\ on}}{=} -\epsilon_{ijk}\widetilde{\mathbf{S}}^k,\end{array}\right\} su(2)_{\mathrm{DEF.}} \otimes \mathrm{su}(2)_{\mathrm{DEF.}}; \tag{22.158}$$

$$\left.\begin{array}{r}\left[\widetilde{\Upsilon}^\nu,\widetilde{\Upsilon}^\mu\right] = 0\ ;\\ \left[\widetilde{\Upsilon}^5,\widetilde{\Upsilon}^\mu\right] = 0\ ,\end{array}\right\} \mathrm{tr.}(1,4)_{\mathrm{STD.}}. \tag{22.159}$$

Since in this case there are 15 independent Killing vectors, the corresponding Riemann spaces, for the e.m. and weak interactions above threshold and for the strong one below threshold, are maximally symmetric and have therefore constant curvature (zero for the e.m. and weak interactions, as it is straightforward to check directly by means of (20.12)).

### 22.4.2 Strong Metric for Violated $\Upsilon$-Hypothesis

Let us consider the case of the strong metric above threshold ($x^5 \geq x^5_{0,\text{strong}}$) when the hypothesis $\Upsilon$ of functional independence is violated $\forall \mu = 0, 1, 2, 3$ (see Sect. 22.2.3). The metric is given by (22.87) and for $c = 0$ takes the form:

$$g_{AB,\text{DR5}}(x^5)$$
$$= \text{diag}\left(\left(\frac{x^5}{x^5_0}\right)^2, -\frac{2}{25}, -\frac{4}{25}, -\left(\frac{x^5}{x^5_0}\right)^2, \pm\frac{1}{\gamma\left(\frac{x^5}{x^5_0}\right)^2}\right). \tag{22.160}$$

The contravariant Killing vector $\xi^A(x, x^5)$ is given by (22.90)–(22.93), which read now:

$$\begin{cases} \xi^0 = 0; \\ \xi^1(x^2) = -\dfrac{25}{2}d_2x^2 - \dfrac{25}{2}T_1 = \alpha^3x^2 + \alpha^1; \\ \xi^2(x^1) = \dfrac{25}{4}d_2x^1 - \dfrac{25}{4}T_2 = \dfrac{1}{2}\alpha^3x^1 + \alpha^2; \\ \xi^3 = 0; \\ \xi^5\left(x^5\right) = \pm\dfrac{\gamma\alpha}{(x^5_0)^2}x^5 = \pm\alpha^4x^5, \end{cases} \tag{22.161}$$

where the dependence on the transformation parameters $\alpha^{\mathcal{A}}$ ($\mathcal{A}$=1,2,3,4) has been made explicit.

The contravariant Killing vectors $\xi^A_{\mathcal{A}}(x)$ are therefore

$$\xi^A_1(x) = \begin{pmatrix} \xi^\mu_{2,\text{SR}}(x) \\ \\ 0 \end{pmatrix} = \begin{pmatrix} 0 \\ 1 \\ 0 \\ 0 \\ 0 \end{pmatrix}; \tag{22.162}$$

$$\xi^A_2(x) = \begin{pmatrix} \xi^\mu_{3,\text{SR}}(x) \\ \\ 0 \end{pmatrix} = \begin{pmatrix} 0 \\ 0 \\ 1 \\ 0 \\ 0 \end{pmatrix}; \tag{22.163}$$

$$\xi_3^A(x^1,x^2) = \begin{pmatrix} 0 \\ x^2 \\ -\dfrac{x^1}{2} \\ 0 \\ 0 \end{pmatrix}; \tag{22.164}$$

$$\xi_{4,\pm}^A(x^5) = \begin{pmatrix} 0 \\ 0 \\ 0 \\ 0 \\ \pm x^5 \end{pmatrix}. \tag{22.165}$$

It is possible to do the following identifications:

$$\xi_1^A = \begin{pmatrix} \Upsilon_{\mathrm{SR}}^{1\mu} \\ 0 \end{pmatrix} = \Upsilon^{1A}, \quad \alpha^1 = \mp T^{1\prime}; \tag{22.166}$$

$$\xi_2^A = \begin{pmatrix} \Upsilon_{\mathrm{SR}}^{2\mu} \\ 0 \end{pmatrix} = \Upsilon^{2A}, \quad \alpha^2 = \mp T^{2\prime}; \tag{22.167}$$

$$\xi_3^A(x^1,x^2) = \begin{pmatrix} S^{3\mu}(x)|_{x^2 \longrightarrow \sqrt{2}x^2} \\ 0 \end{pmatrix} = S^{3A}, \quad \alpha^3 = \mp \Theta^3; \tag{22.168}$$

$$\xi_{4,\pm}^A(x^5) = R(x^5). \tag{22.169}$$

In (22.168), the notation for $S^3$ has to be interpreted in the sense of Sect. 22.2.3, namely it is the generator of rotations in the 2D manifold $\Pi_2 = (x^1, x^2)$ with metric rescaling $x^2 \longrightarrow \sqrt{2}x^2$. One therefore gets the following Killing algebra

$$\begin{cases} \left[\widetilde{\Upsilon}^1, \widetilde{\Upsilon}^2\right] = 0; \\ \left[\widetilde{\Upsilon}^1, \widetilde{\mathbf{S}}^3\right] = -\dfrac{1}{2}\widetilde{\Upsilon}^2; \\ \left[\widetilde{\Upsilon}^2, \widetilde{\mathbf{S}}^3\right] = \widetilde{\Upsilon}^1; \\ \left[\widetilde{\Upsilon}^i, \widetilde{\mathbf{R}}\right] = 0, \quad i = 1,2; \\ \left[\widetilde{\mathbf{S}}^3, \widetilde{\mathbf{R}}\right] = 0. \end{cases} \tag{22.170}$$

### *22.4.3 Power Ansatz Metrics with Violated $\Upsilon$-Hypothesis*

We shall now discuss the infinitesimal algebraic structure of DR5 for the metrics in the Power Ansatz when the hypothesis $\Upsilon$ of functional independence is not satisfied by any $\mu = 0, 1, 2, 3$. According to Appendix A, this occurs in five cases (only three of which are independent). We will consider only the two which correspond to physical metrics.

**Case 1**

This corresponds to the VI class of solutions, characterized by the exponent set $\widetilde{\mathbf{q}}_{\mathrm{VI}} = (p, 0, 0, 0, p-2)$. The 5D metric is given by (A.75). The contravariant Killing five-vector $\xi^A(x, x^5)$ depends on the time-like or space-like nature of the fifth coordinate, and writes (cf. (A.79)–(A.83))

$$\xi^0(x^0, x^5; p)$$

$$= \begin{cases} \text{“}+\text{”}: \\ \left(x_0^5\right)^{p-1}\left[A\cos\left(\dfrac{p}{2}\dfrac{x^0}{x_0^5}\right) - B\sin\left(\dfrac{p}{2}\dfrac{x^0}{x_0^5}\right)\right](x^5)^{-p/2} + \alpha(x_0^5)^p; \\ \text{“}-\text{”}: \\ \left(x_0^5\right)^{p-1}\left[C\cosh\left(\dfrac{p}{2}\dfrac{x^0}{x_0^5}\right) + D\sinh\left(\dfrac{p}{2}\dfrac{x^0}{x_0^5}\right)\right](x^5)^{-p/2} + \alpha(x_0^5)^p; \end{cases}$$

$$\xi^1(x^2, x^3) = -\Theta^3 x^2 - \Theta^2 x^3 + T^1;$$

$$\xi^2(x^1, x^3) = \Theta^3 x^1 - \Theta^1 x^3 + T^2;$$

$$\xi^3(x^1, x^2) = \Theta^2 x^1 + \Theta^1 x^2 + T^3;$$

$$\xi^5(x^0, x^5, p)$$

$$= \begin{cases} \text{“}+\text{”}: \\ \left(x_0^5\right)^{p-2}\left[A\sin\left(\dfrac{p}{2}\dfrac{x^0}{x_0^5}\right) + B\cos\left(\dfrac{p}{2}\dfrac{x^0}{x_0^5}\right)\right](x^5)^{-((p/2)-1)}; \\ \text{“}-\text{”}: \\ -\left(x_0^5\right)^{p-2}\left[C\sinh\left(\dfrac{p}{2}\dfrac{x^0}{x_0^5}\right) + D\cosh\left(\dfrac{p}{2}\dfrac{x^0}{x_0^5}\right)\right](x^5)^{-((p/2)-1)}. \end{cases} \tag{22.171}$$

By the identifications

$$-\Theta^3 = \alpha^5; \quad \Theta^2 = \alpha^6; \quad \Theta^1 = -\alpha^7;$$

$$\alpha(x_0^5)^p = T^0(\alpha, x_0^5, p) = \alpha^1(\alpha, x_0^5, p); T^i = \alpha^{i+1}, i = 1, 2, 3;$$

$$\text{“}+\text{”}\begin{cases} A\left(x_0^5\right)^{(p/2)-2} = \alpha^8 \\ B\left(x_0^5\right)^{(p/2)-2} = \alpha^9 \end{cases}; \quad \text{“}-\text{”}\begin{cases} C\left(x_0^5\right)^{(p/2)-2} = \alpha^8 \\ D\left(x_0^5\right)^{(p/2)-2} = \alpha^9 \end{cases}, \tag{22.172}$$

the Killing vector can be written as[8]

$$\xi^0(x^0, x^5; p)$$
$$= \alpha^1 + \begin{cases} \text{“}+\text{”} : \left(x_0^5\right)^{(p/2)+1} \left[\alpha^8 \cos\left(\frac{p}{2}\frac{x^0}{x_0^5}\right) - \alpha^9 \sin\left(\frac{p}{2}\frac{x^0}{x_0^5}\right)\right] (x^5)^{-(p/2)}; \\ \text{“}-\text{”} : \left(x_0^5\right)^{(p/2)+1} \left[\alpha^8 \cosh\left(\frac{p}{2}\frac{x^0}{x_0^5}\right) + \alpha^9 \sinh\left(\frac{p}{2}\frac{x^0}{x_0^5}\right)\right] (x^5)^{-(p/2)}; \end{cases}$$
$$\xi^1(x^2, x^3) = \alpha^2 + \alpha^5 x^2 - \alpha^6 x^3;$$
$$\xi^2(x^1, x^3) = \alpha^3 - \alpha^5 x^1 + \alpha^7 x^3;$$
$$\xi^3(x^1, x^2) = \alpha^4 + \alpha^6 x^1 - \alpha^7 x^2;$$
$$\xi^5(x^0, x^5, p)$$
$$= \begin{cases} \text{“}+\text{”} : \left(x_0^5\right)^{p/2} \left[\alpha^8 \sin\left(\frac{p}{2}\frac{x^0}{x_0^5}\right) + \alpha^9 \cos\left(\frac{p}{2}\frac{x^0}{x_0^5}\right)\right] (x^5)^{-((p/2)-1)}; \\ \text{“}-\text{”} : -\left(x_0^5\right)^{p/2} \left[\alpha^8 \sinh\left(\frac{p}{2}\frac{x^0}{x_0^5}\right) + \alpha^9 \cosh\left(\frac{p}{2}\frac{x^0}{x_0^5}\right)\right] (x^5)^{-((p/2)-1)}. \end{cases}$$
(22.173)

One gets therefore, for the Killing vectors $\xi_{\mathcal{A}}^A(x)$ ($\mathcal{A} = 1, \ldots, 9$):

$$\xi_{\mathcal{A}}^A(x) = \begin{pmatrix} \xi_{\mathcal{A},\mathrm{SR}}^\mu(x) \\ \\ 0 \end{pmatrix}, \quad \mathcal{A} = 1, 2, 3, 4; \tag{22.174}$$

$$\xi_5^A(x^1, x^2) = \begin{pmatrix} 0 \\ x^2 \\ -x^1 \\ 0 \\ 0 \end{pmatrix} = \begin{pmatrix} \xi_{8,\mathrm{SR}}^\mu(x) \\ \\ 0 \end{pmatrix}; \tag{22.175}$$

$$\xi_6^A(x^1, x^3) = \begin{pmatrix} 0 \\ -x^3 \\ 0 \\ x^1 \\ 0 \end{pmatrix} = \begin{pmatrix} \xi_{9,\mathrm{SR}}^\mu(x) \\ \\ 0 \end{pmatrix}; \tag{22.176}$$

[8]The definitions of $\alpha^{\mathcal{A}}$ for $\mathcal{A} = 8, 9$ have been chosen so to make also these transformation parameters dimensionless.

$$\xi_7^A(x^2, x^3) = \begin{pmatrix} 0 \\ 0 \\ x^3 \\ -x^2 \\ 0 \end{pmatrix} = \begin{pmatrix} \xi_{10,\mathrm{SR}}^{\mu}(x) \\ \\ 0 \end{pmatrix}; \qquad (22.177)$$

$$\xi_{8,\pm}^A(x^0, x^5, p) = \begin{pmatrix} \left(x_0^5\right)^{(p/2)+1} \cosh\left(\dfrac{p}{2}\dfrac{x^0}{x_0^5}\right)(x^5)^{-p/2} \\ 0 \\ 0 \\ 0 \\ \pm\left(x_0^5\right)^{p/2} \sinh\left(\dfrac{p}{2}\dfrac{x^0}{x_0^5}\right)(x^5)^{-((p/2)-1)} \end{pmatrix}; \qquad (22.178)$$

$$\xi_{9,\pm}^A(x^0, x^5, p) = \begin{pmatrix} \mp\left(x_0^5\right)^{(p/2)+1} \sinh\left(\dfrac{p}{2}\dfrac{x^0}{x_0^5}\right)(x^5)^{-p/2} \\ 0 \\ 0 \\ 0 \\ \pm\left(x_0^5\right)^{p/2} \cosh\left(\dfrac{p}{2}\dfrac{x^0}{x_0^5}\right)(x^5)^{-((p/2)-1)} \end{pmatrix}. \qquad (22.179)$$

The earlier Killing vectors can be identified with generators as follows:

$$\xi_1^A = \begin{pmatrix} \Upsilon_{\mathrm{SR}}^{0\mu} \\ \\ 0 \end{pmatrix} = \Upsilon^{1A}; \qquad (22.180)$$

$$\xi_{i+1}^A = \begin{pmatrix} \Upsilon_{\mathrm{SR}}^{i\mu} \\ \\ 0 \end{pmatrix} = -\Upsilon^{iA}, \quad i = 1, 2, 3; \qquad (22.181)$$

$$\xi_{i+4}^A(x) = \begin{pmatrix} \xi_{i+7,\mathrm{SR}}^{\mu}(x) \\ \\ 0 \end{pmatrix} = \begin{pmatrix} \begin{cases} S_{\mathrm{SR}}^{3\mu}(x) \\ S_{\mathrm{SR}}^{2\mu}(x) \\ S_{\mathrm{SR}}^{1\mu}(x) \end{cases} \\ 0 \end{pmatrix} = S^{iA}, \quad i = 1, 2, 3; \qquad (22.182)$$

$$\xi_{8,\pm}^A(x^0, x^5; p) = Z_{\pm}^{1A}(x^0, x^5; p); \qquad (22.183)$$

$$\xi_{9,\pm}^A(x^0, x^5; p) = Z_{\pm}^{2A}(x^0, x^5; p). \qquad (22.184)$$

The Killing algebra is then specified by the following commutation relations:

$$\begin{cases} \left[\widetilde{\Upsilon}^{\mu},\widetilde{\Upsilon}^{\nu}\right]=0 \quad \forall\mu,\nu=0,1,2,3; \\ \left[\widetilde{\mathbf{S}}^{i}(x),\widetilde{\mathbf{S}}^{j}(x)\right] \overset{ESC\ on}{=} \epsilon_{ijk}\widetilde{\mathbf{S}}^{k} \quad \forall i,j=1,2,3; \\ \left[\widetilde{\mathbf{S}}^{i}(x),\widetilde{\Upsilon}^{0}\right]=0 \quad \forall i,j=1,2,3; \\ \left[\widetilde{\mathbf{S}}^{i}(x),\widetilde{\Upsilon}^{j}\right] \overset{ESC\ on}{=} \epsilon_{ijk}\widetilde{\Upsilon}^{k} \quad \forall i,j=1,2,3; \\ \left[\widetilde{\Upsilon}^{0},\widetilde{\mathbf{Z}}^{1}_{\pm}(x^{0},x^{5},p)\right]=\dfrac{p}{2x_0^5}\widetilde{\mathbf{Z}}^{2}_{\pm}(x^{0},x^{5},p); \\ \left[\widetilde{\Upsilon}^{0},\widetilde{\mathbf{Z}}^{2}_{\pm}(x^{0},x^{5},p)\right]=\mp\dfrac{p}{2x_0^5}\widetilde{\mathbf{Z}}^{1}_{\pm}(x^{0},x^{5},p); \\ \left[\widetilde{\Upsilon}^{i},\widetilde{\mathbf{Z}}^{1}_{\pm}(x^{0},x^{5},p)\right]=\left[\widetilde{\Upsilon}^{i},\widetilde{\mathbf{Z}}^{2}_{\pm}(x^{0},x^{5},p)\right]=0; \\ \left[\widetilde{\mathbf{S}}^{i}(x),\widetilde{\mathbf{Z}}^{m}_{\pm}(x^{0},x^{5},p)\right]=0 \quad \forall i=1,2,3,\quad \forall m=1,2. \end{cases} \tag{22.185}$$

It is easily seen that the Killing algebra for this case contains the subalgebra $su(2)_{STD.} \otimes_s tr.(1,3)_{STD.}$.

### Cases 2–4

The metrics belonging to classes II ($\widetilde{\mathbf{q}}_{II} = (0,p,0,0,p-2)$), IV ($\widetilde{\mathbf{q}}_{IV} = (0,0,0,p,p-2)$) and IX ($\widetilde{\mathbf{q}}_{IX} = (0,0,p,0,p-2)$) differ only for an exchange of spatial axes (see (A.86) for the first case). They are discussed in Sects. A.2.2–A.2.4. The Killing algebra for all these three (physically equivalent) metrics can be dealt with by an unitary mathematical approach.

Let us label the three classes II, IV, and IX by $i = 1,2,3$, respectively (according to the space axis involved). The metric coefficients are given, in compact form, by

$$g_{AB,\mathrm{DR5},i}\begin{cases} g_{00}=-g_{jj}=1, \quad j\neq i, \quad j\in\{j_1,j_2\}, j_1<j_2; \\ g_{ii}\left(x^5\right)=-\left(\dfrac{x^5}{x_0^5}\right)^{p}; g_{55}\left(x^5\right)=\pm\left(\dfrac{x^5}{x_0^5}\right)^{p-2}. \end{cases} \tag{22.186}$$

As easily seen, the three indices $i$, $j_1$, $j_2$ take the values $\{123; 213; 312\}$. The contravariant Killing vector has the form (cf. (A.88)–(A.92))

$$\xi^0(x^{j_1},x^{j_2};\alpha)=\alpha^1+\alpha^5x^{j_1}+\alpha^6x^{j_2};$$

$$\xi^i_{\pm}(x^i,x^5;\alpha)=\alpha^{i+1}(\alpha,x_0^5,p)$$

$$+\begin{cases} \text{“}+\text{”}: \left(x_0^5\right)^{(p/2)+1}\left[\alpha^8\cosh\left(\dfrac{p}{2}\dfrac{x^i}{x_0^5}\right)+\alpha^9\sinh\left(\dfrac{p}{2}\dfrac{x^i}{x_0^5}\right)\right]\left(x^5\right)^{-p/2}; \\ \text{“}-\text{”}: \left(x_0^5\right)^{(p/2)+1}\left[\alpha^8\cos\left(\dfrac{p}{2}\dfrac{x^i}{x_0^5}\right)-\alpha^9\sin\left(\dfrac{p}{2}\dfrac{x^i}{x_0^5}\right)\right]\left(x^5\right)^{-p/2}; \end{cases} \tag{22.187}$$

$$\xi^{j_1}(x^0, x^{j_2}; \alpha) = \alpha^{j_1+1} + \alpha^5 x^0 + \alpha^7 x^{j_2};$$
$$\xi^{j_2}(x^0, x^{j_1}; \alpha) = \alpha^{j_2+1} + \alpha^6 x^0 - \alpha^7 x^{j_1};$$
$$\xi^5_{\pm}(x^i, x^5; \alpha) =$$
$$= \begin{cases} \text{“}+\text{”} : (x^5_0)^{p/2}\left[\alpha^8 \sinh\left(\dfrac{p}{2}\dfrac{x^i}{x^5_0}\right) + \alpha^9 \cosh\left(\dfrac{p}{2}\dfrac{x^i}{x^5_0}\right)\right](x^5)^{-((p/2)-1)}; \\ \text{“}-\text{”} : -(x^5_0)^{p/2}\left[\alpha^8 \sin\left(\dfrac{p}{2}\dfrac{x^i}{x^5_0}\right) + \alpha^9 \cos\left(\dfrac{p}{2}\dfrac{x^i}{x^5_0}\right)\right](x^5)^{-((p/2)-1)}, \end{cases} \tag{22.188}$$
(22.189)

where

$$\text{“}+\text{”} \begin{cases} A\,(x^5_0)^{(p/2)-2} = \alpha^8; \\ B\,(x^5_0)^{(p/2)-2} = \alpha^9; \end{cases} \qquad \text{“}-\text{”} \begin{cases} C\,(x^5_0)^{(p/2)-2} = \alpha^8; \\ D\,(x^5_0)^{(p/2)-2} = \alpha^9; \end{cases}$$
$$\alpha^{i+1}(\alpha, x^5_0, p) = -\alpha(x^5_0)^p;$$
$$\alpha^7 = \begin{cases} \mp\Theta, & i = 1; \\ \pm\Theta, & i = 2; \\ \mp\Theta, & i = 3. \end{cases} \tag{22.190}$$

The parameter $\alpha^7$ corresponds to a true infinitesimal rotation in the 2D Euclidean plane $\Pi_{(x^{j_1}, x^{j_2})}$.

Therefore the explicit form of the Killing vectors $\xi^A_{\mathcal{A}}(x)$ ($\mathcal{A} = 1, \ldots, 9$) is

$$\xi^A_{\mathcal{A}}(x) = \begin{pmatrix} \xi^\mu_{\mathcal{A},\mathrm{SR}}(x) \\ 0 \end{pmatrix}, \quad \mathcal{A} = 1, i+1, j_1+1, j_2+1; \tag{22.191}$$

$$\xi^A_5(x^0, x^{j_1}) = \begin{pmatrix} x^{j_1} \\ x^0 \\ (j_1\text{th row}) \\ 0 \end{pmatrix} = \begin{cases} i = 1 : \begin{pmatrix} \xi^\mu_{6,\mathrm{SR}}(x^0, x^2) \\ 0 \end{pmatrix}; \\ i = 2 : \begin{pmatrix} \xi^\mu_{5,\mathrm{SR}}(x^0, x^1) \\ 0 \end{pmatrix}; \end{cases} \tag{22.192}$$

$$\xi^A_6(x^0, x^{j_2}) = \begin{pmatrix} x^{j_2} \\ x^0 \\ (j_2\text{th row}) \\ 0 \end{pmatrix} = \begin{cases} i = 1 : \begin{pmatrix} \xi^\mu_{7,\mathrm{SR}}(x^0, x^3) \\ 0 \end{pmatrix}; \\ i = 2 : \begin{pmatrix} \xi^\mu_{7,\mathrm{SR}}(x^0, x^3) \\ 0 \end{pmatrix}; \\ i = 3 : \begin{pmatrix} \xi^\mu_{6,\mathrm{SR}}(x^0, x^2) \\ 0 \end{pmatrix}; \end{cases} \tag{22.193}$$

$$\xi_7^A(x^{j_1}, x^{j_2}) = \begin{pmatrix} 0 \\ x^{j_2} \\ (j_2\text{th row}) \\ 0 \\ x^{j_1} \\ (j_1\text{th row}) \\ 0 \end{pmatrix} = \begin{cases} i=1: \begin{pmatrix} \xi^{\mu}_{10,\mathrm{SR}}(x^2, x^3) \\ 0 \end{pmatrix}; \\ i=2: \begin{pmatrix} -\xi^{\mu}_{9,\mathrm{SR}}(x^1, x^3) \\ 0 \end{pmatrix}; \\ i=3: \begin{pmatrix} -\xi^{\mu}_{8,\mathrm{SR}}(x^1, x^2) \\ 0 \end{pmatrix}; \end{cases} \tag{22.194}$$

$$\xi^A_{8,\pm}(x^i, x^5; p) = \begin{pmatrix} 0 \\ \begin{cases} \text{“}+\text{”}: \left(x_0^5\right)^{(p/2)+1} \cosh\left(\frac{p}{2}\frac{x^i}{x_0^5}\right)(x^5)^{-p/2} \\ \text{“}-\text{”}: \left(x_0^5\right)^{(p/2)+1} \cos\left(\frac{p}{2}\frac{x^i}{x_0^5}\right)(x^5)^{-p/2} \end{cases} \\ (i\text{th row}) \\ 0 \\ 0 \\ \begin{cases} \left(x_0^5\right)^{p/2} \sinh\left(\frac{p}{2}\frac{x^i}{x_0^5}\right)(x^5)^{-((p/2)-1)} \\ \left(x_0^5\right)^{p/2} \sin\left(\frac{p}{2}\frac{x^i}{x_0^5}\right)(x^5)^{-((p/2)-1)} \end{cases} \end{pmatrix}; \tag{22.195}$$

$$\xi^A_{9,\pm}(x^i, x^5 : p) = \begin{pmatrix} 0 \\ \begin{cases} \text{“}+\text{”}: \left(x_0^5\right)^{(p/2)+1} \sinh\left(\frac{p}{2}\frac{x^i}{x_0^5}\right)(x^5)^{-p/2} \\ \text{“}-\text{”}: -\left(x_0^5\right)^{(p/2)+1} \sin\left(\frac{p}{2}\frac{x^i}{x_0^5}\right)(x^5)^{-(p/2)} \end{cases} \\ (i\text{th row}) \\ 0 \\ 0 \\ \begin{cases} -\left(x_0^5\right)^{p/2} \cosh\left(\frac{p}{2}\frac{x^i}{x_0^5}\right)(x^5)^{-((p/2)-1)} \\ -\left(x_0^5\right)^{p/2} \cos\left(\frac{p}{2}\frac{x^i}{x_0^5}\right)(x^5)^{-((p/2)-1)} \end{cases} \end{pmatrix}. \tag{22.196}$$

The nine Killing vectors can be identified with the generators of the related algebra as follows:

$$\xi_1^A = \begin{pmatrix} \Upsilon^{0\mu}_{\mathrm{SR}} \\ \\ 0 \end{pmatrix} = \Upsilon^{1A}, \quad \alpha^1 = T^0; \tag{22.197}$$

$$\xi^A_{\mathcal{A}} = \begin{pmatrix} \Upsilon^{\mathcal{A}\mu}_{\mathrm{SR}} \\ \\ 0 \end{pmatrix} = \Upsilon^{\mathcal{A}A}, \quad \mathcal{A} = i, j_1, j_2, \begin{cases} \alpha^{i+1}(\alpha, x^5_0, p) = -T^i(\alpha, x^5_0, p), \\ \alpha^{j_1+1}(\alpha, x^5_0, p) = -T^{j_1}(\alpha, x^5_0, p), \\ \alpha^{j_2+1}(\alpha, x^5_0, p) = -T^{j_2}(\alpha, x^5_0, p); \end{cases} \tag{22.198}$$

$$\xi^A_5(x) = \begin{pmatrix} \begin{cases} i=1 : K^{2\mu}_{\mathrm{SR}}(x) \\ i=2,3 : K^{1\mu}_{\mathrm{SR}}(x) \end{cases} \\ \\ 0 \end{pmatrix}, \quad \alpha^5 = \begin{cases} \rho^2, & i=1, \\ \rho^1, & i=2,3; \end{cases} \tag{22.199}$$

$$\xi^A_6(x) = \begin{pmatrix} \begin{cases} i=1,2 : K^{3\mu}_{\mathrm{SR}}(x) \\ i=3 : K^{2\mu}_{\mathrm{SR}}(x) \end{cases} \\ \\ 0 \end{pmatrix}, \quad \alpha^6 = \begin{cases} \rho^3, & i=1,2, \\ \rho^2, & i=3; \end{cases} \tag{22.200}$$

$$\xi^A_7(x) = \begin{pmatrix} \begin{cases} i=1 : S^{1\mu}_{\mathrm{SR}}(x) \\ i=2 : -S^{2\mu}_{\mathrm{SR}}(x) \\ i=3 : S^{3\mu}_{\mathrm{SR}}(x) \end{cases} \\ 0 \end{pmatrix}, \quad \alpha^7 = \begin{cases} -\Theta^1, & i=1, \\ \Theta^2, & i=2, \\ -\Theta^3, & i=3; \end{cases} \tag{22.201}$$

$$\xi^A_{8,\mp}(x^i, x^5; p) = Z^{1A}_{\mp}(x^i, x^5; p); \tag{22.202}$$

$$\xi^A_{9,\mp}(x^i, x^5; p) = Z^{2A}_{\mp}(x^i, x^5; p). \tag{22.203}$$

The commutation relations of the Killing algebra are therefore

$$\left[\widetilde{\Upsilon}^\mu, \widetilde{\Upsilon}^\nu\right] = 0 \quad \forall \mu, \nu = 0,1,2,3;$$

$$\begin{cases} i=1 : \left[\widetilde{\mathbf{K}}^2(x), \widetilde{\mathbf{K}}^3(x)\right] = -\widetilde{\mathbf{S}}^1(x), \\ i=2 : \left[\widetilde{\mathbf{K}}^1(x), \widetilde{\mathbf{K}}^3(x)\right] = \widetilde{\mathbf{S}}^2(x), \\ i=3 : \left[\widetilde{\mathbf{K}}^1(x), \widetilde{\mathbf{K}}^2(x)\right] = -\widetilde{\mathbf{S}}^3(x); \end{cases}$$

$$\left[\widetilde{\mathbf{Z}}^1_{\mp}(x^i, x^5; p), \widetilde{\mathbf{Z}}^2_{\mp}(x^0, x^5; p)\right] = 0;$$

$$\left[\widetilde{\Upsilon}^0, \begin{cases} i=1 : \begin{cases} \widetilde{\mathbf{K}}^2(x) \\ \widetilde{\mathbf{K}}^3(x) \end{cases} \\ i=2 : \begin{cases} \widetilde{\mathbf{K}}^1(x) \\ \widetilde{\mathbf{K}}^3(x) \end{cases} \\ i=3 : \begin{cases} \widetilde{\mathbf{K}}^1(x) \\ \widetilde{\mathbf{K}}^2(x) \end{cases} \end{cases}\right] = \begin{cases} \widetilde{\Upsilon}^{j_1}, \\ \widetilde{\Upsilon}^{j_2}; \end{cases}$$

$$\left[\widetilde{\Upsilon}^0, \begin{cases} i=1: \widetilde{\mathbf{S}}^1(x) \\ i=2: -\widetilde{\mathbf{S}}^2(x) \\ i=3: \widetilde{\mathbf{S}}^3(x) \end{cases}\right] = 0;$$

$$\left[\widetilde{\Upsilon}^i, \begin{cases} i=1: \begin{cases} \widetilde{\mathbf{K}}^2(x) \\ \widetilde{\mathbf{K}}^3(x) \\ \widetilde{\mathbf{S}}^1(x) \end{cases} \\ i=2: \begin{cases} \widetilde{\mathbf{K}}^1(x) \\ \widetilde{\mathbf{K}}^3(x) \\ -\widetilde{\mathbf{S}}^2(x) \end{cases} \\ i=3: \begin{cases} \widetilde{\mathbf{K}}^1(x) \\ \widetilde{\mathbf{K}}^2(x) \\ \widetilde{\mathbf{S}}^3(x) \end{cases} \end{cases}\right] = 0;$$

$$\left[\widetilde{\Upsilon}^{j_1}, \begin{cases} i=1: \widetilde{\mathbf{K}}^2(x) \\ i=2,3: \widetilde{\mathbf{K}}^1(x) \end{cases}\right] = \widetilde{\Upsilon}^0;$$

$$\left[\widetilde{\Upsilon}^{j_1}, \begin{cases} i=1,2: \widetilde{\mathbf{K}}^3(x) \\ i=3: \widetilde{\mathbf{K}}^2(x) \end{cases}\right] = 0;$$

$$\left[\widetilde{\Upsilon}^{j_1}, \begin{cases} i=1: \widetilde{\mathbf{S}}^1(x) \\ i=2: -\widetilde{\mathbf{S}}^2(x) \\ i=3: \widetilde{\mathbf{S}}^3(x) \end{cases}\right] = -\widetilde{\Upsilon}^{j_2};$$

$$\left[\widetilde{\Upsilon}^{j_2}, \begin{cases} i=1: \widetilde{\mathbf{K}}^2(x) \\ i=2,3: \widetilde{\mathbf{K}}^1(x) \end{cases}\right] = 0;$$

$$\left[\widetilde{\Upsilon}^{j_2}, \begin{cases} i=1,2: \widetilde{\mathbf{K}}^3(x) \\ i=3: \widetilde{\mathbf{K}}^2(x) \end{cases}\right] = \widetilde{\Upsilon}^0;$$

$$\left[\widetilde{\Upsilon}^{j_2}, \begin{cases} i=1: \widetilde{\mathbf{S}}^1(x) \\ i=2: -\widetilde{\mathbf{S}}^2(x) \\ i=3: \widetilde{\mathbf{S}}^3(x) \end{cases}\right] = \widetilde{\Upsilon}^{j_1};$$

$$\begin{cases} i=1: \left[\widetilde{\mathbf{K}}^2(x), \widetilde{\mathbf{S}}^1(x)\right] = -\widetilde{\mathbf{K}}^3(x), \\ i=2: \left[\widetilde{\mathbf{K}}^1(x), -\widetilde{\mathbf{S}}^2(x)\right] = -\widetilde{\mathbf{K}}^3(x), \\ i=3: \left[\widetilde{\mathbf{K}}^1(x), \widetilde{\mathbf{S}}^3(x)\right] = -\widetilde{\mathbf{K}}^2(x); \end{cases}$$

$$\begin{cases} i=1: \left[\widetilde{\mathbf{K}}^3(x), \widetilde{\mathbf{S}}^1(x)\right] = \widetilde{\mathbf{K}}^2(x), \\ i=2: \left[\widetilde{\mathbf{K}}^3(x), -\widetilde{\mathbf{S}}^2(x)\right] = \widetilde{\mathbf{K}}^1(x), \\ i=3: \left[\widetilde{\mathbf{K}}^2(x), \widetilde{\mathbf{S}}^3(x)\right] = \widetilde{\mathbf{K}}^1(x); \end{cases}$$

$$\left[\widetilde{\Upsilon}^0, \widetilde{\mathbf{Z}}^1_{\mp}(x;p)\right] = \left[\widetilde{\Upsilon}^0, \widetilde{\mathbf{Z}}^2_{\mp}(x;p)\right] = 0;$$

$$\left[\widetilde{\Upsilon}^i, \widetilde{\mathbf{Z}}^1_{\mp}(x;p)\right] = \frac{p}{2x_0^5}\widetilde{\mathbf{Z}}^2_{\mp}(x;p);$$

$$\left[\widetilde{\Upsilon}^i, \widetilde{\mathbf{Z}}^2_{\mp}(x;p)\right] = \mp\frac{p}{2x_0^5}\widetilde{\mathbf{Z}}^1_{\mp}(x;p);$$

$$\left[\widetilde{\Upsilon}^{j_1}, \widetilde{\mathbf{Z}}^1_{\mp}(x;p)\right] = \left[\widetilde{\Upsilon}^{j_1}, \widetilde{\mathbf{Z}}^2_{\mp}(x;p)\right] = 0;$$

$$\left[\widetilde{\Upsilon}^{j_2}, \widetilde{\mathbf{Z}}^1_{\mp}(x;p)\right] = \left[\widetilde{\Upsilon}^{j_2}, \widetilde{\mathbf{Z}}^2_{\mp}(x;p)\right] = 0;$$

$$\left[\widetilde{\mathbf{Z}}^{1,2}_{\mp}(x;p), \begin{cases} i=1: \begin{cases} \widetilde{\mathbf{K}}^2(x) \\ \widetilde{\mathbf{K}}^3(x) \\ \widetilde{\mathbf{S}}^1(x) \end{cases} \\ i=2: \begin{cases} \widetilde{\mathbf{K}}^1(x) \\ \widetilde{\mathbf{K}}^3(x) \\ -\widetilde{\mathbf{S}}^2(x) \end{cases} \\ i=3: \begin{cases} \widetilde{\mathbf{K}}^1(x) \\ \widetilde{\mathbf{K}}^2(x) \\ \widetilde{\mathbf{S}}^3(x) \end{cases} \end{cases} \right] = 0. \tag{22.204}$$

It is easy to see that in all three cases $i = 1, 2, 3$ the Killing algebra contains the subalgebra so$(1,2) \otimes_s$ tr.$(1,3)$, generated by

$$\underbrace{\widetilde{\Upsilon}^\mu}_{\text{tr.}(1,3)}; \begin{cases} \underbrace{\begin{matrix} i=1: \widetilde{\mathbf{K}}^2(x); \widetilde{\mathbf{K}}^3(x); \widetilde{\mathbf{S}}^1(x); \\ i=2: \widetilde{\mathbf{K}}^1(x); \widetilde{\mathbf{K}}^3(x); -\widetilde{\mathbf{S}}^2(x); \\ i=3: \widetilde{\mathbf{K}}^1(x); \widetilde{\mathbf{K}}^2(x); \widetilde{\mathbf{S}}^3(x). \end{matrix}}_{\text{so}(1,2)} \end{cases} \tag{22.205}$$

Some remarks are in order. For all metrics discussed in this section, it was possible, in general, neither to identify the global algebra obeyed by the Killing generators, nor even ascertain its Lie nature. Such an issue deserves further investigations. However, it is clearly seen from the explicit forms of the generators and of the commutation relations that, even in cases (like that corresponding to metric (22.1)) in which the space–time sector is Minkowskian, the presence of the fifth dimension implies transformations involving the energy coordinate (see definitions (22.150)–(22.153) of the generators $\widetilde{\mathbf{\Sigma}}^i$, $\widetilde{\Gamma}^j$), and therefore entirely new physical symmetries. Moreover, it is easily seen from (22.126), expressing the general infinitesimal form of a metric automorphism in $\Re_5$, and from the explicit form of the

Killing vectors, that the isometric transformations are in general *nonlinear* (in particular in $x^5$). Then, the preliminary results obtained seemingly show that the isometries of $\Re_5$ related to the derived Killing algebras require *an invariance of physical laws under nonlinear coordinate transformations, in which energy is directly involved.*

## 22.5 Features of Killing Isometries in $\Re_5$

The examples of Killing symmetries we just discussed show some peculiar features of $\Re_5$ isometries corresponding to the 5D phenomenological metrics of the four fundamental interactions. Indeed, due to the piecewise nature of such metrics, the respective symmetries are strongly related to the energy range considered. This is at variance with the DSR case, in which the (deformed) isometries of $\widetilde{M}$ are *independent* of the energy parameter $x^5$. This is why we never speak of a symmetry as related to a given interaction in the DSR context. On the contrary, in DR5 the metric nature of $x^5$, and the consequent piecewise structure of the phenomenological metrics, is fundamental in determining the $\Re_5$ isometries. As a matter of fact, for a given interaction, in general *different* Killing symmetries are obtained in the two energy ranges below and above threshold. This is reflected in the discontinuous behavior of the $\Re_5$ Killing vectors at the energy threshold $x^5_{0,\text{int.}}$, namely one has in general

$$\lim_{x^5 \longrightarrow x^{5+}_{0,\text{int.}}} \xi^A_{\text{DR5,int.}}(x^5) \neq \lim_{x^5 \longrightarrow x^{5-}_{0,\text{int.}}} \xi^A_{\text{DR5,int.}}(x^5). \quad (22.206)$$

Conversely, at metric level, there is a continuity in the 5D metric tensor at the energy threshold (as clearly seen by their expressions in terms of the Heaviside function: see Sect. 19.3)

$$\begin{aligned}\lim_{x^5 \longrightarrow x^{5+}_{0,\text{int.}}} g_{AB,\text{DR5,int.}}(x^5) &= \lim_{x^5 \longrightarrow x^{5-}_{0,\text{int.}}} g_{AB,\text{DR5,int.}}(x^5) \\ &= g_{AB,\text{DR5,int.}}(x^5_{0,\text{int.}}). \end{aligned} \quad (22.207)$$

This implies that symmetries present in an energy range in which the space–time sector is standard Minkowskian – or at least its metric coefficients are constant – may no longer hold when (in a different energy range) the space–time of $\Re_5$ becomes Minkowskian deformed, and vice versa.

As already stressed, this is essentially due to the change of nature (from parameter to coordinate) of the energy $x^5$ in the passage DSR→DR5, i.e., in the geometrical embedding of $\widetilde{M}$ in $\Re_5$. In this process, at the metric level, the slicing property (19.10) holds, namely the sections of $\Re_5$ at constant energy $x^5 = \overline{x^5}$ ($dx^5 = 0$) do possess the same metric structure of $\widetilde{M}\left(\overline{x^5}\right)$.

*This is no longer true at the level of the Killing symmetries.* We can write, symbolically:

$$\begin{aligned} &\text{Isometries of } \Re_5|_{dx^5=0 \Leftrightarrow x^5=\overline{x^5}} \\ &\neq \text{Isometries of } \widetilde{M}\left(x^5 = \overline{x^5}\right) = \text{Deformed Poincaré group } P(1,3)^{10}_{\text{DEF.}}. \end{aligned} \tag{22.208}$$

In fact, in increasing the dimension number by taking energy as the fifth coordinate, some of the 10 symmetry degrees of freedom of the maximal Killing group of DSR (i.e., the deformed Poincaré group) are lost.

The Killing isometries are therefore strictly related to the geometrical context considered. This is easily seen on account of the fact that the slicing process is carried out in a genuine Riemannian geometric framework, in which the effect of the fifth coordinate is perceptible even at the 4D level of space–time sections. From the point of view of the algebraic structure, this is reflected by the arising of new generators, associated to true or pseudo rotations involving both the space–time coordinates and the energy one. This situation exactly reminds that occurring in Special Relativity, where the presence of time as a genuine coordinate – no longer a parameter as in classical physics – together with the ensuing geometrical structure of the Minkowski space, does affect the physics in the ordinary, Euclidean 3-space (the "shadow" of the pseudoeuclidean metric of $M$). This is a further evidence that *the embedding of $\widetilde{M}$ in $\Re_5$ is not a mere formal artifact, but has deep physical motivations and implications.*

We shall see in the next Part that similar considerations apply to dynamics, too.

# Part V

# DEFORMED SPACE–TIME IN FIVE DIMENSIONS: DYNAMICS

# 23
# Dynamics in DR5

It is well known that, in a Riemann space, the dynamic laws are actually geometrical laws. As familiar from General Relativity, the motion of a body in a gravitational field is described by the geodesic equations, which are in turn related to the affine and metric properties of the Riemann space–time. Analogously, we expect that, in the framework of DR5, the *local* dynamics of particles ruled by the four fundamental interactions – described by the 5D embeddings of the DSR phenomenological metrics: see Sect. 19.3 – is embodied in the 5D geodesic equations in $\Re_5$.

This has to be compared with the case of DSR, where – as in any special-relativistic theory – the geodesics equations are trivially given by

$$\frac{\mathrm{d}^2 x^\mu(\tau_{\mathrm{DSR}})}{\mathrm{d}\tau_{\mathrm{DSR}}^2} = 0, \tag{23.1}$$

(with $\tau_{\mathrm{DSR}}$ being the proper time in DSR: see Sect. 3.4.1) and yield solutions which are straight word-lines:

$$x^\mu(\tau_{\mathrm{DSR}}) = \alpha_{\mu 1}\tau_{\mathrm{DSR}} + \alpha_{\mu 2} \tag{23.2}$$

($\mu = 0, 1, 2, 3$). Actually, as it was seen in Part II, *the nontrivial dynamics of DSR is related to the intrinsic geometrical structure of the deformed Minkowski space* $\widetilde{M}$ *as Generalized Lagrange space* (see Sect. 9.2).

On the contrary, we shall see that the geodesic equations in $\Re_5$ do possess a nontrivial structure, corresponding to an *extrinsic dynamic behavior* which complements the intrinsic one related to the embedded space–time $\widetilde{M} \subset \Re_5$.

## 23.1 Proper Time in DR5

The proper time in DR5 can be found by a procedure analogous to that followed in DSR. We have

$$\left(\mathrm{d}\tau_{\mathrm{DR5}}(x^5)\right)^2 = \frac{1}{c^2}\left(\mathrm{d}s_{\mathrm{DR5}}(x^5)\right)^2 = \frac{1}{c^2}\left(\mathrm{d}S(x^5)\right)^2$$
$$= \frac{1}{c^2}\left[b_0^2(x^5)c^2\mathrm{d}t^2 - \sum_i b_i^2(x^5)\left(\mathrm{d}x^i\right)^2 \pm f(x^5)\left(\mathrm{d}x^5\right)^2\right]. \quad (23.3)$$

As in DSR, the natural frame is the frame where the particle is at rest with fixed energy, namely[1]

$$\left(x^0, x^1, x^2, x^3, x^5\right)_{\mathrm{DR5,nat}} = \left(x^0, \overline{x^1}, \overline{x^2}, \overline{x^3}, \overline{x^5}\right). \quad (23.4)$$

In other words, due to the slicing structure of $\Re_5$ (cf. (19.10)), the natural frame in DR5 is the 5D, *local* (since the fifth metric coordinate is fixed) generalization of the 4D, *global* inertial (topical) natural frame of DSR.

Then

$$\left(\mathrm{d}\tau_{\mathrm{DR5}}(\overline{x^5})\right)^2 = \frac{1}{c^2}\left(\mathrm{d}S(\overline{x^5})\right)^2\Big|_{nat} = b_0^2(\overline{x^5})\mathrm{d}t^2$$
$$\Longrightarrow \mathrm{d}\tau_{\mathrm{DR5}} = b_0\mathrm{d}t, \quad (23.5)$$

or, in finite form:

$$\tau_{\mathrm{DR5}}(\overline{x^5}) = b_0(\overline{x^5})t \quad (23.6)$$

(where $t$ is the coordinate time in the natural frame (23.4)). One gets therefore the same result of DSR (see Sect. 3.4.1), as expected on physical and mathematical grounds, on account of the embedding $\widetilde{M}(x^5) \subset \Re_5$.

## 23.2 Geodesics Equations in $\Re_5$

Let us therefore consider the geodesics in the 5D space–time–energy Riemannian space $\Re_5$, in order to clarify their possible physical meaning (see [137] for a thorough discussion of the geodesic equation of motion in a general Kaluza–Klein model).

The geodesic equations are

$$\frac{\mathrm{d}^2x^A}{\mathrm{d}\tau^2} + \Gamma^A_{BC}\frac{\mathrm{d}x^B}{\mathrm{d}\tau}\frac{\mathrm{d}x^C}{\mathrm{d}\tau} = 0, \quad (23.7)$$

[1]However, let us notice that, in the DR5 framework, the natural frame is in general noninertial, due to the Riemannian structure of $\Re_5$.

where, for massive particles, $\tau = \tau_{\mathrm{DR5}}$ is the proper time in $\Re_5$ (or another affine parameter – not necessarily invariant – for massless particles).[2] The compatibility condition of the definition (23.6) of $\tau_{\mathrm{DR5}}$ with the geodesic equations (23.7) in $\Re_5$ is straightforward (the components of the connection $\Gamma^A_{BC}$ vanish for $x^5 = \overline{x^5}$: see (20.2)).

On account of the explicit form (20.3) of the affine connection, one gets the following system of five coupled differential equations

$$
\left\{
\begin{array}{c}
\dfrac{\mathrm{d}^2 x^\mu(\tau)}{\mathrm{d}\tau^2} + 2\dfrac{b'_\mu(x^5(\tau))}{b_\mu(x^5(\tau))}\dfrac{\mathrm{d}x^\mu(\tau)}{\mathrm{d}\tau}\dfrac{\mathrm{d}x^5(\tau)}{\mathrm{d}\tau} = 0, \\
\mu = 0,1,2,3 \quad \text{(ESC off)}; \\
\dfrac{\mathrm{d}^2 x^5(\tau)}{\mathrm{d}\tau^2} = \\
= \pm\dfrac{1}{f(x^5(\tau))}\left[g_{\alpha\beta} b_\alpha(x^5(\tau)) b'_\beta(x^5(\tau))\left(\dfrac{\mathrm{d}x^\alpha(\tau)}{\mathrm{d}\tau}\right)^2 \mp f'(x^5(\tau))\left(\dfrac{\mathrm{d}x^5(\tau)}{\mathrm{d}\tau}\right)^2\right] \\
\text{(ESC on)},
\end{array}
\right.
\tag{23.8}
$$

where as by now familiar the prime denotes derivation with respect to $x^5$ and $g_{\alpha\beta}$ is the Minkowskian metric tensor.

System (23.8) does not admit solutions in general. In the following, we shall confine ourselves to look for physically significant solutions in the simpler case of the Power Ansatz for the metric coefficients.

---

[2] Indeed, let us recall that, in a Riemannian space, the geodesic equations – although formally identical in any reference frame – actually do depend on the chosen frame, due to the nontensor nature of the affine connection $\Gamma^A_{BC}$.

For instance, it is possible to parametrize the space–time trajectory of a massless particle in terms of $x^0$, determined by the null-interval condition as follows:

$$
\mathrm{d}s^2_{\mathrm{DR5}} = 0
$$

$$
\Longleftrightarrow b_0^2\left(\mathrm{d}x^0\right)^2 - \sum_i b_i^2(x^5)\left(\mathrm{d}x^i\right)^2 \pm f(x^5)\left(\mathrm{d}x^5\right)^2 = 0
$$

$$
\Longleftrightarrow \mathrm{d}x^0 = \frac{1}{b_0}\sqrt{\left[\sum_i b_i^2(x^5)\,(\mathrm{d}x^i)^2 \mp f(x^5)\,(\mathrm{d}x^5)^2\right]}.
$$

# 24

# Solution of the Geodesic Equations in the Power Ansatz

It has been seen in Sect. 20.4 that the phenomenological metrics for the electromagnetic, weak, gravitational, and strong interactions, derived in the context of DSR, can be recovered as 5D metrics found, in the Power Ansatz, as solutions of the Einstein equations in vacuum and with cosmological constant $\Lambda_{(5)} = 0$. In this chapter we will solve the geodesic equations in the Power Ansatz, and apply the results obtained to the 5D metrics of the four fundamental interactions.

## 24.1 General Solution

In the Power Ansatz for the metric coefficients (see Chap. 20), the system of geodesic equations (23.8) takes the form

$$\begin{cases} \dfrac{d^2x^0}{d\tau^2} + \dfrac{q_0}{x^5}\dfrac{dx^0}{d\tau}\dfrac{dx^5}{d\tau} = 0; \\ \dfrac{d^2x^1}{d\tau^2} + \dfrac{q_1}{x^5}\dfrac{dx^1}{d\tau}\dfrac{dx^5}{d\tau} = 0; \\ \dfrac{d^2x^2}{d\tau^2} + \dfrac{q_2}{x^5}\dfrac{dx^2}{d\tau}\dfrac{dx^5}{d\tau} = 0; \\ \dfrac{d^2x^3}{d\tau^2} + \dfrac{q_3}{x^5}\dfrac{dx^3}{d\tau}\dfrac{dx^5}{d\tau} = 0; \\ \dfrac{d^2x^5}{d\tau^2} \pm \dfrac{r}{2x^5}\left(\dfrac{dx^5}{d\tau}\right)^2 \mp \dfrac{1}{2\left(x^5\right)^{r+1}}\left[q_0\left(\dfrac{x^5}{x_0^5}\right)^{q_0}\left(\dfrac{dx^0}{d\tau}\right)^2 \right. \\ \left. -q_1\left(\dfrac{x^5}{x_0^5}\right)^{q_1}\left(\dfrac{dx^1}{d\tau}\right)^2 - q_2\left(\dfrac{x^5}{x_0^5}\right)^{q_2}\left(\dfrac{dx^2}{d\tau}\right)^2 - q_3\left(\dfrac{x^5}{x_0^5}\right)^{q_3}\left(\dfrac{dx^3}{d\tau}\right)^2\right] = 0. \end{cases} \tag{24.1}$$

It can be synthetically written as

$$\frac{d^2x^\mu}{d\tau^2}+\frac{q_\mu}{x^5}\frac{dx^\mu}{d\tau}\frac{dx^5}{d\tau}=0, \quad \mu=0,1,2,3 \text{ (ESC off)}; \tag{24.2}$$

$$\frac{d^2x^5}{d\tau^2}\mp\frac{1}{2}\left[g_{\alpha\beta}q_\alpha\left(\frac{x^5}{x_0^5}\right)^{q_\alpha-r-1}\frac{1}{x_0^5}\left(\frac{dx^\alpha}{d\tau}\right)^2\mp\frac{r}{2x^5}\left(\frac{dx^5}{d\tau}\right)^2\right]=0 \text{ (ESC on)} \tag{24.3}$$

with $g_{\alpha\beta}$ denoting, as usual, the Minkowskian metric tensor.

The solution of (24.2) in terms of $x^5$ reads (ESC off)

$$x^\mu(\tau)=C_{\mu 1}+C_{\mu 2}\int d\tau\left(x^5(\tau)\right)^{-q_\mu}, \tag{24.4}$$

where $C_{\mu 1}$, $C_{\mu 2}$ are real integration constants. Replacing (24.4) in (24.3) yields (ESC on)

$$\frac{d^2x^5}{d\tau^2}\mp\frac{1}{2}\left(x_0^5\right)^r\left(x^5\right)^{-r-1}g_{\alpha\beta}C_{\alpha 2}^2q_\alpha\left(\frac{x^5}{x_0^5}\right)^{-q_\beta}+\frac{r}{2x^5}\left(\frac{dx^5}{d\tau}\right)^2=0. \tag{24.5}$$

By solving this equation one gets the following implicit functional relation for $x^5(\tau)$:

$$\tau+A_1\mp\left(x_0^5\right)^{-r/2}\int d\zeta F_\pm(\zeta;\tilde{\mathbf{q}},A_2)\Big|_{\zeta=x^5(\tau)}=0, \tag{24.6}$$

($A_1$, $A_2 \in R$), where (ESC on)

$$F_\pm(\zeta;\tilde{\mathbf{q}},A_2)\equiv\left\{\mp\left[g_{\mu\nu}\left(x_0^5\right)^{q_\mu}C_{\mu 2}^2\zeta^{-r-q_\nu}\mp\left(x_0^5\right)^{-r}A_2\zeta^{-r}\right]\right\}^{-1/2} \tag{24.7}$$

and $\tilde{\mathbf{q}}=(q_0,q_1,q_2,q_3,r)$ is the parametric set of exponents of the metric coefficients (in the Power Ansatz) for the class of solutions considered (see Sect. 20.4). Of course, the explicit form of the function $F_\pm(\zeta;\tilde{\mathbf{q}},A_2)$, and therefore of the indefinite integral in $\zeta$ in (24.6), depends on the set $\tilde{\mathbf{q}}$ (and on the integration constant $A_2$); moreover, it determines the geodesic motions in $\Re_5$ for the class of solutions characterized by the exponent set $\tilde{\mathbf{q}}$. This is why we shall refer to it as *the geodesic generating function*. The integral in (24.6) then becomes (ESC on)

$$\int d\zeta F_\pm(\zeta;\tilde{\mathbf{q}},A_2)$$

$$=\int d\zeta\left[\alpha_{\mu,\pm}(q_\mu,C_{\mu 2}^2,(x_0^5))\delta_{\mu\nu}\zeta^{-r-q_\nu}+\alpha_5(r,A_2,(x_0^5))\zeta^{-r}\right]^{-1/2} \tag{24.8}$$

with

$$\begin{aligned}\alpha_{0,\pm}(q_0, C_{02}^2, (x_0^5)) &= \mp (x_0^5)^{q_0} C_{02}^2;\\ \alpha_{1,\pm}(q_1, C_{12}^2, (x_0^5)) &= \pm (x_0^5)^{q_1} C_{12}^2;\\ \alpha_{2,\pm}(q_2, C_{22}^2, (x_0^5)) &= \pm (x_0^5)^{q_2} C_{22}^2;\\ \alpha_{3,\pm}(q_3, C_{32}^2, (x_0^5)) &= (x_0^5)^{q_3} C_{32}^2;\\ \alpha_5(r, A_2, (x_0^5)) &= (x_0^5)^{-r} A_2. \end{aligned} \tag{24.9}$$

Such an integral is of the type (ESC on)

$$\int \mathrm{d}x \frac{x^{r/2}}{\sqrt{a + \delta_{\mu\nu} c_\mu x^{-q_\nu}}} \tag{24.10}$$

with $a$, $c_\mu$ ($\mu = 0, 1, 2, 3$) real constants. For $r \neq -2$, putting $y = x^{\frac{r}{2}+1}$ yields

$$\int \mathrm{d}x \frac{x^{r/2}}{\sqrt{a + \delta_{\mu\nu} c_\mu x^{-q_\nu}}} = \frac{2}{r+2} \int \frac{\mathrm{d}y}{\sqrt{a + \delta_{\mu\nu} c_\mu y^{\gamma_\nu(q_\nu, r)}}}, \tag{24.11}$$

where $\gamma_\nu(q_\nu, r) = -\dfrac{2q_\nu}{r+2}$. For $r = -2$ one gets instead, by the substitution $y = \ln x$ ($x \geq 0$)

$$\int \frac{\mathrm{d}x}{x\sqrt{a + \delta_{\mu\nu} c_\mu x^{-q_\nu}}} = \int \frac{\mathrm{d}y}{\sqrt{a + \delta_{\mu\nu} c_\mu \mathrm{e}^{-q_\nu y}}}. \tag{24.12}$$

The Riemann integrals (24.11) and (24.12) do not exist in literature for generic values of the constants. Then, the general solution of the geodetic equations (24.2), even in the Power Ansatz, can only be expressed in terms of such integrals by means of (24.6)–(24.12), and therefore takes the implicit form

$$x^\mu(\tau) = C_{\mu 1} + C_{\mu 2} \int \mathrm{d}\tau \left(x^5(\tau)\right)^{-q_\mu} \text{ (ESC off)}; \tag{24.13}$$

$$\tau + A_1 \mp (x_0^5)^{-r/2} \int \mathrm{d}\zeta F_\pm(\zeta; \widetilde{\mathbf{q}}, A_2)\Big|_{\zeta = x^5(\tau)} = 0 \tag{24.14}$$

with the generating function given by (24.7). Equation (24.14) for the energy coordinate explicitly reads

$$\tau + A_1$$

$$\stackrel{\text{ESC on}}{=} \pm \frac{2\left(x_0^5\right)^{-r/2}}{r+2} \int \frac{\mathrm{d}y}{\sqrt{\alpha_{\mu,\pm}(q_\mu, C_{\mu 2}^2, (x_0^5))\delta_{\mu\nu} y^{\gamma_\nu(q_\nu, r)} + \alpha_5(r, A_2, (x_0^5))}} = 0,$$

$$\gamma_\nu(q_\nu, r) = -\frac{2q_\nu}{r+2}, \quad y = (x_5(\tau))^{\frac{r}{2}+1}, \quad r \neq -2; \tag{24.15}$$

$$\tau + A_1 \overset{\text{ESC on}}{=} \pm (x_0^5)^{-r/2} \int^{x_5(\tau)} \frac{dx}{\sqrt{\alpha_{\mu,\pm}(q_\mu, C_{\mu 2}^2, (x_0^5))\delta_{\mu\nu} x^{-q_\nu} + \alpha_5(r, A_2, (x_0^5))}} = 0,$$

$$r = 2; \tag{24.16}$$

$$\begin{aligned}
\alpha_{0,\pm}(q_0, C_{02}^2, (x_0^5)) &= \mp (x_0^5)^{q_0} C_{02}^2; \\
\alpha_{1,\pm}(q_1, C_{12}^2, (x_0^5)) &= \pm (x_0^5)^{q_1} C_{12}^2; \\
\alpha_{2,\pm}(q_2, C_{22}^2, (x_0^5)) &= \pm (x_0^5)^{q_2} C_{22}^2; \\
\alpha_{3,\pm}(q_3, C_{32}^2, (x_0^5)) &= (x_0^5)^{q_3} C_{32}^2; \\
\alpha_5(r, A_2, (x_0^5)) &= (x_0^5)^{-r} A_2.
\end{aligned} \tag{24.17}$$

The geodesic motions expressed by (24.13), (24.14) for all classes (I)–(XII) of the Einstein equations in vacuum in the Power Ansatz, discussed in Sect. 20.4, can be found in Appendix C.

## 24.2 Geodesic Motions for the 5D Metrics of Fundamental Interactions

### *24.2.1 Generating Function for Electromagnetic and Weak Metrics*

The metrics for electromagnetic and weak interactions are characterized by the power dependence $\left(\frac{x^5}{x_0^5}\right)^{1/3}$. Then, as shown in Sect. 20.4, they can be obtained from the following classes of solutions of the algebraic Einstein equations (20.19) with $\Lambda_{(5)} = 0$:

(1) *Class II for* $m = \frac{1}{3}$, characterized therefore by the coefficient set $\widetilde{\mathbf{q}}_{\text{II,int.}} = (0, 1/3, 0, 0, -5/3)$, int.=e.m., weak. One gets:

$$g_{AB,\text{DR5,II},m=\frac{1}{3}}(x^5) = \text{diag}\left(1, -\left(\frac{x^5}{x_0^5}\right)^{1/3}, -1, -1, \pm\left(\frac{x^5}{x_0^5}\right)^{-5/3}\right). \tag{24.18}$$

The geodesic generating function $F_\pm$ (24.7) takes the form

$$F_{\pm,\text{II}}\left(\zeta; m = \frac{1}{3}, A_2\right) = \left\{\pm\left[C_{22}^2 + C_{32}^2 - C_{02}^2 \pm A_2 (x_0^5)^{5/3}\right]\zeta^{5/3} \pm C_{12}^2 (x_0^5)^{1/3} \zeta^{4/3}\right\}^{-1/2}. \tag{24.19}$$

(2) *Class IV for* $p = \frac{1}{3}$ $(\widetilde{\mathbf{q}}_{\mathrm{IV,int.}} = (0, 0, 0, 1/3, -5/3))$:

$$g_{AB,\mathrm{DR5,IV},p=\frac{1}{3}}(x^5) = \mathrm{diag}\left(1, , -1, -1, -\left(\frac{x^5}{x_0^5}\right)^{1/3}, \pm\left(\frac{x^5}{x_0^5}\right)^{5/3}\right); \tag{24.20}$$

$$F_{\pm,\mathrm{IV}}\left(\zeta; p = \frac{1}{3}, A_2\right)$$
$$= \left\{\pm\left[C_{12}^2 + C_{22}^2 - C_{02}^2 \pm A_2\left(x_0^5\right)^{5/3}\right]\zeta^{5/3} \pm C_{32}^2\left(x_0^5\right)^{1/3}\zeta^{4/3}\right\}^{-1/2}. \tag{24.21}$$

(3) *Class IX for* $n = \frac{1}{3}$ $(\widetilde{\mathbf{q}}_{\mathrm{IX,int.}} = (0, 0, 1/3, 0, -5/3))$:

$$g_{AB,\mathrm{DR5,IX},n=\frac{1}{3}}(x^5) = \mathrm{diag}\left(1, , -1, -\left(\frac{x^5}{x_0^5}\right)^{1/3}, -1, , \pm\left(\frac{x^5}{x_0^5}\right)^{-5/3}\right); \tag{24.22}$$

$$F_{\pm,\mathrm{IX}}\left(\zeta; n = \frac{1}{3}, A_2\right)$$
$$= \left\{\pm\left[C_{12}^2 + C_{32}^2 - C_{02}^2 \pm A_2\left(x_0^5\right)^{5/3}\right]\zeta^{5/3} \pm C_{22}^2\left(x_0^5\right)^{1/3}\zeta^{4/3}\right\}^{-1/2}. \tag{24.23}$$

It is easily seen that, at level of both the metric structure $g_{AB,\mathrm{DR5}}(x^5)$ and of the integrand function $F_{\pm}(\zeta; \widetilde{\mathbf{q}}, A_2)$, the following relations hold:

$$II)|_{m=\frac{1}{3}} \underset{x^1 \leftrightarrow x^3 (C_{12} \leftrightarrow C_{32})}{\Leftrightarrow} IV)|_{p=\frac{1}{3}} \underset{x^2 \leftrightarrow x^3 (C_{22} \leftrightarrow C_{32})}{\Leftrightarrow} IX)|_{n=\frac{1}{3}} \underset{x^1 \leftrightarrow x^2 (C_{12} \leftrightarrow C_{22})}{\Leftrightarrow} II)|_{m=\frac{1}{3}}. \tag{24.24}$$

This essentially means that (as it is also seen by the expressions of the coefficient sets) the three cases are equivalent – apart from a redenomination of the spatial axes – and it is therefore possible to consider any of them without loss of generality.

The forms (24.19), (24.21), (24.23) of the function $F_{\pm,\mathrm{int.}}(\zeta; \widetilde{\mathbf{q}}, A_2)$ (int. = e.m., weak) for the three classes can be summarized as

$$F_{\pm,\mathrm{int.}}(\zeta; A_2, K_{1,\pm,\mathrm{int.}}, K_{2,\mathrm{int.}}) = \left[\pm\left(K_{1,\pm,\mathrm{int.}}\zeta^{5/3} + K_{2,\mathrm{int.}}\zeta^{4/3}\right)\right]^{-1/2}, \tag{24.25}$$

where the constants $K_{1,\pm,\mathrm{int.}}$, $K_{2,\mathrm{int.}}$ are given by

$$\begin{aligned} II)|_{m=\frac{1}{3}} &: \\ K_{1,\pm,\mathrm{int.}} &= C_{22}^2 + C_{32}^2 - C_{02}^2 \pm A_2\left(x_{0,\mathrm{int.}}^5\right)^{5/3}, \\ K_{2,\mathrm{int.}} &= C_{12}^2\left(x_{0,\mathrm{int.}}^5\right)^{1/3}; \end{aligned} \tag{24.26}$$

$$IV)|_{p=\frac{1}{3}} \quad :$$

$$K_{1,\pm,\text{int.}} = C_{12}^2 + C_{22}^2 - C_{02}^2 \pm A_2 \left(x_{0,\text{int.}}^5\right)^{5/3},$$
$$K_{2,\text{int.}} = C_{32}^2 \left(x_{0,\text{int.}}^5\right)^{1/3}; \qquad (24.27)$$

$$IX)|_{n=\frac{1}{3}} \quad :$$

$$K_{1,\pm,\text{int.}} = C_{12}^2 + C_{32}^2 - C_{02}^2 \pm A_2 \left(x_{0,\text{int.}}^5\right)^{5/3},$$
$$K_{2,\text{int.}} = C_{22}^2 \left(x_{0,\text{int.}}^5\right)^{1/3}. \qquad (24.28)$$

### *24.2.2 Generating Function for Strong and Gravitational Metrics*

The metrics for strong and gravitational interactions are characterized by the power dependence $\left(\frac{x^5}{x_0^5}\right)^2$, and can be therefore obtained from the following classes of solutions:

(1) *Class I for* $n = 2$ $(p = 0)$ $(\widetilde{\mathbf{q}}_{\text{I,int.}} = (2, -1, 2, 0, 1)$, int.=strong, grav.). One gets:

$$g_{AB,\text{DR5,I},n=2}(x^5)$$
$$= \text{diag}\left(\left(\frac{x^5}{x_0^5}\right)^2, -\left(\frac{x^5}{x_0^5}\right)^{-(4+4p)/(4+p)},\right.$$
$$\left. -\left(\frac{x^5}{x_0^5}\right)^2, -\left(\frac{x^5}{x_0^5}\right)^p, \pm\left(\frac{x^5}{x_0^5}\right)^{(p^2+2p+4)/(4+p)}\right); \qquad (24.29)$$

$$g_{AB,\text{DR5,I},n=2,p=0}(x^5)$$
$$= \text{diag}\left(\left(\frac{x^5}{x_0^5}\right)^2, -\left(\frac{x^5}{x_0^5}\right)^{-1}, -\left(\frac{x^5}{x_0^5}\right)^2, -1, \pm\left(\frac{x^5}{x_0^5}\right)\right); \qquad (24.30)$$

whence the following forms of the generating function:

$$F_{\pm,I}(\zeta; n = 2, p, A_2)$$
$$= \left[\pm \left(x_0^5\right)^{-(4+4p)/(4+p)} C_{12}^2 \zeta^{-\frac{p^2-2p}{4+p}}\right.$$
$$\pm \left(x_0^5\right)^2 \left(C_{22}^2 - C_{02}^2\right) \zeta^{-(p^2+4p+12)/(4+p)} +$$
$$\pm \left(x_0^5\right)^p C_{32}^2 \zeta^{-(2p^2+6p+4)/4+p)}$$
$$\left.\pm A_2 \left(x_0^5\right)^{-(p^2+2p+4)/(4+p)} \zeta^{-\frac{p^2+2p+4}{4+p}}\right]^{-1/2}, \qquad (24.31)$$

$$F_{\pm,I}(\zeta; n=2, p=0, A_2)$$
$$= \left\{ \pm C_{12}^2 \left(x_0^5\right)^{-1} + \left(\pm C_{32}^2 + A_2 \left(x_0^5\right)^{-1}\right) \zeta^{-1} \right.$$
$$\left. \pm \left(x_0^5\right)^2 \left(C_{22}^2 - C_{02}^2\right) \zeta^{-3} \right\}^{-1/2} . \tag{24.32}$$

The last equation can be written as

$$F_{\pm,\mathrm{I,int.}}(\zeta; K_{1,\pm,\mathrm{int.}}, K_{2,\pm,\mathrm{int.}}, K_{3,\pm,\mathrm{int.}})$$
$$= \left[K_{1,\pm,\mathrm{int.}} + K_{2,\pm,\mathrm{int.}}\zeta^{-1} + K_{3,\pm,\mathrm{int.}}\zeta^{-3}\right]^{-1/2} \tag{24.33}$$

with

$$\begin{aligned} K_{1,\pm,\mathrm{int.}} &= \pm C_{12}^2 \left(x_0^5\right)^{-1}, \\ K_{2,\pm,\mathrm{int.}} &= \pm C_{32}^2 + A_2 \left(x_0^5\right)^{-1}, \\ K_{3,\pm,\mathrm{int.}} &= \pm \left(x_0^5\right)^2 \left(C_{22}^2 - C_{02}^2\right). \end{aligned} \tag{24.34}$$

(2) *Class X for* $q = 2$, $n = p = 0$ ($\widetilde{\mathbf{q}}_{\mathrm{X,int.}} = (2, 0, 0, 0, 0)$):

$$g_{AB,\mathrm{DR5,X},q=2,n=p=0}(x^5) = \mathrm{diag}\left(\left(\frac{x^5}{x_0^5}\right)^2, -1, -1, -1, \pm 1\right); \tag{24.35}$$

$$F_{\pm,\mathrm{X}}(\zeta; q=2, n=p=0, A_2)$$
$$= \left\{\pm \left(C_{12}^2 + C_{22}^2 + C_{32}^2\right) + A_2 \mp \left(x_0^5\right)^2 C_{02}^2 \zeta^{-2}\right\}^{-1/2} . \tag{24.36}$$

(3) *Class XI for* $q = 2$, $n = 0$ ($\widetilde{\mathbf{q}}_{\mathrm{XI,int.}} = (2, 0, 0, 0, 0)$):

$$g_{AB,\mathrm{DR5,XI},q=2,n=0}(x^5) = \mathrm{diag}\left(\left(\frac{x^5}{x_0^5}\right)^2, -1, -1, -1, \pm 1\right); \tag{24.37}$$

$$F_{\pm,\mathrm{XI}}(\zeta; q=2, n=0, A_2)$$
$$= \left\{\pm \left(C_{12}^2 + C_{22}^2 + C_{32}^2\right) + A_2 \mp \left(x_0^5\right)^2 C_{02}^2 \zeta^{-2}\right\}^{-1/2} . \tag{24.38}$$

(4) *Class XII for* $q = 2$, $n = 0$ ($\widetilde{\mathbf{q}}_{\mathrm{X,int.}} = (2, 0, 0, 0, 0)$):

$$g_{AB,\mathrm{DR5,XII},q=2,n=0}(x^5) = \mathrm{diag}\left(\left(\frac{x^5}{x_0^5}\right)^2, -1, -1, -1, \pm 1\right); \tag{24.39}$$

$$F_{\pm,\mathrm{XII}}(\zeta; q=2, n=0, A_2)$$
$$= \left\{\pm\left(C_{12}^2 + C_{22}^2 + C_{32}^2\right) + A_2 \mp \left(x_0^5\right)^2 C_{02}^2 \zeta^{-2}\right\}^{-1/2}. \tag{24.40}$$

The last three cases are perfectly equivalent at level both of the metric structure $g_{AB,\mathrm{DR5}}(x^5)$ and of the generating function $F_\pm(\zeta; \widetilde{\mathbf{q}}, A_2)$:

$$\mathrm{X})|_{q=2,n=p=0} = \mathrm{XI})|_{q=2,n=0} = \mathrm{XII})|_{q=2,n=0}\,. \tag{24.41}$$

The function $F_\pm(\zeta)$ for all three cases can be therefore written in the compact form

$$F_{\pm,\mathrm{int.}}(\zeta; K_{1,\pm,\mathrm{int.}}, K_{2,\mathrm{int.}}) = \left[\pm\left(K_{1,\pm,\mathrm{int.}} + K_{2,\mathrm{int.}}\zeta^{-2}\right)\right]^{-1/2}, \tag{24.42}$$

where we put

$$\begin{aligned} K_{1,\pm,\mathrm{int.}} &= \left(C_{12}^2 + C_{22}^2 + C_{32}^2\right) \pm A_2, \\ K_{2,\mathrm{int.}} &= -\left(x_{0,\mathrm{int.}}^5\right)^2 C_{02}^2. \end{aligned} \tag{24.43}$$

### *24.2.3 Geodesics for Electromagnetic and Weak Interactions*

It is now possible, on account of the results of Sect. 24.1 and Sect. 24.2.1, to write the explicit expressions of the geodesics in $\Re_5$ corresponding to the electromagnetic and weak metrics. Equation (24.14) for the energy coordinate reads, in this case:

$$\tau + A_1$$
$$= \pm\left(x_0^5\right)^{5/6} \int d\zeta F_{\pm,\mathrm{e.m.,weak}}(\zeta; A_2, K_{1,\pm}, K_2)\Big|_{\zeta = x^5(\tau)}, \tag{24.44}$$

whence, from (24.25)

$$x_{\pm,\mathrm{int.}}^5(\tau) = \pm\frac{1}{K_{1,\pm,\mathrm{int.}}^3}\left[\frac{K_{1,\pm,\mathrm{int.}}^2}{36\left(x_{0,\mathrm{int.}}^5\right)^{5/3}}\left(\tau^2 + 2A_2\tau + A_2^2\right) \mp K_{2,\mathrm{int.}}\right]^3. \tag{24.45}$$

This equation can be also put in the form

$$x_{\pm,\mathrm{int.}}^5(\tau) = \pm a_{1,\pm,\mathrm{int.}}\left(a_{2,\pm,\mathrm{int.}}\tau^2 + a_{3,\pm,\mathrm{int.}}\tau + a_{4,\pm,\mathrm{int.}}\right)^3 \tag{24.46}$$

with

$$\begin{aligned} a_{1,\pm,\mathrm{int.}} &= \frac{1}{K_{1,\pm,\mathrm{int.}}^3}; \quad a_{2,\pm,\mathrm{int.}} = \frac{K_{1,\pm,\mathrm{int.}}^2}{36\left(x_{0,\mathrm{int.}}^5\right)^{5/3}}; \\ a_{3,\pm,\mathrm{int.}} &= \frac{K_{1,\pm,\mathrm{int.}}^2 A_2}{18\left(x_{0,\mathrm{int.}}^5\right)^{5/3}}; \quad a_{4,\pm,\mathrm{int.}} = \frac{K_{1,\pm,\mathrm{int.}}^2 A_2^2}{36\left(x_{0,\mathrm{int.}}^5\right)^{5/3}} \mp K_{2,\mathrm{int.}}. \end{aligned} \tag{24.47}$$

As noted in Sect. 24.2.1, one can, without loss of generality, consider any of the three classes (which only differ by the name of the spatial axes). Taking e.g., class II, we get, for the space–time coordinates of the geodesics (cf. (24.13)):

$$\begin{aligned} x^{\mu}_{\pm,\mathrm{int.}}(\tau) &= C_{\mu 1} + C_{\mu 2} \int \mathrm{d}\tau \left(x^5(\tau)\right)^{-q_\mu} \\ &= C_{\mu 1} + C_{\mu 2}\left(\tau + \chi_\mu\right) = \widetilde{C}_{\mu 1} + C_{\mu 2}\tau, \\ \widetilde{C}_{\mu 1} &= C_{\mu 1} + C_{\mu 2}\chi_\mu, \qquad \mu = 0, 2, 3; \end{aligned} \tag{24.48}$$

$$\begin{aligned} x^1_{\pm,\mathrm{int.}}(\tau) &= C_{11} + C_{12} \int \mathrm{d}\tau \left(x^5(\tau)\right)^{-q_1} = C_{11} + C_{12} \int \mathrm{d}\tau \left(x^5(\tau)\right)^{-1/3} \\ &= C_{11} + C_{12} \int \mathrm{d}\tau \left[\pm a_{1,\pm,\mathrm{int.}} \left(a_{2,\pm,\mathrm{int.}}\tau^2 + a_{3,\pm,\mathrm{int.}}\tau + a_{4,\pm,\mathrm{int.}}\right)^3\right]^{-1/3} \\ &= \begin{cases} C_{11} + C_{12}(\pm a_{1,\pm,\mathrm{int.}})^{-1/3} \dfrac{2}{\sqrt{|\Delta_\pm|}} arctg\left(\dfrac{2a_{2,\pm,\mathrm{int.}}\tau + a_{3,\pm,\mathrm{int.}}}{\sqrt{|\Delta_\pm|}}\right), \\ \qquad \Delta_\pm < 0, \\ C_{11} + C_{12}(\pm a_{1,\pm,\mathrm{int.}})^{-1/3} \dfrac{1}{\sqrt{\Delta_\pm}} arctg\left(\dfrac{2a_{2,\pm,\mathrm{int.}}\tau + a_{3,\pm,\mathrm{int.}} - \sqrt{\Delta_\pm}}{2a_{2,\pm,\mathrm{int.}}\tau + a_{3,\pm,\mathrm{int.}} + \sqrt{\Delta_\pm}}\right), \\ \qquad \Delta_\pm > 0, \\ C_{11} + C_{12}\left(\pm a_{1,\pm,\mathrm{int.}}\right)^{-1/3} \dfrac{2}{2a_{2,\pm,\mathrm{int.}}\tau + a_{3,\pm,\mathrm{int.}}}, \\ \qquad \Delta_\pm - 0, \end{cases} \end{aligned} \tag{24.49}$$

where we put

$$\Delta_\pm = a^2_{3,\pm,\mathrm{int.}} - 4a_{2,\pm,\mathrm{int.}}a_{4,\pm,\mathrm{int.}}. \tag{24.50}$$

### *24.2.4 Geodesics for Strong and Gravitational Interactions*

Analogously to what done in Sect. 24.2.3 for the electromagnetic and the weak interaction, one can exploit the results of Sect. 24.1 and Sect. 24.2.2 in order to get the expressions of the geodesics in $\Re_5$ for particles subjected either to the strong force or to the gravitational one. On account of the fact that the last three metrics discussed in Sect. 24.2.2 are equivalent (they differ only for labeling of the spatial axes), it is possible to consider only two cases (and the related subcases).

**Case 1 (Class I for $n = 2$, $p = 0$)** In this case the energy coordinate is determined by the equation (int.=strong, grav.)

$$\tau + A_1$$
$$= \pm \left(x^5_{0,\text{int.}}\right)^{-r/2} \int d\zeta F_{\pm,I,int.}(\zeta; K_{1,\pm,\text{int.}}, K_{2,\pm,\text{int.}}, K_{3,\pm,\text{int.}})\Big|_{\zeta=x^5(\tau)} \tag{24.51}$$

$$= \pm \left(x^5_{0,\text{int.}}\right)^{-1/2} \int d\zeta \left[K_{1,\pm,\text{int.}} + K_{2,\pm,\text{int.}}\zeta^{-1} + K_{3,\pm,\text{int.}}\zeta^{-3}\right]^{-1/2}\Big|_{\zeta=x^5(\tau)} \tag{24.52}$$

(cf. (24.14),(24.42)). Let us consider the following subcases:

(i) $K_{3,\pm,\text{int.}} = 0$:

(i.1) $K_{1,\pm,\text{int.}} = 0$, $K_{2,\pm,\text{int.}} \neq 0$:

$$x^5_{\pm,\text{int.}}(\tau) = \left[\pm\frac{2}{3}\sqrt{x^5_{0,\text{int.}}K_{2,\pm,\text{int.}}}\,(\tau + A_1)\right]^{2/3}; \tag{24.53}$$

(i.2) $K_{1,\pm,\text{int.}} \neq 0$, $K_{2,\pm,\text{int.}} = 0$:

$$x^5_{\pm,\text{int.}}(\tau) = \pm\sqrt{x^5_{0,\text{int.}}K_{1,\pm,\text{int.}}}\,(\tau + A_1); \tag{24.54}$$

(i.3) $K_{1,\pm,\text{int.}} \neq 0$, $K_{2,\pm,\text{int.}} \neq 0$: The function $x^5_{\pm,\text{int.}}(\tau)$ is determined implicitly by the equation

$$\begin{aligned}
\tau + A_1 = \pm \left(x^5_{0,\text{int.}}\right)^{-1/2} & \left\{ \frac{x^5_{\pm,\text{int.}}(\tau)}{\sqrt{K_{1,\pm,\text{int.}}}}\sqrt{1 + \frac{K_{2,\pm,\text{int.}}}{x^5_{0,\text{int.}}K_{1,\pm,\text{int.}}}} \right. \\
& - \frac{K_{2,\pm,\text{int.}}}{2K^{3/2}_{1,\pm,\text{int.}}}\left[\widehat{\Theta}\left(K_{1,\pm,\text{int.}}\right)\ln\left|\frac{1+\sqrt{1+\frac{K_{2,\pm,\text{int.}}}{x^5_{0,\text{int.}}K_{1,\pm,\text{int.}}}}}{1-\sqrt{1+\frac{K_{2,\pm,\text{int.}}}{x^5_{0,\text{int.}}K_{1,\pm,\text{int.}}}}}\right|\right. \\
& + 2\widehat{\Theta}\left(-K_{1,\pm,\text{int.}}\right)\widehat{\Theta}\left(1-\sqrt{1+\frac{K_{2,\pm,\text{int.}}}{x^5_{0,\text{int.}}K_{1,\pm,\text{int.}}}}\right) \\
\times \widehat{\Theta}\left(1+\sqrt{1+\frac{K_{2,\pm,\text{int.}}}{x^5_{0,\text{int.}}K_{1,\pm,\text{int.}}}}\right) & \left.\left.\operatorname{arctgh}\left(\sqrt{1+\frac{K_{2,\pm,\text{int.}}}{x^5_{0,\text{int.}}K_{1,\pm,\text{int.}}}}\right)\right]\right\};
\end{aligned} \tag{24.55}$$

(ii) $K_{2,\pm,\text{int.}} = 0$:

(ii.1) $K_{1,\pm,\text{int.}} \neq 0$, $K_{3,\pm,\text{int.}} \neq 0$: The Riemann integral at the right-hand side of (24.51) is unknown in this case;

(ii.2) $K_{1,\pm,\text{int.}} \neq 0$, $K_{3,\pm,\text{int.}} = 0$:

$$x^5_{\pm,\text{int.}}(\tau) = \pm\sqrt{x^5_{0,\text{int.}} K_{1,\pm,\text{int.}}}\,(\tau + A_1)\,; \tag{24.56}$$

(ii.3) $K_{1,\pm,\text{int.}} = 0$, $K_{3,\pm,\text{int.}} \neq 0$:

$$x^5_{\pm,\text{int.}}(\tau) = \left[\pm\frac{5}{2}\sqrt{x^5_{0,\text{int.}} K_{3,\pm,\text{int.}}}\,(\tau + A_1)\right]^{2/5}\,; \tag{24.57}$$

(iii) $K_{1,\pm,\text{int.}} = 0$:

(iii.1) $K_{2,\pm,\text{int.}} \neq 0$, $K_{3,\pm,\text{int.}} \neq 0$:

(iii.1.1) $\dfrac{K_{2,\pm,\text{int.}}}{K_{3,\pm,\text{int.}}} > 0$: The Riemann integral at the right-hand side of (24.51) is unknown;

(iii.1.2) $\dfrac{K_{2,\pm,\text{int.}}}{K_{3,\pm,\text{int.}}} < 0$: The function $x^5_{\pm,\text{int.}}(\tau) = x^5(\tau)$ is determined implicitly by the equation

$$\begin{aligned}
&\tau + A_1 \\
&= \pm\left(x^5_{0,\text{int.}}\right)^{-1/2}\left\{\pm i\frac{2\sqrt{K_{3,\pm,\text{int.}}}}{3K_{2,\pm,\text{int.}}}\widehat{\Theta}\left(\sqrt{-\frac{K_{2,\pm,\text{int.}}}{K_{3,\pm,\text{int.}}}} - \frac{1}{x^5_{\pm,\text{int.}}(\tau)}\right)\right. \\
&\quad\times\left[\left(-\frac{K_{2,\pm,\text{int.}}}{K_{3,\pm,\text{int.}}}\right)^{-\frac{1}{4}}\int_0^{\sqrt{1-\frac{1}{x^5_{\pm,\text{int.}}(\tau)}\sqrt{-\frac{K_{3,\pm,\text{int.}}}{K_{2,\pm,\text{int.}}}}}}\frac{dt}{\sqrt{(1-t^2)(2-t^2)}}\right. \\
&\quad\left.+\sqrt{-\frac{K_{2,\pm,\text{int.}}}{K_{3,\pm,\text{int.}}}\left(x^5_{\pm,\text{int.}}(\tau)\right)^3 - x^5_{\pm,\text{int.}}(\tau)}\right] \\
&\quad-\frac{2\sqrt{K_{3,\pm,\text{int.}}}}{3K_{2,\pm,\text{int.}}}\widehat{\Theta}\left(\frac{1}{x^5_{\pm,\text{int.}}(\tau)} - \sqrt{-\frac{K_{2,\pm,\text{int.}}}{K_{3,\pm,\text{int.}}}}\right)
\end{aligned}$$

$$\times \left\{ \left[ -\left(-\frac{K_{2,\pm,\text{int.}}}{K_{3,\pm,\text{int.}}}\right)^{-1/4} \int_0^{\sqrt{1-x^5_{\pm,\text{int.}}(\tau)\sqrt{-\frac{K_{3,\pm,\text{int.}}}{K_{2,\pm,\text{int.}}}}}} \frac{dt}{\sqrt{(1-t^2)(2-t^2)}} \right.\right.$$
$$\left.\left. - \sqrt{-\frac{K_{2,\pm,\text{int.}}}{K_{3,\pm,\text{int.}}}\left(x^5_{\pm,\text{int.}}(\tau)\right)^3 - x^5_{\pm,\text{int.}}(\tau)} \right] \right\}, \tag{24.58}$$

or

$$\times \left\{ \left[ \left(-\frac{K_{2,\pm,\text{int.}}}{K_{3,\pm,\text{int.}}}\right)^{-1/4} \int_0^{\sqrt{\frac{2\sqrt{-\frac{K_{2,\pm,\text{int.}}}{K_{3,\pm,\text{int.}}}}}{\frac{1}{x^5_{\pm,\text{int.}}(\tau)}+\sqrt{-\frac{K_{2,\pm,\text{int.}}}{K_{3,\pm,\text{int.}}}}}}} \frac{dt}{\sqrt{(1-t^2)(2-t^2)}} \right.\right.$$
$$\left.\left. - \sqrt{-\frac{K_{2,\pm,\text{int.}}}{K_{3,\pm,\text{int.}}}\left(x^5_{\pm,\text{int.}}(\tau)\right)^3 - x^5_{\pm,\text{int.}}(\tau)} \right] \right\};$$

(iii.2) $K_{2,\pm,\text{int.}} \neq 0$, $K_{3,\pm,\text{int.}} = 0$:

$$x^5_{\pm,\text{int.}}(\tau) = \left[\pm\frac{2}{3}\sqrt{x^5_{0,\text{int.}}K_{3,\pm,\text{int.}}}\,(\tau + A_1)\right]^{2/3}; \tag{24.59}$$

(iii.3) $K_{2,\pm,\text{int.}} = 0$, $K_{3,\pm,\text{int.}} \neq 0$:

$$x^5_{\pm,\text{int.}}(\tau) = \left[\pm\frac{5}{2}\sqrt{x^5_{0,\text{int.}}K_{3,\pm,\text{int.}}}\,(\tau + A_1)\right]^{2/5}; \tag{24.60}$$

(iv) $K_{1,\pm,\text{int.}}, K_{2,\pm,\text{int.}}, K_{3,\pm,\text{int.}} \neq 0$: The Riemann integral at the right-hand side of (24.51) is unknown.

The space–time coordinates $x^\mu$ are obtained from (24.13) by substituting the expressions (24.52)–(24.55) obtained in the various subcases.

**Case 2 (Classes** $\text{X})|_{q=2,n=p=0} = \text{XI})|_{q=2,n=0} = \text{XII})|_{q=2,n=0}$) As stressed earlier, these three cases are equivalent. The fifth coordinate is determined by the equation (*int.* = *strong*, *grav.*)

$$\begin{aligned}\tau + A_1 &= \pm\left(x^5_{0,\text{int.}}\right)^{-r/2} \int d\zeta F_{\pm,,\text{int.}}(\zeta; K_{1,\pm,\text{int.}}, K_{2,,\mathit{int.}})\Big|_{\zeta=x^5(\tau)} \\ &= \pm \int d\zeta \left[\pm K_{1,\pm,\text{int.}} + K_{2,\text{int.}}\zeta^{-2}\right]^{-1/2}\Big|_{\zeta=x^5(\tau)},\end{aligned} \tag{24.61}$$

whose explicit solution is

$$x^5_{\pm,\text{int.}}(\tau) = \sqrt{\pm\frac{1}{K_{1,\pm,\text{int.}}}\left[K^2_{1,\pm,\text{int.}}(\tau + A_1)^2 \mp K_{2,\text{int.}}\right]} \tag{24.62}$$

(we discarded the solution $x^5_{\pm,\text{int.}}(\tau) < 0$, on account of the physical meaning of the fifth coordinate, energy, in DR5).

Solution (24.62) can be also written as

$$x^5_{\pm,\text{int.}}(\tau) = \sqrt{\pm\alpha_{1,\pm,\text{int.}}\tau^2 \pm \alpha_{2,\pm,\text{int.}}\tau + \alpha_{3,\pm,\text{int.}}}, \tag{24.63}$$

where we put

$$\begin{aligned} \alpha_{1,\pm,\text{int.}} &= K_{1,\pm,\text{int.}};\ \alpha_{2,\pm,\text{int.}} = 2A_1K_{1,\pm,\text{int.}}; \\ \alpha_{3,\pm,\text{int.}} &= \pm A_1^2 K_{1,\pm,\text{int.}} - \frac{K_{2,\text{int.}}}{K_{1,\pm,\text{int.}}}. \end{aligned} \tag{24.64}$$

The space–time coordinates are obtained by replacing (24.63) in (24.13) and using the expressions (24.64) and (24.43) of the constants. One gets, for the time coordinate:

$$\begin{aligned} x^0_{\pm,\text{int.}}(\tau) &= C_{01} + C_{02}\int d\tau\,\left(x^5(\tau)\right)^{-q_0} = C_{01} + C_{02}\int d\tau\,\left(x^5(\tau)\right)^{-2} \\ &= C_{01} + C_{02}\int \frac{d\tau}{\pm\alpha_{1,\pm,\text{int.}}\tau^2 \pm \alpha_{2,\pm,\text{int.}}\tau + \alpha_{3,\pm,\text{int.}}} = \\ &= \begin{cases} C_{01} + C_{02}\dfrac{2}{\sqrt{|\varDelta_\pm|}}arctg\left(\dfrac{\pm 2\alpha_{1,\pm,\text{int.}}\tau \pm \alpha_{2,\pm,\text{int.}}}{\sqrt{|\varDelta_\pm|}}\right), \\ \qquad \varDelta_\pm < 0, \\ C_{01} + C_{02}\dfrac{1}{\sqrt{\varDelta_\pm}}\ln\left|\dfrac{\pm 2\alpha_{1,\pm,\text{int.}}\tau \pm \alpha_{2,\pm,\text{int.}} - \sqrt{\varDelta_\pm}}{\pm 2\alpha_{1,\pm,\text{int.}}\tau \pm \alpha_{2,\pm,\text{int.}} + \sqrt{\varDelta_\pm}}\right|, \\ \qquad \varDelta_\pm > 0, \\ C_{01} - C_{02}\dfrac{2}{\pm 2\alpha_{1,\pm,\text{int.}}\tau \pm \alpha_{2,\pm,\text{int.}}}, \\ \qquad \varDelta_\pm = 0, \end{cases} \end{aligned} \tag{24.65}$$

where

$$\varDelta_\pm = \alpha^2_{2,\pm,\text{int.}} \mp 4\alpha_{1,\pm,\text{int.}}\alpha_{3,\pm,\text{int.}}. \tag{24.66}$$

Finally, the space coordinates read

$$\begin{aligned} x^i_{\pm,\text{int.}}(\tau) &= C_{i1} + C_{i2}\int d\tau\,\left(x^5(\tau)\right)^{-q_i} \\ &= C_{i1} + C_{i2}(\tau + \chi_i) = \widetilde{C}_{i1} + C_{i2}\tau, \\ \widetilde{C}_{i1} &= C_{i1} + C_{i2}\chi_i, \quad i = 1, 2, 3. \end{aligned} \tag{24.67}$$

## 24.3 Gravitational Metric of the Einstein Type

Let us consider the class VI of solutions of the vacuum Einstein equations in the Power Ansatz, characterized by the coefficient set $\widetilde{\mathbf{q}}_{\mathrm{VI}} = (q, 0, 0, 0, q-2)$, to which correspond the 5d metric

$$g_{AB,\mathrm{DR5,VI}}(x^5) = \mathrm{diag}\left(\left(\frac{x^5}{x_0^5}\right)^q, -1, -1, -1, \pm\left(\frac{x^5}{x_0^5}\right)^{q-2}\right) \tag{24.68}$$

and the function

$$\begin{aligned} &F_{\pm,\mathrm{VI}}(\zeta; q, A_2) \\ &= \left\{\pm\left[C_{12}^2 + C_{22}^2 + C_{32}^2 \pm A_2\left(x_0^5\right)^{2-q}\right]\zeta^{2-q} \mp C_{02}^2\left(x_0^5\right)^q \zeta^{2-2q}\right\}^{-1/2}. \end{aligned} \tag{24.69}$$

By putting $q = 1$, these equations become

$$g_{AB,\mathrm{DR5,VI},q=1}(x^5) = \mathrm{diag}\left(\frac{x^5}{x_0^5}, -1, -1, -1, \pm\left(\frac{x^5}{x_0^5}\right)^{-1}\right); \tag{24.70}$$

$$F_{\pm,VI}(\zeta; q = 1, A_2) = \left\{\pm\left[C_{12}^2 + C_{22}^2 + C_{32}^2 \pm A_2 x_0^5\right]\zeta \mp C_{02}^2 x_0^5\right\}^{-1/2}. \tag{24.71}$$

Metric (24.70) has standard Minkowskian structure for the spatial part, whereas the time metric coefficient is linear in the energy coordinate $x^5$. Then, as far as the space–time sector is concerned, it is similar to the 4D gravitational metric (2.21) introduced by Einstein in order to account for the slowing down of clocks in a (weak) gravitational field.

Let us derive the geodesic equations for such a metric. The function $F_{\pm,VI}$ can be written as

$$F_{\pm,VI}(\zeta; K_1, K_{2,\pm}) = \left[\pm\left(K_1 + K_{2,\pm}\zeta\right)\right]^{-1/2}, \tag{24.72}$$

where

$$\begin{aligned} K_1 &= -C_{02}^2 x_0^5; \\ K_{2,\pm} &= C_{12}^2 + C_{22}^2 + C_{32}^2 \pm A_2 x_0^5. \end{aligned} \tag{24.73}$$

On account of the results of Sect. 24.1, one gets therefore, for the fifth coordinate of the geodesic equation:

$$x_{\pm}^5(\tau) = \pm\frac{K_{2,\pm}}{4x_0^5}\tau^2 \pm \frac{K_{2,\pm}A_1}{2x_0^5}\tau \pm \frac{K_{2,\pm}A_1^2}{4x_0^5} \quad K_1, \tag{24.74}$$

or also

$$x_{\pm}^5(\tau) = \pm a_{1,\pm}\tau^2 \pm a_{2,\pm}\tau \pm a_{3,\pm}, \tag{24.75}$$

where

$$a_{1,\pm} = \frac{K_{2,\pm}}{4x_0^5}; \quad a_{2,\pm} = \frac{K_{2,\pm}A_1}{2x_0^5}; \quad a_{3,\pm} = \frac{K_{2,\pm}A_1^2}{4x_0^5} \mp K_1. \tag{24.76}$$

The time coordinate is given by

$$\begin{aligned}
x^0_\pm(\tau) &= C_{01} + C_{02}\int \mathrm{d}\tau\,\left(x^5(\tau)\right)^{-q_0} = C_{01} + C_{02}\int \mathrm{d}\tau\,\left(x^5(\tau)\right)^{-1} \\
&= C_{01} + C_{02}\int \frac{\mathrm{d}\tau}{\pm a_{1,\pm}.\tau^2 \pm a_{2,\pm}\tau + a_{3,\pm}.} \\
&= \begin{cases}
C_{01} + C_{02}\dfrac{2}{\sqrt{|\Delta_\pm|}} arctg\left(\dfrac{\pm 2a_{1,\pm}.\tau \pm a_{2,\pm}}{\sqrt{|\Delta_\pm|}}\right), \\
\qquad \Delta_\pm < 0, \\
C_{01} + C_{02}\dfrac{1}{\sqrt{\Delta_\pm}}\ln\left|\dfrac{\pm 2a_{1,\pm}\tau \pm a_{2,\pm} - \sqrt{\Delta_\pm}}{\pm 2a_{1,\pm}\tau \pm a_{2,\pm} + \sqrt{\Delta_\pm}}\right|, \\
\qquad \Delta_\pm > 0, \\
C_{01} - C_{02}\dfrac{2}{\pm 2a_{1,\pm}\tau \pm a_{2,\pm}}, \\
\qquad \Delta_\pm = 0,
\end{cases}
\end{aligned} \tag{24.77}$$

where

$$\Delta_\pm = a_{2,\pm}^2 \mp 4a_{1,\pm}a_{3,\pm}. \tag{24.78}$$

Finally, the space coordinates read

$$\begin{aligned}
x^i_{\pm,\mathrm{int.}}(\tau) &= C_{i1} + C_{i2}\int \mathrm{d}\tau\,\left(x^5(\tau)\right)^{-q_i} \\
&= C_{i1} + C_{i2}\left(\tau + \chi_i\right) = \widetilde{C}_{i1} + C_{i2}\tau, \\
\widetilde{C}_{i1} &= C_{i1} + C_{i2}\chi_i, \quad i = 1,2,3.
\end{aligned} \tag{24.79}$$

The result obtained is therefore identical to that of classes $\mathrm{X})_{q=2,n=p=0} = \mathrm{XI})|_{q=2,n=0} = \mathrm{XII})|_{q=2,n=0}$ for the strong and gravitational interactions (as expected on physical grounds).

## 24.4 Class VIII and the Heisenberg Time–Energy Uncertainty

As a final case we shall consider the solution of (24.13), (24.14) for the class VIII of solutions ($\widetilde{\mathbf{q}}_{\mathrm{VIII}} = (0,0,0,0,r)$ ). The corresponding metric is

$$g_{AB,\mathrm{DR5,VIII}}(x^5) = \mathrm{diag}\left(1,-1,-1,-1,\pm\left(\frac{x^5}{x_0^5}\right)^r\right), \tag{24.80}$$

whose 4D space–time sector is a standard Minkowski space (which, as by now familiar, represents, in the DSR framework, the electromagnetic interaction), whereas the energy exponent is undetermined.

The generating function $F_{\pm,VIII}$ reads

$$\begin{aligned} F_{\pm,VI}(\zeta;r,A_2) &= \left\{\pm\left[C_{12}^2+C_{22}^2+C_{32}^2-C_{02}^2\pm A_2\left(x_0^5\right)^{-r}\right]\right\}^{-1/2}\zeta^{r/2} \\ &= K_{1,\pm}\zeta^{r/2}, \end{aligned} \tag{24.81}$$

where we put

$$K_{1,\pm}=\left\{\pm\left[C_{12}^2+C_{22}^2+C_{32}^2-C_{02}^2\pm A_2\left(x_0^5\right)^{-r}\right]\right\}^{-1/2}. \tag{24.82}$$

It is therefore easily got for the fifth coordinate

$$x_\pm^5(\tau;r)=\left[\pm\frac{r+2}{2K_{1,\pm}}\left(x_0^5\right)^{r/2}(\tau+A_1)\right]^{2/(r+2)} \tag{24.83}$$

or

$$x_\pm^5(\tau;r)=\lambda_\pm\frac{r+2}{2}(\tau+A_1)^{2/(r+2)}, \tag{24.84}$$

where

$$\lambda_\pm=\left[\pm\frac{\left(x_0^5\right)^{r/2}}{K_{1,\pm}}\right]^{2/(r+2)}. \tag{24.85}$$

As to the space–time coordinates, they read (ESC off)

$$\begin{aligned} x_\pm^\mu(\tau) &= C_{\mu1}+C_{\mu2}\int \mathrm{d}\tau\,\left(x^5(\tau)\right)^{-q_\mu} \\ &= C_{\mu1}+C_{\mu2}\left(\tau+\chi_\mu\right)=\widetilde{C}_{\mu1}+C_{\mu2}\tau, \\ &\mu=0,1,2,3,\ \ \widetilde{C}_{\mu1}=C_{\mu1}+C_{\mu2}\chi_\mu. \end{aligned} \tag{24.86}$$

Let us consider the expression (24.84) of the energy coordinate. Putting $A_1=0$ and $r=-4$, one gets

$$x_\pm^5(\tau;r=4)=-\lambda_\pm\left(C_{02}^2,C_{12}^2,C_{22}^2,C_{32}^2,A_2,\left(x_0^5\right),r=-4\right)\tau^{-1}, \tag{24.87}$$

where the parametric dependence of $\lambda_\pm$ has been made explicit.

Taking

$$\lambda_\pm=-\hbar\ell_0 \tag{24.88}$$

with $\hbar$ being the Planck constant, one obtains

$$x^5(\tau)=\hbar\ell_0\tau^{-1}\Longleftrightarrow\frac{x^5}{\ell_0}\tau=\hbar\Longleftrightarrow E\tau=\hbar. \tag{24.89}$$

From (24.86) one gets

$$x^0(\tau) = ct(\tau) = \widetilde{C}_{01} + C_{02}\tau \Leftrightarrow \tau(t) = \frac{1}{C_{02}}\left(ct - \widetilde{C}_{01}\right), \quad (24.90)$$

where $t$ is the time coordinate. Therefore, by putting $\widetilde{C}_{01} = 0$, $C_{02} = c$,

$$Et = \hbar \quad (24.91)$$

namely *(24.84) takes a form which reminds the quantum-mechanical, Heisenberg uncertainty relation for time and energy.* Otherwise stated, we can say that *the geodesics in a 5D space–time, endowed with the 5D metric (24.80)* (with suitable values of the coefficient $r$ and of the constants), *embedding a standard 4D Minkowski space* (i.e., whose 4D slices at $dx^5 = 0$ coincide with $M$), *correspond to trajectories (5D world lines) of minimal time–energy uncertainty.* This result seemingly indicates that the 5D scheme of DR5 may play a role toward understanding certain aspects of quantum mechanics *in purely classical (geometrical) terms.* Similar conclusions on the connection between Heisenberg's principle and Kaluza–Klein theory can be drawn also in the context of the Space–Time–Matter model [138–140].

Let us remark that actually this result – showing that a seemingly strict quantum effect like the uncertainty relation can be derived on a mere classical basis – is not surprising, on account of the experimental findings on the anomalous behavior of photon systems discussed in Chap. 13. Indeed, we have seen that such results cannot be explained in the framework of the Copenhagen interpretation of quantum mechanics, and are more in favor of the Einstein–de Broglie–Bohm quantum theory, based on the hollow wave. The latter can be interpreted as a deformation of space–time, described by the DSR formalism. In DR5, the propagation of such a deformation in the 4D space–time can possibly be associated to a geodesic motion in $\Re_5$. If this hypothesis is right, then the $\Re_5$ geodesics would assume the role, in a mere classical framework, of the quantum Feynman paths, and a presumed probabilistic phenomenon in four dimensions would be thus brought back to a deterministic one in five dimensions.

# 25

# Complete Solutions of Geodesic Equations for the 5D Metrics

We discussed in Chap. 24 the solutions of the geodesic equations for the four phenomenological metrics of the fundamental interactions, obtained as special cases of the classes of solutions of the vacuum Einstein equations in the Power Ansatz. However, it is easily seen that they hold only for the energy ranges where the metrics are not Minkowskian (namely below threshold for the electromagnetic and weak metrics, and above threshold for the strong and gravitational ones). Moreover, in most cases the value of the parameter $r$ was fixed (as functions of the other coefficients $q_\mu$, $\mu = 0, 1, 2, 3$) by the structure of the Einstein equations. We want now to give the general solutions of the geodesic equations for the four interactions, starting from the general form of the metrics (20.21)–(20.23), obtained by the 5D embedding of the 4D DSR phenomenological metrics in the DR5 framework. As already stressed, such a procedure leaves undetermined the fifth metric coefficient $f(x^5)$, and therefore yields $r$-parametrized metrics.

The general expression of the geodesic generating function $F_\pm(\zeta; \widetilde{\mathbf{q}}, A_2)$, which determines the geodesic motions in the Power Ansatz, is given by (24.7). On account of it, and of the exponent sets $\widetilde{\mathbf{q}}_{\text{int.}}$ (int.=e.m.,weak,strong,grav.), (20.41)–(20.43), one gets, in correspondence to the 5D metrics of the four fundamental interactions in the Power Ansatz:

$$F_{\pm,\text{e.m./weak}}(\zeta; r, A_2, x^5_{0,\text{e.m./weak}})$$
$$= \zeta^{r/2}\left\{A_2\left(x^5_{0,\text{e.m./weak}}\right)^{-r} \mp C^2_{02}\right.$$
$$\left.\pm\left[\left(x^5_{0,\text{e.m./weak}}\right)\zeta^{-1}\right]^{1/3\widehat{\Theta}_{\text{L}}\left(x^5_{0,\text{e.m./weak}}-x^5\right)}\left(C^2_{12}+C^2_{32}\right)\right\}^{-1/2}; \quad (25.1)$$

$$F_{\pm,\text{strong}}(\zeta; r, A_2, x^5_{0,\text{strong}})$$
$$= \zeta^{r/2}\left\{A_2\left(x^5_{0,\text{strong}}\right)^{-r} \pm \left(C^2_{12}+C^2_{22}\right)+\right.$$
$$\left.\pm\left[\left(x^5_{0,\text{strong}}\right)\zeta^{-1}\right]^{2\widehat{\Theta}_{\text{L}}\left(x^5-x^5_{0,\text{strong}}\right)}\left(C^2_{32}-C^2_{02}\right)\right\}^{-1/2}; \quad (25.2)$$

$$F_{\pm,\text{grav.}}(\zeta; r, A_2, x^5_{0,\text{grav.}})$$
$$= \zeta^{r/2}\left\{\left(x^5_{0,\text{grav.}}\right)^{-r}A_2 \pm \left(x^5_{0,\text{grav.}}\right)^{?}\zeta^{-?}\left(C^2_{12}+C^2_{22}\right)+\right.$$
$$\left.\pm\left[\left(x^5_{0,\text{grav.}}\right)\zeta^{-1}\right]^{\widetilde{2\widehat{\Theta}_{\text{L}}\left(x^5-x^5_{0,\text{grav.}}\right)}}\left(C^2_{32}-C^2_{02}\right)\right\}^{-1/2}, \quad (25.3)$$

$$(25.4)$$

where the tilde and the question marks in $F_{\pm,\text{grav.}}(\zeta; r, A_2, x^5_{0,\text{grav.}})$ have the meaning clarified in Sect. 20.2.3.

The solutions for the geodesic equations are still given by (24.4), (24.6). Let us distinguish the two cases of Minkowskian and non-Minkowskian behavior.

## 25.1 Minkowskian Behavior

This is the case of the electromagnetic and weak interactions above threshold, i.e., for $x^5 \geq x^5_{0,\text{e.m./weak}}$, and of the strong and gravitational interactions below threshold, i.e., for $x^5 \leq x^5_{0,\text{strong}}$ and $x^5 \leq x^5_{0,\text{grav.}}$ In these cases, the exponent sets of the metrics reduce to

$$\widetilde{\mathbf{q}}_{\text{e.m./weak}}\left(x^5 \geq x^5_{0,\text{e.m./weak}}\right) = (0,0,0,0,r); \quad (25.5)$$

$$\widetilde{\mathbf{q}}_{\text{strong}}\left(0 < x^5 \leq x^5_{0,\text{strong}}\right) = (0,(0,0),0,r); \quad (25.6)$$

$$\widetilde{\mathbf{q}}_{\text{grav.}}\left(0 < x^5 \leq x^5_{0,\text{grav.}}\right) = (0,(0,0),0,r). \quad (25.7)$$

These coefficient sets are identical to that of Class VIII we already discussed in Sect. 24.4. The corresponding solution for the fifth coordinate is therefore

$$x^5_{\pm,\mathrm{int.}}(\tau; x^5_{0,\mathrm{int.}}, r_{\mathrm{int.}}) = \left[\pm\frac{r_{\mathrm{int.}}+2}{2K_{1,\pm,\mathrm{int.}}}\left(x^5_{0,\mathrm{int.}}\right)^{r_{\mathrm{int}}/2}(\tau + A_1)\right]^{2/(r_{\mathrm{int}}+2)}, \tag{25.8}$$

$$K_{1,\pm,\mathrm{int.}} = \left\{\pm\left[C^2_{12} + C^2_{22} + C^2_{32} - C^2_{02} \pm A_2\left(x^5_{0,\mathrm{int.}}\right)^{-r_{\mathrm{int.}}}\right]\right\}^{-1/2}, \tag{25.9}$$

whereas the space–time coordinates are given by (ESC off)

$$\begin{aligned} x^\mu_{\pm,\mathrm{int.}}(\tau) &= C_{\mu 1} + C_{\mu 2}\int \mathrm{d}\tau \left(x^5(\tau)\right)^{-q_\mu} \\ &= C_{\mu 1} + C_{\mu 2}\left(\tau + \chi_\mu\right) = \widetilde{C}_{\mu 1} + C_{\mu 2}\tau, \\ \mu &= 0,1,2,3, \quad \widetilde{C}_{\mu 1} = C_{\mu 1} + C_{\mu 2}\chi_\mu. \end{aligned} \tag{25.10}$$

## 25.2 Non-Minkowskian Behavior

Let us consider separately the different cases.

### *25.2.1 Electromagnetic and Weak Interactions under Threshold*

(a.1) $\mp C^2_{02} + A_2\left(x^5_{0,\mathrm{e.m./weak}}\right)^{-r} = 0,\ C^2_{12} + C^2_{22} + C^2_{32} \neq 0$:

(a.1.1) $3r + 7 \neq 0$:

$$\begin{aligned} &x^5_{\pm,\mathrm{e.m./weak.}}(\tau) \\ &= \left[\pm\frac{3r+7}{6}\sqrt{\pm\left(C^2_{12} + C^2_{22} + C^2_{32}\right)}\left(x^5_{0,\mathrm{e.m./weak}}\right)^{(3r+1)/6}(\tau + A_1)\right]^{6/(3r+7)}; \end{aligned} \tag{25.11}$$

(a.1.2) $r = -\dfrac{7}{3}$:

$$\begin{aligned} &x^5_{\pm,\mathrm{e.m./weak.}}(\tau) \\ &= \exp\left[\pm\sqrt{\pm\left(C^2_{12} + C^2_{22} + C^2_{32}\right)}\left(x^5_{0,\mathrm{e.m./weak}}\right)^{(3r+1)/6}(\tau + A_1)\right]; \end{aligned} \tag{25.12}$$

(a.2) $\mp C_{02}^2 + A_2\left(x^5_{0,\text{e.m./weak}}\right)^{-r} \neq 0,\ C_{12}^2 + C_{22}^2 + C_{32}^2 = 0$:

(a.2.1) $r \neq -2$:

$$x^5_{\pm,\text{e.m./weak.}}(\tau) = \left[\pm\frac{r+2}{2}\sqrt{\mp C_{02}^2 + A_2(x^5_{0,\text{e.m./weak}})^{-r}}(x^5_{0,\text{e.m./weak}})^{r/2}(\tau + A_1)\right]^{2/(r+2)}; \tag{25.13}$$

(a.2.2) $r = -2$:

$$x^5_{\pm,\text{e.m./weak.}}(\tau) = \exp\left[\pm\sqrt{\mp C_{02}^2 + A_2\left(x^5_{0,\text{e.m./weak}}\right)^{-r}}\left(x^5_{0,\text{e.m./weak}}\right)^{r/2}(\tau + A_1)\right]; \tag{25.14}$$

(a.3) $\mp C_{02}^2 + A_2\left(x^5_{0,\text{e.m./weak}}\right)^{-r} \neq 0,\ C_{12}^2 + C_{22}^2 + C_{32}^2 \neq 0$:

(a.3.1) $3r + 7 \neq 0$:

$$\begin{aligned}\tau + A_1 &= \pm\frac{6\left(x^5_{0,\text{e.m./weak}}\right)^{(-3r+2)/9}}{(3r+7)\sqrt{\pm\left(C_{12}^2 + C_{22}^2 + C_{32}^2\right)}} \\ &\quad \times \left[\frac{-\left(C_{12}^2 + C_{22}^2 + C_{32}^2\right)}{C_{02}^2 \mp A_2\left(x^5_{0,\text{e.m./weak}}\right)^{-r}}\right]^{(3r+7)/6} \int \frac{dt}{\sqrt{t^{\frac{2}{3r+7}} + 1}},\end{aligned} \tag{25.15}$$

where

$$t = \left[\frac{-C_{02}^2 \pm A_2(x^5_{0,\text{e.m./weak}})^{-r}}{(x^5_{0,\text{e.m./weak}})^{1/3}(C_{12}^2 + C_{22}^2 + C_{32}^2)}\right]^{(3r+7)/2} (x^5_{\pm,\text{e.m./weak.}}(\tau))^{(3r+7)/6}; \tag{25.16}$$

(a.3.2) $r = -\dfrac{7}{3}$:

$$x^5_{\pm,\text{e.m./weak.}}(\tau) = x^5_{0,\text{e.m./weak}}\left[\frac{C_{12}^2 + C_{22}^2 + C_{32}^2}{-C_{02}^2 \pm A_2\left(x^5_{0,\text{e.m./weak}}\right)^{-r}}\right]^3 \tag{25.17}$$

$$\times \sinh^{-6}\left[\pm\frac{1}{6}\sqrt{\pm\left(C_{12}^2 + C_{22}^2 + C_{32}^2\right)}\left(x^5_{0,\text{e.m./weak}}\right)^{(3r+1)/6}(\tau + A_1)\right]. \tag{25.18}$$

### 25.2.2 Strong Interaction above Threshold

(b.1) $\pm\left(C_{12}^2+C_{22}^2\right)+A_2\left(x^5_{0,\text{strong}}\right)^{-r}=0,\ C_{02}^2-C_{32}^2\neq 0$:

(b.1.1) $r\neq -4$:

$$x^5_{\pm,\text{strong}}(\tau)=\left[\pm\frac{r+4}{2}(\tau+A_1)(x^5_{0,\text{strong}})^{(r+2)/2}\sqrt{\mp(C_{02}^2-C_{32}^2)}\right]^{2/(r+4)}; \tag{25.19}$$

(b.1.2) $r=-4$:

$$x^5_{\pm,\text{strong}}(\tau)=\exp\left[\pm(\tau+A_1)\left(x^5_{0,\text{strong}}\right)^{(r+2)/2}\sqrt{\mp\left(C_{02}^2-C_{32}^2\right)}\right]; \tag{25.20}$$

(b.2) $\pm\left(C_{12}^2+C_{22}^2\right)+A_2\left(x^5_{0,\text{strong}}\right)^{-r}\neq 0,\ C_{02}^2-C_{32}^2=0$:

(b.2.1) $r\neq -2$:

$$x^5_{\pm,\text{strong}}(\tau)$$
$$=\left[\pm\frac{r+2}{2}\sqrt{\pm(C_{12}^2+C_{22}^2)+A_2(x^5_{0,\text{strong}})^{-r}}(x^5_{0,\text{strong}})^{r/2}(\tau+A_1)\right]^{2/(r+2)}; \tag{25.21}$$

(b.2.2) $r=-2$:

$$x^5_{\pm,\text{strong}}(\tau)$$
$$=\exp\left[\pm\sqrt{\pm(C_{12}^2+C_{22}^2)+A_2(x^5_{0,\text{strong}})^{-r}}(x^5_{0,\text{strong}})^{r/2}(\tau+A_1)\right]; \tag{25.22}$$

(b.3) $\pm\left(C_{12}^2+C_{22}^2\right)+A_2\left(x^5_{0,\text{strong}}\right)^{-r}\neq 0,\ C_{02}^2-C_{32}^2\neq 0$:

(b.3.1) $r\neq -4$:

$$\tau+A_1=\pm\frac{2\left[\mp\left(C_{02}^2-C_{32}^2\right)\right]^{(r-2)/4}}{(r+4)\left[\pm\left(C_{12}^2+C_{22}^2\right)+A_2\left(x^5_{0,\text{strong}}\right)^{-r}\right]^{(r+4)/4}} \times\int\frac{\mathrm{d}t}{\sqrt{t^{4/(r+4)}+1}}, \tag{25.23}$$

where

$$t=\left[\frac{\pm\left(C_{12}^2+C_{22}^2\right)+A_2\left(x^5_{0,\text{strong}}\right)^{-r}}{\left(x^5_{0,\text{strong}}\right)^2\left(\mp C_{02}^2\pm C_{32}^2\right)}\right]^{(r+4)/4}\left(x^5_{\pm,\text{strong}}(\tau)\right)^{(r+4)/2}; \tag{25.24}$$

(b.3.2) $r = -4$:

$$x^5_{\pm,\mathrm{strong}}(\tau) = x^5_{0,\mathrm{strong}} \sqrt{\frac{-C^2_{02} + C^2_{32}}{C^2_{12} + C^2_{22} \pm A_2 \left(x^5_{0,\mathrm{strong}}\right)^{-r}}} \times \sinh^{-1}\left[\pm\sqrt{\mp\left(C^2_{02} - C^2_{32}\right)} \left(x^5_{0,\mathrm{strong}}\right)^{(r+2)/2} (\tau + A_1)\right] . \tag{25.25}$$

## *25.2.3 Gravitational Interaction above Threshold*

(I) $? = 0$:

(c.I.1) $\pm\left(C^2_{12} + C^2_{22}\right) + A_2 \left(x^5_{0,\mathrm{grav}}\right)^{-r} = 0$, $C^2_{02} - C^2_{32} \neq 0$:

(c.I.1.1) $r \neq -4$:

$$x^5_{\pm,\mathrm{grav}}(\tau) = \left[\pm\frac{r+4}{2} (\tau + A_1) \left(x^5_{0,\mathrm{grav}}\right)^{(r+2)/2} \sqrt{\mp\left(C^2_{02} - C^2_{32}\right)}\right]^{2/(r+4)} ; \tag{25.26}$$

(c.I.1.2) $r = -4$:

$$x^5_{\pm,\mathrm{grav}}(\tau) = \exp\left[\pm(\tau + A_1) \left(x^5_{0,\mathrm{grav}}\right)^{(r+2)/2} \sqrt{\mp\left(C^2_{02} - C^2_{32}\right)}\right] ; \tag{25.27}$$

(c.I.2) $\pm(C^2_{12} + C^2_{22}) + A_2(x^5_{0,\mathrm{grav}})^{-r} \neq 0$, $C^2_{02} - C^2_{32} = 0$:

(c.I.2.1) $r \neq -2$:

$$x^5_{\pm,\mathrm{grav}}(\tau) = \left[\pm\frac{r+2}{2}\sqrt{\pm\left(C^2_{12} + C^2_{22}\right) + A_2 \left(x^5_{0,\mathrm{grav}}\right)^{-r}} \left(x^5_{0,\mathrm{grav}}\right)^{r/2} (\tau + A_1)\right]^{2/(r+2)} ; \tag{25.28}$$

(c.I.2.2) $r = -2$:

$$x^5_{\pm,\mathrm{grav}}(\tau) = \exp\left[\pm\sqrt{\pm\left(C^2_{12} + C^2_{22}\right) + A_2 \left(x^5_{0,\mathrm{grav}}\right)^{-r}} \left(x^5_{0,\mathrm{grav}}\right)^{r/2} (\tau + A_1)\right] ; \tag{25.29}$$

(c.I.2) $\pm\left(C^2_{12} + C^2_{22}\right) + A_2 \left(x^5_{0,\mathrm{grav}}\right)^{-r} \neq 0$, $C^2_{02} - C^2_{32} \neq 0$:

(c.I.2.1) $r \neq -4$:

$$\tau + A_1 = \pm \frac{2\left[\mp\left(C_{02}^2 - C_{32}^2\right)\right]^{(r-2)/4}}{(r+4)\left[\pm\left(C_{12}^2 + C_{22}^2\right) + A_2(x_{0,\mathrm{grav}}^5)^{-r}\right]^{(r+4)/4}} \times \int \frac{\mathrm{d}t}{\sqrt{t^{4/(r+4)}+1}} \tag{25.30}$$

where

$$t = \left[\frac{\pm\left(C_{12}^2 + C_{22}^2\right) + A_2\left(x_{0,\mathrm{grav}}^5\right)^{-r}}{\left(x_{0,\mathrm{grav}}^5\right)^2\left(\mp C_{02}^2 \pm C_{32}^2\right)}\right]^{(r+4)/4} \left(x_{\pm,\mathrm{grav}}^5(\tau)\right)^{(r+4)/2}; \tag{25.31}$$

(c.I.2.2) $r = -4$:

$$x_{\pm,\mathrm{grav}}^5(\tau) = x_{0,\mathrm{grav}}^5 \sqrt{\frac{-C_{02}^2 + C_{32}^2}{C_{12}^2 + C_{22}^2 \pm A_2\left(x_{0,\mathrm{grav}}^5\right)^{-r}}} \times \sinh^{-1}\left[\pm\sqrt{\mp\left(C_{02}^2 - C_{32}^2\right)}\left(x_{0,\mathrm{grav}}^5\right)^{(r+2)/2}(\tau + A_1)\right]. \tag{25.32}$$

(II) $? = \tilde{2}$:

(c.II.1) $A_2\left(x_{0,\mathrm{grav}}^5\right)^{-r} = 0, \ \mp\left(x_{0,\mathrm{grav}}^5\right)^{\tilde{2}}\left(C_{02}^2 - C_{12}^2 - C_{22}^2 - C_{32}^2\right) \neq 0$:

(c.II.1.1) $r \neq -4$:

$$x_{\pm,\mathrm{grav}}^5(\tau) = \left[\pm\frac{r+4}{2}(\tau + A_1)(x_{0,\mathrm{grav}}^5)^{(r+2)/2} \times \sqrt{\mp(x_{0,\mathrm{grav}}^5)^{\tilde{2}}(C_{02}^2 - C_{12}^2 - C_{22}^2 - C_{32}^2)}\right]^{2/(r+4)}; \tag{25.33}$$

(c.II.1.2) $r = -4$:

$$x_{\pm,\mathrm{grav}}^5(\tau) = \exp\left[\pm(\tau + A_1)\left(x_{0,\mathrm{grav}}^5\right)^{(r+2)/2}\sqrt{\mp\left(C_{02}^2 - C_{32}^2\right)}\right]; \tag{25.34}$$

(c.II.2) $A_2(x_{0,\mathrm{grav}}^5)^{-r} \neq 0, \ \mp(x_{0,\mathrm{grav}}^5)^{\tilde{2}}\left(C_{02}^2 - C_{12}^2 - C_{22}^2 - C_{32}^2\right) = 0$:

(c.II.2.1) $r \neq -2$:

$$x^5_{\pm,\text{grav}}(\tau) = \left[\pm\frac{r+2}{2}\sqrt{A_2\left(x^5_{0,\text{grav}}\right)^{-r}}\left(x^5_{0,\text{grav}}\right)^{r/2}(\tau + A_1)\right]^{2/(r+2)}; \tag{25.35}$$

(c.II.2.2) $r = -2$:

$$x^5_{\pm,\text{grav}}(\tau) = \exp\left[\pm\sqrt{A_2\left(x^5_{0,\text{grav}}\right)^{-r}}\left(x^5_{0,\text{grav}}\right)^{r/2}(\tau + A_1)\right]; \tag{25.36}$$

(c.II.3) $A_2\left(x^5_{0,\text{grav}}\right)^{-r} \neq 0,\ \mp\left(x^5_{0,\text{grav}}\right)^{\tilde{2}}\left(C^2_{02} - C^2_{12} - C^2_{22} - C^2_{32}\right) \neq 0$:

(c.II.3.1) $r \neq -4$:

$$\frac{\tau + A_1}{2} = \pm\left(x^5_{0,\text{grav}}\right)^{-r/2} \frac{2}{(r+4)\sqrt{\mp\left(x^5_{0,\text{grav}}\right)^{\tilde{2}}\left(C^2_{02} - C^2_{12} - C^2_{22} - C^2_{32}\right)}}$$
$$\times\left[\frac{\mp\left(x^5_{0,\text{grav}}\right)^{\tilde{2}}\left(C^2_{02} - C^2_{12} - C^2_{22} - C^2_{32}\right)}{A_2\left(x^5_{0,\text{grav}}\right)^{-r}}\right]^{(r+4)/4}\int\frac{dt}{\sqrt{t^{4/(r+4)}+1}}, \tag{25.37}$$

where

$$t = \left[\frac{A_2\left(x^5_{0,\text{grav}}\right)^{-r}}{\mp\left(x^5_{0,\text{grav}}\right)^{\tilde{2}}\left(C^2_{02} - C^2_{12} - C^2_{22} - C^2_{32}\right)}\right]^{(r+4)/4}\left(x^5_{\pm,\text{grav}}(\tau)\right)^{(r+4)/2}; \tag{25.38}$$

(c.II.3.2) $r = -4$:

$$x^5_{\pm,\text{grav}}(\tau) = \sqrt{\frac{\mp\left(x^5_{0,\text{grav}}\right)^{\tilde{2}}\left(C^2_{02} - C^2_{12} - C^2_{22} - C^2_{32}\right)}{A_2\left(x^5_{0,\text{grav}}\right)^{-r}}}$$
$$\times\sinh^{-1}\left[\pm\sqrt{\mp\left(x^5_{0,\text{grav}}\right)^{\tilde{2}}\left(C^2_{02} - C^2_{12} - C^2_{22} - C^2_{32}\right)}\left(x^5_{0,\text{grav}}\right)^{r/2}(\tau + A_1)\right]. \tag{25.39}$$

Let us stress that, in both special cases ? $= 0, \tilde{2}$, the treatment is quite analogous to those of the strong interaction above energy threshold.

## 25.3 Slicing and Dynamics

It must be by now clear, from the examples discussed in Sect. 25.2, that the explicit form of the geodesics in $\Re_5$ (namely, its dynamics) strictly depends on the sets of metric exponents $\widetilde{\mathbf{q}}_{\text{int.}}$ ((20.24)–(20.26)), which determine $x^A(\tau)$ through the knowledge of the generating function $F_{\pm}(\zeta; \widetilde{\mathbf{q}}, A_2)$.

But – as is easily seen from their expressions – the exponent sets $\widetilde{\mathbf{q}}_{\text{int.}}$ are discontinuous at the threshold energy $x^5_{0,\text{int.}}$:

$$\lim_{x^5 \longrightarrow x^{5+}_{0,\text{int.}}} \widetilde{\mathbf{q}}_{\text{int.}}(x^5) \neq \lim_{x^5 \longrightarrow x^{5-}_{0,\text{int.}}} \widetilde{\mathbf{q}}_{\text{int.}}(x^5), \tag{25.40}$$

namely, for a given interaction, different sets are obtained in the two different energy ranges (below and above threshold). This entails among the others that, as done in Sect. 20.2, it is necessary to use the right and left specifications of the Heaviside function in order to write $\widetilde{\mathbf{q}}_{\text{int.}}$ in the compact form (20.41)–(20.43), valid on the whole energy range. In turn, such a discontinuity in $\widetilde{\mathbf{q}}_{\text{int.}}$ at $x^5_{0,\text{int.}}$ causes an analogous behavior in the geodesic motions. In fact, let $x^5_{\text{int.},<}(\tau)$, $x^5_{\text{int.},>}(\tau)$ denote the solutions of the geodesic equation (24.3) for the fifth coordinate under and above threshold, respectively. Then, it is possible to impose e.g., $x^5_{\text{int.},<}(\overline{\tau}) = x^5_{0,\text{int.}}$ and find the corresponding value $\overline{\tau} \in R$. However, if such a value is replaced in the geodesic solution corresponding to the other energy range, one finds in general $x^5_{\text{int.},>}(\overline{\tau}) = \overline{x^5_{\text{int.}}} \neq x^5_{0,\text{int.}}$.

The situation is exactly analogous to that we encountered in the case of the Killing symmetries (Sect. 22.4). The nontrivial "bifurcation of dynamics" in the two energy ranges is clearly related to the nature change (from parameter to coordinate) the variable $x^5$ undergoes in the passage DSR→DR5. Therefore, dynamic structures present in an energy range in which the space–time sector is standard Minkowskian – or at least its metric coefficients are constant – may no longer occur when (in a different energy range) the space–time of $\Re_5$ becomes Minkowskian deformed, and vice versa.

Such a change of role of energy in the geometrical embedding of $\widetilde{M}$ in $\Re_5$ implies also, in full analogy with the case of the Killing isometries, that the dynamics in a given 4D space $\widetilde{M}\left(x^5 = \overline{x^5}\right)$ is *different* from the dynamics obtained for the slice of $\Re_5$ at constant energy $x^5 = \overline{x^5}$ with space–time sector coinciding with $\widetilde{M}\left(x^5 = \overline{x^5}\right)$. Symbolically one has:

$$\text{Dynamics in } \Re_5|_{\mathrm{d}x^5=0 \Leftrightarrow x^5=\overline{x^5}} \neq \text{Dynamics in } \widetilde{M}\left(x^5 = \overline{x^5}\right). \tag{25.41}$$

In fact the change of role of $x^5$ causes the destruction of the nonhomogeneous linearity in $\tau$ of the geodesic motions in DSR, which is no longer recovered in the inverse process of slicing of $\Re_5$ at $\mathrm{d}x^5 = 0$. This is again at variance with the metric level, where (see (19.10)) the constant-energy sections of $\Re_5$ at $x^5 = \overline{x^5}$ are endowed with the same metric structure of $\widetilde{M}\left(\overline{x^5}\right)$. Again, as in the case of the Killing symmetries, it is possible to understand this point by remembering that one is considering sections of a genuine Riemannian space, which therefore do keep memory of the fifth coordinate.

An explicit example of the key dynamic role played by the fifth coordinate in the embedding process is provided by the results of Sect. 24.4 for the geodesics relevant to class VIII of solutions of the 5D Einstein's equations. As already noted, the 5D metric (24.80) corresponding to the exponent set $\widetilde{\mathbf{q}}_{\mathrm{VIII}} = (0, 0, 0, 0, r)$ has a standard Minkowski structure for its space–time sector. In spite of this, the embedding of such a Minkowski space in $\Re_5$ (i.e., the presence and the form of the fifth metric coefficient) makes the dynamic behavior genuinely nontrivial, because to the standard geodesic motion of $M$ it is added the further condition that the geodesics must correspond to a minimal value of the time–energy uncertainty.

We can therefore conclude that not only Killing isometries, but dynamics, too, depends on the geometrical framework. This further supports the deep physical (not only mathematical) significance of the geometrical embedding of $\widetilde{M}$ in $\Re_5$.

# 26
# Conclusions and Perspectives

*"Two roads diverged in a wood, and I –*
*I took the one less traveled by,*
*And that has made the difference."*
(R. Frost: "The road not taken,"
from *Mountain interval*)

After the in-depth exposition of Deformed Relativity (DR) in four and five dimensions carried out in this book, it is worth making some remarks, drawing a few conclusions and outlining possible developments.

The implementation of the Finzi principle of solidarity – our cornerstone, a real first principle – for all the fundamental interactions (electromagnetic, weak, gravitational and strong) led us to generalize Special Relativity by building up DSR and its 5D extension, DR5. Their formalism allows one to geometrically represent interactions as space–time deformations (flat in the DSR case and curved in the DR5 one), described by metrics depending on the energy of the process considered.

Such a geometrization of interactions, in agreement with (and consequence of) the solidarity principle, was attained by paying a price for. The paid price was abandoning two unifying principles of (Einsteinian) Special Relativity , namely the invariance of the light speed (and its feature of maximal causal velocity for all interactions) and the uniqueness of the transformations connecting inertial frames (Lorentz transformations). Indeed, we have seen that DSR (and therefore DR5) implies a different maximal causal velocity for each interaction – and even unlike velocities for a given interaction, depending on the physical process and/or the space direction considered – and different sets of space–time coordinate transformations. Notice that, on the basis of the axiomatic formulation of SR, discussed in Sect. 1.2, it is implicit in the Principle of Relativity that one has to specify the class of physical phenomena to which it must apply (as first stressed in [2]). As a result, there exist a priori more relativities, depending on the class of phenomena considered and the interaction(s) ruling them.

This scenario is not new at all. As already said in Sect. 3.3.7, both these features of DSR, the noninvariance of the light speed and the existence of diverse sets of coordinate transformations for different interactions, are already present in the Lorentz–Poincaré version of Special Relativity. We can therefore state that *Deformed Relativity inherits the legacy of Lorentzian Relativity* .

This aspect of DSR is by no means a drawback of our formalism. Actually, the structure of Lorentzian Relativity is richer and in a sense more flexible than that of the Einsteinian one. Its great merit lies in its ability of fitting physical phenomena, by allowing maximal causal speeds and coordinate transformations suited to the interactions involved. Apparently this occurs at the price of loosing the unifying tissue provided, in Einstein's view to relativity, by the uniqueness of light speed and of the coordinate transformations. However, in Lorentzian and even more in Deformed Relativity there is a perhaps more profound and fundamental unifying principle: symmetry. Indeed, whereas the too rigid structure of Einsteinian Relativity is forced to take into account the possibility of breakdown of Lorentz invariance, when facing with interactions different from the electromagnetic one, Deformed Relativity in its role of heir of LR is able to adjust its geometric structure in order to suit the other interactions (or their nonlocal parts), by deforming its space–time and then its isometric transformations. In other words, *the unifying principle of both LR and DR is Lorentz invariance*, which, although apparently broken by some physical phenomena (see Part III), is actually recovered as deformed Lorentz invariance (see Sect. 3.3.5).

Contrarily to Einstein's view, both LR and DR oust the speed of light $c$ from its privileged position of absolute velocity and scale it down to an ordinary physical quantity, relative to the observer and/or the interaction. In this way, they seemingly give up the second unifying feature of Einsteinian Relativity, i.e., the existence of an universal constant. However, it must be stressed that the formalism of Deformed Relativity actually contains a constant with an universal character. Indeed, it follows from the discussion of Sect. 19.1 that in DR5 the energy–length conversion constant $l_0$ (introduced for dimensional consistency in the 5D interval in $\Re_5$: see (19.3)) can be identified with the standard gravitational constant $\kappa$, and be interpreted as determining the deformation of space–time for all interactions. Then, $\kappa$

is an universal constant in DR (the deformation constant) and plays the unifying role of the light speed $c$ in Einsteinian Relativity .

In this sense, we can maintain that Einstein was right in looking for unifying principles in Relativity, and that the main difference between ER on one side, and LR and DR on the other side, is simply in the choice of such principles. In our opinion, the choice originally done by Lorentz and Poincaré (and inherited by DR) is more fundamental, and perhaps – as we tried among the others to show in this book – more open to significant developments.

An important point concerns the relations between DSR and DR5. It has been shown that the formalism of Deformed Relativity, in either number of dimensions, provides us with a geometrization of interactions ruling physical phenomena for electromagnetic, weak, strong and gravitational processes. This is accomplished in terms of a deformed Minkowskian (DSR) or Riemannian (DR5) metric structure, by allowing for the dependence of metric on the energy. The connection between DSR and DR5 is apparently a mere mathematical one, implemented by embedding the deformed, 4D Minkowski space $\widetilde{M}(x^5)$ in the 5D Riemannian space $\Re_5$ with energy as fifth dimension. This seemingly entails that energy-constant slices of $\Re_5$ at $x^5 = \overline{x^5}$ would have to be equivalent in all respects to $\widetilde{M}\left(x^5 = \overline{x^5}\right)$. Although this is in fact true at metric level, actually this equivalence is lost as far as isometries and dynamics are concerned (see Sects. 22.4 and 25.3, respectively). We have also seen that in some cases, like the geodesic motions for Class VIII of solutions of Einstein equations (Sect. 24.4), which implies an Heisenberg-type time–energy uncertainty relation, the presence of the fifth coordinate affects the behavior of geodesics in the standard Minkowski space in a nontrivial way.

This physical influence of the extra dimension present in DR5 on the physics in the standard 4D space–time has another implication, of which some inklings have been already come into light in the experimental review of Part III. In fact, it can be seen that (as already noted) some of the experiments discussed involve physical processes occurring at fixed energy, like the double-slit and the coil experiments. In these cases, a treatment in terms of DSR (or of constant-energy sections of DR5) is enough. For DR5, this means leaving aside the fifth metric coefficient $f(x^5)$. On the contrary, in the Cavendish-like experiment to measure the gravity speed, and in the cavitation experiments (at least at microscopic level), energy changes during the involved processes. Then, a treatment in terms of DSR is not enough, and one has to make recourse to the whole 5D structure of DR5. In particular, the energy coefficient $f(x^5)$ is expected to play a basic role in the description of dynamic processes. Let us notice that, in this framework, the terms "dynamic" and "static" refer to the energy behavior. By "static" we mean a process occurring at constant energy, described by a fixed deformation of space–time. It corresponds just to zero speed in energy – i.e., to a static situation in the 5D space – like in the cases of the double-slit and coil experiments . The geometrical picture of such a process is gotten in terms of a fixed-energy metric, either in DSR or in (constant-energy slices of) DR5. A "dynamic" process occurs instead with variable energy, and in this context amounts to the establishing of the space–time deformation. In $\Re_5$, dynamic processes may occur with a constant speed in energy (like in cavitation: see Sect. 16.3) or with a variable one (the case of the gravity speed experiment, Sect. 15.3). Accounting for such a dynamic case (in which the space–time deformation is changing too) requires the full machinery of the 5D DR5 formalism, in particular by taking into account the energy coefficient $f(x^5)$ and its functional form. In other words, the experiments of Part III teach us that, *whereas DSR is able to geometrically describe interactions only in a static ("frozen") case, DR5 permits to describe the dynamic settling of the space–time deformation brought about by the interaction considered.* In this connection, we recall that in the DR5 framework room is allowed also for metrics with space coefficients depending either functionally or parametrically on $f(x^5)$ (see for instance the gravitational case in Sect. 22.3). This permits the energy dimension to affect the space deformation in a *direct* way.

Still concerning the influence of the extra dimension on the physics in the 4D deformed space–time, other points worth investigating are the possible connection between Lorentz invariance in DR5 and the usual gauge invariance, and the occurrence of parity violation as consequence of space anisotropy when viewed from the standpoint of the space–time–energy manifold $\Re_5$.

A further basic topic deserving study in DSR is the extension to the non-abelian case of the results obtained for the abelian gauge fields (like the e.m. one), based on the structure of the deformed Minkowski space $\widetilde{M}$ as Generalized Lagrange Space (see Sect. 9.4). In other words, it would be worth verifying if also non-abelian internal gauge fields can exist in absence of external fields, due to the intrinsic geometry of $\widetilde{M}$.

As to DR5, its formalism lends itself to a number of possible, future developments. These include e.g., solving the general Einstein equations with a nonzero cosmological constant, $\Lambda_{(5)} \neq 0$. Further improvements of the predictive power of the theory may come from the explicit introduction of a space–time-coordinate dependence in the fifth metric coefficient $f$ and/or in the cosmological constant $\Lambda_{(5)}$, i.e., assuming $f = f(E, x)$, $\Lambda_{(5)} = \Lambda_{(5)}(E, x)$. As is easily seen, this amounts to taking into account also the presence of matter in the DR5 scheme. In some cases, it comes out possible to relate mass (and therefore matter) to the LLI breakdown in four dimensions (see [6]). However, solving the 5D Einstein equations with matter sources is expected to be a quite formidable task.

The Killing symmetries of DR5 deserve further investigation on many respects. Let us quote, for instance: The Lie nature of the infinitesimal symmetries derived; the passage from the infinitesimal level to the finite one; and, last but not least, the physical meaning of the symmetries obtained. As far as this last point is concerned, the results obtained seemingly show an invariance of physical laws under nonlinear coordinate transformations (in particular in time and energy).

Besides the above "classical" problems, there are also what we may call the possible "quantum" aspects of the formalism. The basic question is whether the extra dimension energy can classically account for quantum features in four dimensions. A first result in this direction is provided by the Heisenberg-type time–energy uncertainty relation obtained from geodesic motion in $\Re_5$ for the class VIII of solutions of vacuum Einstein equations. This agrees with similar results derived within other noncompactified Kaluza–Klein-like models [138–140]. Actually, 5D schemes seem to provide a classical framework where to deal with (4D) quantum properties, including not only uncertainty relations but also quantization rules [123]. In DR5, the fact that energy is the fifth dimension is apparently a further complication factor, since actually, in most systems of physical interest at a microscopical level, energy is quantized. The two basic issues to be faced are to determine how energy quantization matches in this scheme, and to account for energy jumps within an apparently completely classical framework. A possible working hypothesis on heuristic ground is to explore the topology induced, for each interaction, by its own metric in the 5D space $\Re_5$. Such topologies may exhibit discontinuities or even singularities. Indeed, whereas in DSR the geometrical properties of the deformed Minkowski space for a given interaction are caused by the "frozen" 4D metric at fixed energy, it is the *whole structure* of the interaction 5D metric on the *whole energy range* which determines the $\Re_5$ geometry in DR5.[1] As we have seen in Sect. 19.3, the phenomenological 5D metrics of the four fundamental interactions are discontinuous at the threshold energy $E_{0,\text{int.}}$ (int.—e.m., weak, grav., strong). The metric discontinuities may well affect the corresponding topologies in a neighborhood of $E_{0,\text{int.}}$. By means of suitable mathematical tools (like the Mordell conjecture , exploiting diophantine equations [141], or the homology theory [142]), such "holes" in topologies can be associated to integers. These latter, in turn, can be possibly connected to quantized physical quantities like energy, charge, and other "charges" and quantum numbers related to the interaction considered.

If the above conjecture will reveal itself feasible, in the DR5 formalism quantization rules will arise from *geometry* of the space $\Re_5$ (topological structure), whereas uncertainty relations will be determined by its

---

[1] We can say, somewhat loosely, that $\Re_5$ is a functional of the metric:

$$\Re_5 = \Re_5\,[g]\,.$$

*dynamics* (geodesic motions). The task is far from easy, and the journey expected to be long, but we hope to have convinced the reader that exploring the space–time–energy land, by both theoretical and experimental instruments, is well worth the effort.

# Appendix A

# Reductivity of the $\Upsilon$-Hypothesis for the 12 Classes of the Vacuum Einstein Equations in the Power Ansatz

We shall discuss here the possible reductivity of the $\Upsilon$-hypothesis of functional independence, stated in Sect. 21.2, for the 12 classes of Power Ansatz solutions of the Einstein equations in vacuum (derived in Sect. 20.2), and solve explicitly the Killing equations in the five cases in which this hypothesis is violated. In the notation of Sect. 20.2, each class will be specified by an exponent set $\tilde{\mathbf{q}} \equiv (q_0, q_1, q_2, q_3, r)$.

## A.1 Analysis of Reductivity of the $\Upsilon$-Hypothesis

### A.1.1 Class (I)

$$\tilde{\mathbf{q}}_{\mathrm{I}} = \left( n, -n\left(\frac{2p+n}{2n+p}\right), n, p, \frac{p^2-2p+2np-4n+3n^2}{2n+p} \right).$$

One gets

$$r+2 = \frac{p^2+2np+3n^2}{2n+p} \tag{A.1}$$

$$\begin{cases} q_0 - r - 2 = q_2 - r - 2 = \dfrac{3n^3-7n^2-4np+np^2+2n^2p-p^2}{2n+p}; \\ q_1 - r - 2 = -(2n+p); \\ q_3 - r - 2 = \dfrac{p^3-3p^2-6np+2np^2+3n^2p-3n^2}{2n+p}. \end{cases} \tag{A.2}$$

The condition of nonvanishing denominators yields

$$\text{Den.} \neq 0 \Longleftrightarrow 2n + p \neq 0 \Longleftrightarrow q_1 - r - 2 \neq 0. \tag{A.3}$$

Therefore, under the further assumptions

$$\begin{cases} q_1 = -n\left(\dfrac{2p+n}{2n+p}\right) \neq 0 \Longleftrightarrow \begin{cases} n \neq 0, \\ 2p+n \neq 0, \end{cases} \\ r+2 = \dfrac{p^2+2np+3n^2}{2n+p} \neq 0 \Longleftrightarrow p^2+2np+3n^2 \neq 0, \end{cases} \tag{A.4}$$

one finds that the $\Upsilon$-hypothesis is satisfied at least by $\mu = 1$.

Moreover, we have the following possible *degenerate cases* (i.e., those in which the $\Upsilon$-hypothesis is violated for any value of $\mu$):

(Ia) $n = 0 \Rightarrow p \neq 0$. Then

$$\begin{cases} q_0 - r - 2 = q_2 - r - 2 = -p \neq 0; \\ q_1 - r - 2 = -(2n+p) \neq 0; \\ q_3 - r - 2 = p(p-3), \end{cases} \tag{A.5}$$

whence for $p \neq 3$ the $\Upsilon$-hypothesisis satisfied only by $\mu = 3$, and the case considered is not a degenerate one.

Therefore the true degenerate case of this class is characterized by

$$n = 0, \quad p = 3 \tag{A.6}$$

and corresponds to the 5D metric

$$g_{AB,\text{DR5power}}(x^5) = \text{diag}\left(1, -1, -1, -\left(\frac{x^5}{x_0^5}\right)^3, \pm\frac{x^5}{x_0^5}\right), \tag{A.7}$$

special case for $p = 3$ of the metric

$$g_{AB,\text{DR5power}}(x^5) = \text{diag}\left(1, -1, -1, -\left(\frac{x^5}{x_0^5}\right)^p, \pm\left(\frac{x^5}{x_0^5}\right)^{p-2}\right), \tag{A.8}$$

that will be discussed in Sect. A.2.4.

(Ib) $2n + p \Longleftrightarrow n = -2p$. From (A.3) it follows:

$$2n + p = -3p \neq 0 \tag{A.9}$$

and therefore

$$\begin{cases} q_0 - r - 2 = q_2 - r - 2 = p(6p+7); \\ q_1 - r - 2 = 3p \neq 0; \\ q_3 - r - 2 = -3p\left(p - \dfrac{1}{3}\right). \end{cases} \tag{A.10}$$

Then, the $\Upsilon$-hypothesis is satisfied for $p \neq -\frac{7}{6}$ by $\mu = 0, 2$, and $p \neq -\frac{1}{3}$ by $\mu = 3$.

For $p = -\frac{7}{6}$ one gets

$$\begin{cases} q_0 - r - 2 = q_2 - r - 2 = 0; \\ q_1 - r - 2 = -\dfrac{7}{2}; \\ q_3 - r - 2 = -\dfrac{21}{4}. \end{cases} \tag{A.11}$$

and the $\Upsilon$-hypothesisis still satisfied by $\mu = 3$.

For $p = \frac{1}{3}$ it is

$$\begin{cases} q_0 - r - 2 = q_2 - r - 2 = 3; \\ q_1 - r - 2 = 1; \\ q_3 - r - 2 = 0. \end{cases} \tag{A.12}$$

and the $\Upsilon$-hypothesis is still satisfied by $\mu = 0, 2$.

(Ic) $p^2 + 2np + 3n^2 = 0$. The only pair of real solutions of this equation is $(n, p) = (0, 0)$, that must be discarded because it entails the vanishing of the denominators.

### *A.1.2 Class (II)*

$$\widetilde{\mathbf{q}}_{\mathrm{II}} = (0, m, 0, 0, m - 2)\,.$$

We have

$$r + 2 = m; \tag{A.13}$$

$$\begin{cases} q_0 - r - 2 = q_2 - r - 2 = q_3 - r - 2 = -m, \\ q_1 - r - 2 = 0. \end{cases} \tag{A.14}$$

The $\Upsilon$-hypothesis is violated $\forall m \in R$ and $\forall \mu \in \{0, 1, 2, 3\}$.

In general Class (II) corresponds to the 5D metric

$$g_{AB,\mathrm{DR5power}}(x^5) = \mathrm{diag}\left(1, -\left(\frac{x^5}{x_0^5}\right)^m, -1, -1, \pm\left(\frac{x^5}{x_0^5}\right)^{m-2}\right), \tag{A.15}$$

we shall consider in Sect. A.2.2.

In particular, for $m = 0$ one gets

$$g_{AB,\mathrm{DR5power}}(x^5) = \mathrm{diag}\left(1, -1, -1, -1, \pm\left(\frac{x^5}{x_0^5}\right)^{-2}\right), \tag{A.16}$$

whose space–time part is the same as the standard Minkowski space $M$. This metric is a special case of

$$g_{AB,\text{DR5power}}(x^5) = \text{diag}\left(a, -b, -c, -d, \pm f\left(x^5\right)\right), \tag{A.17}$$

whose Killing equations coincide with those of the metric with $a = b = c = d = 1$, solved in Sect. 22.1.1.

### *A.1.3 Class (III)*

$$\widetilde{\mathbf{q}}_{\text{III}} = (n, -n, n, n, -2(1-n)).$$

It is

$$r + 2 = 2n; \tag{A.18}$$

$$\begin{cases} q_0 - r - 2 = q_2 - r - 2 = q_3 - r - 2 = -n, \\ q_1 - r - 2 = -3n. \end{cases} \tag{A.19}$$

The $\Upsilon$-hypothesis is satisfied for $n \neq 0\ \forall \mu \in \{0,1,2,3\}$.

The degenerate case is characterized by $n = 0$ and corresponds to metric (A.16).

### *A.1.4 Class (IV)*

$$\widetilde{\mathbf{q}}_{\text{IV}} = (0, 0, 0, p, p-2).$$

One gets

$$r + 2 = m; \tag{A.20}$$

$$\begin{cases} q_0 - r - 2 = q_1 - r - 2 = q_3 - r - 2 = -p, \\ q_3 - r - 2 = 0. \end{cases} \tag{A.21}$$

The $\Upsilon$-hypothesis is violated $\forall p \in R$ and $\forall \mu \in \{0,1,2,3\}$.

In general this class corresponds to the 5D metric (A.8), whose special case $p = 0$ is given by metric (A.16).

### *A.1.5 Class (V)*

$$\widetilde{\mathbf{q}}_{\text{V}} = (-p, -p, -p, p, -(1+p)).$$

We have

$$r + 2 = 1 - p; \tag{A.22}$$

$$\begin{cases} q_0 - r - 2 = q_1 - r - 2 = q_2 - r - 2 = -1, \\ q_3 - r - 2 = 2p - 1. \end{cases} \tag{A.23}$$

Therefore the $\Upsilon$-hypothesis is satisfied for $p \neq 0$, $p \neq 1$ by $\mu = 0, 1, 2$ (and for $p \neq 0$, $p \neq 1$, $p \neq \frac{1}{2}$ also by $\mu = 3$).

There are three possible degenerate cases:

(Va) $p = 1$. One has

$$\begin{cases} q_0 = q_1 = q_2 = -1; \\ q_3 = 1; \\ r = -2. \end{cases} \tag{A.24}$$

The $\Upsilon$-hypothesis is not satisfied by any value of $\mu$. The corresponding 5D metric is

$$g_{AB,\mathrm{DR5power}}(x^5) = \mathrm{diag}\left(\left(\frac{x^5}{x_0^5}\right)^{-1}, -\left(\frac{x^5}{x_0^5}\right)^{-1}, -\left(\frac{x^5}{x_0^5}\right)^{-1}, -\frac{x^5}{x_0^5}, \pm\left(\frac{x^5}{x_0^5}\right)^{-2}\right) \tag{A.25}$$

and is discussed in Sect. A.2.5.

(Vb) $p = 0$. It is

$$\begin{cases} q_0 = q_1 = q_2 = q_3 = 0; \\ r = -1. \end{cases} \tag{A.26}$$

and the $\Upsilon$-hypothesis is violated by any value of $\mu$. The corresponding 5D metric is

$$g_{AB,\mathrm{DR5power}}(x^5) = \mathrm{diag}\left(1, -1, -1, -1, \pm\left(\frac{x^5}{x_0^5}\right)^{-1}\right), \tag{A.27}$$

special case of the metric (A.17).

(Vc) $p = \frac{1}{2}$. One gets

$$r + 2 = \frac{1}{2}; \tag{A.28}$$

$$\begin{cases} q_0 = q_1 = q_2 = -\dfrac{1}{2}, \\ q_3 = \dfrac{1}{2}. \end{cases} \tag{A.29}$$

Therefore the $\Upsilon$-hypothesis is satisfied by $\mu = 0, 1, 2$.

### *A.1.6 Class (VI)*

$$\widetilde{\mathbf{q}}_{\mathrm{VI}} = (q, 0, 0, 0, q - 2)$$

One has

$$r + 2 = q; \tag{A.30}$$

$$\begin{cases} q_0 - r - 2 = 0, \\ q_1 - r - 2 = q_2 - r - 2 = q_3 - r - 2 = -q. \end{cases} \tag{A.31}$$

The $\Upsilon$-hypothesis is not satisfied $\forall q \in R$ and $\forall \mu \in \{0,1,2,3\}$.

The corresponding 5D metric reads

$$g_{AB,\mathrm{DR5power}}(x^5) = \mathrm{diag}\left(\left(\frac{x^5}{x_0^5}\right)^q, -1, -1, -1, \pm\left(\frac{x^5}{x_0^5}\right)^{q-2}\right) \tag{A.32}$$

(with special case $q = 0$ given by (A.16)), and is discussed in Sect. A.2.1.

### A.1.7 Class (VII)

$$\widetilde{\mathbf{q}}_{\mathrm{VII}} = (q, -q, -q, -q, -q, -2)$$

It is

$$r + 2 = -q; \tag{A.33}$$

$$\begin{cases} q_0 - r - 2 = 2q, \\ q_1 - r - 2 = q_2 - r - 2 = q_3 - r - 2 = 0. \end{cases} \tag{A.34}$$

The $\Upsilon$-hypothesis is satisfied for $q \neq 0$ by $\mu = 0$.

The degenerate case $q = 0$ corresponds to the 5D metric (A.16).

### A.1.8 Class (VIII)

$\widetilde{\mathbf{q}}_{\mathrm{VIII}} = (0, 0, 0, 0, r \in R)$

One has

$$q_0 - r - 2 = q_1 - r - 2 = q_3 - r - 2 = q_3 - r - 2 = -r - 2. \tag{A.35}$$

The $\Upsilon$-hypothesis is violated $\forall r \in R$ and $\forall \mu \in \{0,1,2,3\}$.

The Class (VIII) corresponds to the 5D metric

$$g_{AB,\mathrm{DR5power}}(x^5) = \mathrm{diag}\left(1, -1, -1, -1, \pm\left(\frac{x^5}{x_0^5}\right)^{-r}\right), \tag{A.36}$$

that generalizes metric (A.16) and is a special case of the metric (A.17).

### A.1.9 Class (IX)

$\widetilde{\mathbf{q}}_{\mathrm{IX}} = (0, 0, n, 0, n - 2)$

We have

$$r + 2 = n; \tag{A.37}$$

$$\begin{cases} q_0 - r - 2 = q_1 - r - 2 = q_3 - r - 2 = -n, \\ q_2 - r - 2 = 0. \end{cases} \tag{A.38}$$

The $\Upsilon$-hypothesis is not satisfied $\forall n \in R$ and $\forall \mu \in \{0,1,2,3\}$.

The corresponding 5D metric is

$$g_{AB,\mathrm{DR5power}}(x^5) = \mathrm{diag}\left(1, -1, -\left(\frac{x^5}{x_0^5}\right)^n, -1, \pm\left(\frac{x^5}{x_0^5}\right)^{n-2}\right) \tag{A.39}$$

and is discussed in Sect. A.2.3 (whereas $n = 0$ gives metric (A.16)).

### A.1.10 Class (X)

$$\widetilde{\mathbf{q}}_{\mathrm{X}} = \left(q, -\frac{pq+np+nq}{n+p+q}, n, p, \frac{(n+p+q)(n+p+q-2)-(pq+np+nq)}{n+p+q}\right).$$

One gets

$$r+2 = \frac{(n+p+q)^2-(pq+np+nq)}{n+p+q}; \tag{A.40}$$

$$\begin{cases} q_0 - r - 2 = -\dfrac{p^2+n^2+np}{n+p+q}; \\ q_1 - r - 2 = -(n+p+q); \\ q_2 - r - 2 = -\dfrac{q^2+p^2+pq}{n+p+q}; \\ q_3 - r - 2 = -\dfrac{q^2+n^2+nq}{n+p+q}. \end{cases} \tag{A.41}$$

The condition of nonvanishing denominators yields

$$\text{Den.} \neq 0 \Longleftrightarrow n+p+q \neq 0 \Longleftrightarrow q_1 - r - 2 \neq 0. \tag{A.42}$$

Therefore, under the further assumptions

$$\begin{cases} q_1 = -\dfrac{pq+np+nq}{n+p+q} \neq 0 \Longleftrightarrow pq+np+nq \neq 0; \\ r+2 = \dfrac{(n+p+q)^2-(pq+np+nq)}{n+p+q} \neq 0 \\ \qquad \Longleftrightarrow p^2+n^2+q^2+pq+np+nq \neq 0, \end{cases} \tag{A.43}$$

one finds that the $\Upsilon$-hypothesis is satisfied at least by $\mu = 1$.

Then, we have the following possible degenerate cases:

(Xa) $pq+np+nq = 0$. Therefore

$$r+2 = n+p+q \neq 0; \tag{A.44}$$

$$\begin{cases} q_0 - r - 2 = -\dfrac{p^2+n^2+np}{n+p+q}; \\ q_1 - r - 2 = -(n+p+q) \neq 0; \\ q_2 - r - 2 = -\dfrac{q^2+p^2+pq}{n+p+q}; \\ q_3 - r - 2 = -\dfrac{q^2+n^2+nq}{n+p+q}. \end{cases} \tag{A.45}$$

One has to consider the following subcases:

(Xa.1) $q = 0$. Then

$$np = 0, n + p \neq 0 \Longleftrightarrow \begin{cases} \text{(Xa.1.1)} & n = 0,\ p \neq 0 \\ \text{or} & \\ \text{(Xa.1.2)} & n \neq 0,\ p = 0. \end{cases} \tag{A.46}$$

(Xa.1.1) One gets

$$r = p - 2; \tag{A.47}$$

$$\begin{cases} q_0 - r - 2 = -p \neq 0; \\ q_1 - r - 2 = p \neq 0; \\ q_2 - r - 2 = -p \neq 0; \\ q_3 - r - 2 = 0. \end{cases} \tag{A.48}$$

Therefore the $\Upsilon$-hypothesis is satisfied by no value of $\mu$. The corresponding metric is given by (A.8) (Class (IV)).

(Xa.1.2) One finds

$$r = n - 2; \tag{A.49}$$

$$\begin{cases} q_0 - r - 2 = q_1 - r - 2 = -n \neq 0; \\ q_2 - r - 2 = 0; \\ q_3 - r - 2 = -n \neq 0. \end{cases} \tag{A.50}$$

Again, the $\Upsilon$-hypothesis is not satisfied by any value of $\mu$. The corresponding 5D metric is given by (A.39) (Class (IX)).

(Xa.2) $n = 0$. Then

$$pq = 0, p + q \neq 0 \Longleftrightarrow \begin{cases} \text{(Xa.2.1)} & p = 0,\ q \neq 0 \\ \text{or} & \\ \text{(Xa.2.2)} & p \neq 0,\ q = 0. \end{cases} \tag{A.51}$$

(Xa.2.1) One gets

$$r = q - 2; \tag{A.52}$$

$$\begin{cases} q_0 - r - 2 = 0; \\ q_1 - r - 2 = q_2 - r - 2 = q_3 - r - 2 = -q \neq 0. \end{cases} \tag{A.53}$$

Therefore the $\Upsilon$-hypothesis is satisfied by no value of $\mu$. The corresponding metric coincides with that of Class (VI), (A.32).

The case (Xa.2.2) coincides with the case (X a.1.1).

(Xa.3) $p = 0$. It is

$$nq = 0, n + q \neq 0 \Longleftrightarrow \begin{cases} \text{(Xa.3.1)} & n = 0,\ q \neq 0 \\ \text{or} & \\ \text{(Xa.3.2)} & n \neq 0,\ q = 0, \end{cases} \tag{A.54}$$

and therefore the subcases (Xa.3.1) and (Xa.3.2) coincide with subcases (Xa.2.1) and (Xa.1.2), respectively.

(Xa.4) $q_0 - r - 2 = 0 \Leftrightarrow p^2 + n^2 + np = 0$. The only possible pair of real solutions of this equation is $(p, n) = (0, 0)$. From the condition of nonvanishing denominators it then follows $q \neq 0$. Therefore such a case coincides with (Xa.2.1).

(Xa.5) $q_1 - r - 2 = 0 \Leftrightarrow n + p + q = 0$. This condition expresses the vanishing of the denominators, and therefore this case is impossible.

(Xa.6) $q_2 - r - 2 = 0 \Leftrightarrow q^2 + p^2 + pq = 0$. The only real solution of this equation is $(q, p) = (0, 0)$. The condition of nonvanishing denominators entails $n \neq 0$, and then this case coincides with subcase (Xa.1.2).

(Xa.7) $q_3 - r - 2 = 0 \Leftrightarrow q^2 + n^2 + nq = 0$. The only real solution is $(q, n) = (0, 0)$. The condition of nonvanishing denominators entails $p \neq 0$. Therefore this case coincides with subcase (Xa.1.1).

(Xb) $p^2 + n^2 + q^2 + pq + np + nq = 0$. The only possible solution of such equation is $(p, n, q) = (0, 0, 0)$, which contradicts the nonvanishing condition of denominators. This case is impossible, too.

### *A.1.11 Class (XI)*

$$\widetilde{\mathbf{q}}_{\mathrm{XI}} = \left( q, -\frac{n\,(2q + n)}{2n + q}, n, n, \frac{3n^2 - 4n + 2nq - 2q + q^2}{2n + q} \right)$$

One gets

$$r + 2 = \frac{3n^2 + 2nq + q^2}{2n + q}; \tag{A.55}$$

$$\begin{cases} q_0 - r - 2 = -\dfrac{3n^2}{2n + q}; \\ q_1 - r - 2 = -\,(2n + q)\,; \\ q_2 - r - 2 = q_3 - r - 2 = -\dfrac{q^2 + n^2 + nq}{2n + q}. \end{cases} \tag{A.56}$$

The condition of nonvanishing denominators yields

$$\text{Den.} \neq 0 \Longleftrightarrow 2n + q \neq 0 \Longleftrightarrow q_1 - r - 2 \neq 0. \tag{A.57}$$

Then, by assuming further

$$\left\{ \begin{array}{l} q_1 = -\dfrac{n(2q+n)}{2n+q} \neq 0 \Longleftrightarrow \left\{ \begin{array}{c} n \neq 0; \\ 2q+n \neq 0, \end{array} \right. ; \\ r+2 = \dfrac{3n^2+2nq+q^2}{2n+q} \neq 0 \Longleftrightarrow 3n^2+2nq+q^2 \neq 0 \end{array} \right. , \tag{A.58}$$

one gets that the $\Upsilon$-hypothesisis satisfied at least by $\mu = 1$.

We have also the following possible degenerate cases:

(XIa) $n = 0 \Rightarrow q \neq 0$. Then

$$r = q - 2; \tag{A.59}$$

$$\left\{ \begin{array}{c} q_0 - r - 2 = 0; \\ q_1 - r - 2 = q_2 - r - 2 = q_3 - r - 2 = -q \neq 0. \end{array} \right. \tag{A.60}$$

Therefore the $\Upsilon$-hypothesis is satisfied by no value of $\mu$. The corresponding metric coincides with that of Class (VI).

(XIb) $2q + n = 0$ (which, together with $2n + q \neq 0$, entails $q \neq 0$). Then

$$r = -3q - 2; \tag{A.61}$$

$$\left\{ \begin{array}{c} q_0 - r - 2 = 4q \neq 0; \\ q_1 - r - 2 = 3q \neq 0; \\ q_2 - r - 2 = q_3 - r - 2 = q \neq 0. \end{array} \right. \tag{A.62}$$

Therefore the $\Upsilon$-hypothesis is satisfied by $\mu = 0, 2, 3$.

(XIc) $3n^2 + 2nq + q^2 = 0$. The only possible solution of such equation is $(n, q) = (0, 0)$, contradicting the nonvanishing condition of denominators, whence the impossibility of this case.

### *A.1.12 Class (XII)*

$$\widetilde{\mathfrak{q}}_{\mathrm{XII}} = \left( q, n, n, -\frac{n(2q+n)}{2n+q}, \frac{p^2+pq-2p+np-2n+nq+n^2-2q+q^2}{n+p+q} \right)$$

We have

$$r + 2 = \frac{p^2+n^2+q^2+pq+np+nq}{n+p+q}; \tag{A.63}$$

$$\left\{ \begin{array}{c} q_0 - r - 2 = -\dfrac{p^2+n^2+np}{n+p+q}; \\ q_1 - r - 2 = q_2 - r - 2 = -\dfrac{p^2+q^2+pq}{n+p+q}; \\ q_3 - r - 2 = \\ = -\dfrac{3n^3+q^3+6n^2q+5nq^2+2np^2+3pn^2+q^2p+p^2q+5npq}{(n+p+q)(2n+q)}. \end{array} \right. \tag{A.64}$$

The condition of nonzero denominators yields

$$\text{Den.} \neq 0 \Longleftrightarrow \begin{cases} 2n+q \neq 0; \\ n+p+q \neq 0. \end{cases} \tag{A.65}$$

Let us put

$$\left.\begin{array}{c} n+p+q = a \in R_0 \\ 2n+q = b \in R_0 \end{array}\right\} \Rightarrow p = a - \frac{q+b}{2}. \tag{A.66}$$

Considering e.g., $\mu = 0$, one has

$$q_0 - r - 2 = -\frac{p^2+n^2+np}{n+p+q}$$

$$= \frac{1}{4a}\left[3(q-a)^2 + (a-b)^2\right] \overset{q \neq a \,\text{or}\, a \neq b}{\overbrace{\gtrless}} 0, \quad \text{sgn}(a) = \begin{cases} 1, \\ -1. \end{cases} \tag{A.67}$$

By assuming $q \neq a$ and/or $a \neq b$, the vanishing of the denominators entails $q_0 - r - 2 \neq 0$. Then, under the further assumptions

$$\begin{cases} q_0 = q \neq 0; \\ r+2 = \dfrac{p^2+n^2+q^2+pq+np+nq}{n+p+q} \neq 0 \\ \Longleftrightarrow p^2+n^2+q^2+pq+np+nq \neq 0, \end{cases} \tag{A.68}$$

one gets that the $\Upsilon$-hypothesis is satisfied at least by $\mu = 0$.

The possible degenerate cases are:

(XIIa) $q = 0 \Rightarrow \begin{cases} n \neq 0, \\ n+p \neq 0. \end{cases}$ Then

$$r+2 = \frac{p^2+n^2+np}{n+p}; \tag{A.69}$$

$$\begin{cases} q_0 - r - 2 = -\dfrac{p^2+n^2+np}{n+p}; \\ q_1 - r - 2 = q_2 - r - 2 = -\dfrac{p^2}{n+p}; \\ q_3 - r - 2 = -\dfrac{3n^2+2p^2+3pn}{2(n+p)}. \end{cases} \tag{A.70}$$

It is $r+2 \neq 0$ and $q_0 - r - 2 \neq 0$ because the only possible solution of $p^2+n^2+np = 0$ is the pair $(p,n) = (0,0)$, incompatible with the nonvanishing of the denominators. Therefore this case is impossible.

We have the following subcases:

(XIIa.1) $p = 0$. Then

$$r = n - 2; \tag{A.71}$$

$$\begin{cases} q_0 - r - 2 = -n \neq 0; \\ q_1 - r - 2 = q_2 - r - 2 = 0; \\ q_3 - r - 2 = -\frac{3}{2}n \neq 0. \end{cases} \tag{A.72}$$

Therefore the $\Upsilon$-hypothesis is satisfied by $\mu = 3$.

(XIIa.2) $3n^2 + 2p^2 + 3pn = 0$. This equation admits as only real solution the pair $(p, n) = (0, 0)$, incompatible with the nonvanishing of the denominators. Therefore, this case is impossible.

(XIIb) $p^2 + n^2 + q^2 + pq + np + nq = 0$. This equation has the only real solution $(p, n, q) = (0, 0, 0)$, contradicting the nonvanishing of the denominators, and then this case has to be discarded.

(XIIc) $q = a = b \Longleftrightarrow n = p = 0$. We have

$$r = a - 2; \tag{A.73}$$

$$\begin{cases} q_0 - r - 2 = 0; \\ q_1 - r - 2 = q_2 - r - 2 = q_3 - r - 2 = -a \neq 0. \end{cases} \tag{A.74}$$

Consequently the $\Upsilon$-hypothesis is satisfied by no value of $\mu$. The metric obtained is the same of Class (VI).

## A.2 Solution of the 5D Killing Equations for Totally Violated $\Upsilon$-Hypothesis

The analysis of the previous section has shown that, in the framework of the Power Ansatz, there exist five cases (actually only three of them are independent) in which the hypothesis $\Upsilon$ of functional independence is violated $\forall \mu = 0, 1, 2, 3$. In the following, we shall explicitly solve the Killing equations in such cases.

### *A.2.1 Case 1*

In the framework of the Power Ansatz, the first case we shall consider in which the $\Upsilon$-hypothesis is not satisfied by any value of $\mu$ corresponds to the 5D metric belonging to the VI class ($p \in R$)

$$g_{AB,\mathrm{DR5},1}(x^5) = \mathrm{diag}\left(\left(\frac{x^5}{x_0^5}\right)^p, -1, -1, -1, \pm\left(\frac{x^5}{x_0^5}\right)^{p-2}\right). \tag{A.75}$$

For $p = 0$ one gets the metric

$$g_{AB,\mathrm{DR5}}(x^5) = \mathrm{diag}\left(1, -1, -1, -1, \pm\left(\frac{x^5}{x_0^5}\right)^{-2}\right) \tag{A.76}$$

that is a special case of the metric

$$g_{AB,\mathrm{DR5}}(x^5) = \mathrm{diag}\left(a, -b, -c, -d, \pm f\left(x^5\right)\right), \tag{A.77}$$

whose Killing equations (coincident with those relevant to the metric with $a = b = c = d = 1$) have been solved in Sect. 22.1.1. Therefore, one can assume $p \in R_0$.

The Killing system (21.6)–(21.17) in this case reads

$$\left\{\begin{array}{c}
\pm\xi_{0,0}(x^A) + \dfrac{p}{2}\dfrac{x^5}{(x_0^5)^2}\xi_5(x^A) = 0; \\
\xi_{0,1}(x^A) + \xi_{1,0}(x^A) = 0; \\
\xi_{0,2}(x^A) + \xi_{2,0}(x^A) = 0; \\
\xi_{0,3}(x^A) + \xi_{3,0}(x^A) = 0; \\
\xi_{0,5}(x^A)x^5 - p\xi_0(x^A) + \xi_{5,0}(x^A)x^5 = 0; \\
\xi_{1,1}(x^A) = 0; \\
\xi_{1,2}(x^A) + \xi_{2,1}(x^A) = 0; \\
\xi_{1,3}(x^A) + \xi_{3,1}(x^A) = 0; \\
\xi_{1,5}(x^A) + \xi_{5,1}(x^A) = 0; \\
\xi_{2,2}(x^A) = 0; \\
\xi_{2,3}(x^A) + \xi_{3,2}(x^A) = 0; \\
\xi_{2,5}(x^A) + \xi_{5,2}(x^A) = 0; \\
\xi_{3,3}(x^A) = 0; \\
\xi_{3,5}(x^A) + \xi_{5,3}(x^A) = 0; \\
-2\xi_{5,5}(x^A)x^5 + (p-2)\,\xi_5(x^A) = 0.
\end{array}\right. \tag{A.78}$$

Its solution depends on the signature (time-like or space-like) of $x^5$. One gets, for the covariant Killing five-vector:

$$\xi_0(x^0, x^5; p)$$

$$= \left\{\begin{array}{c}
\text{“}+\text{”}: \\
\left(x_0^5\right)^{-1}\left[A\cos\left(\dfrac{p}{2}\dfrac{x^0}{x_0^5}\right) - B\sin\left(\dfrac{p}{2}\dfrac{x^0}{x_0^5}\right)\right](x^5)^{p/2} + \alpha(x^5)^p; \\
\text{“}-\text{”}: \\
\left(x_0^5\right)^{-1}\left[C\cosh\left(\dfrac{p}{2}\dfrac{x^0}{x_0^5}\right) - D\sinh\left(\dfrac{p}{2}\dfrac{x^0}{x_0^5}\right)\right](x^5)^{p/2} + \alpha(x^5)^p;
\end{array}\right. \tag{A.79}$$

$$\xi_1(x^2, x^3) = \Theta_3 x^2 + \Theta_2 x^3 - T_1; \tag{A.80}$$

$$\xi_2(x^1, x^3) = -\Theta_3 x^1 + \Theta_1 x^3 - T_2; \tag{A.81}$$

$$\xi_3(x^1, x^2) = -\Theta_2 x^1 + \Theta_1 x^2 - T_3; \tag{A.82}$$

$$\xi_5(x^0, x^5; p) = \begin{cases} \text{“}+\text{”}: \\ \left[A\cos\left(\frac{p}{2}\frac{x^0}{x_0^5}\right) + B\sin\left(\frac{p}{2}\frac{x^0}{x_0^5}\right)\right](x^5)^{(p/2)-1}; \\ \text{“}-\text{”}: \\ \left[C\cosh\left(\frac{p}{2}\frac{x^0}{x_0^5}\right) + D\sinh\left(\frac{p}{2}\frac{x^0}{x_0^5}\right)\right](x^5)^{(p/2)-1} \end{cases} \tag{A.83}$$

with $\Theta_i$,$T_i$ $(i = 1, 2, 3)$, $A$, $B$, $C$, $D$, $\alpha \in R$. Since $\left[x_0^5\right] = l$, the dimensions of the transformation parameters are

$$[A] = [B] = [C] = [D] = l^{-((p/2)-1)}, \quad [\alpha] = l^{-p}, \quad [T_i] = l\ , \quad [\ \Theta_i] = l^0\ \forall i. \tag{A.84}$$

The 3D Killing group(of the Euclidean sections at $dx^5 = 0$, $dx^0 = 0$) is trivially the group of rototranslations of the Euclidean space $E_3$ with metric $\delta_{ij} = \text{diag}(-1, -1, -1)$ $((i, j) \in \{1, 2, 3\}^2)$:

$$\text{SO}(3)_{\text{STD.}} \otimes_s \text{Tr.}(3)_{\text{STD.}}. \tag{A.85}$$

### A.2.2 Case 2

The 5D metric for this case is

$$g_{AB,\text{DR5},2}(x^5) = \text{diag}\left(1, -\left(\frac{x^5}{x_0^5}\right)^p, -1, -1, \pm\left(\frac{x^5}{x_0^5}\right)^{p-2}\right). \tag{A.86}$$

The corresponding Killing equations are

$$\begin{cases} \xi_{0,0}(x^A) = 0; \\ \xi_{0,1}(x^A) + \xi_{1,0}(x^A) = 0; \\ \xi_{0,2}(x^A) + \xi_{2,0}(x^A) = 0; \\ \xi_{0,3}(x^A) + \xi_{3,0}(x^A) = 0; \\ \xi_{0,5}(x^A) + \xi_{5,0}(x^A) = 0; \\ \mp\xi_{1,1}(x^A) + \frac{p}{2}\frac{x^5}{(x_0^5)^2}\xi_5(x^A) = 0; \\ \xi_{1,2}(x^A) + \xi_{2,1}(x^A) = 0; \\ \xi_{1,3}(x^A) + \xi_{3,1}(x^A) = 0; \\ \xi_{1,5}(x^A)x^5 - p\xi_1(x^A) + \xi_{5,1}(x^A)x^5 = 0; \\ \xi_{2,2}(x^A) = 0; \\ \xi_{2,3}(x^A) + \xi_{3,2}(x^A) = 0; \\ \xi_{2,5}(x^A) + \xi_{5,2}(x^A) = 0; \\ \xi_{3,3}(x^A) = 0; \\ \xi_{3,5}(x^A) + \xi_{5,3}(x^A) = 0; \\ -2\xi_{5,5}(x^A)x^5 + (p-2)\,\xi_5(x^A) = 0. \end{cases} \tag{A.87}$$

Having as solution the covariant Killing vector

$$\xi_0(x^2, x^3) = \zeta_2 x^2 + \zeta_3 x^3 + T_0; \tag{A.88}$$

$$\xi_1(x^1, x^5; p) = \begin{cases} \text{“}+\text{”}: \\ \left(x_0^5\right)^{-1}\left[A\cosh\left(\dfrac{p}{2}\dfrac{x^1}{x_0^5}\right) + B\sinh\left(\dfrac{p}{2}\dfrac{x^1}{x_0^5}\right)\right](x^5)^{p/2} + \alpha(x^5)^p; \\ \text{“}-\text{”}: \\ \left(x_0^5\right)^{-1}\left[C\cos\left(\dfrac{p}{2}\dfrac{x^1}{x_0^5}\right) - D\sin\left(\dfrac{p}{2}\dfrac{x^1}{x_0^5}\right)\right](x^5)^{p/2} + \alpha(x^5)^p; \end{cases} \tag{A.89}$$

$$\xi_2(x^0, x^3) = -\zeta_2 x^1 + \Theta_1 x^3 - T_2; \tag{A.90}$$

$$\xi_3(x^0, x^2) = -\zeta_3 x^1 - \Theta_1 x^2 - T_3; \tag{A.91}$$

$$\xi_5(x^1, x^5; p) = \begin{cases} \text{“}+\text{”}: \\ \left[A\sinh\left(\dfrac{p}{2}\dfrac{x^1}{x_0^5}\right) + B\cosh\left(\dfrac{p}{2}\dfrac{x^1}{x_0^5}\right)\right](x^5)^{(p/2)-1}; \\ \text{“}-\text{”}: \\ \left[C\sin\left(\dfrac{p}{2}\dfrac{x^1}{x_0^5}\right) + D\cos\left(\dfrac{p}{2}\dfrac{x^1}{x_0^5}\right)\right](x^5)^{(p/2)-1} \end{cases} \tag{A.92}$$

with $\zeta_k (k = 2, 3)$, $\Theta_1$,$T_\nu$ $(\nu = 0, 2, 3)$, $A$, $B$, $C$, $D$, $\alpha \in R$. The dimensions of the transformation parameters are

$$[A] = [B] = [C] = [D] = l^{-((p/2)-1)},\ [\alpha] = l^{-p},\ [T_\nu] = l\ ,\ [\ \zeta_i] = [\Theta_1] = l^0. \tag{A.93}$$

The 3D Killing group(of the sections at $\mathrm{d}x^5 = 0$, $\mathrm{d}x^1 = 0$) is trivially the group of rototranslations of the pseudoeuclidean space $E_3'$ with metric $g_{\mu\nu} = \mathrm{diag}(1, -1, -1)$ $((\mu, \nu) \in \{0, 2, 3\}^2)$:

$$\mathrm{SO}(2,1)_{\mathrm{STD.}} \otimes_s \mathrm{Tr.}(2,1)_{\mathrm{STD.}}. \tag{A.94}$$

### *A.2.3 Case 3*

In this case the 5D metric reads

$$g_{AB,\mathrm{DR5},3}(x^5) = \mathrm{diag}\left(1, , -1, -\left(\frac{x^5}{x_0^5}\right)^p, -1, \pm\left(\frac{x^5}{x_0^5}\right)^{p-2}\right) \tag{A.95}$$

which is the same as Case 2, apart from an exchange of the space axes $x$ and $y$. The Killing vector is therefore obtained from the previous solution (A.88)–(A.92) by the exchange $1 \leftrightarrow 2$.

### *A.2.4 Case 4*

The 5D metric of this case

$$g_{AB,\mathrm{DR5},4}(x^5) = \mathrm{diag}\left(1, -1, -1, -\left(\frac{x^5}{x_0^5}\right)^p, \pm\left(\frac{x^5}{x_0^5}\right)^{p-2}\right) \tag{A.96}$$

amounts again to an exchange of space axes ($1 \leftrightarrow 3$) with respect to case 2. Accordingly, the solution for the Killing vector is obtained by such an exchange in the relevant equations.

### *A.2.5 Case 5*

The 5D metric of this case is given by

$$g_{AB,\mathrm{DR5},5}(x^5) = \mathrm{diag}\left(\left(\frac{x^5}{x_0^5}\right)^{-1}, -\left(\frac{x^5}{x_0^5}\right)^{-1}, -\left(\frac{x^5}{x_0^5}\right)^{-1}, -\frac{x^5}{x_0^5}, \pm\left(\frac{x^5}{x_0^5}\right)^{-2}\right) \tag{A.97}$$

to which corresponds the Killing system

$$\begin{cases} 2\xi_{0,0}(x^A)x_0^5 \mp \xi_5(x^A) = 0; \\ \xi_{0,1}(x^A) + \xi_{1,0}(x^A) = 0; \\ \xi_{0,2}(x^A) + \xi_{2,0}(x^A) = 0; \\ \xi_{0,3}(x^A) + \xi_{3,0}(x^A) = 0; \\ \xi_{0,5}(x^A)x^5 + \xi_0(x^A) + \xi_{5,0}(x^A)x^5 = 0; \\ 2\xi_{1,1}(x^A)x_0^5 \pm \xi_5(x^A) = 0; \\ \xi_{1,2}(x^A) + \xi_{2,1}(x^A) = 0; \\ \xi_{1,3}(x^A) + \xi_{3,1}(x^A) = 0; \\ \xi_{1,5}(x^A)x^5 + \xi_1(x^A) + \xi_{5,1}(x^A)x^5 = 0; \\ \xi_{2,2}(x^A)x_0^5 \pm \xi_5(x^A) = 0; \\ \xi_{2,3}(x^A) + \xi_{3,2}(x^A) = 0; \\ \xi_{2,5}(x^A)x^5 + \xi_2(x^A) + \xi_{5,2}(x^A)x^5 = 0; \\ \mp 2\xi_{3,3}(x^A)\left(x_0^5\right)^3 + \left(x^5\right)^2 \xi_5(x^A) = 0; \\ \xi_{3,5}(x^A)x^5 - \xi_3(x^A) + \xi_{5,3}(x^A)x^5 = 0; \\ \xi_{5,5}(x^A)x^5 + \xi_5(x^A) = 0. \end{cases} \tag{A.98}$$

Solving this system yields the covariant Killing vector

$$\xi_0(x^1, x^2, x^5) = \eta_1\frac{x^1}{x^5} + \eta_2\frac{x^2}{x^5} + \tau_0\frac{1}{x^5}; \tag{A.99}$$

$$\xi_1(x^0, x^2, x^5) = -\eta_1\frac{x^0}{x^5} + \Theta_3\frac{x^2}{x^5} - \tau_1\frac{1}{x^5}; \tag{A.100}$$

$$\xi_2(x^0, x^1, x^5) = -\eta_2\frac{x^0}{x^5} - \Theta_3\frac{x^1}{x^5} - \tau_2\frac{1}{x^5}; \tag{A.101}$$

$$\xi_3 = 0; \tag{A.102}$$

$$\xi_5 = 0 \tag{A.103}$$

with $\eta_k (k = 1, 2)$, $\Theta_3$, $\tau_\nu$ $(\nu = 0, 1, 2) \in R$. The dimensions of the transformation parameters are

$$[\tau_\nu] = l^2 \ , \quad [\ \eta_k] = [\Theta_3] = l \ . \tag{A.104}$$

In this case the Killing group is the group of rototranslations of the pseudoeuclidean space $M_3$ with metric $g_{\mu\nu} = \operatorname{diag}\left(\frac{x^5}{x_0^5}\right)^{-1} (1, -1, -1)$ $((\mu, \nu) \in \{0, 1, 2\}^2)$:

$$\mathrm{SO}(2,1)_{\mathrm{STD}.M_3} \otimes_s \mathrm{Tr}.(2,1)_{\mathrm{STD}.M_3}. \tag{A.105}$$

In all five cases discussed, the 5D contravariant Killing vectors $\xi^A(x, x^5)$ are obtained by means of the contravariant deformed metric tensor $g_{\mathrm{DR5}}^{AB}(x^5)$ as

$$\xi^A(x, x^5) = g_{\mathrm{DR55}}^{AB}(x^5)\xi_B(x, x^5). \tag{A.106}$$

For instance, in case 4, the contravariant metric tensor is

$$g_{\mathrm{DR5,4}}^{AB}(x^5) = \operatorname{diag}\left(1, -1, -1, -\left(\frac{x^5}{x_0^5}\right)^{-p}, \pm\left(\frac{x^5}{x_0^5}\right)^{-p+2}\right) \tag{A.107}$$

and therefore the contravariant components of the Killing vector read

$$\xi^0(x^1, x^2) = \zeta_1 x^1 + \zeta_2 x^2 + T_0; \tag{A.108}$$

$$\xi^1(x^0, x^2) = \zeta_1 x^0 - \Theta_3 x^2 + T_1; \tag{A.109}$$

$$\xi^2(x^0, x^1) = \zeta_2 x^0 + \Theta_3 x^1 + T_2; \tag{A.110}$$

$$\xi^3(x^3, x^5; p) = \begin{cases} \text{“}+\text{”}: \\ -\left(x_0^5\right)^{p-1}\left[A\cosh\left(\dfrac{p}{2}\dfrac{x^3}{x_0^5}\right) + B\sinh\left(\dfrac{p}{2}\dfrac{x^3}{x_0^5}\right)\right]\left(x^5\right)^{-p/2} + \alpha\left(x_0^5\right)^p; \\ \text{“}-\text{”}: \\ -\left(x_0^5\right)^{p-1}\left[C\cos\left(\dfrac{p}{2}\dfrac{x^3}{x_0^5}\right) - D\sin\left(\dfrac{p}{2}\dfrac{x^3}{x_0^5}\right)\right]\left(x^5\right)^{-p/2} + \alpha\left(x_0^5\right)^p; \end{cases} \tag{A.111}$$

$$\xi^5(x^3, x^5; p) = \begin{cases} \text{“}+\text{”}: \\ \left(x_0^5\right)^{p-2}\left[A\sinh\left(\dfrac{p}{2}\dfrac{x^3}{x_0^5}\right) + B\cosh\left(\dfrac{p}{2}\dfrac{x^3}{x_0^5}\right)\right]\left(x^5\right)^{-(p/2)+1}; \\ \text{“}-\text{”}: \\ -\left(x_0^5\right)^{p-2}\left[C\sin\left(\dfrac{p}{2}\dfrac{x^3}{x_0^5}\right) + D\cos\left(\dfrac{p}{2}\dfrac{x^3}{x_0^5}\right)\right]\left(x^5\right)^{-(p/2)+1}. \end{cases} \tag{A.112}$$

# Appendix B

# Gravitational Killing Symmetries for Special Forms of $b_1^2(x^5)$ and $b_2^2(x^5)$

In this appendix, we shall investigate the integrability of the Killing system for the gravitational interaction in different cases by assuming special forms for the spatial metric coefficients $b_1^2(x^5)$ and $b_2^2(x^5)$. For each case, we will consider the two energy ranges $0 < x^5 \leqslant x_0^5$ (subcase a)) and $x^5 > x_0^5$ (subcase b)).

## B.1 Form I

The phenomenological metric 5D is assumed to be

$$
\begin{aligned}
& g_{AB,\mathrm{DR5,grav.}}(x^5) \\
&= \mathrm{diag}\left(1 + \Theta(x^5 - x^5_{0,\mathrm{grav.}})\left[\frac{1}{4}\left(1 + \frac{x^5}{x^5_{0,\mathrm{grav.}}}\right)^2 - 1\right], -c_1, -c_2,\right. \\
&\left. -\left\{1 + \Theta(x^5 - x^5_{0,\mathrm{grav.}})\left[\frac{1}{4}\left(1 + \frac{x^5}{x^5_{0,\mathrm{grav.}}}\right)^2 - 1\right]\right\}, \pm f(x^5)\right), \\
& c_1, c_2 \in R_0^+,\ (\text{in gen.: } c_1 \neq 1, c_2 \neq 1, c_1 \neq c_2).
\end{aligned}
\tag{B.1}
$$

### *B.1.1 (Ia)*

Metric (B.1) becomes:

$$g_{AB,\text{DR5}}(x^5) = \text{diag}\left(1, -c_1, -c_2, -1, \pm f(x^5)\right) = \text{diag}\left(g_{\mu\nu,\overline{M_4}}(x^5), \pm f(x^5)\right), \tag{B.2}$$

where $\overline{M_4}$ is a standard 4D Minkowskian space with the following coordinate rescaling:

$$\begin{aligned} x^1 &\longrightarrow \sqrt{c_1}x^1 \underset{\substack{(\text{in gen.})\\ \not\Leftarrow}}{\Rightarrow} \mathrm{d}x^1 \longrightarrow \sqrt{c_1}\mathrm{d}x^1; \\ x^2 &\longrightarrow \sqrt{c_2}x^2 \underset{\substack{(\text{in gen.})\\ \not\Leftarrow}}{\Rightarrow} \mathrm{d}x^2 \longrightarrow \sqrt{c_2}\mathrm{d}x^2. \end{aligned} \tag{B.3}$$

This case is therefore the same of the e.m. and weak interactions in the energy range $x^5 \geqslant x_0^5$ (Sect. 22.1.1) and of the strong interaction in the range $0 < x^5 \leqslant x_0^5$ (Sect. 22.2.1). Thus, the $\Upsilon$-hypothesis of functional independence is violated for any $\mu \in \{0, 1, 2, 3\}$, and the contravariant Killing five-vector $\xi^A(x, x^5)$ is given by (22.5)–(22.10).

The Killing group of the sections at $\mathrm{d}x^5 = 0$ of $\Re_5$ is therefore the standard Poincaré group, suitably rescaled:

$$\left[P(1,3)_{\text{STD.}} = \text{SO}(1,3)_{\text{STD.}} \otimes_s \text{Tr.}(1,3)_{\text{STD.}}\right]\Big|_{x^1 \longrightarrow \sqrt{c_1}x^1, x^2 \longrightarrow \sqrt{c_2}x^2} . \tag{B.4}$$

### *B.1.2 (Ib)*

The metric takes the form

$$g_{AB,\text{DR5}}(x^5) = \text{diag}\left(\frac{1}{4}\left(1 + \frac{x^5}{x^5_{0,\text{grav.}}}\right)^2, -c_1, -c_2, -\frac{1}{4}\left(1 + \frac{x^5}{x^5_{0,\text{grav.}}}\right)^2, \pm f(x^5)\right) \tag{B.5}$$

and from (21.22)–(21.23) it follows:

$$\begin{aligned} &A_0(x^5) = -A_3(x^5) \\ &= \tfrac{1}{8}\left(1 + \frac{x^5}{x_0^5}\right)\frac{x^5}{(x_0^5)^2}\left(f(x^5)\right)^{-1/2}\left[\frac{1}{x^5} + \frac{1}{2}\frac{f'(x^5)}{f(x^5)} + \frac{1}{2}\frac{x_0^5}{x^5}\frac{f'(x^5)}{f(x^5)}\right]; \\ &A_1(x^5) = A_2(x^5) = 0; \end{aligned} \tag{B.6}$$

$$\begin{aligned} B_0(x^5) &= B_3(x^5) = \frac{1}{2}\left(1 + \frac{x_0^5}{x^5}\right)\left(f(x^5)\right)^{1/2}; \\ \frac{1}{\sqrt{c_1}}B_1(x^5) &= \frac{1}{\sqrt{c_2}}B_2(x^5) = \left(f(x^5)\right)^{1/2}, \end{aligned} \tag{B.7}$$

whence

$$\frac{\pm A_0(x^5)}{B_0(x^5)} = \frac{\mp A_3(x^5)}{B_3(x^5)} = \pm\frac{1}{4}\frac{1}{f(x^5)}\frac{x^5}{(x_0^5)^2}\left[\frac{1}{x^5} + \frac{1}{2}\frac{f'(x^5)}{f(x^5)} + \frac{1}{2}\frac{x_0^5}{x^5}\frac{f'(x^5)}{f(x^5)}\right]. \tag{B.8}$$

Therefore, the $\Upsilon$-hypothesis is satisfied only for $\mu = 0, 3$ under condition

$$\frac{1}{x^5} + \frac{1}{2}\frac{f'(x^5)}{f(x^5)} + \frac{1}{2}\frac{x_0^5}{x^5}\frac{f'(x^5)}{f(x^5)} \neq c\frac{f(x^5)}{x^5}, c \in R. \tag{B.9}$$

Then, on the basis of the results of Sect. 21.3, the components of the contravariant Killing vector $\xi^A(x, x^5)$ in this case are given by (21.41)–(21.45), in which (some of) the real parameters are constrained by the following system (cf. (21.46)):

$$\left\{\begin{array}{l}
(01)\ \frac{1}{4}\left(1+\frac{x^5}{x^5_{0,\text{grav.}}}\right)^2 \left[d_8x^2x^3 + d_7x^2 + d_6x^3 + (d_5 + a_2)\right] \\
\quad +c_1\left[h_2x^2x^3 + h_1x^2 + h_8x^3 + (h_7 + e_2)\right] = 0; \\
(02)\ \frac{1}{4}\left(1+\frac{x^5}{x^5_{0,\text{grav.}}}\right)^2 \left(d_8x^1x^3 + d_7x^1 + d_4x^3 + d_3\right) \\
\quad +c_2\left[l_2x^1x^3 + l_1x^1 + l_6x^3 + (l_5 + e_4)\right] = 0; \\
(03)\ \frac{1}{4}\left(1+\frac{x^5}{x^5_{0,\text{grav.}}}\right)^2 \left(d_8x^1x^2 + d_6x^1 + d_4x^2 + d_2\right) \\
\quad +\frac{1}{4}\left(1+\frac{x^5}{x^5_{0,\text{grav.}}}\right)^2 \left[m_8x^1x^2 + m_7x^1 + m_6x^2 + (m_5 + g_2)\right] = 0; \\
(12)\ c_1\left(h_2x^0x^3 + h_1x^0 + h_4x^3 + h_3\right) \\
\quad +c_2\left(l_2x^0x^3 + l_1x^0 + l_4x^3 + l_3\right) = 0; \\
(13)\ c_1\left(h_2x^0x^2 + h_8x^0 + h_4x^2 + h_6\right) \\
\quad +\frac{1}{4}\left(1+\frac{x^5}{x^5_{0,\text{grav.}}}\right)^2 \left(m_8x^0x^2 + m_7x^0 + m_4x^2 + m_3\right) = 0; \\
(23)\ c_2\left(l_2x^0x^1 + l_6x^0 + l_4x^1 + l_8\right) \\
\quad +\frac{1}{4}\left(1+\frac{x^5}{x^5_{0,\text{grav.}}}\right)^2 \left(m_8x^0x^1 + m_6x^0 + m_4x^1 + m_2\right) = 0.
\end{array}\right. \tag{B.10}$$

Solving system (B.10) one finally gets for $\xi^A(x, x^5)$:

$$\xi^0(x^3) = \widetilde{F_0}(x^3) = d_2x^3 + (a_1 + d_1 + K_0); \tag{B.11}$$

$$\xi^1(x^2) = -\widetilde{F_1}(x^2) = \frac{c_2}{c_1}l_3x^2 - (K_1 + h_5 + e_1); \tag{B.12}$$

$$\xi^2(x^1) = -\widetilde{F_2}(x^1) = -l_3x^1 - (l_7 + K_2 + e_3); \tag{B.13}$$

$$\xi^3(x^0) = -\widetilde{F_3}(x^0) = d_2 x^0 - (m_1 + g_1 + c); \tag{B.14}$$

$$\xi^5 = 0. \tag{B.15}$$

It follows from the earlier equations that the 5D Killing group in the range considered is

$$\left(\mathrm{SO}(2)_{\mathrm{STD.},\Pi(x^1,x^2\longrightarrow\sqrt{\frac{c_2}{c_1}}x^2)} \otimes B_{x^3,\mathrm{STD.}}\right) \otimes_s \mathrm{Tr.}(1,3)_{\mathrm{STD.}}, \tag{B.16}$$

where $\mathrm{SO}(2)_{\mathrm{STD.},\Pi(x^1,x^2\longrightarrow\sqrt{\frac{c_2}{c_1}}x^2)} = \mathrm{SO}(2)_{\mathrm{STD.},\Pi(x^1\longrightarrow\sqrt{c_1}x^1,x^2\longrightarrow\sqrt{c_2}x^2)}$ is the one-parameter group (generated by $S^3_{\mathrm{SR}}\big|_{x^2\longrightarrow\sqrt{\frac{c_2}{c_1}}x^2}$) of the 2D rotations in the plane $\Pi(x^1,x^2)$ characterized by the scale transformation (B.3), $B_{x^3,STD.}$ is the usual one-parameter group (generated by $K^3_{\mathrm{SR}}$) of the standard Lorentzian boosts along $\widehat{x^3}$ and $\mathrm{Tr.}(1,3)_{\mathrm{STD.}}$ is the usual space–time translation group.

Notice that, by introducing the right distribution $\widehat{\Theta}_{\mathrm{R}}(x_0^5 - x^5)$, putting

$$\begin{aligned}
&\frac{B^1}{c_1} \equiv \zeta^1, \frac{B^2}{c_2} \equiv \zeta^2, B^3 \equiv \zeta^3,\\
&\frac{\Theta^1}{c_2} \equiv \theta^1, \frac{\Theta^2}{c_1} \equiv \theta^2, \frac{\Theta^3}{c_2} \equiv \theta^3,\\
&\Xi^0 \equiv \zeta^5, \frac{\Xi^1}{c_1} \equiv \Xi^{1\prime}, \frac{\Xi^2}{c_2} \equiv \Xi^{2\prime},\\
&\frac{T^1}{c_1} \equiv T^{1\prime}, \frac{T^2}{c_2} \equiv T^{2\prime},
\end{aligned} \tag{B.17}$$

and making the identifications

$$\begin{aligned}
&(a_1 + d_1 + K_0) = T^0;\\
&-(K_1 + h_5 + e_1) = \frac{1}{c_1}T^1;\\
&-(l_7 + K_2 + e_3) = \frac{1}{c_2}T^2;\\
&-(m_1 + g_1 + c) = T^3;\\
&l_3 = \frac{1}{c_2}\Theta^3;\\
&d_2 = -B^3,
\end{aligned} \tag{B.18}$$

it is possible to express the contravariant five-vector $\xi^A(x, x^5)$ for the gravitational interaction in case (I) in the following form, valid in the whole energy range ($x^5 \in R_0^+$):

$$\xi^0(x^1,x^2,x^3,x^5) = \widehat{\Theta}_{\mathrm{R}}(x_0^5 - x^5)\left[-c_1\zeta^1 x^1 - c_2\zeta^2 x^2 + \zeta^5 F(x^5)\right] - \zeta^3 x^3 + T^0; \tag{B.19}$$

$$\xi^1(x^0,x^2,x^3,x^5) = \widehat{\Theta}_{\mathrm{R}}(x_0^5 - x^5)\left[-\zeta^1 x^0 - \theta^2 x^3 - \Xi^1 F(x^5)\right] + \frac{c_2}{c_1}\theta^3 x^2 + T^1; \tag{B.20}$$

$$\xi^2(x^0,x^1,x^3,x^5) = \widehat{\Theta}_{\rm R}(x_0^5 - x^5)\left[-\zeta^2 x^0 + \theta^1 x^3 - \Xi^2 F(x^5)\right] - \theta^3 x^1 + T^2; \tag{B.21}$$

$$\xi^3(x^0,x^1,x^2,x^5) = \widehat{\Theta}_{\rm R}(x_0^5 - x^5)\left[c_1\theta^2 x^1 - c_2\theta^1 x^2 - \Xi^3 F(x^5)\right] - \zeta^3 x^0 + T^3; \tag{B.22}$$

$$\begin{aligned}&\xi^5(x,x^5)\\ &= \widehat{\Theta}_{\rm R}(x_0^5 - x^5)\left\{\mp\left(f(x^5)\right)^{-1/2}\left[\zeta^5 x^0 + c_1\Xi^1 x^1 + c_2\Xi^2 x^2 + \Xi^3 x^3 - T^5\right]\right\}.\end{aligned} \tag{B.23}$$

## B.2 Form II

$$\begin{aligned}&g_{AB,\mathrm{DR5,grav.}}(x^5)\\ &= \mathrm{diag}\left(1 + \Theta(x^5 - x^5_{0,\mathrm{grav.}})\left[\frac{1}{4}\left(1 + \frac{x^5}{x^5_{0,\mathrm{grav.}}}\right)^2 - 1\right],\right.\\ &\quad -\left\{c_1 + \Theta(x^5 - x^5_{0,\mathrm{grav.}})\left[\beta_1^2(x^5) - c_1\right]\right\},\\ &\quad -\left\{c_2 + \Theta(x^5 - x^5_{0,\mathrm{grav.}})\left[\beta_2^2(x^5) - c_2\right]\right\},\\ &\left.-\left\{1 + \Theta(x^5 - x^5_{0,\mathrm{grav.}})\left[\frac{1}{4}\left(1 + \frac{x^5}{x^5_{0,\mathrm{grav.}}}\right)^2 - 1\right]\right\}, \pm f(x^5)\right);\end{aligned} \tag{B.24}$$

$$c_1, c_2 \in R_0^+,\ (\text{in gen.: } c_1 \neq 1, c_2 \neq 1, c_1 \neq c_2),$$

where the functions $\beta_1^2(x^5)$ and $\beta_2^2(x^5)$ have the properties:

$$\begin{aligned}&\beta_1^2(x^5), \beta_2^2(x^5) \in R_0^+,\ \forall x^5 \in \left([x_0^5, \infty)\right) \subset R_0^+;\\ &\beta_1^2(x^5) \neq \beta_2^2(x^5);\\ &\beta_1^2(x_0^5) = c_1, \quad \beta_2^2(x_0^5) = c_2;\\ &\beta_1^2(x^5) \neq \frac{1}{4}\left(1 + \frac{x^5}{x^5_{0,\mathrm{grav.}}}\right)^2, \quad \beta_2^2(x^5) \neq \frac{1}{4}\left(1 + \frac{x^5}{x^5_{0,\mathrm{grav.}}}\right)^2.\end{aligned} \tag{B.25}$$

### *B.2.1 (IIa)*

In the energy range $0 < x^5 \leqslant x_0^5$, the 5D metric has the same form (B.2) of case I a), and therefore the same results of that case hold true. In particular the $\Upsilon$-hypothesis is not satisfied by any value of $\mu$, and the contravariant vector $\xi^A(x,x^5)$ is still given by (22.5)–(22.10).

### *B.2.2 (IIb)*

In this case the 5D gravitational metric reads

$$g_{AB,\mathrm{DR5}}(x^5) = \mathrm{diag}\left(\frac{1}{4}\left(1+\frac{x^5}{x^5_{0,\mathrm{grav.}}}\right)^2, -\beta_1^2(x^5), -\beta_2^2(x^5), -\frac{1}{4}\left(1+\frac{x^5}{x^5_{0,\mathrm{grav.}}}\right)^2, \pm f(x^5)\right) \tag{B.26}$$

The fake vectors (21.22), (21.23) become (ESC off):

$$\begin{aligned} A_0(x^5) &= -A_3(x^5) \\ &= \frac{1}{8}\left(1+\frac{x^5}{x^5_0}\right)\frac{x^5}{(x^5_0)^2}\left(f(x^5)\right)^{-1/2} \\ &\quad\times\left[\frac{1}{x^5}+\frac{1}{2}\frac{f'(x^5)}{f(x^5)}+\frac{1}{2}\frac{x^5_0}{x^5}\frac{f'(x^5)}{f(x^5)}\right]; \end{aligned} \tag{B.27}$$

$$A_i(x^5) \equiv \beta_i(x^5)(f(x^5))^{-1/2} \times\left[-\left(\beta_i'(x^5)\right)^2+\beta_i(x^5)\beta_i''(x^5)-\frac{1}{2}\beta_i(x^5)\beta_i'(x^5)f'(x^5)(f(x^5))^{-1}\right],\quad i=1,2; \tag{B.28}$$

$$B_0(x^5) = B_3(x^5) = \frac{1}{2}\left(1+\frac{x^5_0}{x^5}\right)\left(f(x^5)\right)^{1/2}; \tag{B.29}$$

$$B_i(x^5) \equiv \beta_i(x^5)(f(x^5))^{1/2}, i=1,2, \tag{B.30}$$

whence

$$\frac{\pm A_0(x^5)}{B_0(x^5)} = \frac{\mp A_3(x^5)}{B_3(x^5)} = \pm\frac{1}{4}\frac{1}{f(x^5)}\frac{x^5}{(x^5_0)^2}\left[\frac{1}{x^5}+\frac{1}{2}\frac{f'(x^5)}{f(x^5)}+\frac{1}{2}\frac{x^5_0}{x^5}\frac{f'(x^5)}{f(x^5)}\right], \tag{B.31}$$

$$\begin{aligned} \frac{\pm A_i(x^5)}{B_i(x^5)} &= \pm(f(x^5))^{-1} \\ &\quad\times\left[-\left(\beta_i'(x^5)\right)^2+\beta_i(x^5)\beta_i''(x^5)-\frac{1}{2}\beta_i(x^5)\beta_i'(x^5)f'(x^5)(f(x^5))^{-1}\right], \\ i &= 1,2. \end{aligned} \tag{B.32}$$

Then, the $\Upsilon$-hypothesis is satisfied for $\mu = 0, 3$ under constraint (B.9), and for $\mu = 1$ and/or 2 under the conditions

$$\left.\begin{array}{l}\left[-\left(\beta_i'(x^5)\right)^2 + \beta_i(x^5)\beta_i''(x^5) - \frac{1}{2}\beta_i(x^5)\beta_i'(x^5)f'(x^5)(f(x^5))^{-1}\right] \neq 0;\\ (f(x^5))^{-1}\left[-\left(\beta_i'(x^5)\right)^2 + \beta_i(x^5)\beta_i''(x^5)\right.\\ \left.-\frac{1}{2}\beta_i(x^5)\beta_i'(x^5)f'(x^5)(f(x^5))^{-1}\right] \neq c,\\ c \in R_0,\end{array}\right\}$$

$$\Leftrightarrow \begin{cases}-\left(\beta_i'(x^5)\right)^2 + \beta_i(x^5)\beta_i''(x^5) - \frac{1}{2}\beta_i(x^5)\beta_i'(x^5)f'(x^5)(f(x^5))^{-1} \neq cf(x^5),\\ \qquad c \in R, \quad i = 1 \text{ and/or } 2.\end{cases} \tag{B.33}$$

By exploiting the results of Sect. 21.3, the components of the contravariant Killing vector $\xi^A(x, x^5)$ corresponding to form II) of the 5D gravitational metric over threshold are given by (21.41)–(21.45), in which (some of) the real parameters are constrained by the following system:

$$\begin{cases}(01)\ \frac{1}{4}\left(1+\frac{x^5}{x^5_{0,\text{grav.}}}\right)^2\left[d_8x^2x^3 + d_7x^2 + d_6x^3 + (d_5 + a_2)\right]\\ \quad +\beta_1^2(x^5)\left[h_2x^2x^3 + h_1x^2 + h_8x^3 + (h_7 + e_2)\right] = 0;\\ (02)\ \frac{1}{4}\left(1+\frac{x^5}{x^5_{0,\text{grav.}}}\right)^2\left(d_8x^1x^3 + d_7x^1 + d_4x^3 + d_3\right)\\ \quad +\beta_2^2(x^5)\left[l_2x^1x^3 + l_1x^1 + l_6x^3 + (l_5 + e_4)\right] = 0;\\ (03)\ \frac{1}{4}\left(1+\frac{x^5}{x^5_{0,\text{grav.}}}\right)^2\left[d_8x^1x^2 + d_6x^1 + d_4x^2 + d_2\right.\\ \quad \left.+m_8x^1x^2 + m_7x^1 + m_6x^2 + (m_5 + g_2)\right] = 0;\\ (12)\ \beta_1^2(x^5)\left(h_2x^0x^3 + h_1x^0 + h_4x^3 + h_3\right)\\ \quad +\beta_2^2(x^5)\left(l_2x^0x^3 + l_1x^0 + l_4x^3 + l_3\right) = 0;\\ (13)\ \frac{1}{4}\left(1+\frac{x^5}{x^5_{0,\text{grav.}}}\right)^2\left(m_8x^0x^2 + m_7x^0 + m_4x^2 + m_3\right)\\ \quad +\beta_1^2(x^5)\left(h_2x^0x^2 + h_8x^0 + h_4x^2 + h_6\right) = 0;\\ (23)\ \frac{1}{4}\left(1+\frac{x^5}{x^5_{0,\text{grav.}}}\right)^2\left(m_8x^0x^1 + m_6x^0 + m_4x^1 + m_2\right)\\ \quad +\beta_2^2(x^5)\left(l_2x^0x^1 + l_6x^0 + l_4x^1 + l_8\right) = 0.\end{cases} \tag{B.34}$$

Solving system (B.34) yields for $\xi^A(x, x^5)$ in this case

$$\xi^0(x^3) = \widetilde{F_0}(x^3) = d_2x^3 + (a_1 + d_1 + K_0); \tag{B.35}$$

$$\xi^1 = -\widetilde{F_1} = -\left(K_1 + h_5 + e_1\right); \tag{B.36}$$

$$\xi^2 = -\widetilde{F_2} = -(l_7 + K_2 + e_3); \tag{B.37}$$

$$\xi^3(x^0) = -\widetilde{F_3}(x^0) = d_2x^0 - (m_1 + g_1 + c); \tag{B.38}$$

$$\xi^5 = 0. \tag{B.39}$$

The Killing group in this range is

$$B_{x^3,\mathrm{STD.}} \otimes_s \mathrm{Tr.}(1,3)_{\mathrm{STD.}}. \tag{B.40}$$

By means of the distribution $\widehat{\Theta}_{\mathrm{R}}(x_0^5 - x^5)$, by the redenominations (B.17) of case (I) and putting

$$\begin{aligned}
&(a_1 + d_1 + K_0) = T^0;\\
&-(K_1 + h_5 + e_1) = \frac{1}{c_1}T^1;\\
&-(l_7 + K_2 + e_3) = \frac{1}{c_2}T^2;\\
&-(m_1 + g_1 + c) = T^3;\\
&d_2 = -B^3,
\end{aligned} \tag{B.41}$$

one gets the following expression of the contravariant vector for the form II of the gravitational metric in the whole energy range:

$$\xi^0(x^1,x^2,x^3,x^5) = \widehat{\Theta}_{\mathrm{R}}(x_0^5 - x^5)\left[-c_1\zeta^1x^1 - c_2\zeta^2x^2 + \zeta^5F(x^5)\right] - \zeta^3x^3 + T^0; \tag{B.42}$$

$$\xi^1(x^0,x^2,x^3,x^5) = \widehat{\Theta}_{\mathrm{R}}(x_0^5 - x^5)\left[-\zeta^1x^0 + \frac{c_2}{c_1}\theta^3x^2 - \theta^2x^3 - \Xi^1F(x^5)\right] + T^1; \tag{B.43}$$

$$\xi^2(x^0,x^1,x^3,x^5) = \widehat{\Theta}_{\mathrm{R}}(x_0^5 - x^5)\left[-\zeta^2x^0 - \theta^3x^1 + \theta^1x^3 - \Xi^2F(x^5)\right] + T^2; \tag{B.44}$$

$$\xi^3(x^0,x^1,x^2,x^5) = \widehat{\Theta}_{\mathrm{R}}(x_0^5 - x^5)\left[c_1\theta^2x^1 - c_2\theta^1x^2 - \Xi^3F(x^5)\right] - \zeta^3x^0 + T^3; \tag{B.45}$$

$$\begin{aligned}
&\xi^5(x,x^5)\\
&= \widehat{\Theta}_{\mathrm{R}}(x_0^5 - x^5)\left\{\mp\left(f(x^5)\right)^{-\frac{1}{2}}\left[\zeta^5x^0 + c_1\Xi^1x^1 + c_2\Xi^2x^2 + \Xi^3x^3 - T^5\right]\right\}.
\end{aligned} \tag{B.46}$$

## B.3 Form III

$$
\begin{aligned}
&g_{AB,\mathrm{DR5,grav.}}(x^5)\\
&= \mathrm{diag}\left(1+\Theta(x^5-x^5_{0,\mathrm{grav.}})\left[\frac{1}{4}\left(1+\frac{x^5}{x^5_{0,\mathrm{grav.}}}\right)^2-1\right],\right.\\
&\quad -\beta_1^2(x^5), -\beta_2^2(x^5),\\
&\left.-\left\{1+\Theta(x^5-x^5_{0,\mathrm{grav.}})\left[\frac{1}{4}\left(1+\frac{x^5}{x^5_{0,\mathrm{grav.}}}\right)^2-1\right]\right\}, \pm f(x^5)\right), \qquad \text{(B.47)}
\end{aligned}
$$

where the functions $\beta_1^2(x^5)$ and $\beta_2^2(x^5)$ have in general the properties:

$$
\begin{aligned}
&\beta_1^2(x^5), \beta_2^2(x^5) \in R_0^+, \quad \forall x^5 \in R_0^+;\\
&\beta_1^2(x^5) \neq \beta_2^2(x^5);\\
&\beta_1^2(x^5) \neq \frac{1}{4}\left(1+\frac{x^5}{x^5_{0,\mathrm{grav.}}}\right)^2, \quad \beta_2^2(x^5) \neq \frac{1}{4}\left(1+\frac{x^5}{x^5_{0,\mathrm{grav.}}}\right)^2.
\end{aligned}
\qquad \text{(B.48)}
$$

Therefore, the present case (III) differs from the previous case II) for the nature strictly functional (and not composite, namely expressible in terms of one or more Heaviside functions) of $\beta_1^2(x^5)$ and $\beta_2^2(x^5)$.

### *B.3.1 (IIIa)*

The 5D metric reads

$$
g_{AB,\mathrm{DR5}}(x^5) = \mathrm{diag}\left(1, -\beta_1^2(x^5), -\beta_2^2(x^5), -1, \pm f(x^5)\right). \qquad \text{(B.49)}
$$

Then, from definitions (21.22) and (21.23), there follow (B.28),(B.30) and

$$
\begin{aligned}
A_0(x^5) &= A_3(x^5) = 0;\\
B_0(x^5) &= B_3(x^5) = \left(f(x^5)\right)^{1/2}.
\end{aligned}
\qquad \text{(B.50)}
$$

Therefore, the $\Upsilon$-hypothesis is satisfied only for $\mu = 1$ and/or 2 under condition (22.105). From Sect. 21.3, the contravariant five-vector $\xi^A(x, x^5)$ corresponding to form (III) of the 5D gravitational metric below threshold is given by (21.41)–(21.45), in which (some of) the real parameters are constrained by the system:

$$\begin{cases} (01) \ \left[d_8x^2x^3 + d_7x^2 + d_6x^3 + (d_5 + a_2)\right] \\ \quad +\beta_1^2(x^5)\left[h_2x^2x^3 + h_1x^2 + h_8x^3 + (h_7 + e_2)\right] = 0; \\ (02) \ \left(d_8x^1x^3 + d_7x^1 + d_4x^3 + d_3\right) \\ \quad +\beta_2^2(x^5)\left[l_2x^1x^3 + l_1x^1 + l_6x^3 + (l_5 + e_4)\right] = 0; \\ (03) \ \left(d_8x^1x^2 + d_6x^1 + d_4x^2 + d_2\right) \\ \quad + \left[m_8x^1x^2 + m_7x^1 + m_6x^2 + (m_5 + g_2)\right] = 0; \\ (12) \ \beta_1^2(x^5)\left(h_2x^0x^3 + h_1x^0 + h_4x^3 + h_3\right) \\ \quad +\beta_2^2(x^5)\left(l_2x^0x^3 + l_1x^0 + l_4x^3 + l_3\right) = 0; \\ (13) \ \beta_1^2(x^5)\left(h_2x^0x^2 + h_8x^0 + h_4x^2 + h_6\right) \\ \quad + \left(m_8x^0x^2 + m_7x^0 + m_4x^2 + m_3\right) = 0; \\ (23) \ \beta_2^2(x^5)\left(l_2x^0x^1 + l_6x^0 + l_4x^1 + l_8\right) \\ \quad + \left(m_8x^0x^1 + m_6x^0 + m_4x^1 + m_2\right) = 0. \end{cases} \tag{B.51}$$

Then, from the solutions of the above system, one finds:

$$\xi^0(x^3) = \widetilde{F_0}(x^3) = d_2x^3 + (a_1 + d_1 + K_0); \tag{B.52}$$

$$\xi^1 = -\widetilde{F_1} = -\left(K_1 + h_5 + e_1\right); \tag{B.53}$$

$$\xi^2 = -\widetilde{F_2} = -(l_7 + K_2 + e_3); \tag{B.54}$$

$$\xi^3(x^0) = -\widetilde{F_3}(x^0) = d_2x^0 - (m_1 + g_1 + c); \tag{B.55}$$

$$\xi^5 = 0. \tag{B.56}$$

Let us notice that the result obtained for $\xi^A(x, x^5)$ coincides with that of case (IIb), (B.35)–(B.39).

### *B.3.2 (IIIb)*

The form of the 5D metric is identical to that of case (IIb):

$$\begin{aligned} &g_{AB,\mathrm{DR5}}(x^5) \\ &= \mathrm{diag}\left(\frac{1}{4}\left(1 + \frac{x^5}{x^5_{0,\mathrm{grav.}}}\right)^2, -\beta_1^2(x^5),\right. \\ &\left. -\beta_2^2(x^5), -\frac{1}{4}\left(1 + \frac{x^5}{x^5_{0,\mathrm{grav.}}}\right)^2, \pm f(x^5)\right). \end{aligned} \tag{B.57}$$

Therefore, the same results of Sect. B.2.2 hold. Moreover, we have just noted that case (IIIa) yields the same results of case (IIb), and thus of case (IIIb) too. Consequently, for the form III of the gravitational metric, the contravariant Killing vector of the gravitational metric is independent of the energetic range considered. In conclusion, putting

$$\begin{aligned}&(a_1 + d_1 + K_0) \equiv T^0;\\ &-(K_1 + h_5 + e_1) \equiv T^1;\\ &-(l_7 + K_2 + e_3) \equiv T^2;\\ &-(m_1 + g_1 + c) = T^3;\\ &d_2 = -\zeta^3,\end{aligned} \tag{B.58}$$

one gets the following general form for $\xi^A(x, x^5)$ for form III of the gravitational metric ($\forall x^5 \in R_0^+$):

$$\xi^0(x^3) = -\zeta^3 x^3 + T^0; \tag{B.59}$$

$$\xi^1 = +T^1; \tag{B.60}$$

$$\xi^2 = +T^2; \tag{B.61}$$

$$\xi^3(x^0) = -\zeta^3 x^0 + T^3; \tag{B.62}$$

$$\xi^5 = 0. \tag{B.63}$$

Moreover, $\forall x^5 \in R_0^+$ the 5D Killing group is

$$B_{x^3,\text{STD.}} \otimes_s \text{Tr.}(1,3)_{\text{STD.}}. \tag{B.64}$$

## B.4 Form IV

$$\begin{aligned}&g_{AB,\text{DR5,grav.}}(x^5)\\ &= \operatorname{diag}\left(1 + \Theta(x^5 - x^5_{0,\text{grav.}})\left[\frac{1}{4}\left(1 + \frac{x^5}{x^5_{0,\text{grav.}}}\right)^2 - 1\right],\right.\\ &\quad -\left\{c + \Theta(x^5 - x^5_{0,\text{grav.}})\left[\beta^2(x^5) - c\right]\right\},\\ &\quad -\left\{c + \Theta(x^5 - x^5_{0,\text{grav.}})\left[\beta^2(x^5) - c\right]\right\},\\ &\left. -\left\{1 + \Theta(x^5 - x^5_{0,\text{grav.}})\left[\frac{1}{4}\left(1 + \frac{x^5}{x^5_{0,\text{grav.}}}\right)^2 - 1\right]\right\}, \pm f(x^5)\right)\end{aligned} \tag{B.65}$$

($c \in R_0^+$, $c \neq 1$), where the function $\beta^2(x^5)$ has the following properties:

$$\begin{aligned}&\beta^2(x^5) \in R_0^+, \forall x^5 \in \left([x_0^5, \infty)\right) \subset R_0^+;\\ &\beta^2(x_0^5) = c;\\ &\beta^2(x^5) \neq \frac{1}{4}\left(1 + \frac{x^5}{x^5_{0,\text{grav.}}}\right)^2.\end{aligned} \tag{B.66}$$

Therefore this case is a special case of case (II) with $\beta_1^2(x^5) = \beta_2^2(x^5)$.

### *B.4.1 (IVa)*

The 5D metric in this case coincides with that of case (Ia) with $c_1 = c_2 = c$:

$$g_{AB,\mathrm{DR5}}(x^5) = \mathrm{diag}\left(1, -c, -c, -1, \pm f(x^5)\right). \tag{B.67}$$

and therefore all the results of Sect. B.1.1 still hold with $c_1 = c_2 = c$.

### *B.4.2 (IVb)*

The 5D metric is

$$\begin{aligned} & g_{AB,\mathrm{DR5}}(x^5) \\ & = \left(\mathrm{diag}\frac{1}{4}\left(1+\frac{x^5}{x^5_{0,\mathrm{grav.}}}\right)^2, -\beta^2(x^5), -\beta^2(x^5), \right. \\ & \left. -\frac{1}{4}\left(1+\frac{x^5}{x^5_{0,\mathrm{grav.}}}\right)^2, \pm f(x^5)\right). \end{aligned} \tag{B.68}$$

The results are the same of case II b) with $\beta_1^2(x^5) = \beta_2^2(x^5)$ and $c_1 = c_2 = c$.

## B.5 Form V

$$\begin{aligned} & g_{AB,\mathrm{DR5,grav.}}(x^5) \\ & = \mathrm{diag}\left(1+\Theta(x^5 - x^5_{0,\mathrm{grav.}})\left[\frac{1}{4}\left(1+\frac{x^5}{x^5_{0,\mathrm{grav.}}}\right)^2 - 1\right],\right. \\ & -\beta^2(x^5), -\beta^2(x^5), \\ & \left. -\left\{1+\Theta(x^5 - x^5_{0,\mathrm{grav.}})\left[\frac{1}{4}\left(1+\frac{x^5}{x^5_{0,\mathrm{grav.}}}\right)^2 - 1\right]\right\}, \pm f(x^5)\right), \end{aligned} \tag{B.69}$$

where the function $\beta^2(x^5)$ has the following properties:

$$\begin{aligned} & \beta^2(x^5) \in R_0^+, \forall x^5 \in R_0^+; \\ & \beta^2(x^5) \neq \frac{1}{4}\left(1+\frac{x^5}{x^5_{0,\mathrm{grav.}}}\right)^2. \end{aligned} \tag{B.70}$$

Therefore this case is a special case of case (III) with $\beta_1^2(x^5) = \beta_2^2(x^5)$.

### B.5.1 (Va)

In the energy range considered the 5D metric (B.69) becomes

$$g_{AB,\mathrm{DR5}}(x^5) = \mathrm{diag}\left(1, -\beta^2(x^5), -\beta^2(x^5), -1, \pm f(x^5)\right). \tag{B.71}$$

Then, from definitions (21.22) and (21.23), one gets (B.28), (B.30) and

$$\begin{aligned} A_0(x^5) &= A_3(x^5) = 0; \\ B_0(x^5) &= B_3(x^5) = \left(f(x^5)\right)^{1/2}. \end{aligned} \tag{B.72}$$

Therefore, the $\Upsilon$-hypothesis is satisfied only for $\mu = 1{,}2$ under condition (22.105). From Sect. 21.3, the contravariant Killing five-vector $\xi^A(x, x^5)$ corresponding to form V) of the 5D gravitational metric below threshold is given by (21.41)–(21.45), in which (some of) the real parameters are constrained by the system:

$$\begin{aligned}
&(01)\ \ d_8x^2x^3 + d_7x^2 + d_6x^3 + (d_5 + a_2) \\
&\quad +\beta^2(x^5)\left[h_2x^2x^3 + h_1x^2 + h_8x^3 + (h_7 + e_2)\right] = 0; \\
&(02)\ \ d_8x^1x^3 + d_7x^1 + d_4x^3 + d_3 \\
&\quad +\beta^2(x^5)\left[l_2x^1x^3 + l_1x^1 + l_6x^3 + (l_5 + e_4)\right] = 0; \\
&(03)\ \ d_8x^1x^2 + d_6x^1 + d_4x^2 + d_2 \\
&\quad +m_8x^1x^2 + m_7x^1 + m_6x^2 + (m_5 + g_2) = 0; \\
&(12)\ \ \beta^2(x^5)\left(h_2x^0x^3 + h_1x^0 + h_4x^3 + h_3\right) \\
&\quad +\beta^2(x^5)\left(l_2x^0x^3 + l_1x^0 + l_4x^3 + l_3\right) = 0; \\
&(13)\ \ \beta^2(x^5)\left(h_2x^0x^2 + h_8x^0 + h_4x^2 + h_6\right) \\
&\quad +m_8x^0x^2 + m_7x^0 + m_4x^2 + m_3 = 0; \\
&(23)\ \ \beta^2(x^5)\left(l_2x^0x^1 + l_6x^0 + l_4x^1 + l_8\right) \\
&\quad +m_8x^0x^1 + m_6x^0 + m_4x^1 + m_2 = 0.
\end{aligned} \tag{B.73}$$

The solution of system (B.73) yields therefore, for $\xi^A(x, x^5)$:

$$\xi^0(x^3) = \widetilde{F_0}(x^3) = d_2x^3 + (a_1 + d_1 + K_0); \tag{B.74}$$

$$\xi^1(x^2) = -\widetilde{F_1}(x^2) = l_3x^2 - (K_1 + h_5 + e_1); \tag{B.75}$$

$$\xi^2(x^1) = -\widetilde{F_2}(x^1) = -l_3x^1 - (l_7 + K_2 + e_3); \tag{B.76}$$

$$\xi^3(x^0) = -\widetilde{F_3}(x^0) = d_2x^0 - (m_1 + g_1 + c); \tag{B.77}$$

$$\xi^5 = 0. \tag{B.78}$$

### B.5.2 (Vb)

The 5D metric has the same form of case (IVb), (B.68). Consequently, all the results of Sect. B.4.2 hold. Moreover, the Killing vector has the same expression of case (Va), and it is therefore independent of the energy range. By the following redenomination of the parameters:

$$\begin{aligned}
&(a_1 + d_1 + K_0) \equiv T^0;\\
&-(K_1 + h_5 + e_1) \equiv T^1;\\
&-(l_7 + K_2 + e_3) \equiv T^2;\\
&-(m_1 + g_1 + c) = T^3;\\
&d_2 = -\zeta^3;\\
&l_3 = -\theta^3,
\end{aligned} \tag{B.79}$$

the contravariant Killing vector $\xi^A(x, x^5)$ of the gravitational metric (B.68) can be written in the form (valid $\forall x^5 \in R_0^+$):

$$\xi^0(x^3) = -\zeta^3 x^3 + T^0; \tag{B.80}$$

$$\xi^1(x^2) = \theta^3 x^2 + T^1; \tag{B.81}$$

$$\xi^2(x^1) = -\theta^3 x^1 + T^2; \tag{B.82}$$

$$\xi^3(x^0) = -\zeta^3 x^0 + T^3; \tag{B.83}$$

$$\xi^5 = 0. \tag{B.84}$$

Then, $\forall x^5 \in R_0^+$, the 5D Killing group is

$$\left(SO(2)_{\mathrm{STD.},\Pi(x^1,x^2)} \otimes B_{x^3,\mathrm{STD.}}\right) \otimes_s \mathrm{Tr.}(1,3)_{\mathrm{STD.}}. \tag{B.85}$$

## B.6 Form VI

$$\begin{aligned}
&g_{AB,\mathrm{DR5,grav.}}(x^5)\\
&= \mathrm{diag}\left(1 + \Theta(x^5 - x^5_{0,\mathrm{grav.}})\left[\frac{1}{4}\left(1 + \frac{x^5}{x^5_{0,\mathrm{grav.}}}\right)^2 - 1\right],\right.\\
&-\left\{1 + \Theta(x^5 - x^5_{0,\mathrm{grav.}})\left[\frac{1}{4}\left(1 + \frac{x^5}{x^5_{0,\mathrm{grav.}}}\right)^2 - 1\right]\right\},\\
&-\left\{c + \Theta(x^5 - x^5_{0,\mathrm{grav.}})\left[\beta^2(x^5) - c\right]\right\},\\
&\left.-\left\{1 + \Theta(x^5 - x^5_{0,\mathrm{grav.}})\left[\frac{1}{4}\left(1 + \frac{x^5}{x^5_{0,\mathrm{grav.}}}\right)^2 - 1\right]\right\}, \pm f(x^5)\right),
\end{aligned} \tag{B.86}$$

($c \in R_0^+$, $c \neq 1$) where the function $\beta^2(x^5)$ has in general the properties

$$\begin{aligned}
&\beta^2(x^5) \in R_0^+, \forall x^5 \in \left([x_0^5, \infty)\right) \subset R_0^+;\\
&\beta^2(x_0^5) = c;\\
&\beta^2(x^5) \neq \frac{1}{4}\left(1 + \frac{x^5}{x^5_{0,\mathrm{grav.}}}\right)^2.
\end{aligned} \tag{B.87}$$

## *B.6.1 (VIa)*

The form of the 5D metric is

$$g_{AB,\mathrm{DR5}}(x^5) = \mathrm{diag}\left(1, -1, -c, -1, \pm f(x^5)\right). \tag{B.88}$$

This is exactly case (Ia) with $c_1 = 1$, $c_2 = c$, and therefore all the results of Sect. B.1.1 hold.

## *B.6.2 (VIb)*

The 5D metric is

$$\begin{aligned}
&g_{AB,\mathrm{DR5}}(x^5)\\
&= \mathrm{diag}\left(\frac{1}{4}\left(1 + \frac{x^5}{x^5_{0,\mathrm{grav.}}}\right)^2, -\frac{1}{4}\left(1 + \frac{x^5}{x^5_{0,\mathrm{grav.}}}\right)^2,\right.\\
&\left. -\beta^2(x^5), -\frac{1}{4}\left(1 + \frac{x^5}{x^5_{0,\mathrm{grav.}}}\right)^2, \pm f(x^5)\right)
\end{aligned} \tag{B.89}$$

and from definitions (21.22), (21.23) it follows:

$$\begin{aligned}
&A_0(x^5) = -A_1(x^5) = -A_3(x^5)\\
&= \frac{1}{8}\left(1 + \frac{x^5}{x_0^5}\right)\frac{x^5}{(x_0^5)^2}\left(f(x^5)\right)^{-1/2}\left[\frac{1}{x^5} + \frac{1}{2}\frac{f'(x^5)}{f(x^5)} + \frac{1}{2}\frac{x_0^5}{x^5}\frac{f'(x^5)}{f(x^5)}\right];\\
&B_0(x^5) = B_1(x^5) = B_3(x^5) = \frac{1}{2}\left(1 + \frac{x^5}{x_0^5}\right)\left(f(x^5)\right)^{1/2};
\end{aligned} \tag{B.90}$$

$$\begin{aligned}
&\frac{\pm A_0(x^5)}{B_0(x^5)} = \frac{\mp A_1(x^5)}{B_1(x^5)} = \frac{\mp A_3(x^5)}{B_3(x^5)}\\
&= \pm\frac{1}{4}\frac{1}{f(x^5)}\frac{x^5}{(x_0^5)^2}\left[\frac{1}{x^5} + \frac{1}{2}\frac{f'(x^5)}{f(x^5)} + \frac{1}{2}\frac{x_0^5}{x^5}\frac{f'(x^5)}{f(x^5)}\right];
\end{aligned} \tag{B.91}$$

$$A_2(x^5) \equiv \beta(x^5)(f(x^5))^{-1/2}$$
$$\times \left[ -\left(\beta'(x^5)\right)^2 + \beta(x^5)\beta''(x^5) - \frac{1}{2}\beta(x^5)\beta'(x^5)f'(x^5)(f(x^5))^{-1} \right]; \quad \text{(B.92)}$$

$$B_2(x^5) \equiv \beta(x^5)(f(x^5))^{1/2}; \quad \text{(B.93)}$$

$$\frac{\pm A_2(x^5)}{B_2(x^5)}$$
$$= \pm(f(x^5))^{-1} \left[ -\left(\beta'(x^5)\right)^2 + \beta(x^5)\beta''(x^5) - \frac{1}{2}\beta(x^5)\beta'(x^5)f'(x^5)(f(x^5))^{-1} \right]. \quad \text{(B.94)}$$

Then, the $\Upsilon$-hypothesis is satisfied for $\mu = 0, 1, 3$ under condition (B.9), and for $\mu = 2$ under condition:

$$-\left(\beta'(x^5)\right)^2 + \beta(x^5)\beta''(x^5) - \frac{1}{2}\beta(x^5)\beta'(x^5)f'(x^5)(f(x^5))^{-1} \neq \lambda f(x^5), \quad \lambda \in R. \quad \text{(B.95)}$$

So, under at least one of the conditions (B.9), (B.94), the contravariant Killing five-vector $\xi^A(x, x^5)$ of the gravitational metric in case (VIb) is given by (21.41)–(21.45), with (some of) the real parameters being constrained by the system:

$$\begin{aligned}
&(01)\ \ d_8x^2x^3 + d_7x^2 + d_6x^3 + (d_5 + a_2) \\
&\quad +h_2x^2x^3 + h_1x^2 + h_8x^3 + (h_7 + e_2) = 0; \\
&(02)\ \ \frac{1}{4}\left(1 + \frac{x^5}{x^5_{0,\text{grav.}}}\right)^2 (d_8x^1x^3 + d_7x^1 + d_4x^3 + d_3) \\
&\quad +\beta^2(x^5)\left[l_2x^1x^3 + l_1x^1 + l_6x^3 + (l_5 + e_4)\right] = 0; \\
&(03)\ \ d_8x^1x^2 + d_6x^1 + d_4x^2 + d_2 \\
&\quad +m_8x^1x^2 + m_7x^1 + m_6x^2 + (m_5 + g_2) = 0; \\
&(12)\ \ \frac{1}{4}\left(1 + \frac{x^5}{x^5_{0,\text{grav.}}}\right)^2 (h_2x^0x^3 + h_1x^0 + h_4x^3 + h_3) \\
&\quad +\beta^2(x^5)\left(l_2x^0x^3 + l_1x^0 + l_4x^3 + l_3\right) = 0; \\
&(13)\ \ h_2x^0x^2 + h_8x^0 + h_4x^2 + h_6 \\
&\quad +m_8x^0x^2 + m_7x^0 + m_4x^2 + m_3 = 0; \\
&(23)\ \ \beta^2(x^5)\left(l_2x^0x^1 + l_6x^0 + l_4x^1 + l_8\right) \\
&\quad +\frac{1}{4}\left(1 + \frac{x^5}{x^5_{0,\text{grav.}}}\right)^2 (m_8x^0x^1 + m_6x^0 + m_4x^1 + m_2) = 0.
\end{aligned} \quad \text{(B.96)}$$

Its solution yields the following explicit form of $\xi^A(x, x^5)$:

$$\xi^0(x^1, x^3) = \widetilde{F_0}(x^1, x^3) = -(h_7 + e_2)\,x^1 - (m_5 + g_2)\,x^3 + (a_1 + d_1 + K_0); \quad \text{(B.97)}$$

$$\xi^1(x^0,x^3) = -\widetilde{F_1}(x^0,x^3) = -(h_7+e_2)\,x^0 - h_6x^3 - (K_1+h_5+e_1)\,; \quad \text{(B.98)}$$

$$\xi^2 = -\widetilde{F_2} = -(l_7+K_2+e_3); \quad \text{(B.99)}$$

$$\xi^3(x^0,x^1) = -\widetilde{F_3}(x^0,x^1) = -(m_5+g_2)\,x^0 + h_6x^1 - (m_1+g_1+c); \quad \text{(B.100)}$$

$$\xi^5 = 0. \quad \text{(B.101)}$$

Then, it is easily seen that the 5D Killing group in this subcase is

$$\mathrm{SO}(2,1)_{\mathrm{STD}.M_3} \otimes_s \mathrm{Tr}.(1,3)_{\mathrm{STD}.}. \quad \text{(B.102)}$$

Here, $\mathrm{SO}(2,1)_{\mathrm{STD}.M_3}$ is the 3-parameter, homogeneous Lorentz group (generated by $S^2_{\mathrm{SR}}$, $K^1_{\mathrm{SR}}$, $K^3_{\mathrm{SR}}$) of the 3D space $M_3$ endowed with the metric interval

$$\mathrm{d}s^2_{M_3} = \left(\frac{1}{4}\left(1+\frac{x^5}{x^5_{0,\mathrm{grav.}}}\right)^2\right)\left(\left(\mathrm{d}x^0\right)^2 - \left(\mathrm{d}x^1\right)^2 - \left(\mathrm{d}x^3\right)^2\right) \quad \text{(B.103)}$$

and $\mathrm{Tr}.(1,3)_{\mathrm{STD}.}$ is the usual space–time translation group. Equation (B.101) can be rewritten as

$$P(1,2)_{\mathrm{STD}.M_3} \otimes_s \mathrm{Tr}._{\mathrm{STD}.,x^2} \quad \text{(B.104)}$$

where $P(1,2)_{\mathrm{STD}.M_3} = \mathrm{SO}(1,2)_{\mathrm{STD}.M_3} \otimes_s \mathrm{Tr}.(1)_{\mathrm{STD}.M_3}$ is the Poincaré group of $M_3$ and $\mathrm{Tr}.(1)_{\mathrm{STD}.,x^2}$ is the one-parameter group (generated by $\Upsilon^2_{\mathrm{SR}}$) of the translations along $\widehat{x^2}$.

In case VI, too, it is possible to write the Killing vector $\xi^A(x,x^5)$ of the gravitational metric (B.86) in a form valid on the whole energy range by means of the (right) Heaviside distribution $\widehat{\Theta}_{\mathrm{R}}(x^5_0 - x^5)$, by putting

$$\begin{aligned}
&B^1 \equiv \zeta^1, \frac{B^2}{c} \equiv \zeta^2, B^3 \equiv \zeta^3;\\
&\frac{\Theta^1}{c} \equiv \theta^1, \Theta^2 \equiv \theta^2, \frac{\Theta^3}{c} \equiv \theta^3;\\
&\Xi^0 \equiv \zeta^5, \frac{\Xi^2}{c} \equiv \Xi^{2\prime};\\
&\frac{T^2}{c} \equiv T^{2\prime}
\end{aligned} \quad \text{(B.105)}$$

and identifying

$$\begin{aligned}
&(a_1+d_1+K_0) = T^0;\\
&-(K_1+h_5+e_1) = T^1;\\
&-(l_7+K_2+e_3) = \frac{1}{c}T^2;\\
&-(m_1+g_1+c) = T^3;\\
&m_5+g_2 = B^3;\\
&h_6 = \Theta^2;\\
&h_7+e_2 = B^1.
\end{aligned} \quad \text{(B.106)}$$

One gets

$$\xi^0(x^1,x^2,x^3,x^5) = \widehat{\Theta}_{\mathrm{R}}(x_0^5 - x^5)\left[-c\zeta^2 x^2 + \zeta^5 F(x^5)\right] - \zeta^1 x^1 - \zeta^3 x^3 + T^0; \tag{B.107}$$

$$\xi^1(x^0,x^2,x^3,x^5) = \widehat{\Theta}_{\mathrm{R}}(x_0^5 - x^5)\left[c\theta^3 x^2 - \Xi^1 F(x^5)\right] - \zeta^1 x^0 - \theta^2 x^3 + T^1; \tag{B.108}$$

$$\xi^2(x^0,x^1,x^3,x^5) = \widehat{\Theta}_{\mathrm{R}}(x_0^5 - x^5)\left[-\zeta^2 x^0 + \theta^1 x^3 - \theta^3 x^1 - \Xi^2 F(x^5)\right] + T^2; \tag{B.109}$$

$$\xi^3(x^0,x^1,x^2,x^5) = \widehat{\Theta}_{\mathrm{R}}(x_0^5 - x^5)\left[-c\theta^1 x^2 - \Xi^3 F(x^5)\right] - \zeta^3 x^0 + \theta^2 x^1 + T^3; \tag{B.110}$$

$$\begin{aligned}&\xi^5(x,x^5)\\ &= \widehat{\Theta}_{\mathrm{R}}(x_0^5 - x^5)\left\{\mp\left(f(x^5)\right)^{-\frac{1}{2}}\left[\zeta^5 x^0 + \Xi^1 x^1 + c\Xi^2 x^2 + \Xi^3 x^3 - T^5\right]\right\}.\end{aligned} \tag{B.111}$$

## B.7 Form VII

$$\begin{aligned}&g_{AB,\mathrm{DR5,grav.}}(x^5)\\ &= \mathrm{diag}\left(1 + \Theta(x^5 - x^5_{0,\mathrm{grav.}})\left[\frac{1}{4}\left(1 + \frac{x^5}{x^5_{0,\mathrm{grav.}}}\right)^2 - 1\right],\right.\\ &-\left\{1 + \Theta(x^5 - x^5_{0,\mathrm{grav.}})\left[\frac{1}{4}\left(1 + \frac{x^5}{x^5_{0,\mathrm{grav.}}}\right)^2 - 1\right]\right\}, \beta^2(x^5),\\ &\left.-\left\{1 + \Theta(x^5 - x^5_{0,\mathrm{grav.}})\left[\frac{1}{4}\left(1 + \frac{x^5}{x^5_{0,\mathrm{grav.}}}\right)^2 - 1\right]\right\}, \pm f(x^5)\right)\end{aligned} \tag{B.112}$$

with the function $\beta^2(x^5)$ having in general the properties

$$\begin{aligned}&\beta^2(x^5) \in R_0^+, \forall x^5 \in R_0^+;\\ &\beta^2(x^5) \neq \frac{1}{4}\left(1 + \frac{x^5}{x^5_{0,\mathrm{grav.}}}\right)^2.\end{aligned} \tag{B.113}$$

### *B.7.1 (VIIa)*

The 5D metric (B.112) reads

$$g_{AB,\mathrm{DR5}}(x^5) = \mathrm{diag}\left(1, -1, -\beta^2(x^5), -1, \pm f(x^5)\right). \tag{B.114}$$

Definitions (21.22) and (21.23) yield:

$$A_0(x^5) = A_1(x^5) = A_3(x^5) = 0;$$
$$A_2(x^5) \equiv \beta(x^5)(f(x^5))^{-1/2}\cdot$$
$$\cdot\Big[-\left(\beta'(x^5)\right)^2 + \beta(x^5)\beta''(x^5) - \frac{1}{2}\beta(x^5)\beta'(x^5)f'(x^5)(f(x^5))^{-1}\Big]; \quad \text{(B.115)}$$

$$B_0(x^5) = B_1(x^5) = B_3(x^5) = \left(f(x^5)\right)^{1/2};$$
$$B_2(x^5) \equiv \beta(x^5)(f(x^5))^{1/2}; \quad \text{(B.116)}$$

$$\frac{\pm A_2(x^5)}{B_2(x^5)}$$
$$= \pm(f(x^5))^{-1}\Big[-\left(\beta'(x^5)\right)^2 + \beta(x^5)\beta''(x^5) - \frac{1}{2}\beta(x^5)\beta'(x^5)f'(x^5)(f(x^5))^{-1}\Big]. \quad \text{(B.117)}$$

So, the $\Upsilon$-hypothesis is satisfied only for $\mu = 2$ under condition (B.94). The contravariant Killing vector $\xi^A(x, x^5)$ for the gravitational metric in this subcase is still given by (21.41)–(21.45), in which (some of) the real parameters satisfy the following constraint system:

$$\begin{aligned}
&(01)\ \ d_8x^2x^3 + d_7x^2 + d_6x^3 + (d_5 + a_2)\\
&\quad +h_2x^2x^3 + h_1x^2 + h_8x^3 + (h_7 + e_2) = 0;\\
&(02)\ \ d_8x^1x^3 + d_7x^1 + d_4x^3 + d_3\\
&\quad +\beta^2(x^5)\left[l_2x^1x^3 + l_1x^1 + l_6x^3 + (l_5 + e_4)\right] = 0;\\
&(03)\ \ d_8x^1x^2 + d_6x^1 + d_4x^2 + d_2\\
&\quad +m_8x^1x^2 + m_7x^1 + m_6x^2 + (m_5 + g_2) = 0;\\
&(12)\ \ h_2x^0x^3 + h_1x^0 + h_4x^3 + h_3\\
&\quad +\beta^2(x^5)\left(l_2x^0x^3 + l_1x^0 + l_4x^3 + l_3\right) = 0;\\
&(13)\ \ h_2x^0x^2 + h_8x^0 + h_4x^2 + h_6\\
&\quad +m_8x^0x^2 + m_7x^0 + m_4x^2 + m_3 = 0;\\
&(23)\ \ \beta^2(x^5)\left(l_2x^0x^1 + l_6x^0 + l_4x^1 + l_8\right)\\
&\quad +m_8x^0x^1 + m_6x^0 + m_4x^1 + m_2 = 0,
\end{aligned} \quad \text{(B.118)}$$

whose solution can be shown to coincide with that of subcase (VIb). One gets therefore

$$\xi^0(x^1, x^3) = \widetilde{F_0}(x^1, x^3) = -\left(h_7 + e_2\right)x^1 - \left(m_5 + g_2\right)x^3 + \left(a_1 + d_1 + K_0\right); \quad \text{(B.119)}$$

$$\xi^1(x^0, x^3) = -\widetilde{F_1}(x^0, x^3) = -\left(h_7 + e_2\right)x^0 - h_6x^3 - \left(K_1 + h_5 + e_1\right); \quad \text{(B.120)}$$

$$\xi^2 = -\widetilde{F_2} = -(l_7 + K_2 + e_3); \quad \text{(B.121)}$$

$$\xi^3(x^0, x^1) = -\widetilde{F_3}(x^0, x^1) = -\left(m_5 + g_2\right)x^0 + h_6x^1 - (m_1 + g_1 + c); \quad \text{(B.122)}$$

$$\xi^5 = 0. \quad \text{(B.123)}$$

### *B.7.2 (VIIb)*

In the energy range $x^5 > x_0^5$ the form of the metric coincides with that of the case VI in the same range, (B.89), and therefore all results of Sect. B.6.2 still hold.

By putting

$$\begin{aligned}
&(a_1 + d_1 + K_0) = T^0; \\
&-(K_1 + h_5 + e_1) = T^1; \\
&-(l_7 + K_2 + e_3) = T^2; \\
&-(m_1 + g_1 + c) = T^3; \\
&m_5 + g_2 = \zeta^3; \\
&h_7 + e_2 = \zeta^1,
\end{aligned} \tag{B.124}$$

one finds that the general form of $\xi^A(x, x^5)$ for the gravitational metric in form VII is independent of the energy range and coincides with that obtained in case VI in the range $x^5 > x_0^5$. It reads explicitly (cf. (B.107)–(B.111))

$$\xi^0(x^1, x^3) = -\zeta^1 x^1 - \zeta^3 x^3 + T^0; \tag{B.125}$$

$$\xi^1(x^0 x^3) = -\zeta^1 x^0 - \theta^2 x^3 + T^1; \tag{B.126}$$

$$\xi^2 = +T^2; \tag{B.127}$$

$$\xi^3(x^0, x^1) = -\zeta^3 x^0 + \theta^2 x^1 + T^3; \tag{B.128}$$

$$\xi^5(x, x^5) = 0. \tag{B.129}$$

Consequently, the 5D Killing group is, $\forall x^5 \in R_0^+$ (see case VI):

$$\mathrm{SO}(1,2)_{\mathrm{STD}.M_3} \otimes_s \mathrm{Tr.}(1,3)_{\mathrm{STD.}} = P(1,2)_{\mathrm{STD}.M_3} \otimes_s \mathrm{Tr.}(1)_{\mathrm{STD.},x^2}. \tag{B.130}$$

## B.8 Form VIII

$$\begin{aligned}
&g_{AB,\mathrm{DR5,grav.}}(x^5) \\
&= \mathrm{diag}\left(1 + \Theta(x^5 - x^5_{0,\mathrm{grav.}})\left[\frac{1}{4}\left(1 + \frac{x^5}{x^5_{0,\mathrm{grav.}}}\right)^2 - 1\right],\right. \\
&\quad - \left\{c + \Theta(x^5 - x^5_{0,\mathrm{grav.}})\left[\beta^2(x^5) - c\right]\right\}, \\
&\quad - \left\{1 + \Theta(x^5 - x^5_{0,\mathrm{grav.}})\left[\frac{1}{4}\left(1 + \frac{x^5}{x^5_{0,\mathrm{grav.}}}\right)^2 - 1\right]\right\}, \\
&\quad \left. - \left\{1 + \Theta(x^5 - x^5_{0,\mathrm{grav.}})\left[\frac{1}{4}\left(1 + \frac{x^5}{x^5_{0,\mathrm{grav.}}}\right)^2 - 1\right]\right\}, \pm f(x^5)\right)
\end{aligned} \tag{B.131}$$

($c \in R_0^+$, $c \neq 1$) with

$$\begin{aligned} &\beta^2(x^5) \in R_0^+, \forall x^5 \in \left([x_0^5, \infty)\right) \subset R_0^+; \\ &\beta^2(x_0^5) = c; \\ &\beta^2(x^5) \neq \frac{1}{4}\left(1 + \frac{x^5}{x^5_{0,\text{grav.}}}\right)^2 . \end{aligned} \tag{B.132}$$

### B.8.1 (VIIIa)

In this subcase metric (B.131) becomes

$$g_{AB,\text{DR5}}(x^5) = \text{diag}\left(1, -c, -1, -1, \pm f(x^5)\right), \tag{B.133}$$

the same as case (Ia) with $c_1 = c$, $c_2 = 1$ (see (B.2)). Then, the results of Sect. B.1.1 are valid.

### B.8.2 (VIIIb)

The metric (B.131) reads

$$\begin{aligned} &g_{AB,\text{DR5}}(x^5) \\ &= \text{diag}\left(\frac{1}{4}\left(1 + \frac{x^5}{x^5_{0,\text{grav.}}}\right)^2, -\beta^2(x^5), -\frac{1}{4}\left(1 + \frac{x^5}{x^5_{0,\text{grav.}}}\right)^2, \right. \\ &\qquad \left. -\frac{1}{4}\left(1 + \frac{x^5}{x^5_{0,\text{grav.}}}\right)^2, \pm f(x^5)\right) \end{aligned} \tag{B.134}$$

and coincides with metric (B.89) of case (VIb) except for the exchange of the space axes $1 \rightleftarrows 2$. By the usual procedure it is found that the $\Upsilon$-hypothesis is satisfied for $\mu = 0, 2, 3$ under condition (B.9) and for $\mu = 1$ under condition (B.94). Therefore, under at least one of these two conditions, the contravariant Killing vector $\xi^A(x, x^5)$ of the gravitational metric in subcase VIII b) are given by (21.41)–(21.45), in which (at least some of) the real parameters are constrained by the following system:

$$\begin{cases} (01)\ \frac{1}{4}\left(1+\frac{x^5}{x^5_{0,\text{grav.}}}\right)^2\left[d_8x^2x^3+d_7x^2+d_6x^3+(d_5+a_2)\right] \\ \quad +\beta^2(x^5)\left[h_2x^2x^3+h_1x^2+h_8x^3+(h_7+e_2)\right]=0; \\ (02)\ d_8x^1x^3+d_7x^1+d_4x^3+d_3 \\ \quad +l_2x^1x^3+l_1x^1+l_6x^3+(l_5+e_4)=0; \\ (03)\ d_8x^1x^2+d_6x^1+d_4x^2+d_2 \\ \quad +m_8x^1x^2+m_7x^1+m_6x^2+(m_5+g_2)=0; \\ (12)\ \beta^2(x^5)\left(h_2x^0x^3+h_1x^0+h_4x^3+h_3\right) \\ \quad +\frac{1}{4}\left(1+\frac{x^5}{x^5_{0,\text{grav.}}}\right)^2\left(l_2x^0x^3+l_1x^0+l_4x^3+l_3\right)=0; \\ (13)\ \beta^2(x^5)\left(h_2x^0x^2+h_8x^0+h_4x^2+h_6\right) \\ \quad +\frac{1}{4}\left(1+\frac{x^5}{x^5_{0,\text{grav.}}}\right)^2\left(m_8x^0x^2+m_7x^0+m_4x^2+m_3\right)=0; \\ (23)\ l_2x^0x^1+l_6x^0+l_4x^1+l_8 \\ \quad +m_8x^0x^1+m_6x^0+m_4x^1+m_2=0. \end{cases} \tag{B.135}$$

Then, $\xi^A(x,x^5)$ is given by

$$\xi^0(x^2,x^3)=\widetilde{F_0}(x^2,x^3)=-\left(l_5+e_4\right)x^2-\left(m_5+g_2\right)x^3+\left(a_1+d_1+K_0\right); \tag{B.136}$$

$$\xi^1=-\widetilde{F_1}=-\left(K_1+h_5+e_1\right); \tag{B.137}$$

$$\xi^2(x^0,x^3)=-\widetilde{F_2}(x^0,x^3)=-\left(l_5+e_4\right)x^0-l_8x^3-\left(l_7+K_2+e_3\right); \tag{B.138}$$

$$\xi^3(x^0,x^2)=-\widetilde{F_3}(x^0,x^2)=-\left(m_5+g_2\right)x^0+l_8x^2-\left(m_1+g_1+c\right); \tag{B.139}$$

$$\xi^5=0. \tag{B.140}$$

It follows that the 5D Killing group is:

$$\text{SO}(1,2)_{\text{STD}.\overline{M_3}}\otimes_s \text{Tr.}(1,3)_{\text{STD.}} \tag{B.141}$$

Here $\text{SO}(1,2)_{\text{STD}.\overline{M_3}}$ is the 3-parameter Lorentz group (generated by $S^1_{\text{SR}}$, $K^2_{\text{SR}}$ , $K^3_{\text{SR}}$) of the 3D manifold $\overline{M_3}$ with metric interval

$$\text{d}s^2_{\overline{M_3}}=\left(\frac{1}{4}\left(1+\frac{x^5}{x^5_{0,\text{grav.}}}\right)^2\right)\left(\left(\text{d}x^0\right)^2-\left(\text{d}x^2\right)^2-\left(\text{d}x^3\right)^2\right) \tag{B.142}$$

and $\text{Tr.}(1,3)_{\text{STD.}}$ is the usual space–time translation group. Equation (B.141) can be rewritten in the form

$$P(1,2)_{\text{STD}.\overline{M_3}}\otimes_s \text{Tr.}(1)_{\text{STD.},x^1}, \tag{B.143}$$

where we introduced the six-parameter Poincaré group of $\overline{M_3}$, $P(1,2)_{\mathrm{STD}.\overline{M_3}} = \mathrm{SO}(1,2)_{\mathrm{STD}.\overline{M_3}} \otimes_s \mathrm{Tr}.(1)_{\mathrm{STD}.\overline{M_3}}$, and $\mathrm{Tr}.(1)_{\mathrm{STD}.,x^1}$ is the one-parameter translation group along $\widehat{x^1}$ (generated by $\Upsilon^1_{\mathrm{SR}}$).

Moreover, it is easy to see that the compact form of the contravariant Killing vector for the gravitational metric in case VIII, valid $\forall x^5 \in R_0^+$, is still given by (B.107)–(B.111), with the exchange $1 \rightleftarrows 2$.

## B.9 Form IX

$$
\begin{aligned}
& g_{AB,\mathrm{DR5,grav.}}(x^5) \\
&= \mathrm{diag}\left(1 + \Theta(x^5 - x^5_{0,\mathrm{grav.}})\left[\frac{1}{4}\left(1 + \frac{x^5}{x^5_{0,\mathrm{grav.}}}\right)^2 - 1\right],\right. \\
&\quad -\beta^2(x^5), -\left\{1 + \Theta(x^5 - x^5_{0,\mathrm{grav.}})\left[\frac{1}{4}\left(1 + \frac{x^5}{x^5_{0,\mathrm{grav.}}}\right)^2 - 1\right]\right\}, \\
&\quad \left. -\left\{1 + \Theta(x^5 - x^5_{0,\mathrm{grav.}})\left[\frac{1}{4}\left(1 + \frac{x^5}{x^5_{0,\mathrm{grav.}}}\right)^2 - 1\right]\right\}, \pm f(x^5)\right)
\end{aligned}
\tag{B.144}
$$

where

$$
\begin{aligned}
& \beta^2(x^5) \in R_0^+ \quad \forall x^5 \in R_0^+; \\
& \beta^2(x^5) \neq \frac{1}{4}\left(1 + \frac{x^5}{x^5_{0,\mathrm{grav.}}}\right)^2 .
\end{aligned}
\tag{B.145}
$$

### *B.9.1 (IXa)*

In the range $0 < x^5 \leqslant x_0^5$ metric (B.144) becomes

$$
g_{AB,\mathrm{DR5}}(x^5) = \mathrm{diag}\left(1, -\beta^2(x^5), -1, -1, \pm f(x^5)\right), \tag{B.146}
$$

namely the same of case (VIIa) with the exchange $2 \rightarrow 1$. Therefore, performing such an exchange in (B.113)–(B.116), one finds that the hypothesis $\Upsilon$ of functional independence is satisfied only for $\mu = 1$ under condition (B.94). Under such a condition, the contravariant Killing five-vector $\xi^A(x, x^5)$ of the gravitational metric in the subcase IX a) is given by (21.41)–(21.45), with (some of) the real parameters being constrained by the system:

$$\begin{aligned}
&(01)\ \left[d_8x^2x^3+d_7x^2+d_6x^3+(d_5+a_2)\right]\\
&\quad+\beta^2(x^5)\left[h_2x^2x^3+h_1x^2+h_8x^3+(h_7+e_2)\right]=0;\\
&(02)\ d_8x^1x^3+d_7x^1+d_4x^3+d_3\\
&\quad+l_2x^1x^3+l_1x^1+l_6x^3+(l_5+e_4)=0;\\
&(03)\ d_8x^1x^2+d_6x^1+d_4x^2+d_2\\
&\quad+m_8x^1x^2+m_7x^1+m_6x^2+m_5+g_2=0;\\
&(12)\ \beta^2(x^5)\left(h_2x^0x^3+h_1x^0+h_4x^3+h_3\right)\\
&\quad+l_2x^0x^3+l_1x^0+l_4x^3+l_3=0;\\
&(13)\ \beta^2(x^5)\left(h_2x^0x^2+h_8x^0+h_4x^2+h_6\right)\\
&\quad+m_8x^0x^2+m_7x^0+m_4x^2+m_3=0;\\
&(23)\ l_2x^0x^1+l_6x^0+l_4x^1+l_8\\
&\quad+m_8x^0x^1+m_6x^0+m_4x^1+m_2=0.
\end{aligned} \qquad (B.147)$$

The solution of the above system is the same obtained in subcase (VIIIb). So, the explicit form of $\xi^A(x,x^5)$ for the gravitational metric in form IX under threshold coincides with that of case VIII over threshold, (B.136)–(B.140).

### B.9.2 (IXb)

In the energy range $x^5 > x_0^5$ the form of the metric coincides with that of case VIII in the same range:

$$\begin{aligned}
&g_{AB,\mathrm{DR5}}(x^5)\\
&=\mathrm{diag}\left(\frac{1}{4}\left(1+\frac{x^5}{x^5_{0,\mathrm{grav.}}}\right)^2,-\beta^2(x^5),\right.\\
&\left.-\frac{1}{4}\left(1+\frac{x^5}{x^5_{0,\mathrm{grav.}}}\right)^2,-\frac{1}{4}\left(1+\frac{x^5}{x^5_{0,\mathrm{grav.}}}\right)^2,\pm f(x^5)\right),
\end{aligned} \qquad (B.148)$$

so all the results of Sect. B.8.2 hold. The contravariant Killing vector $\xi^A(x,x^5)$ is therefore independent of the energy range, and (on account of the discussion of Sect. B.8.2) its expression is obtained from (B.124)–(B.129) by the exchange $1 \leftrightarrows 2$.

As to the 5D Killing group, it is therefore, $\forall x^5 \in R_0^+$ (cf. case (VIIIb))

$$\mathrm{SO}(1,2)_{\mathrm{STD.}\overline{M_3}} \otimes_s \mathrm{Tr.}(1,3)_{\mathrm{STD.}} = P(1,2)_{\mathrm{STD.}\overline{M_3}} \otimes_s \mathrm{Tr.}(1)_{\mathrm{STD.},x^2}. \qquad (B.149)$$

## B.10 Form X

$$
\begin{aligned}
&g_{AB,\mathrm{DR5,grav.}}(x^5)\\
&= \mathrm{diag}\left(1+\Theta(x^5-x^5_{0,\mathrm{grav.}})\left[\frac{1}{4}\left(1+\frac{x^5}{x^5_{0,\mathrm{grav.}}}\right)^2-1\right],\right.\\
&-\left\{c_1+\Theta(x^5-x^5_{0,\mathrm{grav.}})\left[\frac{c_1}{4}\left(1+\frac{x^5}{x^5_{0,\mathrm{grav.}}}\right)^2-c_1\right]\right\}\\
&-\left\{c_2+\Theta(x^5-x^5_{0,\mathrm{grav.}})\left[\frac{c_2}{4}\left(1+\frac{x^5}{x^5_{0,\mathrm{grav.}}}\right)^2-c_2\right]\right\},\\
&\left.-\left\{1+\Theta(x^5-x^5_{0,\mathrm{grav.}})\left[\frac{1}{4}\left(1+\frac{x^5}{x^5_{0,\mathrm{grav.}}}\right)^2-1\right]\right\},\pm f(x^5)\right)
\end{aligned}
\tag{B.150}
$$

($c_1, c_2 \in R_0^+$, in gen.: $c_1 \neq 1,\ c_2 \neq 1,\ c_1 \neq c_2$).

### *B.10.1 (Xa)*

In the energy range $0 < x^5 \leqslant x_0^5$ the form of the 5D metric is identical to that of cases (I) and (II) in the same range, (B.2). All the results of Sects. B.1.1 and B.2.1 are valid. The $\Upsilon$-hypothesis is violated $\forall \mu \in \{0,1,2,3\}$. The Killing equations, and therefore the Killing vector, coincide with those corresponding to the e.m. and weak metrics above threshold (Sect. 22.1.1) and to the strong metric below threshold (Sect. 22.2.1). Therefore the contravariant Killing five-vector $\xi^A(x, x^5)$ is given by (22.5)–(22.10), and the Killing group of the sections at $\mathrm{d}x^5 = 0$ of $\Re_5$ is of course the rescaled Poincaré group (B.4).

### *B.10.2 (Xb)*

The 5D metric becomes

$$
\begin{aligned}
&g_{AB,\mathrm{DR5}}(x^5)\\
&= \mathrm{diag}\left(\frac{1}{4}\left(1+\frac{x^5}{x^5_{0,\mathrm{grav.}}}\right)^2, -\frac{c_1}{4}\left(1+\frac{x^5}{x^5_{0,\mathrm{grav.}}}\right)^2,\right.\\
&\left.-\frac{c_2}{4}\left(1+\frac{x^5}{x^5_{0,\mathrm{grav.}}}\right)^2, -\frac{1}{4}\left(1+\frac{x^5}{x^5_{0,\mathrm{grav.}}}\right)^2, \pm f(x^5)\right).
\end{aligned}
\tag{B.151}
$$

Therefore

$$A_0(x^5) = -A_3(x^5) = = \frac{1}{8}\left(1+\frac{x^5}{x_0^5}\right)\frac{x^5}{(x_0^5)^2}\left(f(x^5)\right)^{-1/2}\left[\frac{1}{x^5}+\frac{1}{2}\frac{f'(x^5)}{f(x^5)}+\frac{1}{2}\frac{x_0^5}{x^5}\frac{f'(x^5)}{f(x^5)}\right];$$

$$B_0(x^5) = B_3(x^5) = \frac{1}{2}\left(1+\frac{x^5}{x_0^5}\right)\left(f(x^5)\right)^{1/2}; \tag{B.152}$$

$$\frac{\pm A_0(x^5)}{B_0(x^5)} = \frac{\mp A_3(x^5)}{B_3(x^5)} = \pm\frac{1}{4}\frac{1}{f(x^5)}\frac{x^5}{(x_0^5)^2}\left[\frac{1}{x^5}+\frac{1}{2}\frac{f'(x^5)}{f(x^5)}+\frac{1}{2}\frac{x_0^5}{x^5}\frac{f'(x^5)}{f(x^5)}\right]; \tag{B.153}$$

$$A_i(x^5) = -\frac{(c_i)^{3/2}}{8}\left(1+\frac{x^5}{x_0^5}\right)\frac{x^5}{(x_0^5)^2}\left(f(x^5)\right)^{-1/2}\left[\frac{1}{x^5}+\frac{1}{2}\frac{f'(x^5)}{f(x^5)}+\frac{1}{2}\frac{x_0^5}{x^5}\frac{f'(x^5)}{f(x^5)}\right];$$

$$B_i(x^5) = \frac{(c_i)^{1/2}}{2}\left(1+\frac{x_0^5}{x^5}\right)\left(f(x^5)\right)^{1/2}, \quad i=1,2; \tag{B.154}$$

$$\frac{\pm A_i(x^5)}{B_i(x^5)} = \mp\frac{c_i}{4}\frac{1}{f(x^5)}\frac{x^5}{(x_0^5)^2}\left[\frac{1}{x^5}+\frac{1}{2}\frac{f'(x^5)}{f(x^5)}+\frac{1}{2}\frac{x_0^5}{x^5}\frac{f'(x^5)}{f(x^5)}\right], \quad i=1,2. \tag{B.155}$$

It follows that the $\Upsilon$-hypothesis is satisfied by any $\mu = 0,1,2,3$ under condition (B.9). Then, under such condition, the contravariant Killing vector $\xi^A(x,x^5)$ for the gravitational metric in this subcase is given by (21.41)–(21.45), in which (some of) the real parameters satisfy the constraint system:

$$\begin{aligned}
&(01)\ \ d_8x^2x^3 + d_7x^2 + d_6x^3 + (d_5+a_2)\\
&\quad +c_1\left[h_2x^2x^3 + h_1x^2 + h_8x^3 + (h_7+e_2)\right] = 0;\\
&(02)\ \ d_8x^1x^3 + d_7x^1 + d_4x^3 + d_3\\
&\quad +c_2\left[l_2x^1x^3 + l_1x^1 + l_6x^3 + (l_5+e_4)\right] = 0;\\
&(03)\ \ d_8x^1x^2 + d_6x^1 + d_4x^2 + d_2\\
&\quad +m_8x^1x^2 + m_7x^1 + m_6x^2 + (m_5+g_2) = 0;\\
&(12)\ \ c_1\left(h_2x^0x^3 + h_1x^0 + h_4x^3 + h_3\right)\\
&\quad +c_2\left(l_2x^0x^3 + l_1x^0 + l_4x^3 + l_3\right) = 0;\\
&(13)\ \ c_1\left(h_2x^0x^2 + h_8x^0 + h_4x^2 + h_6\right)\\
&\quad +m_8x^0x^2 + m_7x^0 + m_4x^2 + m_3 = 0;\\
&(23)\ \ c_2\left(l_2x^0x^1 + l_6x^0 + l_4x^1 + l_8\right)\\
&\quad +m_8x^0x^1 + m_6x^0 + m_4x^1 + m_2 = 0.
\end{aligned} \tag{B.156}$$

Then, $\xi^A(x,x^5)$ explicitly reads, in this subcase:

$$\xi^0(x^1,x^2,x^3) = \widetilde{F_0}(x^1,x^2,x^3) = -c_1(h_7+e_2)x^1 - c_2(l_5+e_4)x^2 - (m_5+g_2)x^3 + (a_1+d_1+K_0); \tag{B.157}$$

$$\xi^1(x^0, x^2, x^3) = -\widetilde{F_1}(x^0, x^2, x^3)$$
$$= -(h_7 + e_2)x^0 - h_3 x^2 - h_6 x^3 - (K_1 + h_5 + e_1)\,; \qquad \text{(B.158)}$$

$$\xi^2(x^0, x^1, x^3) = -\widetilde{F_2}(x^0, x^1, x^3)$$
$$= -(l_5 + e_4)\,x^0 + \frac{c_1}{c_2} h_3 x^1 - l_8 x^3 - (l_7 + K_2 + e_3); \qquad \text{(B.159)}$$

$$\xi^3(x^0, x^1, x^2) = -\widetilde{F_3}(x^0, x^1, x^2)$$
$$= -(m_5 + g_2)\,x^0 + c_1 h_6 x^1 + c_2 l_8 x^2 - (m_1 + g_1 + c); \qquad \text{(B.160)}$$

$$\xi^5 = 0. \qquad \text{(B.161)}$$

By introducing as usual the distribution $\widehat{\Theta}_{\mathrm{R}}(x_0^5 - x^5)$, putting

$$\begin{aligned}
&\frac{B^1}{c_1} \equiv \zeta^1, \frac{B^2}{c_2} \equiv \zeta^2, B^3 \equiv \zeta^3;\\
&\frac{\Theta^1}{c_2} \equiv \theta^1, \frac{\Theta^2}{c_1} \equiv \theta^2, \frac{\Theta^3}{c_2} \equiv \theta^3;\\
&\Xi^0 \equiv \zeta^5, \frac{\Xi^1}{c_1} \equiv \Xi^{1\prime}, \frac{\Xi^2}{c_2} \equiv \Xi^{2\prime};\\
&\frac{T^1}{c_1} \equiv T^{1\prime}, \frac{T^2}{c_2} \equiv T^{2\prime}
\end{aligned} \qquad \text{(B.162)}$$

and identifying

$$\begin{aligned}
&(a_1 + d_1 + K_0) = T^0;\\
&-(K_1 + h_5 + e_1) = \frac{1}{c_1} T^1;\\
&-(l_7 + K_2 + e_3) = \frac{1}{c_2} T^2;\\
&-(m_1 + g_1 + c) = T^3;\\
&h_7 + e_2 = \frac{B_1}{c_1}; \quad l_5 + e_4 = \frac{B^2}{c_2}; \quad m_5 + g_2 = B^3;\\
&l_8 = -\frac{\Theta^1}{c_2}; \quad h_6 = \frac{\Theta^2}{c_1}; \quad h_3 = -\frac{\Theta^3}{c_1},
\end{aligned} \qquad \text{(B.163)}$$

one gets the following general form of the Killing vector in case X, valid $\forall x^5 \in R_0^+$:

$$\xi^0(x^1, x^2, x^3, x^5) = \widehat{\Theta}_{\mathrm{R}}(x_0^5 - x^5)\left[\zeta^5 F(x^5)\right] - c_1\zeta^1 x^1 - c_2\zeta^2 x^2 - \zeta^3 x^3 + T^0; \qquad \text{(B.164)}$$

$$\xi^1(x^0, x^2, x^3, x^5) = \widehat{\Theta}_{\mathrm{R}}(x_0^5 - x^5)\left[-\Xi^1 F(x^5)\right] - \zeta^1 x^0 + \frac{c_2}{c_1}\theta^3 x^2 - \theta^2 x^3 + T^1; \qquad \text{(B.165)}$$

$$\xi^2(x^0, x^1, x^3, x^5) = \widehat{\Theta}_{\mathrm{R}}(x_0^5 - x^5)\left[-\Xi^2 F(x^5)\right] - \zeta^2 x^0 - \theta^3 x^1 + \theta^1 x^3 + T^2; \qquad \text{(B.166)}$$

$$\xi^3(x^0,x^1,x^2,x^5) = \widehat{\Theta}_{\mathrm{R}}(x_0^5 - x^5)\left[-\Xi^3 F(x^5)\right] - \zeta^3 x^0 + c_1\theta^2 x^1 - c_2\theta^1 x^2 + T^3; \tag{B.167}$$

$$\begin{aligned}&\xi^5(x,x^5)\\ &= \widehat{\Theta}_{\mathrm{R}}(x_0^5 - x^5)\left\{\mp\left(f(x^5)\right)^{-1/2}\left[\zeta^5 x^0 + c_1\Xi^1 x^1 + c_2\Xi^2 x^2 + \Xi^3 x^3 - T^5\right]\right\}.\end{aligned} \tag{B.168}$$

The 5D Killing group in this range is the Poincaré one, suitably rescaled

$$[P(1,3)_{\mathrm{STD.}} = \mathrm{SO}(1,3)_{\mathrm{STD.}} \otimes_s \mathrm{Tr.}(1,3)_{\mathrm{STD.}}]|_{x^1 \longrightarrow \sqrt{c_1}x^1, x^2 \longrightarrow \sqrt{c_2}x^2} \tag{B.169}$$

as in the energy range $0 < x^5 \leqslant x_0^5$.

## B.11 Form XI

$$\begin{aligned}&g_{AB,\mathrm{DR5,grav.}}(x^5)\\ &= \mathrm{diag}\left(1 + \Theta(x^5 - x^5_{0,\mathrm{grav.}})\left[\frac{1}{4}\left(1 + \frac{x^5}{x^5_{0,\mathrm{grav.}}}\right)^2 - 1\right],\right.\\ &\quad -\frac{c_1}{4}\left(1 + \frac{x^5}{x^5_{0,\mathrm{grav.}}}\right)^2, -\frac{c_2}{4}\left(1 + \frac{x^5}{x^5_{0,\mathrm{grav.}}}\right)^2,\\ &\quad \left. -\left\{1 + \Theta(x^5 - x^5_{0,\mathrm{grav.}})\left[\frac{1}{4}\left(1 + \frac{x^5}{x^5_{0,\mathrm{grav.}}}\right)^2 - 1\right]\right\}, \pm f(x^5)\right)\end{aligned} \tag{B.170}$$

($c_1, c_2 \in R_0^+$. In gen.: $c_1 \neq 1, c_2 \neq 1, c_1 \neq c_2$).

### *B.11.1 (XIa)*

Metric (B.170) reads

$$\begin{aligned}&g_{AB,\mathrm{DR5}}(x^5)\\ &= \mathrm{diag}\left(1, -\frac{c_1}{4}\left(1 + \frac{x^5}{x^5_{0,\mathrm{grav.}}}\right)^2, -\frac{c_2}{4}\left(1 + \frac{x^5}{x^5_{0,\mathrm{grav.}}}\right)^2, -1, \pm f(x^5)\right).\end{aligned} \tag{B.171}$$

Therefore

$$\begin{aligned} A_0(x^5) &= A_3(x^5) = 0; \\ B_0(x^5) &= B_3(x^5) = \left(f(x^5)\right)^{1/2}; \end{aligned} \tag{B.172}$$

$$A_i(x^5) =$$
$$-\frac{(c_i)^{3/2}}{8}\left(1+\frac{x^5}{x_0^5}\right)\frac{x^5}{(x_0^5)^2}\left(f(x^5)\right)^{-1/2}\left[\frac{1}{x^5}+\frac{1}{2}\frac{f'(x^5)}{f(x^5)}+\frac{1}{2}\frac{x_0^5}{x^5}\frac{f'(x^5)}{f(x^5)}\right];$$
$$B_i(x^5) = \frac{(c_i)^{1/2}}{2}\left(1+\frac{x_0^5}{x^5}\right)\left(f(x^5)\right)^{1/2}, \quad i=1,2; \tag{B.173}$$

$$\frac{\pm A_i(x^5)}{B_i(x^5)} = \mp\frac{c_i}{4}\frac{1}{f(x^5)}\frac{x^5}{(x_0^5)^2}\left[\frac{1}{x^5}+\frac{1}{2}\frac{f'(x^5)}{f(x^5)}+\frac{1}{2}\frac{x_0^5}{x^5}\frac{f'(x^5)}{f(x^5)}\right], \quad i=1,2. \tag{B.174}$$

So the $\Upsilon$-hypothesis is satisfied for $\mu = 1,2$ under condition (B.9), which ensures that the contravariant Killing vector $\xi^A(x,x^5)$ for the gravitational metric in this subcase is given by (21.41)–(21.45), where (some of) the real parameters are constrained by the system

$$\begin{aligned}
&(01)\ \ d_8x^2x^3 + d_7x^2 + d_6x^3 + (d_5+a_2)\\
&\quad +\frac{c_1}{4}\left(1+\frac{x^5}{x^5_{0,\text{grav.}}}\right)^2\left[h_2x^2x^3 + h_1x^2 + h_8x^3 + (h_7+e_2)\right] = 0;\\
&(02)\ \ \left(d_8x^1x^3 + d_7x^1 + d_4x^3 + d_3\right)\\
&\quad +\frac{c_2}{4}\left(1+\frac{x^5}{x^5_{0,\text{grav.}}}\right)^2\left[l_2x^1x^3 + l_1x^1 + l_6x^3 + (l_5+e_4)\right] = 0;\\
&(03)\ \ \left(d_8x^1x^2 + d_6x^1 + d_4x^2 + d_2\right)\\
&\quad +\left[m_8x^1x^2 + m_7x^1 + m_6x^2 + (m_5+g_2)\right] = 0;\\
&(12)\ \ c_1\left(h_2x^0x^3 + h_1x^0 + h_4x^3 + h_3\right)\\
&\quad +c_2\left(l_2x^0x^3 + l_1x^0 + l_4x^3 + l_3\right) = 0;\\
&(13)\ \ \frac{c_1}{4}\left(1+\frac{x^5}{x^5_{0,\text{grav.}}}\right)^2\left(h_2x^0x^2 + h_8x^0 + h_4x^2 + h_6\right)\\
&\quad +\left(m_8x^0x^2 + m_7x^0 + m_4x^2 + m_3\right) = 0;\\
&(23)\ \ \frac{c_2}{4}\left(1+\frac{x^5}{x^5_{0,\text{grav.}}}\right)^2\left(l_2x^0x^1 + l_6x^0 + l_4x^1 + l_8\right)\\
&\quad +\left(m_8x^0x^1 + m_6x^0 + m_4x^1 + m_2\right) = 0.
\end{aligned}$$

From the solution of this system one gets the explicit form of the Killing five-vector

$$\xi^0(x^3) = \widetilde{F_0}(x^3) = -\left(m_5+g_2\right)x^3 + (a_1+d_1+K_0); \tag{B.175}$$

$$\xi^1(x^2) = -\widetilde{F_1}(x^2) = -h_3x^2 - (K_1+h_5+e_1); \tag{B.176}$$

$$\xi^2(x^1) = -\widetilde{F_2}(x^1) = \frac{c_1}{c_2}h_3x^1 - (l_7+K_2+e_3); \tag{B.177}$$

$$\xi^3(x^0) = -\widetilde{F_3}(x^0) = -\left(m_5+g_2\right)x^0 - (m_1+g_1+c); \tag{B.178}$$

$$\xi^5 = 0. \tag{B.179}$$

The metric (B.170) can be rewritten as

$$g_{AB,\mathrm{DR5,grav.}}(x^5) = \mathrm{diag}\left(\frac{1}{4}\left(1+\frac{x^5}{x^5_{0,\mathrm{grav.}}}\right)^2 g_{\mu\nu,\overline{M_4}}(x^5), \pm f(x^5)\right) \tag{B.180}$$

where $\overline{M_4}$ is the standard 4D Minkowski space with the coordinate scale transformation (B.3) introduced in Sect. B.1.1. Therefore, in the energy range $0 < x^5 \leqslant x^5_0$ the 5D Killing group is the rescaled Poincaré group

$$[P(1,3)_{\mathrm{STD.}} = \mathrm{SO}(1,3)_{\mathrm{STD.}} \otimes_s \mathrm{Tr.}(1,3)_{\mathrm{STD.}}]|_{x^1 \longrightarrow \sqrt{c_1}x^1, x^2 \longrightarrow \sqrt{c_2}x^2} \,. \tag{B.181}$$

### *B.11.2 (XIb)*

In the energy range $x^5 > x^5_0$ the metric (B.170) coincides with that of case (X) in the same range, (B.151). Therefore, the results for this case coincide with those obtained in Sect. B.10.2. In particular, solution (B.157)–(B.161) holds for the contravariant Killing vector $\xi^A(x, x^5)$.

The 5D Killing group in this range is

$$\left(\mathrm{SO}(2)_{\mathrm{STD.}(x^1 \longrightarrow \sqrt{c_1}x^1, x^2 \longrightarrow \sqrt{c_2}x^2)} \otimes B_{x^3,STD.}\right) \otimes_s \mathrm{Tr.}(1,3)_{\mathrm{STD.}}, \tag{B.182}$$

where $\mathrm{SO}(2)_{\mathrm{STD.}(x^1 \longrightarrow \sqrt{c_1}x^1, x^2 \longrightarrow \sqrt{c_2}x^2)} = \mathrm{SO}(2)_{\mathrm{STD.},\Pi(x^1 \longrightarrow \sqrt{\frac{c_1}{c_2}}x^1)}$ is the one-parameter group (generated by $S^3_{\mathrm{SR}}$) of the 2D rotations in the plane $\Pi(x^1, x^2)$ characterized by the scale transformations (B.3), $B_{x^3,STD.}$ is the usual one-parameter group (generated by $K^3_{\mathrm{SR}}$) of the Lorentz boosts along $\widehat{x^3}$ and $\mathrm{Tr.}(1,3)_{\mathrm{STD.}}$ is the group of standard space–time translations.

Since

$$\begin{aligned}&\left(\mathrm{SO}(2)_{\mathrm{STD.}(x^1 \longrightarrow \sqrt{c_1}x^1, x^2 \longrightarrow \sqrt{c_2}x^2)} \otimes B_{x^3,STD.}\right) \otimes_s \mathrm{Tr.}(1,3)_{\mathrm{STD.}} \subsetneq \\ &\subsetneq [P(1,3)_{\mathrm{STD.}} = \mathrm{SO}(1,3)_{\mathrm{STD.}} \otimes_s \mathrm{Tr.}(1,3)_{\mathrm{STD.}}]|_{x^1 \longrightarrow \sqrt{c_1}x^1, x^2 \longrightarrow \sqrt{c_2}x^2}\end{aligned}$$

we can state that the present case XI is the only one – in the framework of the Ansatz of solution of Killing equations for the 5D phenomenological metric of gravitational interaction – in which the 5D Killing group in the range $0 < x^5 \leqslant x^5_0$ is a proper (non-abelian) subgroup of the Killing group in the range $x^5 > x^5_0$.

By introducing the left distribution $\widehat{\Theta}_{\mathrm{L}}(x^5 - x_0^5)$ (see (20.38))[1] and putting

$$\begin{aligned}
&(a_1 + d_1 + K_0) \equiv T^0;\\
&-(K_1 + h_5 + e_1) \equiv T^1;\\
&-(l_7 + K_2 + e_3) \equiv T^2;\\
&-(m_1 + g_1 + c) \equiv T^3;\\
&m_5 + g_2 \equiv \zeta^3; \quad l_5 + e_4 \equiv \zeta^2; \quad h_7 + e_2 \equiv \zeta^1;\\
&l_8 \equiv -\theta^1; \quad h_6 \equiv \theta^2; \quad h_3 \equiv -\frac{c_2}{c_1}\theta^3,
\end{aligned} \tag{B.183}$$

the contravariant Killing vector $\xi^A(x, x^5)$ of the gravitational metric in case XI can be written in the following form (valid $\forall x^5 \in R_0^+$):

$$\xi^0(x^1, x^2, x^3, x^5) = \widehat{\Theta}_{\mathrm{L}}(x^5 - x_0^5)\left[-c_1\zeta^1 x^1 - c_2\zeta^2 x^2\right] - \zeta^3 x^3 + T^0; \tag{B.184}$$

$$\xi^1(x^0, x^2, x^3, x^5) = \widehat{\Theta}_{\mathrm{L}}(x^5 - x_0^5)\left[-\zeta^1 x^0 - \theta^2 x^3\right] + \frac{c_2}{c_1}\theta^3 x^2 + T^1; \tag{B.185}$$

$$\xi^2(x^0, x^1, x^3, x^5) = \widehat{\Theta}_{\mathrm{L}}(x^5 - x_0^5)\left[-\zeta^2 x^0 + \theta^1 x^3\right] - \theta^3 x^1 + T^2; \tag{B.186}$$

$$\xi^3(x^0, x^1, x^2, x^5) = \widehat{\Theta}_{\mathrm{L}}(x^5 - x_0^5)\left[c_1\theta^2 x^1 - c_2\theta^1 x^2\right] - \zeta^3 x^0 + T^3; \tag{B.187}$$

$$\xi^5 = 0. \tag{B.188}$$

Notice that in this case the dependence of (B.184)–(B.188) on $x^5$ is fictitious, because actually $\xi^\mu$ depends on $x^5$ through the distribution $\widehat{\Theta}_{\mathrm{L}}(x^5 - x_0^5)$ only.

[1]The use of the left distribution $\widehat{\Theta}_{\mathrm{L}}(x^5 - x_0^5)$ (instead of the right one $\widehat{\Theta}_{\mathrm{R}}(x_0^5 - x^5)$ used in all the other cases) is due to the fact already stressed that in the present case the 5D Killing group in the range $0 < x^5 \leqslant x_0^5$ is a proper (non-abelian) subgroup of the Killing group in the range $x^5 > x_0^5$ . Indeed, let us recall the complementary nature of the two distributions, expressed by (20.40).

# Appendix C

# Explicit and Implicit Forms of Geodesics for the 12 Classes of Solutions of Einstein's Equations in Vacuum in the Power Ansatz

We shall give in the following the (explicit or implicit) solutions of the geodesic equations corresponding to the 12 classes of metrics, solutions of the 5D Einstein equations in vacuum in the Power Ansatz. The reader is referred to [127] for calculation details.

## C.1 Class (I)

$$\widetilde{\mathbf{q}}_{\mathrm{I}} = \left( n, -n\left(\frac{2p+n}{2n+p}\right), n, p, \frac{p^2-2p+2np-4n+3n^2}{2n+p} \right).$$

The geodesic generating function $F_{\pm,I}(\zeta; n, p, A_2)$ (24.7) is given by

$$\begin{aligned} &F_{\pm,I}(\zeta; n, p, A_2) \\ &= \Big[ \pm (x_0^5)^{-(2np+n^2)/(2n+p)} C_{12}^2 \zeta^{-(p^2-2p-4n+2n^2)/(2n+p)} \\ &\quad \pm (x_0^5)^n (C_{22}^2 - C_{02}^2) \zeta^{-(p^2-2p+3np-4n+5n^2)/(2n+p)} \\ &\quad \pm (x_0^5)^p C_{32}^2 \zeta^{-(2p^2-2p+4np-4n+3n^2)(2n+p)} \\ &+ A_2 (x_0^5)^{-(p^2-2p+2np-4n+3n^2)/(2n+p)} \zeta^{-(p^2-2p+2np-4n+3n^2)/(2n+p)} \Big]^{-1/2}. \end{aligned} \tag{C.1}$$

Therefore, the integral (24.8) can be put in the form $(2n+p\neq 0)$:

$$\frac{4n+p}{3n^2+4np+2p^2}\int\frac{\mathrm{d}y}{\sqrt{c_0y^{\alpha(n,p)}+c_1y^{\beta(n,p)}+c_3y^{\gamma(n,p)}+c_2}} \tag{C.2}$$

with

$$\begin{cases} \alpha(n,p)=\dfrac{2p^2+2n^2+8np}{3n^2+4np+2p^2}; \\ \beta(n,p)=\dfrac{2p^2-4n^2+2np}{3n^2+4np+2p^2}; \\ \gamma(n,p)=\dfrac{2p^2+4np}{3n^2+4np+2p^2}; \\ y=(x_5(\tau))^{3n^2/(4n+2p)+p}, \quad x_5(\tau)>0; \end{cases} \tag{C.3}$$

$$\begin{cases} c_0=\pm(x_0^5)^{-(2np+n^2)/(2n+p)}C_{12}^2; \\ c_1=\pm(x_0^5)^n(C_{22}^2-C_{02}^2); \\ c_2=\pm(x_0^5)^pC_{32}^2; \\ c_3=A_2(x_0^5)^{-(p^2-2p+2np-4n+3n^2)/(2n+p)}, \end{cases} \tag{C.4}$$

and cannot be evaluated for arbitrary values of the parameters. Therefore, the solution for Class (I) can only be given in implicit form by replacing (C.3)–(C.5) in the general solution (24.13)–(24.14).

## C.2 Class (II)

$$\widetilde{\mathbf{q}}_{\mathrm{II}}=(0,m,0,0,m-2)\,.$$

The function $F_\pm(\zeta;q,A_2)$ reads, in this case

$$F_{\pm,\mathrm{II}}(\zeta;m,A_2)$$
$$=\left\{\pm\left[C_{22}^2+C_{32}^2-C_{02}^2\pm A_2(x_0^5)^{-m+2}\right]\zeta^{-m+2}\pm(x_0^5)^mC_{12}^2\zeta^{-2m+2}\right\}^{-1/2}. \tag{C.5}$$

The Riemann indefinite integrals of the generating functions $F_{\pm,i}(\zeta;\widetilde{\mathbf{q}},A_2)$ ($i$=II,IV,VI,IX) for the classes (II), (IV), (VI), and (IX) can be put in the following form (ESC off):

$$\int\mathrm{d}\zeta F_{\pm,i}(\zeta;\widetilde{\mathbf{q}},A_2)=\int\mathrm{d}\zeta\left[a_i\zeta^{-k_i+2}+b_i\zeta^{-2k_i+2}\right]^{-1/2},$$
$$k_{\mathrm{II}}=m,\quad k_{\mathrm{IV}}=p,\quad k_{\mathrm{VI}}=q,\quad k_{\mathrm{IX}}=n, \tag{C.6}$$

where $a_i = a_i(C_{02}^2, C_{12}^2, C_{22}^2, C_{32}^2, \widetilde{\mathbf{q}}, A_2, x_0^5)$, $b_i = b_i(C_{02}^2, C_{12}^2, C_{22}^2, C_{32}^2, \widetilde{\mathbf{q}}, A_2, x_0^5)$. They can be evaluated as follows ($y = x^k$):

$$\int \frac{\mathrm{d}x}{\sqrt{ax^{-k+2} + bx^{-2k+2}}} = \begin{cases} \dfrac{2\sqrt{b}}{ka}\sqrt{\dfrac{a}{b}y + 1}, & a \neq 0,\ b \neq 0; \\ \dfrac{1}{\sqrt{b}}\dfrac{x^k}{k}, & a = 0,\ b \neq 0; \\ \dfrac{2}{\sqrt{a}}\dfrac{x^{k/2}}{k}, & a \neq 0,\ b = 0; \\ \dfrac{1}{\sqrt{a+b}}\ln|x|, & k = 0. \end{cases} \tag{C.7}$$

The solution therefore reads:

$$x^\mu(\tau) = C_{\mu 1} + C_{\mu 2}\tau, \quad \mu = 0, 2, 3; \tag{C.8}$$

$$x^1(\tau) = C_{11} + C_{12}\int \mathrm{d}\tau \left(x^5(\tau)\right)^{-m}; \tag{C.9}$$

As to $x^5(\tau)$, we have the following cases:

(1) $m \neq 0$:

(1.1) $C_{22}^2 + C_{32}^2 - C_{02}^2 \pm A_2(x_0^5)^{-m+2} \neq 0$, $C_{12} \neq 0$:

$$x^5(\tau) = \left\{\frac{(x_0^5)^m C_{12}^2}{[C_{22}^2 + C_{32}^2 - C_{02}^2 \pm A_2(x_0^5)^{-m+2}]} \times \left[\frac{m^2\left(C_{22}^2 + C_{32}^2 - C_{02}^2 \pm A_2(x_0^5)^{-m+2}\right)^2}{4(x_0^5)^2 C_{12}^2}(\tau + A_1)^2 - 1\right]\right\}^{1/m}; \tag{C.10}$$

(1.2) $C_{22}^2 + C_{32}^2 - C_{02}^2 \pm A_2(x_0^5)^{-m+2} = 0$, $C_{12} \neq 0$:

$$x^5(\tau) = \left[\pm m\sqrt{\pm(x_0^5)^m C_{12}^2}(x_0^5)^{\frac{m-2}{2}}(\tau + A_1)\right]^{1/m}; \tag{C.11}$$

(1.3) $C_{22}^2 + C_{32}^2 - C_{02}^2 \pm A_2(x_0^5)^{-m+2} \neq 0$, $C_{12} = 0$:

$$x^5(\tau) = \left[\pm m\frac{\sqrt{\pm\left[C_{22}^2 + C_{32}^2 - C_{02}^2 \pm A_2(x_0^5)^{-m+2}\right]}}{2}(x_0^5)^{(m-2)/2}(\tau + A_1)\right]^{2/m}. \tag{C.12}$$

(2) $m = 0$:

$$x^5(\tau)$$
$$= \exp\left\{\pm\sqrt{\pm\left[C_{22}^2 + C_{32}^2 - C_{02}^2 \pm A_2(x_0^5)^{-m+2} + (x_0^5)^m C_{12}^2\right]}(x_0^5)^{(m-2)/2}(\tau + A_1)\right\}. \tag{C.13}$$

## C.3 Class (III)

$$\widetilde{\mathbf{q}}_{\mathrm{III}} = (n, -n, n, n, -2(1-n)).$$

One gets

$$F_{\pm,\mathrm{III}}(\zeta; n, A_2) = \{\pm[C_{22}^2 + C_{32}^2 - C_{02}^2](x_0^5)^n \zeta^{-3n+2} + \pm A_2(x_0^5)^{-2n+2}\zeta^{-2n+2} \pm (x_0^5)^{-n} C_{12}^2 \zeta^{-n+2}\}^{-1/2}. \tag{C.14}$$

The integral of $F_{\pm,\mathrm{III}}(\zeta; n, A_2)$ is then of the kind

$$\int \frac{\mathrm{d}x}{\sqrt{ax^{2-n} + bx^{2-2n} + cx^{2-3n}}}, \tag{C.15}$$

which admits an explicit solution at least for some values of the parameters.

The geodesic solution therefore reads:

$$x^\mu(\tau) = C_{\mu 1} + C_{\mu 2}\int \mathrm{d}\tau\,\left(x^5(\tau)\right)^{-n}, \quad \mu = 0, 2, 3; \tag{C.16}$$

$$x^1(\tau) = C_{11} + C_{12}\int \mathrm{d}\tau\,\left(x^5(\tau)\right)^{n}; \tag{C.17}$$

where $x^5(\tau)$ is given (explicitly or implicitly) by:

(1) $n \neq 0$:

(1.1) $C_{22}^2 + C_{32}^2 - C_{02}^2 = 0$:

(1.1.1) $C_{12} \neq 0$, $A_2 \neq 0$:

$$x^5(\tau) = \left\{\pm\frac{A_2(x_0^5)^{-n+2}}{C_{12}}\left[\pm(x_0^5)^{2n-4}\frac{n^2}{4}\frac{C_{12}^4}{A_2}(\tau + A_1)^2 - 1\right]\right\}^{1/n}; \tag{C.18}$$

(1.1.2) $C_{12} = 0$, $A_2 \neq 0$:

$$x^5(\tau) = \left[\pm n\sqrt{A_2}\,(\tau + A_1)\right]^{1/n}; \tag{C.19}$$

(1.1.3) $C_{12} \neq 0,\ A_2 = 0$:

$$x^5(\tau) = \left[ \pm \frac{n C_{12} \sqrt{\pm 1}}{2} \left(x_0^5\right)^{n/2-1} (\tau + A_1) - 1 \right]^{2/n} . \quad \text{(C.20)}$$

(1.2) $A_2 = 0$:

(1.2.1) $C_{12} \neq 0,\ C_{22}^2 + C_{32}^2 - C_{02}^2 \neq 0,\ C_{12}^2 \neq \left(C_{22}^2 + C_{32}^2 - C_{02}^2\right) \left(x_0^5\right)^{2n}$:

(1.2.1a) $\left(C_{22}^2 + C_{32}^2 - C_{02}^2\right) < 0$ :

$$\begin{aligned}
(\tau + A_1) = {} & \pm \frac{\left(x_0^5\right)^{-(n/2)+1}}{n C_{12} \sqrt{\mp 1}} \\
& \times \left\{ \widehat{\Theta} \left( \left(x_0^5\right)^n \sqrt{\frac{C_{02}^2 - C_{22}^2 - C_{32}^2}{C_{12}^2}} \right.\right. \\
& \left. - \ \left(x^5(\tau)\right)^n \right) \widehat{\Theta} \left( \left(x^5(\tau)\right)^n \right)
\end{aligned}$$

$$\times \left\{ \begin{array}{l}
\left[ \sqrt{2} \left(x_0^5\right)^{n/2} \sqrt[4]{\frac{C_{02}^2 - C_{22}^2 - C_{32}^2}{C_{12}^2}} \int_0^{\sqrt{\frac{2}{\left(\frac{x_0^5}{x^5(\tau)}\right)^n \sqrt{\frac{C_{02}^2 - C_{22}^2 - C_{32}^2}{C_{12}^2}} - 1}}} \mathrm{d}t \frac{\sqrt{1 - t^2}}{\sqrt{1 - \frac{t^2}{2}}} \right. \\
\left. \qquad -2 \sqrt{\left(x^5(\tau)\right)^n \frac{\sqrt{\frac{C_{02}^2 - C_{22}^2 - C_{32}^2}{C_{12}^2}} - \left(x^5(\tau)\right)^n}{\sqrt{\frac{C_{02}^2 - C_{22}^2 - C_{32}^2}{C_{12}^2}} + \left(x^5(\tau)\right)^n}} \right] \\
\qquad\qquad\qquad \text{or} \\
\sqrt{2} \left(x_0^5\right)^{n/2} \sqrt[4]{\frac{C_{02}^2 - C_{22}^2 - C_{32}^2}{C_{12}^2}} \int_0^{\sqrt{1 - \sqrt{\frac{C_{02}^2 - C_{22}^2 - C_{32}^2}{C_{12}^2}} \left(\frac{x^5(\tau)}{x_0^5}\right)^n}} \mathrm{d}t \frac{\sqrt{1 - t^2}}{\sqrt{1 - \frac{t^2}{2}}}
\end{array} \right.$$

$$\mp i \widehat{\Theta} \left( \left(x^5(\tau)\right)^n - \left(x_0^5\right)^n \sqrt{\frac{C_{02}^2 - C_{22}^2 - C_{32}^2}{C_{12}^2}} \right)$$

$$\times \left[ -\sqrt{2} \left(x_0^5\right)^{\frac{n}{2}} \sqrt[4]{\frac{C_{02}^2 - C_{22}^2 - C_{32}^2}{C_{12}^2}} \int_0^{\sqrt{1 - \sqrt{\frac{C_{02}^2 - C_{22}^2 - C_{32}^2}{C_{12}^2}} \left(\frac{x^5(\tau)}{x_0^5}\right)^n}} \mathrm{d}t \frac{\sqrt{1 - t^2}}{\sqrt{1 - \frac{t^2}{2}}} \right.$$

$$\left. \left. + 2 \sqrt{\left(x^5(\tau)\right)^n - \frac{C_{02}^2 - C_{22}^2 - C_{32}^2}{C_{12}^2} \left(\frac{x_0^5}{x^5(\tau)}\right)^n \left(x_0^5\right)^n} \right] \right\} . \quad \text{(C.21)}$$

(1.2.1b) $(C_{22}^2 + C_{32}^2 - C_{02}^2) > 0$ :
In this case, the integral

$$\int^{x^5(\tau)} \frac{dx}{\sqrt{\pm(x_0^5)^{-n} C_{12}^2 x^{2-n} \pm (C_{22}^2 + C_{32}^2 - C_{02}^2)(x_0^5)^n x^{2-3n}}} \tag{C.22}$$

is unknown, and therefore no solution can be given even in implicit form.

(1.2.2) $C_{12} \neq 0$, $C_{22}^2 + C_{32}^2 - C_{02}^2 \neq 0$, $C_{12}^2 = (C_{22}^2 + C_{32}^2 - C_{02}^2)(x_0^5)^{2n}$:

$$\left(x^5(\tau)\right)^n \mp n^2 C_{12}^2 (x_0^5)^{n-2} (\tau + A_1)^2 + \left(x^5(\tau)\right)^{-n} = 0. \tag{C.23}$$

(1.2.3) $C_{12} \neq 0$, $C_{22}^2 + C_{32}^2 - C_{02}^2 = 0$:

$$x^5(\tau) = \left[\pm\sqrt{\pm 1}\frac{n}{2} C_{12} (x_0^5)^{\frac{n}{2}-1} (\tau + A_1)\right]^{2/n}. \tag{C.24}$$

(1.2.4) $C_{12} = 0$, $C_{22}^2 + C_{32}^2 - C_{02}^2 \neq 0$:

$$x^5(\tau) = \left[\pm\frac{3n}{2}\sqrt{\pm(C_{22}^2 + C_{32}^2 - C_{02}^2)}(x_0^5)^{(3n/2)-1}(\tau + A_1)\right]^{2/3n}. \tag{C.25}$$

(1.3) $C_{12} = 0$:

(1.3.1) $A_2 \neq 0$, $C_{22}^2 + C_{32}^2 - C_{02}^2 \neq 0$:

$$\begin{aligned}
\tau + A_1 = \pm (x_0^5)^{1-n} \Bigg\{ & \frac{(C_{22}^2+C_{32}^2-C_{02}^2)(x_0^5)^{4n-3}}{n A_2^{3/2}} \widehat{\Theta}\left(\mp\frac{A_2}{C_{22}^2+C_{32}^2-C_{02}^2}\right) \\
& \times \widehat{\Theta}\left(\mp\frac{A_2 (x_0^5)^{2-2n}}{C_{22}^2+C_{32}^2-C_{02}^2}\left(\frac{x^5(\tau)}{x_0^5}\right)^n - 1\right) \frac{A_2 (x_0^5)^{2-2n}}{C_{22}^2+C_{32}^2-C_{02}^2}\left(\frac{x^5(\tau)}{x_0^5}\right)^n \\
& \times {}_2F_1\left(\frac{1}{2}, -1; 0; \mp\left(\frac{x_0^5}{x^5(\tau)}\right)^n \frac{C_{22}^2 + C_{32}^2 - C_{02}^2}{A_2}(x_0^5)^{2n-2}\right) \\
& + \widehat{\Theta}\left(\pm\frac{A_2 (x_0^5)^{2-2n}}{C_{22}^2+C_{32}^2-C_{02}^2}\left(\frac{x^5(\tau)}{x_0^5}\right)^n + 1\right)\widehat{\Theta}\left(\mp\frac{A_2}{C_{22}^2+C_{32}^2-C_{02}^2}\right) \\
\times & \begin{cases}
\dfrac{2}{3}\dfrac{A_2 (x_0^5)^{2-(3/2)n}}{C_{22}^2 + C_{32}^2 - C_{02}^2}\left(\dfrac{x^5(\tau)}{x_0^5}\right)^{(3/2)n} \\
\quad \times {}_2F_1\left(\dfrac{1}{2}, \dfrac{3}{2}; \dfrac{5}{2}; \mp\left(\dfrac{x^5(\tau)}{x_0^5}\right)^n \dfrac{A_2 (x_0^5)^{2-2n}}{C_{22}^2 + C_{32}^2 - C_{02}^2}\right) \\
\text{or} \\
\dfrac{A_2 (x_0^5)^{2-(3/2)n}}{C_{22}^2 + C_{32}^2 - C_{02}^2}\left(\dfrac{x^5(\tau)}{x_0^5}\right)^n \\
\quad \times {}_2F_1\left(\dfrac{1}{2}, -1; 0; \mp\left(\dfrac{x_0^5}{x^5(\tau)}\right)^n \dfrac{C_{22}^2 + C_{32}^2 - C_{02}^2}{A_2}(x_0^5)^{2-2n}\right)
\end{cases}
\end{aligned}$$

$$\pm\frac{(C_{22}^2+C_{32}^2-C_{02}^2)(x_0^5)^{4n-3}}{nA_2^{3/2}}\widehat{\Theta}\left(\pm\frac{A_2}{C_{22}^2+C_{32}^2-C_{02}^2}\right)$$

$$\times\left\{\begin{array}{c}\left[\begin{array}{c}\pm\frac{2}{3}\frac{A_2(x_0^5)^{2-(3/2)n}}{C_{22}^2+C_{32}^2-C_{02}^2}\left(\frac{x^5(\tau)}{x_0^5}\right)^{(3/2)n}\\ \times\,{}_2F_1\left(\frac{1}{2},\frac{3}{2};\frac{5}{2};\mp\left(\frac{x^5(\tau)}{x_0^5}\right)^n\frac{A_2(x_0^5)^{2-2n}}{C_{22}^2+C_{32}^2-C_{02}^2}\right)\end{array}\right]\Bigg\}, \\ \text{or} \\ \left[\begin{array}{c}\pm\frac{A_2(x_0^5)^{2-2n}}{C_{22}^2+C_{32}^2-C_{02}^2}\left(\frac{x^5(\tau)}{x_0^5}\right)^n\\ \times\,{}_2F_1\left(\frac{1}{2},-1;0;\mp\left(\frac{x_0^5}{x^5(\tau)}\right)^n\frac{C_{22}^2+C_{32}^2-C_{02}^2}{A_2}(x_0^5)^{2n-2}\right)\end{array}\right]\Bigg\}, \\ \text{or} \\ \left[\sqrt{\pm\frac{A_2(x_0^5)^{2-2n}}{C_{22}^2+C_{32}^2-C_{02}^2}\left(\frac{x^5(\tau)}{x_0^5}\right)^n\left[\pm\frac{A_2(x_0^5)^{2-2n}}{C_{22}^2+C_{32}^2-C_{02}^2}\left(\frac{x^5(\tau)}{x_0^5}\right)^n+1\right]}\right. \\ \left.-\ln\left|\begin{array}{c}\sqrt{\pm\frac{A_2(x_0^5)^{2-2n}}{C_{22}^2+C_{32}^2-C_{02}^2}\left(\frac{x^5(\tau)}{x_0^5}\right)^n}\\ +\sqrt{\pm\frac{A_2(x_0^5)^{2-2n}}{C_{22}^2+C_{32}^2-C_{02}^2}\left(\frac{x^5(\tau)}{x_0^5}\right)^n+1}\end{array}\right|\right]\Bigg\}.\end{array}\right. \tag{C.26}$$

Here, ${}_2F_1$ is the generalized hypergeometric function of class (2,1), defined by

$$ {}_2F_1(\alpha,\beta;\gamma;z)\equiv\sum_{k=0}^{\infty}\frac{(\alpha)_k\,(\beta)_k}{(\gamma)_k}\frac{z^k}{k!} \tag{C.27}$$

with $(\alpha)_k$ being the Pochhammer symbols

$$(\alpha)_k\equiv\frac{\Gamma(\alpha+k)}{\Gamma(\alpha)} \tag{C.28}$$

and

$$\Gamma(z)\equiv\int_0^\infty \mathrm{e}^{-t}t^{z-1}\mathrm{d}t \qquad (\mathrm{Re}(z)>0) \tag{C.29}$$

denotes the Euler gamma function (or Euler integral of second kind).

(1.3.2) $A_2\neq 0,\ C_{22}^2+C_{32}^2-C_{02}^2=0$:

$$x^5(\tau)=\left[\pm n\sqrt{A_2}\,(\tau+A_1)\right]^{1/n}. \tag{C.30}$$

(1.3.3) $A_2 = 0$, $C_{22}^2 + C_{32}^2 - C_{02}^2 \neq 0$:

$$x^5(\tau) = \left[\pm\frac{3n}{2}\sqrt{\pm(C_{22}^2+C_{32}^2-C_{02}^2)}(x_0^5)^{\frac{3n}{2}-1}(\tau+A_1)\right]^{2/3n}. \tag{C.31}$$

(1.4) $C_{12}$, $A_2$, $C_{22}^2 + C_{32}^2 - C_{02}^2 \neq 0$:

(1.4.1) $C_{12}^2 \neq \left(C_{22}^2 + C_{32}^2 - C_{02}^2\right)(x_0^5)^{2n}$:

(1.4.1.a) $\left(C_{22}^2 + C_{32}^2 - C_{02}^2\right) < 0$:

$$(\tau + A_1) = \pm(x_0^5)^{1-n}\left\{\widehat{\Theta}\left(\pm\frac{A_2}{C_{22}^2+C_{32}^2-C_{02}^2}\right)\left[\mp\frac{1}{n}\frac{\sqrt{A_2}}{C_{12}^2}x_0^5\right.\right.$$

$$\times\widehat{\Theta}\left(\begin{array}{c}-\frac{A_2^2(x_0^5)^{4-4n}}{2C_{12}^2\left(C_{22}^2+C_{32}^2-C_{02}^2\right)}\left(1+\sqrt{1-4\frac{C_{12}^2\left(C_{22}^2+C_{32}^2-C_{02}^2\right)}{A_2^2(x_0^5)^{4-4n}}}\right)\\ \mp\frac{A_2^2(x_0^5)^{2-2n}}{C_{22}^2+C_{32}^2-C_{02}^2}\left(\frac{x^5(\tau)}{x_0^5}\right)^n\end{array}\right)$$

$$\times\left\{\begin{array}{l}\left[2\sqrt[4]{1-4\frac{C_{12}^2\left(C_{22}^2+C_{32}^2-C_{02}^2\right)}{A_2^2(x_0^5)^{4-4n}}}E(A,B) - \frac{-1+\sqrt{1-4\frac{C_{12}^2\left(C_{22}^2+C_{32}^2-C_{02}^2\right)}{A_2^2(x_0^5)^{4-4n}}}}{\sqrt[4]{1-4\frac{C_{12}^2\left(C_{22}^2+C_{32}^2-C_{02}^2\right)}{A_2^2(x_0^5)^{4-4n}}}}F(A,B)\right.\\ \left. -2\sqrt{\mp\frac{A_2^2(x_0^5)^{2-2n}}{C_{22}^2+C_{32}^2-C_{02}^2}}\left(\frac{x^5(\tau)}{x_0^5}\right)^n \times\frac{\left(x^5(\tau)\right)^n \pm\frac{A_2^2(x_0^5)^{2-n}}{2C_{12}^2}\left[1+\sqrt{1-4\frac{C_{12}^2\left(C_{22}^2+C_{32}^2-C_{02}^2\right)}{A_2^2(x_0^5)^{4-4n}}}\right]}{\left(x^5(\tau)\right)^n \pm\frac{A_2^2(x_0^5)^{2-n}}{2C_{12}^2}\left[1-\sqrt{1-4\frac{C_{12}^2\left(C_{22}^2+C_{32}^2-C_{02}^2\right)}{A_2^2(x_0^5)^{4-4n}}}\right]}\right]\\ \text{or}\\ \left[-2\sqrt[4]{1-4\frac{C_{12}^2\left(C_{22}^2+C_{32}^2-C_{02}^2\right)}{A_2^2(x_0^5)^{4-4n}}}E(C,B) + \frac{-1+\sqrt{1-4\frac{C_{12}^2\left(C_{22}^2+C_{32}^2-C_{02}^2\right)}{A_2^2(x_0^5)^{4-4n}}}}{\sqrt[4]{1-4\frac{C_{12}^2\left(C_{22}^2+C_{32}^2-C_{02}^2\right)}{A_2^2(x_0^5)^{4-4n}}}}F(C,B)\right]\end{array}\right.$$

$$
\begin{aligned}
&\frac{i}{n}\frac{\sqrt{A_2}}{C_{12}^2}x_0^5\widehat{\Theta}\\
&\times\left(\begin{array}{c}\dfrac{A_2^2(x_0^5)^{4-4n}}{2C_{12}^2\left(C_{22}^2+C_{32}^2-C_{02}^2\right)}\left(1+\sqrt{1-4\dfrac{C_{12}^2\left(C_{22}^2+C_{32}^2-C_{02}^2\right)}{A_2^2(x_0^5)^{4-4n}}}\right)\\
\pm\dfrac{A_2^2(x_0^5)^{2-2n}}{C_{22}^2+C_{32}^2-C_{02}^2}\left(\dfrac{x^5(\tau)}{x_0^5}\right)^n\end{array}\right)\\
&\times\left[\begin{array}{l}2\sqrt[4]{1-4\frac{C_{12}^2\left(C_{22}^2+C_{32}^2-C_{02}^2\right)}{A_2^2(x_0^5)^{4-4n}}}E\left(D,G\right)\\
+\dfrac{-1+\sqrt{1-4\frac{C_{12}^2\left(C_{22}^2+C_{32}^2-C_{02}^2\right)}{A_2^2(x_0^5)^{4-4n}}}}{\sqrt[4]{1-4\frac{C_{12}^2\left(C_{22}^2+C_{32}^2-C_{02}^2\right)}{A_2^2(x_0^5)^{4-4n}}}}F\left(D,G\right)\\
-2\sqrt{\dfrac{-\frac{C_{12}^2(x_0^5)^{3n-2}}{A_2}\left(\frac{x^5(\tau)}{x_0^5}\right)^{2n}-\frac{A_2(x_0^5)^{2-2n}}{C_{22}^2+C_{32}^2-C_{02}^2}\left(\frac{x^5(\tau)}{x_0^5}\right)^{n}\mp 1}{\frac{A_2(x_0^5)^{2-2n}}{C_{22}^2+C_{32}^2-C_{02}^2}\left(\frac{x^5(\tau)}{x_0^5}\right)^{n}}}\end{array}\right]\\
&+\widehat{\Theta}\left(\mp\frac{A_2}{C_{22}^2+C_{32}^2-C_{02}^2}\right)
\end{aligned}
$$

$$
\begin{aligned}
&\times\left\{\left[-\frac{i}{n}\frac{\sqrt{A_2}}{C_{12}^2}x_0^5\widehat{\Theta}\left(\frac{A_2^2(x_0^5)^{4-4n}}{2C_{12}^2\left(C_{22}^2+C_{32}^2-C_{02}^2\right)}\left(-1+\sqrt{1-4\frac{C_{12}^2\left(C_{22}^2+C_{32}^2-C_{02}^2\right)}{A_2^2(x_0^5)^{4-4n}}}\right)\right.\right.\right.\\
&\qquad\left.\left.\mp\frac{A_2^2(x_0^5)^{2-2n}}{C_{22}^2+C_{32}^2-C_{02}^2}\left(\frac{x^5(\tau)}{x_0^5}\right)^n\right)\right]\\
&\times\left[\begin{array}{l}\dfrac{-1+\sqrt{1-4\dfrac{C_{12}^2\left(C_{22}^2+C_{32}^2-C_{02}^2\right)}{A_2^2(x_0^5)^{4-4n}}}}{\sqrt[4]{1-4\dfrac{C_{12}^2\left(C_{22}^2+C_{32}^2-C_{02}^2\right)}{A_2^2(x_0^5)^{4-4n}}}}F(H,B)\\
-2\sqrt[4]{1-4\dfrac{C_{12}^2\left(C_{22}^2+C_{32}^2-C_{02}^2\right)}{A_2^2(x_0^5)^{4-4n}}}E(H,B)\\
+2\sqrt{\dfrac{\frac{C_{12}^2(x_0^5)^{3n-2}}{A_2}\left(\frac{x^5(\tau)}{x_0^5}\right)^{2n}-\frac{A_2(x_0^5)^{2-2n}}{C_{22}^2+C_{32}^2-C_{02}^2}\left(\frac{x^5(\tau)}{x_0^5}\right)^{n}\pm 1}{\frac{A_2(x_0^5)^{2-2n}}{C_{22}^2+C_{32}^2-C_{02}^2}\left(\frac{x^5(\tau)}{x_0^5}\right)^{n}}}\end{array}\right]
\end{aligned}
$$

$$+\frac{i}{n}\frac{\sqrt{A_2}}{C_{12}^2}x_0^5\widehat{\Theta}$$

$$\times\left(-\frac{A_2^2(x_0^5)^{4-4n}}{2C_{12}^2\left(C_{22}^2+C_{32}^2-C_{02}^2\right)}\left(-1+\sqrt{1-4\frac{C_{12}^2\left(C_{22}^2+C_{32}^2-C_{02}^2\right)}{A_2^2(x_0^5)^{4-4n}}}\right)\pm\frac{A_2^2(x_0^5)^{2-2n}}{C_{22}^2+C_{32}^2-C_{02}^2}\left(\frac{x^5(\tau)}{x_0^5}\right)^n\right)$$

$$\times\left\{\left[2\sqrt[4]{1-4\frac{C_{12}^2\left(C_{22}^2+C_{32}^2-C_{02}^2\right)}{A_2^2(x_0^5)^{4-4n}}}\,E(L,B)-\frac{1+\sqrt{1-4\frac{C_{12}^2\left(C_{22}^2+C_{32}^2-C_{02}^2\right)}{A_2^2(x_0^5)^{4-4n}}}}{\sqrt[4]{1-4\frac{C_{12}^2\left(C_{22}^2+C_{32}^2-C_{02}^2\right)}{A_2^2(x_0^5)^{4-4n}}}}\,F(L,B)\right]\right.$$

or

$$\left[-2\sqrt[4]{1-4\frac{C_{12}^2\left(C_{22}^2+C_{32}^2-C_{02}^2\right)}{A_2^2(x_0^5)^{4-4n}}}\,E(M,G)+\frac{1+\sqrt{1-4\frac{C_{12}^2\left(C_{22}^2+C_{32}^2-C_{02}^2\right)}{A_2^2(x_0^5)^{4-4n}}}}{\sqrt[4]{1-4\frac{C_{12}^2\left(C_{22}^2+C_{32}^2-C_{02}^2\right)}{A_2^2(x_0^5)^{4-4n}}}}\,F(M,G)\right.$$

$$\left.\left.+2\sqrt{\frac{\pm\frac{C_{12}^2(x_0^5)^{2n-2}}{A_2}\left(\frac{x^5(\tau)}{x_0^5}\right)^{2n}+\frac{1}{2}\left[1-\sqrt{1-4\frac{C_{12}^2\left(C_{22}^2+C_{32}^2-C_{02}^2\right)}{A_2^2(x_0^5)^{4-4n}}}\right]}{-1\mp\frac{A_2(x_0^5)^{2-2n}}{2C_{12}^2}\left(\frac{x_0^5}{x^5(\tau)}\right)^n\left[1+\sqrt{1-4\frac{C_{12}^2\left(C_{22}^2+C_{32}^2-C_{02}^2\right)}{A_2^2(x_0^5)^{4-4n}}}\right]}}\right]\right\},\qquad \text{(C.32)}$$

where $F(z,k)$ and $E(z,k)$ are the elliptic integrals of first and second kind, respectively, in the normal Legendre form

$$F(z,k)\equiv\int_0^z\frac{\mathrm{d}x}{\sqrt{1-k^2\sin^2x}}=\int_0^{\sin z}\frac{\mathrm{d}t}{\sqrt{1-t^2}\sqrt{1-k^2t^2}},\qquad z,k\in C;\qquad \text{(C.33)}$$

$$E(z,k)\equiv\int_0^z\mathrm{d}x\sqrt{1-k^2\sin^2x}=\int_0^{\sin z}\frac{\mathrm{d}t\sqrt{1-k^2t^2}}{\sqrt{1-t^2}},\qquad z,k\in C,\qquad \text{(C.34)}$$

and we put

$$A = \arcsin\left\{\left[\pm 4\frac{C_{12}^2}{A_2(x_0^5)^{2n-2}}\left(\frac{x^5(\tau)}{x_0^5}\right)^n\sqrt{1-4\frac{C_{12}^2\left(C_{22}^2+C_{32}^2-C_{02}^2\right)}{A_2^2(x_0^5)^{4-4n}}}\right]\right.$$
$$\times\left[1-\sqrt{1-4\frac{C_{12}^2\left(C_{22}^2+C_{32}^2-C_{02}^2\right)}{A_2^2(x_0^5)^{4-4n}}}\right]^{-1}$$
$$\left.\times\left[1+\sqrt{1-4\frac{C_{12}^2\left(C_{22}^2+C_{32}^2-C_{02}^2\right)}{A_2^2(x_0^5)^{4-4n}}}\pm 2\frac{C_{12}^2}{A_2(x_0^5)^{2n-2}}\left(\frac{x^5(\tau)}{x_0^5}\right)^n\right]^{-1}\right\}^{1/2}; \quad (C.35)$$

$$B = \sqrt{\frac{1+\sqrt{1-4\frac{C_{12}^2\left(C_{22}^2+C_{32}^2-C_{02}^2\right)}{A_2^2(x_0^5)^{4-4n}}}}{2\sqrt{1-4\frac{C_{12}^2\left(C_{22}^2+C_{32}^2-C_{02}^2\right)}{A_2^2(x_0^5)^{4-4n}}}}}; \quad (C.36)$$

$$C = \arcsin\sqrt{\frac{1+\sqrt{1-4\frac{C_{12}^2\left(C_{22}^2+C_{32}^2-C_{02}^2\right)}{A_2^2(x_0^5)^{4-4n}}}\pm 2\frac{C_{12}^2}{A_2(x_0^5)^{2n-2}}\left(\frac{x^5(\tau)}{x_0^5}\right)^n}{1+\sqrt{1-4\frac{C_{12}^2\left(C_{22}^2+C_{32}^2-C_{02}^2\right)}{A_2^2(x_0^5)^{4-4n}}}}}; \quad (C.37)$$

$$D = \arcsin\sqrt{\frac{1+\sqrt{1-4\frac{C_{12}^2\left(C_{22}^2+C_{32}^2-C_{02}^2\right)}{A_2^2(x_0^5)^{4-4n}}}\pm 2\frac{C_{12}^2}{A_2(x_0^5)^{2n-2}}\left(\frac{x^5(\tau)}{x_0^5}\right)^n}{\pm 2\frac{(x_0^5)^{n-1}C_{12}^2}{A_2}\left(\frac{x^5(\tau)}{x_0^5}\right)^n}}; \quad (C.38)$$

$$G = \sqrt{\frac{-1+\sqrt{1-4\frac{C_{12}^2\left(C_{22}^2+C_{32}^2-C_{02}^2\right)}{A_2^2(x_0^5)^{4-4n}}}}{2\sqrt{1-4\frac{C_{12}^2\left(C_{22}^2+C_{32}^2-C_{02}^2\right)}{A_2^2(x_0^5)^{4-4n}}}}}; \quad (C.39)$$

$$H = \arcsin\sqrt{\frac{\left[1-\sqrt{1-4\frac{C_{12}^2\left(C_{22}^2+C_{32}^2-C_{02}^2\right)}{A_2^2(x_0^5)^{4-4n}}}\pm 2\frac{(x_0^5)^{2n-2}C_{12}^2}{A_2}\left(\frac{x^5(\tau)}{x_0^5}\right)^n\right]}{\pm 2\frac{(x_0^5)^{2n-2}C_{12}^2}{A_2}\left(\frac{x^5(\tau)}{x_0^5}\right)^n}}; \quad (C.40)$$

$$L = \arcsin\sqrt{\frac{\left[1-\sqrt{1-4\frac{C_{12}^2\left(C_{22}^2+C_{32}^2-C_{02}^2\right)}{A_2^2(x_0^5)^{4-4n}}}\pm 2\frac{(x_0^5)^{2n-2}C_{12}^2}{A_2}\left(\frac{x^5(\tau)}{x_0^5}\right)^n\right]}{1-\sqrt{1-4\frac{C_{12}^2\left(C_{22}^2+C_{32}^2-C_{02}^2\right)}{A_2^2(x_0^5)^{4-4n}}}}}; \quad (C.41)$$

$$M = \arcsin\left\{\left[\pm 4\frac{C_{12}^2(x_0^5)^{2n-2}}{A_2}\left(\frac{x^5(\tau)}{x_0^5}\right)^n\sqrt{1-4\frac{C_{12}^2(C_{22}^2+C_{32}^2-C_{02}^2)}{A_2^2(x_0^5)^{4-4n}}}\right]\right.$$
$$\times\left[-1+\sqrt{1-4\frac{C_{12}^2(C_{22}^2+C_{32}^2-C_{02}^2)}{A_2^2(x_0^5)^{4-4n}}}\right]^{-1}$$
$$\left.\times\left[1+\sqrt{1-4\frac{C_{12}^2(C_{22}^2+C_{32}^2-C_{02}^2)}{A_2^2(x_0^5)^{4-4n}}}\pm 2\frac{C_{12}^2(x_0^5)^{2n-2}}{A_2}\left(\frac{x^5(\tau)}{x_0^5}\right)^n\right]^{-1}\right\}^{1/2}. \quad \text{(C.42)}$$

(1.4.1.b) $(C_{22}^2+C_{32}^2-C_{02}^2)>0$:

In this case, the Riemann integral

$$\int^{x^5(\tau)}\frac{\mathrm{d}x}{\sqrt{\pm(x_0^5)^{-n}C_{12}^2x^{2-n}+A_2(x_0^5)^{2-2n}x^{2-2n}\pm(C_{22}^2+C_{32}^2-C_{02}^2)(x_0^5)^nx^{2-3n}}} \quad \text{(C.43)}$$

is unknown, and therefore no solution can be obtained for $x^5(\tau)$.

(1.4.2) $C_{12}^2=(C_{22}^2+C_{32}^2-C_{02}^2)(x_0^5)^{2n}$:

$$\left(x^5(\tau)\right)^n\mp C_{12}^2(x_0^5)^{n-2}(\tau+A_1)^2+\left(x^5(\tau)\right)^{-n}\pm\frac{A_2(x_0^5)^{2n-2}}{C_{12}^2}=0. \quad \text{(C.44)}$$

(2) $n=0$:

$$x^5(\tau)=\exp\left\{\pm\sqrt{\pm(C_{22}^2+C_{32}^2-C_{02}^2)(x_0^5)^{3n-2}\pm C_{12}^2(x_0^5)^{n-2}+A_2}\,(\tau+A_1)\right\}. \quad \text{(C.45)}$$

## C.4 Class (IV)

$\tilde{\mathfrak{q}}_{\mathrm{IV}}=(0,0,0,p,p-2)$

One gets:

$$F_{\pm,IV}(\zeta;p,A_2)$$
$$=\left\{\pm[C_{12}^2+C_{22}^2-C_{02}^2\pm A_2(x_0^5)^{2-p}]\zeta^{2-p}\pm(x_0^5)^pC_{32}^2\zeta^{2-2p}\right\}^{-1/2}. \quad \text{(C.46)}$$

From the results obtained for Class II, we can write the geodesic solution for Class IV as

$$x^\mu(\tau)=C_{\mu 1}+C_{\mu 2}\tau,\quad \mu=0,1,2; \quad \text{(C.47)}$$

$$x^3(\tau)=C_{31}+C_{32}\int \mathrm{d}\tau\,\left(x^5(\tau)\right)^{-p} \quad \text{(C.48)}$$

with $x^5(\tau)$ given by:

(1) $p \neq 0$:

(1.1) $C_{12}^2 + C_{22}^2 - C_{02}^2 \pm A_2(x_0^5)^{2-p} \neq 0,\ C_{32} \neq 0$:

$$x^5(\tau) = \left\{ \frac{(x_0^5)^p C_{32}^2}{C_{12}^2 + C_{22}^2 - C_{02}^2 \pm A_2(x_0^5)^{2-p}} \right.$$
$$\left. \times \left[ \frac{p^2 \left[C_{12}^2 + C_{22}^2 - C_{02}^2 \pm A_2(x_0^5)^{2-p}\right]^2}{4(x_0^5)^2 C_{32}^2} (\tau + A_1)^2 - 1 \right] \right\}^{1/p}; \tag{C.49}$$

(1.2) $C_{12}^2 + C_{22}^2 - C_{02}^2 \pm A_2(x_0^5)^{2-p} = 0,\ C_{32} \neq 0$:

$$x^5(\tau) = \left[ \pm p \sqrt{\pm (x_0^5)^p C_{32}^2} (x_0^5)^{\frac{p-2}{2}} (\tau + A_1) \right]^{1/p}; \tag{C.50}$$

(1.3) $C_{12}^2 + C_{22}^2 - C_{02}^2 \pm A_2(x_0^5)^{2-p} \neq 0,\ C_{32} = 0$:

$$x^5(\tau) = \left[ \pm p \sqrt{\frac{\pm \left[C_{12}^2 + C_{22}^2 - C_{02}^2 \pm A_2(x_0^5)^{2-p}\right]}{2}} (x_0^5)^{(p-2)/2} (\tau + A_1) \right]^{2/p}; \tag{C.51}$$

(2) $p = 0$:

$$x^5(\tau) = \exp \left[ \pm p (x_0^5)^{(p-2)/2} (\tau + A_1) \right.$$
$$\left. \times \sqrt{\pm \left[C_{12}^2 + C_{22}^2 - C_{02}^2 \pm A_2(x_0^5)^{2-p} + (x_0^5)^p C_{32}^2\right]} \right]. \tag{C.52}$$

## C.5 Class (V)

$\widetilde{\mathbf{q}}_{\mathrm{V}} = (-p, -p, -p, p, -(1+p))$

One has

$$F_{\pm,V}(\zeta; p, A_2)$$
$$= \left\{ \pm (x_0^5)^{-p} [C_{12}^2 + C_{22}^2 - C_{02}^2] \zeta^{1+2p} + A_2 (x_0^5)^{1+p} \zeta^{1+p} \pm (x_0^5)^p C_{32}^2 \zeta^{1+p} \right\}^{-1/2}. \tag{C.53}$$

The solution writes

$$x^\mu(\tau) = C_{\mu 1} + C_{\mu 2} \int d\tau \left(x^5(\tau)\right)^p, \quad \mu = 0, 1, 2; \tag{C.54}$$

$$x^3(\tau) = C_{31} + C_{32} \int d\tau \left(x^5(\tau)\right)^{-p} \tag{C.55}$$

with $x^5(\tau)$ given by:

(1) $p \neq 0$:

(1.1) $C_{12}^2 + C_{22}^2 - C_{02}^2 = 0$:

(1.1.1) $C_{32} \neq 0,\ A_2 \neq 0$:

$$0 = \tau + A_1 \mp (x_0^5)^{(1+p)/2} \left\{ (\pm 1)^{\frac{1-p}{2p}} \frac{C_{32}^{(1-p)/p}}{pA_2^{1/2p}} (x_0^5)^{-(p^2+1)/2p} \right.$$
$$\times \left\{ \delta \left( \frac{1}{2p} + m \right) \widehat{\Theta}_{\mathrm{R}} \left( \pm \frac{A_2 (x_0^5)^{1+p}}{C_{32}^2} \left( \frac{x^5(\tau)}{x_0^5} \right)^p + 1 \right) \right.$$

$$\times \left[ \begin{array}{l} \dfrac{2(-1)^{\frac{1}{2p}-1} \sqrt{\pm \dfrac{A_2 (x_0^5)^{1+p}}{C_{32}^2} \left( \dfrac{x^5(\tau)}{x_0^5} \right)^p + 1}}{2\left|\dfrac{1}{2p} - 1\right| - 1} \\ \times \displaystyle\sum_{k=1}^{\left|\frac{1}{2p}-1\right|-1} \frac{\left(2\left|\frac{1}{2p}-1\right|-1\right)\left(2\left|\frac{1}{2p}-1\right|-3\right)\left(2\left|\frac{1}{2p}-1\right|-5\right)\cdots\left(2\left|\frac{1}{2p}-1\right|-2k+1\right)}{2^k\left(\left|\frac{1}{2p}-1\right|-1\right)\left(\left|\frac{1}{2p}-1\right|-2\right)\cdots\left(\left|\frac{1}{2p}-1\right|-k\right)\left[\mp\frac{A_2(x_0^5)^{1+p}}{C_{32}^2}\left(\frac{x^5(\tau)}{x_0^5}\right)\right]^{p|(1/2p)-1|-k}} \\ + \dfrac{\left(2\left|\frac{1}{2p}-1\right|-3\right)!!}{2^{\left|\frac{1}{2p}-1\right|}\left(\left|\frac{1}{2p}-1\right|-1\right)!} \ln \left| \dfrac{1 + \sqrt{\pm \frac{A_2 (x_0^5)^{1+p}}{C_{32}^2} \left( \frac{x^5(\tau)}{x_0^5} \right)^p + 1}}{1 - \sqrt{\pm \frac{A_2 (x_0^5)^{1+p}}{C_{32}^2} \left( \frac{x^5(\tau)}{x_0^5} \right)^p + 1}} \right| \end{array} \right]$$

or

$$\times 2(-1)^{(1/2p)-1} \widehat{\Theta} \left( 1 - \sqrt{\pm \frac{A_2 (x_0^5)^{1+p}}{C_{32}^2} \left( \frac{x^5(\tau)}{x_0^5} \right)^p + 1} \right)$$

$$\times \left[ \begin{array}{l} \dfrac{\sqrt{\pm \dfrac{A_2 (x_0^5)^{1+p}}{C_{32}^2} \left( \dfrac{x^5(\tau)}{x_0^5} \right)^p + 1}}{2\left|\dfrac{1}{2p} - 1\right| - 1} \\ \times \displaystyle\sum_{k=1}^{\left|\frac{1}{2p}-1\right|-1} \frac{\left(2\left|\frac{1}{2p}-1\right|-1\right)\left(2\left|\frac{1}{2p}-1\right|-3\right)\left(2\left|\frac{1}{2p}-1\right|-5\right)\cdots\left(2\left|\frac{1}{2p}-1\right|-2k+1\right)}{2^k\left(\left|\frac{1}{2p}-1\right|-1\right)\left(\left|\frac{1}{2p}-1\right|-2\right)\cdots\left(\left|\frac{1}{2p}-1\right|-k\right)\left[\mp\frac{A_2(x_0^5)^{1+p}}{C_{32}^2}\left(\frac{x^5(\tau)}{x_0^5}\right)\right]^{p|(1/2p)-1|-k}} \\ + \dfrac{\left(2\left|\frac{1}{2p}-1\right|-3\right)!!}{2^{|(1/2p)-1|}\left(\left|\frac{1}{2p}-1\right|-1\right)!} \operatorname{arctgh} \left[ \sqrt{\pm \frac{A_2 (x_0^5)^{1+p}}{C_{32}^2} \left( \frac{x^5(\tau)}{x_0^5} \right)^p + 1} \right] \end{array} \right]$$

$$\pm 2i\widehat{\Theta}_{\mathrm{R}}\left(\pm\frac{A_2(x_0^5)^{1+p}}{C_{32}^2}\left(\frac{x^5(\tau)}{x_0^5}\right)^p+1\right)\left\{\delta\left(\frac{1}{2p}+\frac{1}{2}\right)\sqrt{\frac{\pm\frac{A_2(x_0^5)^{1+p}}{C_{32}^2}\left(\frac{x^5(\tau)}{x_0^5}\right)^p+1}{\mp\frac{A_2(x_0^5)^{1+p}}{C_{32}^2}\left(\frac{x^5(\tau)}{x_0^5}\right)^p}}\right.$$

$$+\delta\left(\frac{1}{2p}+\frac{2m+1}{2}\right)\frac{\sqrt{\pm\frac{A_2(x_0^5)^{1+p}}{C_{32}^2}\left(\frac{x^5(\tau)}{x_0^5}\right)^p+1}}{2m-1}\left[\mp\frac{A_2(x_0^5)^{1+p}}{C_{32}^2}\left(\frac{x^5(\tau)}{x_0^5}\right)^p\right]^{-m+(1/2)}$$

$$\left.\times\left[1+\sum_{k=1}^{m-1}\frac{2^k(m-1)(m-2)\cdots(m-k)}{(2m-3)(2m-5)\cdots(2m-2k-1)}\left(\pm\frac{A_2(x_0^5)^{1+p}}{C_{32}^2}\left(\frac{x^5(\tau)}{x_0^5}\right)^p\right)^k\right]\right\}$$

$$+\widehat{\Theta}(p)\widehat{\Theta}\left(\pm\frac{A_2(x_0^5)^{1+p}}{C_{32}^2}\left(\frac{x^5(\tau)}{x_0^5}\right)^p-1\right)\widehat{\Theta}\left(\frac{1}{2p}-\frac{1}{2}\right)\frac{2p}{1-p}$$

$$\times\left[\pm\frac{A_2(x_0^5)^{1+p}}{C_{32}^2}\left(\frac{x^5(\tau)}{x_0^5}\right)^p\right]^{1/2p-(1/2)}$$

$$\times\ {}_2F_1\left(\frac{1}{2},-\left(\frac{1}{2p}-\frac{1}{2}\right);-\left(\frac{1}{2p}-\frac{3}{2}\right);\left(\mp\frac{A_2(x_0^5)^{1+p}}{C_{32}^2}\left(\frac{x^5(\tau)}{x_0^5}\right)^p\right)^{-1}\right)$$

$$+\widehat{\Theta}\left(\mp\frac{A_2}{C_{32}^2}\right)\widehat{\Theta}\left(\pm\frac{A_2(x_0^5)^{1+p}}{C_{32}^2}\left(\frac{x^5(\tau)}{x_0^5}\right)^p+1\right)$$

$$\times\left\{\begin{array}{c}2p\left[\pm\frac{A_2(x_0^5)^{1+p}}{C_{32}^2}\left(\frac{x^5(\tau)}{x_0^5}\right)^p\right]^{1/2p}\ {}_2F_1\left(\frac{1}{2},\frac{1}{2p};\frac{1}{2p}+1;\mp\frac{A_2(x_0^5)^{1+p}}{C_{32}^2}\left(\frac{x^5(\tau)}{x_0^5}\right)^p\right)\\ \text{or}\\ \left\{\begin{array}{c}\widehat{\Theta}\left(\frac{1}{2p}-\frac{1}{2}\right)\frac{2p}{1-p}\left[\pm\frac{A_2(x_0^5)^{1+p}}{C_{32}^2}\left(\frac{x^5(\tau)}{x_0^5}\right)^p\right]^{(1/2p)-(1/2)}\\ \times\ {}_2F_1\left(\frac{1}{2},-\left(\frac{1}{2p}-\frac{1}{2}\right);-\left(\frac{1}{2p}-\frac{3}{2}\right);\left(\mp\frac{A_2(x_0^5)^{1+p}}{C_{32}^2}\left(\frac{x^5(\tau)}{x_0^5}\right)^p\right)^{-1}\right)\end{array}\right.\end{array}\right.$$

or

$$+\widehat{\Theta}\left(\pm\frac{A_2}{C_{32}^2}\right)\left\{\begin{array}{c}2p\left[\pm\frac{A_2(x_0^5)^{1+p}}{C_{32}^2}\left(\frac{x^5(\tau)}{x_0^5}\right)^p\right]^{1/2p}\\ \times\ {}_2F_1\left(\frac{1}{2},\frac{1}{2p};\frac{1}{2p}+1;\mp\frac{A_2(x_0^5)^{1+p}}{C_{32}^2}\left(\frac{x^5(\tau)}{x_0^5}\right)^p\right)\\ \text{or}\\ \left\{\begin{array}{c}\widehat{\Theta}\left(\frac{1}{2p}-\frac{1}{2}\right)\frac{2p}{1-p}\left[\pm\frac{A_2(x_0^5)^{1+p}}{C_{32}^2}\left(\frac{x^5(\tau)}{x_0^5}\right)^p\right]^{(1/2p)-(1/2)}\\ \times\ {}_2F_1\left(\frac{1}{2},-\left(\frac{1}{2p}-\frac{1}{2}\right);-\left(\frac{1}{2p}-\frac{3}{2}\right);\left(\mp\frac{A_2(x_0^5)^{1+p}}{C_{32}^2}\left(\frac{x^5(\tau)}{x_0^5}\right)^p\right)^{-1}\right)\end{array}\right.\end{array}\right.$$

$$\left.\begin{array}{c}\mp 2i\widehat{\Theta}_{\mathrm{R}}\left(\pm\frac{A_2(x_0^5)^{1+p}}{C_{32}^2}\left(\frac{x^5(\tau)}{x_0^5}\right)^p+1\right)\delta\left(\frac{1}{2p}-\frac{1}{2}\right)\\ \times\arcsin\left[-\sqrt{\pm\frac{A_2(x_0^5)^{1+p}}{C_{32}^2}\left(\frac{x^5(\tau)}{x_0^5}\right)^p+1}\right]\end{array}\right\};$$

or

$$\left.+2\delta\left(\frac{1}{2p}-1\right)\sqrt{\pm\frac{A_2(x_0^5)^{1+p}}{C_{32}^2}\left(\frac{x^5(\tau)}{x_0^5}\right)^p+1}\right\};$$

or

$$\mp i\widehat{\Theta}_{\mathrm{R}}\left(\pm\frac{A_2(x_0^5)^{1+p}}{C_{32}^2}\left(\frac{x^5(\tau)}{x_0^5}\right)^p+1\right)\delta\left(\frac{1}{2p}-\frac{3}{2}\right)$$
$$\times\left\{\sqrt{\mp\frac{A_2(x_0^5)^{1+p}}{C_{32}^2}\left(\frac{x^5(\tau)}{x_0^5}\right)^p\left[\pm\frac{A_2(x_0^5)^{1+p}}{C_{32}^2}\left(\frac{x^5(\tau)}{x_0^5}\right)^p+1\right]}\right.$$
$$\left.+\arcsin\left[-\sqrt{\pm\frac{A_2(x_0^5)^{1+p}}{C_{32}^2}\left(\frac{x^5(\tau)}{x_0^5}\right)^p+1}\right]\right\}\Bigg\};$$

or

$$\pm 2i\widehat{\Theta}_{\mathrm{R}}\left(\pm\frac{A_2(x_0^5)^{1+p}}{C_{32}^2}\left(\frac{x^5(\tau)}{x_0^5}\right)^p+1\right)\delta\left(\frac{1}{2p}-m-\frac{3}{2}\right)$$
$$\times\frac{\sqrt{\mp\frac{A_2(x_0^5)^{1+p}}{C_{32}^2}\left(\frac{x^5(\tau)}{x_0^5}\right)^p\left[\pm\frac{A_2(x_0^5)^{1+p}}{C_{32}^2}\left(\frac{x^5(\tau)}{x_0^5}\right)^p+1\right]}}{2(m+1)}$$
$$\times\left\{\left[\mp\frac{A_2(x_0^5)^{1+p}}{C_{32}^2}\left(\frac{x^5(\tau)}{x_0^5}\right)^p\right]^m\right.$$
$$\left.+\sum_{k=1}^{m-1}\frac{(2m+1)(2m-1)\cdots(2m-2k+1)}{2^{k+1}m(m-1)(m-2)\cdots(m-k)}\left(\mp\frac{A_2(x_0^5)^{1+p}}{C_{32}^2}\left(\frac{x^5(\tau)}{x_0^5}\right)^p\right)^{m-k-1}\right\}$$
$$-\frac{(2m+1)!!}{2^{m+1}(m+1)!}\arcsin\left[-\sqrt{\pm\frac{A_2(x_0^5)^{1+p}}{C_{32}^2}\left(\frac{x^5(\tau)}{x_0^5}\right)^p+1}\right]\Bigg\};$$

or

$$\pm 2(-1)^{(1/2p)-1}\widehat{\Theta}_{\mathrm{R}}\left(\pm\frac{A_2(x_0^5)^{1+p}}{C_{32}^2}\left(\frac{x^5(\tau)}{x_0^5}\right)^p+1\right)\delta\left(\frac{1}{2p}-m-1\right)$$
$$\times\sum_{k=0}^{\frac{1}{2p}-1}\binom{\frac{1}{2p}-1}{k}(-1)^k\frac{\left[\pm\frac{A_2(x_0^5)^{1+p}}{C_{32}^2}\left(\frac{x^5(\tau)}{x_0^5}\right)^p+1\right]^{(2k+1)/2}}{2k+1}\Bigg\};\tag{C.57}$$

(1.1.2) $C_{32}=0,\ A_2\neq 0$:

(1.1.2.1) $p\neq 1$:

$$x^5(\tau)=\left[\pm\frac{2}{1-p}\sqrt{A_2}\,(\tau+A_1)\right]^{(1-p)/2};\tag{C.58}$$

(1.1.2.2) $p = 1$:

$$x^5(\tau) = \exp\left[\pm\sqrt{A_2}\,(\tau + A_1)\right]; \qquad \text{(C.59)}$$

(1.1.3) $C_{32} \neq 0$, $A_2 = 0$:

$$x^5(\tau) = \pm\frac{1}{4}\frac{C_{32}^2}{x_0^5}(\tau + A_1).$$

(1.2) $A_2 = 0$:

(1.2.1) $C_{32} \neq 0$, $C_{12}^2 + C_{22}^2 - C_{02}^2 \neq 0$:

$$0 = \tau + A_1 \mp \left(x_0^5\right)^{(1+p)/2} \left\{\widehat{\Theta}(p)\right.$$

$$\times \left\{ \begin{array}{c} \left[-\widehat{\Theta}\left(-\frac{1}{2p}+1\right) \frac{\left(x^5(\tau)\right)^{(1/2)-p}\left(x_0^5\right)^{p/2}}{p\sqrt{\pm\left(C_{12}^2+C_{22}^2-C_{02}^2\right)}} \frac{\Gamma\left(-\frac{1}{2p}+1\right)}{\Gamma\left(-\frac{1}{2p}\right)}\right] \left[\widehat{\Theta}\left(C_{12}^2+C_{22}^2-C_{02}^2\right)\right. \\ \left. +\widehat{\Theta}\left(C_{02}^2-C_{12}^2-C_{22}^2\right)\widehat{\Theta}\left(\left(x^5(\tau)\right)^p - \left(x_0^5\right)^p\sqrt{\frac{C_{32}^2}{C_{02}^2-C_{12}^2-C_{22}^2}}\right)\right] \\ \times\ {}_3F_2\left(\frac{1}{2}, -\frac{1}{4p}+\frac{1}{2}, -\frac{1}{4p}+1; -\frac{1}{4p}+1, -\frac{1}{4p}+\frac{3}{2}; \frac{C_{32}^2}{C_{02}^2-C_{12}^2-C_{22}^2}\left(\frac{x^5(\tau)}{x_0^5}\right)^{2p}\right) \end{array}\right.$$

or

$$\times \left\{ \begin{array}{c} \widehat{\Theta}\left(C_{12}^2+C_{22}^2-C_{02}^2\right) \frac{\sqrt{x^5(\tau)}\left(x_0^5\right)^{-p/2}}{pC_{32}\sqrt{\pm 1}} B\left(\frac{1}{2p}, 1\right) \\ \times\ {}_3F_2\left(\frac{1}{2}, \frac{1}{4p}, \frac{1}{4p}+\frac{1}{2}; \frac{1}{4p}+\frac{1}{2}, \frac{1}{4p}+1; \frac{C_{02}^2-C_{12}^2-C_{22}^2}{C_{32}^2}\left(\frac{x^5(\tau)}{x_0^5}\right)^{2p}\right) \end{array}\right.$$

$$\times\delta\left(\frac{1}{2p}-1\right)\frac{\left(x_0^5\right)^{(p/2)}}{p\sqrt{\pm\left(C_{12}^2+C_{22}^2-C_{02}^2\right)}}$$

$$\times \left\{ \begin{array}{c} \widehat{\Theta}\left(C_{12}^2+C_{22}^2-C_{02}^2\right) \operatorname{arcsinh}\left[\sqrt{\frac{C_{12}^2+C_{22}^2-C_{02}^2}{C_{32}^2}}\left(\frac{x^5(\tau)}{x_0^5}\right)^p\right] \\ +\widehat{\Theta}\left(C_{02}^2-C_{12}^2-C_{22}^2\right) \\ \times\ln\left\{\left(x^5(\tau)\right)^p\left[1+\sqrt{1+\frac{C_{32}^2}{C_{12}^2+C_{22}^2-C_{02}^2}\left(\frac{x_0^5}{x^5(\tau)}\right)^{2p}}\right]\right\} \end{array}\right\}$$

or

$$\times\delta\left(\frac{1}{2p}-m-1\right)$$

$$\times \left\{ \begin{array}{c} \times\frac{\sqrt{\left(x^5(\tau)\right)^{2p}+\frac{\left(x_0^5\right)^{2p}C_{32}^2}{C_{12}^2+C_{22}^2-C_{02}^2}}}{p\sqrt{\pm\left(x_0^5\right)^{-p}\left(C_{12}^2+C_{22}^2-C_{02}^2\right)}} \\ \times Q\left(x^5(\tau); C_{02}, C_{12}, C_{22}, C_{32}, p, x_0^5\right) + \frac{\kappa\left(x_0^5\right)^{p/2}}{p\sqrt{\pm\left(C_{12}^2+C_{22}^2-C_{02}^2\right)}} \\ \times \left\{ \begin{array}{c} \widehat{\Theta}\left(C_{12}^2+C_{22}^2-C_{02}^2\right) \operatorname{arcsinh}\left[\sqrt{\frac{C_{12}^2+C_{22}^2-C_{02}^2}{C_{32}^2}}\left(\frac{x^5(\tau)}{x_0^5}\right)^p\right] \\ +\widehat{\Theta}\left(C_{02}^2-C_{12}^2-C_{22}^2\right) \\ \times\ln\left\{\left(x^5(\tau)\right)^p\left[1+\sqrt{1+\frac{C_{32}^2}{C_{12}^2+C_{22}^2-C_{02}^2}\left(\frac{x_0^5}{x^5(\tau)}\right)^{2p}}\right]\right\} \end{array}\right\} \end{array}\right.$$

$$+\left\{2\widehat{\Theta}\left(C_{12}^2+C_{22}^2-C_{02}^2\right)\right.$$

$$\left.+\widehat{\Theta}\left(C_{02}^2-C_{12}^2-C_{22}^2\right)\widehat{\Theta}\left(\left(x^5(\tau)\right)^p-\left(x_0^5\right)^p\sqrt{\frac{C_{32}^2}{C_{02}^2-C_{12}^2-C_{22}^2}}\right)\right\}$$

$$\times\frac{\left(x^5(\tau)\right)^{(1/2)-p}\left(x_0^5\right)^{p/2}}{2p^2\sqrt{\pm\left(C_{12}^2+C_{22}^2-C_{02}^2\right)}}$$

$$\times\ {}_3F_2\left(\tfrac{1}{2},-\tfrac{1}{4p}+\tfrac{1}{2},-\tfrac{1}{4p}+1;-\tfrac{1}{4p}+1,-\tfrac{1}{4p}+\tfrac{3}{2};\tfrac{C_{32}^2}{C_{02}^2-C_{12}^2-C_{22}^2}\left(\tfrac{x^5(\tau)}{x_0^5}\right)^{2p}\right),$$

or

$$+\delta\left(\frac{1}{2p}+m\right)\left\{-\frac{\sqrt{\frac{C_{32}^2}{C_{12}^2+C_{22}^2-C_{02}^2}\left(\frac{x_0^5}{x^5(\tau)}\right)^{2p}-1}}{p\sqrt{\pm\left(x_0^5\right)^{-p}\left(C_{12}^2+C_{22}^2-C_{02}^2\right)}}\right.$$

$$\times\widetilde{Q}\left(x^5(\tau);C_{02},C_{12},C_{22},C_{32},p,x_0^5\right)-\frac{\lambda\left(x_0^5\right)^{p/2}}{p\sqrt{\pm\left(C_{12}^2+C_{22}^2-C_{02}^2\right)}}$$

$$\times\left[\widehat{\Theta}\left(C_{12}^2+C_{22}^2-C_{02}^2\right)\left(x_0^5\right)^{-p/2}\sqrt{\frac{C_{12}^2+C_{22}^2-C_{02}^2}{C_{32}^2}}\right.$$

$$\times\operatorname{arcsinh}\left(\sqrt{\frac{C_{32}^2}{C_{12}^2+C_{22}^2-C_{02}^2}\left(\frac{x_0^5}{x^5(\tau)}\right)^{2p}}\right)$$

$$\left.\left.-\begin{array}{c}\widehat{\Theta}\left(C_{02}^2-C_{12}^2-C_{22}^2\right)\sqrt{\frac{C_{02}^2-C_{12}^2-C_{22}^2}{C_{32}^2}}\\ \times\arcsin\left(\sqrt{\frac{C_{32}^2}{C_{02}^2-C_{12}^2-C_{22}^2}\left(\frac{x_0^5}{x^5(\tau)}\right)^{2p}}\right)\end{array}\right\}\right], \tag{C.60}$$

where ${}_3F_2$ is the generalized hypergeometric function of class (3,2)

$$ {}_3F_2(\alpha,\beta,\gamma;\delta,\varepsilon;z)\equiv\sum_{k=0}^{\infty}\frac{(\alpha)_k(\beta)_k(\gamma)_k}{(\delta)_k(\varepsilon)_k}\frac{z^k}{k!}, \tag{C.61}$$

$B(x,y)$ is the Euler beta function (or Euler integral of first kind), defined by

$$B(x,y)\equiv\int_0^1 t^{x-1}(1-t)^{y-1}\,dt,\quad \operatorname{Re}(x)>0\ ,\quad \operatorname{Re}(y)>0 \tag{C.62}$$

and we put

$$\widetilde{Q}\left(x^5(\tau);C_{02},C_{12},C_{22},C_{32},p,x_0^5\right)\equiv$$

$$\equiv q_{-\frac{1}{2p}-1}\left(x^5(\tau)\right)^{(1/2)+p}+\sum_{n=2}^{-\frac{1}{2p}-1}q_{-\frac{1}{2p}--n-1}\left(x^5(\tau)\right)^{(1/2)+(n+1)p};$$

$$q_{-\frac{1}{2p}-1}\equiv-\frac{2p\left(C_{12}^2+C_{22}^2-C_{02}^2\right)}{\left(x_0^5\right)^{2p}C_{32}^2};$$

$$q_{-\frac{1}{2p}-n-1}\equiv-\frac{\left(C_{12}^2+C_{22}^2-C_{02}^2\right)\left(\dfrac{1}{2p}+n+1\right)}{\left(x_0^5\right)^{2p}C_{32}^2\left(\dfrac{1}{2p}+n\right)}q_{-\frac{1}{2p}-n+1},$$

$$n=2,\ldots,-\frac{1}{2p}-1;$$

$$\lambda=-q_1. \tag{C.63}$$

(1.2.2) $C_{32}\neq 0,\ C_{12}^2+C_{22}^2-C_{02}^2=0$:

$$x^5(\tau)=\pm\frac{1}{4}\frac{C_{32}^2}{x_0^5}\left(\tau+A_1\right). \tag{C.64}$$

(1.2.3) $C_{32}=0,\ C_{12}^2+C_{22}^2-C_{02}^2\neq 0$:

(1.2.3.1) $p\neq\dfrac{1}{2}$:

$$x^5(\tau)=\left[\pm\left(-p+\frac{1}{2}\right)\sqrt{\pm\left(C_{12}^2+C_{22}^2-C_{02}^2\right)}\left(x_0^5\right)^{-(1+2p)/2}\left(\tau+A_1\right)\right]^{2/(1-2p)}. \tag{C.65}$$

(1.2.3.2) $p=\dfrac{1}{2}$:

$$x^5(\tau)=\exp\left[\pm\sqrt{\pm\left(C_{12}^2+C_{22}^2-C_{02}^2\right)}\left(x_0^5\right)^{-(1+2p)/2}\left(\tau+A_1\right)\right]. \tag{C.66}$$

(1.3) $C_{32}=0$:

(1.3.1) $A_2\neq 0,\ C_{12}^2+C_{22}^2-C_{02}^2\neq 0$:

$$0=\tau+A_1\mp\left(x_0^5\right)^{(1+p)/2}\left\{\frac{\left(x_0^5\right)^{-(1+p)/2}}{p\sqrt{A_2}}\left(\frac{A_2\left(x_0^5\right)^{1+2p}}{\pm\left(C_{12}^2+C_{22}^2-C_{02}^2\right)}\right)^{(1/2p)-(1/2)}\right.$$

$$\times\left\{2\left(-1\right)^{((1/2p)-(3/2))}\delta\left(\frac{1}{2p}+m-\frac{1}{2}\right)\widehat{\Theta}_{\mathrm{R}}\left[\frac{\pm\left(C_{12}^2+C_{22}^2-C_{02}^2\right)}{\left(x_0^5\right)^{1+p}A_2}\left(\frac{x^5(\tau)}{x_0^5}\right)^p+1\right]\right.$$

$$
\times\left\{\begin{array}{l}
\left[\dfrac{\sqrt{\dfrac{\pm\left(C_{12}^2+C_{22}^2-C_{02}^2\right)}{\left(x_0^5\right)^{1+p}A_2}\left(\dfrac{x^5(\tau)}{x_0^5}\right)^p+1}}{2\left|\dfrac{1}{2p}-\dfrac{3}{2}\right|-1}\right. \\
\times\displaystyle\sum_{k=1}^{\left|\frac{1}{2p}-\frac{3}{2}\right|-1}\frac{\left(2\left|\frac{1}{2p}-\frac{3}{2}\right|-1\right)\left(2\left|\frac{1}{2p}-\frac{3}{2}\right|-3\right)\left(2\left|\frac{1}{2p}-\frac{3}{2}\right|-5\right)\cdots\left(2\left|\frac{1}{2p}-\frac{3}{2}\right|-2k+1\right)}{2^k\left(\left|\frac{1}{2p}-\frac{3}{2}\right|-1\right)\left(\left|\frac{1}{2p}-\frac{3}{2}\right|-2\right)\cdots\left(\left|\frac{1}{2p}-\frac{3}{2}\right|-k\right)\left[\frac{\mp\left(C_{12}^2+C_{22}^2-C_{02}^2\right)}{\left(x_0^5\right)^{1+p}A_2}\left(\frac{x^5(\tau)}{x_0^5}\right)^p\right]^{|(1/2p)-(3/2)|-k}} \\
\left.+\dfrac{\left(2\left|\frac{1}{2p}-\frac{3}{2}\right|-3\right)!!}{2^{\left|\frac{1}{2p}-\frac{3}{2}\right|}\left(\left|\frac{1}{2p}-\frac{3}{2}\right|-1\right)!}\ln\left|\dfrac{1+\sqrt{\frac{\pm\left(C_{12}^2+C_{22}^2-C_{02}^2\right)}{\left(x_0^5\right)^{1+p}A_2}\left(\frac{x^5(\tau)}{x_0^5}\right)^p+1}}{1-\sqrt{\frac{\pm\left(C_{12}^2+C_{22}^2-C_{02}^2\right)}{\left(x_0^5\right)^{1+p}A_2}\left(\frac{x^5(\tau)}{x_0^5}\right)^p+1}}\right|\right] \\
\text{or} \\
\times\widehat{\Theta}\left(1-\sqrt{\dfrac{\pm\left(C_{12}^2+C_{22}^2-C_{02}^2\right)}{\left(x_0^5\right)^{1+p}A_2}\left(\dfrac{x^5(\tau)}{x_0^5}\right)^p+1}\right) \\
\times\left[\dfrac{\sqrt{\dfrac{\pm\left(C_{12}^2+C_{22}^2-C_{02}^2\right)}{\left(x_0^5\right)^{1+p}A_2}\left(\dfrac{x^5(\tau)}{x_0^5}\right)^p+1}}{2\left|\dfrac{1}{2p}-\dfrac{3}{2}\right|-1}\right. \\
\times\displaystyle\sum_{k=1}^{\left|\frac{1}{2p}-\frac{3}{2}\right|-1}\frac{\left(2\left|\frac{1}{2p}-\frac{3}{2}\right|-1\right)\left(2\left|\frac{1}{2p}-\frac{3}{2}\right|-3\right)\left(2\left|\frac{1}{2p}-\frac{3}{2}\right|-5\right)\cdots\left(2\left|\frac{1}{2p}-\frac{3}{2}\right|-2k+1\right)}{2^k\left(\left|\frac{1}{2p}-\frac{3}{2}\right|-1\right)\left(\left|\frac{1}{2p}-\frac{3}{2}\right|-2\right)\cdots\left(\left|\frac{1}{2p}-\frac{3}{2}\right|-k\right)\left[\frac{\mp\left(C_{12}^2+C_{22}^2-C_{02}^2\right)}{\left(x_0^5\right)^{1+p}A_2}\left(\frac{x^5(\tau)}{x_0^5}\right)^p\right]^{|(1/2p)-(3/2)|-k}} \\
\left.+\dfrac{\left(2\left|\frac{1}{2p}-\frac{3}{2}\right|-3\right)!!}{2^{|(1/2p)-(3/2)|-1}\left(\left|\frac{1}{2p}-\frac{3}{2}\right|-1\right)!}\operatorname{arctgh}\left[\sqrt{\dfrac{\pm\left(C_{12}^2+C_{22}^2-C_{02}^2\right)}{\left(x_0^5\right)^{1+p}A_2}\left(\dfrac{x^5(\tau)}{x_0^5}\right)^p+1}\right]\right]
\end{array}\right.
$$

$$
\pm 2i\delta\left(\frac{1}{2p}+m\right)\widehat{\Theta}_{\mathrm{R}}\left[\frac{\pm\left(C_{12}^2+C_{22}^2-C_{02}^2\right)}{\left(x_0^5\right)^{1+p}A_2}\left(\frac{x^5(\tau)}{x_0^5}\right)^p+1\right]
$$

$$
\times\frac{\sqrt{\frac{\pm\left(C_{12}^2+C_{22}^2-C_{02}^2\right)}{\left(x_0^5\right)^{1+p}A_2}\left(\frac{x^5(\tau)}{x_0^5}\right)^p+1}}{(2m-1)}\left[\frac{\mp\left(C_{12}^2+C_{22}^2-C_{02}^2\right)}{\left(x_0^5\right)^{1+p}A_2}\left(\frac{x^5(\tau)}{x_0^5}\right)^p\right]^{-m+(1/2)}
$$

$$
\times\left\{1+\sum_{k=1}^{m-1}\frac{2^k(m-1)(m-2)\cdots(m-k)}{(2m-3)(2m-5)\cdots(2m-2k-1)}\left[\frac{\mp\left(C_{12}^2+C_{22}^2-C_{02}^2\right)}{\left(x_0^5\right)^{1+p}A_2}\left(\frac{x^5(\tau)}{x_0^5}\right)^p\right]^k\right\}
$$

$$
-2\delta\left(\frac{1}{2p}-\frac{1}{2}\right)\widehat{\Theta}_{\mathrm{R}}\left(\frac{\pm\left(C_{12}^2+C_{22}^2-C_{02}^2\right)}{\left(x_0^5\right)^{1+p}A_2}\left(\frac{x^5(\tau)}{x_0^5}\right)^p+1\right)
$$

$$
\times\left\{\begin{array}{l}
\frac{1}{2}\ln\left|\dfrac{1+\sqrt{\frac{\pm\left(C_{12}^2+C_{22}^2-C_{02}^2\right)}{\left(x_0^5\right)^{1+p}A_2}\left(\frac{x^5(\tau)}{x_0^5}\right)^p+1}}{1-\sqrt{\frac{\pm\left(C_{12}^2+C_{22}^2-C_{02}^2\right)}{\left(x_0^5\right)^{1+p}A_2}\left(\frac{x^5(\tau)}{x_0^5}\right)^p+1}}\right| \\
\text{or} \\
\widehat{\Theta}\left(1-\sqrt{\dfrac{\pm\left(C_{12}^2+C_{22}^2-C_{02}^2\right)}{\left(x_0^5\right)^{1+p}A_2}\left(\dfrac{x^5(\tau)}{x_0^5}\right)^p+1}\right) \\
\times\operatorname{arctgh}\left[\sqrt{\dfrac{\pm\left(C_{12}^2+C_{22}^2-C_{02}^2\right)}{\left(x_0^5\right)^{1+p}A_2}\left(\dfrac{x^5(\tau)}{x_0^5}\right)^p+1}\right]
\end{array}\right.
$$

$$
+\delta\left(\frac{1}{2p}-\frac{1}{2}\right)
$$

$$
\times \left\{ \begin{array}{l}
\widehat{\Theta}\left( \frac{\pm\left(C_{12}^2+C_{22}^2-C_{02}^2\right)}{\left(x_0^5\right)^{1+p} A_2} \left(\frac{x^5(\tau)}{x_0^5}\right)^p - 1 \right) \widehat{\Theta}\left(\frac{1}{2p}-1\right) \\
\times \frac{\left[ \frac{\pm\left(C_{12}^2+C_{22}^2-C_{02}^2\right)}{\left(x_0^5\right)^{1+p} A_2} \left(\frac{x^5(\tau)}{x_0^5}\right)^p \right]^{(1/2p)-1}}{\left(\frac{1}{2p}-1\right)} \\
\times {}_2F_1\left( \frac{1}{2}, -\frac{1}{2p}+1; -\frac{1}{2p}+2; \left( \frac{\mp\left(C_{12}^2+C_{22}^2-C_{02}^2\right)}{\left(x_0^5\right)^{1+p} A_2} \left(\frac{x^5(\tau)}{x_0^5}\right)^p \right)^{-1} \right) \\
+\widehat{\Theta}\left( \frac{\pm\left(C_{12}^2+C_{22}^2-C_{02}^2\right)}{\left(x_0^5\right)^{1+p} A_2} \left(\frac{x^5(\tau)}{x_0^5}\right)^p + 1 \right) \widehat{\Theta}\left( \frac{\mp\left(C_{12}^2+C_{22}^2-C_{02}^2\right)}{A_2} \right) \\
\times \left\{ \begin{array}{c}
\frac{\left[ \frac{\pm\left(C_{12}^2+C_{22}^2-C_{02}^2\right)}{\left(x_0^5\right)^{1+p} A_2} \left(\frac{x^5(\tau)}{x_0^5}\right)^p \right]^{(1/2p)-(1/2)}}{\left(\frac{1}{2p}-\frac{1}{2}\right)} \\
\times {}_2F_1\left( \frac{1}{2}, \frac{1}{2p}-\frac{1}{2}; \frac{1}{2p}+\frac{1}{2}; \frac{\mp\left(C_{12}^2+C_{22}^2-C_{02}^2\right)}{\left(x_0^5\right)^{1+p} A_2} \left(\frac{x^5(\tau)}{x_0^5}\right)^p \right) \\
\text{or} \\
\widehat{\Theta}\left(\frac{1}{2p}-1\right) \frac{\left[ \frac{\pm\left(C_{12}^2+C_{22}^2-C_{02}^2\right)}{\left(x_0^5\right)^{1+p} A_2} \left(\frac{x^5(\tau)}{x_0^5}\right)^p \right]^{(1/2p)-1}}{\left(\frac{1}{2p}-1\right)} \\
\times {}_2F_1\left( \frac{1}{2}, -\frac{1}{2p}+1; -\frac{1}{2p}+2; \left( \frac{\mp\left(C_{12}^2+C_{22}^2-C_{02}^2\right)}{\left(x_0^5\right)^{1+p} A_2} \left(\frac{x^5(\tau)}{x_0^5}\right)^p \right)^{-1} \right)
\end{array} \right. \\
+\widehat{\Theta}\left( \frac{\pm\left(C_{12}^2+C_{22}^2-C_{02}^2\right)}{A_2} \right) \\
\times \left\{ \begin{array}{c}
\frac{\left[ \frac{\pm\left(C_{12}^2+C_{22}^2-C_{02}^2\right)}{\left(x_0^5\right)^{1+p} A_2} \left(\frac{x^5(\tau)}{x_0^5}\right)^p \right]^{(1/2p)-(1/2)}}{\left(\frac{1}{2p}-\frac{1}{2}\right)} \\
\times {}_2F_1\left( \frac{1}{2}, \frac{1}{2p}-\frac{1}{2}; \frac{1}{2p}+\frac{1}{2}; \frac{\mp\left(C_{12}^2+C_{22}^2-C_{02}^2\right)}{\left(x_0^5\right)^{1+p} A_2} \left(\frac{x^5(\tau)}{x_0^5}\right)^p \right) \\
\text{or} \\
\widehat{\Theta}\left(\frac{1}{2p}-1\right) \frac{\left[ \frac{\pm\left(C_{12}^2+C_{22}^2-C_{02}^2\right)}{\left(x_0^5\right)^{1+p} A_2} \left(\frac{x^5(\tau)}{x_0^5}\right)^p \right]^{(1/2p)-1}}{\left(\frac{1}{2p}-1\right)} \\
\times {}_2F_1\left( \frac{1}{2}, -\frac{1}{2p}+1; -\frac{1}{2p}+2; \left( \frac{\mp\left(C_{12}^2+C_{22}^2-C_{02}^2\right)}{\left(x_0^5\right)^{1+p} A_2} \left(\frac{x^5(\tau)}{x_0^5}\right)^p \right)^{-1} \right)
\end{array} \right.
\end{array} \right.
$$

$$
\mp 2\delta\left(\frac{1}{2p}-1\right) \widehat{\Theta}_{\mathrm{R}}\left( \frac{\pm\left(C_{12}^2+C_{22}^2-C_{02}^2\right)}{\left(x_0^5\right)^{1+p} A_2} \left(\frac{x^5(\tau)}{x_0^5}\right)^p + 1 \right)
$$

$$
\times 2i \arcsin\left[ -\sqrt{ \frac{\pm\left(C_{12}^2+C_{22}^2-C_{02}^2\right)}{\left(x_0^5\right)^{1+p} A_2} \left(\frac{x^5(\tau)}{x_0^5}\right)^p + 1 } \right] \Bigg\}
$$

$$\mp 2\delta\left(\frac{1}{2p}-1\right)\widehat{\Theta}_{\mathrm{R}}\left(\frac{\pm\left(C_{12}^2+C_{22}^2-C_{02}^2\right)}{\left(x_0^5\right)^{1+p}A_2}\left(\frac{x^5(\tau)}{x_0^5}\right)^p+1\right)$$
$$\times 2i\arcsin\left[-\sqrt{\frac{\pm\left(C_{12}^2+C_{22}^2-C_{02}^2\right)}{\left(x_0^5\right)^{1+p}A_2}\left(\frac{x^5(\tau)}{x_0^5}\right)^p+1}\right]\Bigg\}$$

$$\mp 2\delta\left(\frac{1}{2p}-1\right)\widehat{\Theta}_{\mathrm{R}}\left(\frac{\pm\left(C_{12}^2+C_{22}^2-C_{02}^2\right)}{\left(x_0^5\right)^{1+p}A_2}\left(\frac{x^5(\tau)}{x_0^5}\right)^p+1\right)$$
$$\times 2i\arcsin\left[-\sqrt{\frac{\pm\left(C_{12}^2+C_{22}^2-C_{02}^2\right)}{\left(x_0^5\right)^{1+p}A_2}\left(\frac{x^5(\tau)}{x_0^5}\right)^p+1}\right]\Bigg\}$$

$$\mp 2\delta\left(\frac{1}{2p}-1\right)\widehat{\Theta}_{\mathrm{R}}\left(\frac{\pm\left(C_{12}^2+C_{22}^2-C_{02}^2\right)}{\left(x_0^5\right)^{1+p}A_2}\left(\frac{x^5(\tau)}{x_0^5}\right)^p+1\right)$$
$$\times 2i\arcsin\left[-\sqrt{\frac{\pm\left(C_{12}^2+C_{22}^2-C_{02}^2\right)}{\left(x_0^5\right)^{1+p}A_2}\left(\frac{x^5(\tau)}{x_0^5}\right)^p+1}\right]\Bigg\}$$

$$\mp 2\delta\left(\frac{1}{2p}-1\right)\widehat{\Theta}_{\mathrm{R}}\left(\frac{\pm\left(C_{12}^2+C_{22}^2-C_{02}^2\right)}{\left(x_0^5\right)^{1+p}A_2}\left(\frac{x^5(\tau)}{x_0^5}\right)^p+1\right)$$
$$\times 2i\arcsin\left[-\sqrt{\frac{\pm\left(C_{12}^2+C_{22}^2-C_{02}^2\right)}{\left(x_0^5\right)^{1+p}A_2}\left(\frac{x^5(\tau)}{x_0^5}\right)^p+1}\right]\Bigg\};$$

or

$$\pm 2\delta\left(\frac{1}{2p}-\frac{3}{2}\right)\sqrt{\frac{\pm\left(C_{12}^2+C_{22}^2-C_{02}^2\right)}{\left(x_0^5\right)^{1+p}A_2}\left(\frac{x^5(\tau)}{x_0^5}\right)^p+1}\Bigg\};$$

or

$$\mp i\delta\left(\frac{1}{2p}-2\right)\widehat{\Theta}_{\mathrm{R}}\left(\frac{\pm\left(C_{12}^2+C_{22}^2-C_{02}^2\right)}{\left(x_0^5\right)^{1+p}A_2}\left(\frac{x^5(\tau)}{x_0^5}\right)^p+1\right)$$
$$\times\left\{\sqrt{\frac{\mp\left(C_{12}^2+C_{22}^2-C_{02}^2\right)}{\left(x_0^5\right)^{1+p}A_2}\left(\frac{x^5(\tau)}{x_0^5}\right)^p\left[\frac{\pm\left(C_{12}^2+C_{22}^2-C_{02}^2\right)}{\left(x_0^5\right)^{1+p}A_2}\left(\frac{x^5(\tau)}{x_0^5}\right)^p+1\right]}\right.$$
$$\left.+\arcsin\left[-\sqrt{\frac{\pm\left(C_{12}^2+C_{22}^2-C_{02}^2\right)}{\left(x_0^5\right)^{1+p}A_2}\left(\frac{x^5(\tau)}{x_0^5}\right)^p+1}\right]\right\}\Bigg\};$$

or

$$\pm 2i\delta\left(\frac{1}{2p}-m-2\right)\widehat{\Theta}_{\mathrm{R}}\left(\frac{\pm\left(C_{12}^2+C_{22}^2-C_{02}^2\right)}{\left(x_0^5\right)^{1+p}A_2}\left(\frac{x^5(\tau)}{x_0^5}\right)^p+1\right)$$

$$\times\frac{\sqrt{\frac{\mp\left(C_{12}^2+C_{22}^2-C_{02}^2\right)}{\left(x_0^5\right)^{1+p}A_2}\left(\frac{x^5(\tau)}{x_0^5}\right)^p\left[\frac{\pm\left(C_{12}^2+C_{22}^2-C_{02}^2\right)}{\left(x_0^5\right)^{1+p}A_2}\left(\frac{x^5(\tau)}{x_0^5}\right)^p+1\right]}}{(2m+1)}$$

$$\times\left\{\left[\frac{\pm\left(C_{12}^2+C_{22}^2-C_{02}^2\right)}{\left(x_0^5\right)^{1+p}A_2}\left(\frac{x^5(\tau)}{x_0^5}\right)^p\right]^m+\right.$$

$$\left.+\sum_{k=0}^{m-1}\frac{(2m+1)(2m-1)\cdots(2m-2k+1)}{2^{k+1}m(m-1)(m-2)\cdots(m-k)}\left[\frac{\mp\left(C_{12}^2+C_{22}^2-C_{02}^2\right)}{\left(x_0^5\right)^{1+p}A_2}\left(\frac{x^5(\tau)}{x_0^5}\right)^p\right]^{m-k-1}\right\}$$

$$-\frac{(2m+1)!!}{2^{m+1}(m+1)!}\arcsin\left[-\sqrt{\frac{\pm\left(C_{12}^2+C_{22}^2-C_{02}^2\right)}{\left(x_0^5\right)^{1+p}A_2}\left(\frac{x^5(\tau)}{x_0^5}\right)^p+1}\right]\Bigg\};$$

or

$$+2\,(-1)^{((1/2p)-(3/2))}\,\delta\left(\frac{1}{2p}-m-\frac{3}{2}\right)\widehat{\Theta}_{\mathrm{R}}\left(\frac{\pm\left(C_{12}^2+C_{22}^2-C_{02}^2\right)}{\left(x_0^5\right)^{1+p}A_2}\left(\frac{x^5(\tau)}{x_0^5}\right)^p+1\right)$$

$$\times\sum_{k=0}^{\frac{1}{2p}-\frac{3}{2}}\binom{\frac{1}{2p}-\frac{3}{2}}{k}(-1)^k\frac{\left[\frac{\pm\left(C_{12}^2+C_{22}^2-C_{02}^2\right)}{\left(x_0^5\right)^{1+p}A_2}\left(\frac{x^5(\tau)}{x_0^5}\right)^p+1\right]^{(2k+1)/2}}{(2k+1)}\Bigg\}\Bigg\}. \tag{C.67}$$

(1.3.2) $A_2\neq 0,\ C_{12}^2+C_{22}^2-C_{02}^2=0$:

(1.3.2.1) $p\neq 1$:

$$x^5(\tau)=\left[\pm\frac{1-p}{2}\sqrt{A_2}\,(\tau+A_1)\right]^{2/(1-p)}. \tag{C.68}$$

(1.3.2.2) $p=1$:

$$x^5(\tau)=\exp\left[\pm\sqrt{A_2}\,(\tau+A_1)\right]. \tag{C.69}$$

(1.3.3) $A_2=0,\ C_{12}^2+C_{22}^2-C_{02}^2\neq 0$:

(1.3.3.1) $p\neq\frac{1}{2}$:

$$x^5(\tau)=\left[\pm\left(-p+\frac{1}{2}\right)\sqrt{\pm(C_{12}^2+C_{22}^2-C_{02}^2)}\left(x_0^5\right)^{-(1+2p)/2}(\tau+A_1)\right]^{2/(1-2p)}. \tag{C.70}$$

(1.3.3.2) $p = \frac{1}{2}$:

$$x^5(\tau) = \exp\left[\pm\sqrt{\pm\left(C_{12}^2 + C_{22}^2 - C_{02}^2\right)}\left(x_0^5\right)^{-(1+2p)/2}(\tau + A_1)\right]. \tag{C.71}$$

(1.4) $C_{32},\ A_2,\ C_{12}^2 + C_{22}^2 - C_{02}^2 \neq 0$:

(1.4.1) $\left(\frac{1}{2p} - 1\right) \in N \Longleftrightarrow p = \frac{1}{2(m+1)},\ m \in N$:

$$\begin{aligned}
0 &= \tau + A_1 \mp \left(x_0^5\right)^{(1+p)/2} \\
&\times \Bigg\{\sqrt{\frac{c}{b}\left(x^5(\tau)\right)^{2p} + \frac{b}{a}\left(x^5(\tau)\right)^p + 1}\,Q(x^5(\tau); a, b, c, p) \\
&\quad + \lambda\left[\widehat{\Theta}\left(\frac{ac}{b^2}\right)\widehat{\Theta}\left(\frac{1}{4} - \frac{ac}{b^2}\right)\frac{b}{\sqrt{ac}}\right. \\
&\times \ln\left|2\sqrt{\left(\frac{ac^2}{b^2}\left(x^5(\tau)\right)^{2p} + \frac{c}{b}\left(x^5(\tau)\right)^p + \frac{ac}{b^2}\right)} + 2\frac{c}{b}\left(x^5(\tau)\right)^p + 1\right| \\
&\quad + \delta\left(\frac{ac}{b^2} - \frac{1}{4}\right) 2\ln\left|\frac{b}{2a}\left(x^5(\tau)\right)^p + 1\right| \\
&\quad + \widehat{\Theta}\left(\frac{ac}{b^2} - \frac{1}{4}\right)\frac{b}{\sqrt{ac}}\operatorname{arcsinh}\left(\frac{2\frac{c}{b}\left(x^5(\tau)\right)^p + 1}{\sqrt{4\frac{ac}{b^2} - 1}}\right) \\
&\left. - \widehat{\Theta}\left(-\frac{ac}{b^2}\right)\frac{b}{\sqrt{-ac}}\arcsin\left(\frac{2\frac{c}{b}\left(x^5(\tau)\right)^p + 1}{\sqrt{-4\frac{ac}{b^2} - 1}}\right)\right]\Bigg\},
\end{aligned} \tag{C.72}$$

where

$$\begin{aligned}
&Q(x^5(\tau); a, b, c, p) \\
&\equiv q_{\frac{1}{2p}-2}\left(\frac{b}{a}\right)^{((1/2p)-2)}\left(x^5(\tau)\right)^{((1/2)-2p)} \\
&\quad + q_{\frac{1}{2p}-3}\left(\frac{b}{a}\right)^{((1/2p)-3)}\left(x^5(\tau)\right)^{((1/2)-3p)}
\end{aligned}$$

$$+\sum_{n=2}^{\frac{1}{2p}-2} q_{\frac{1}{2p}-n-2}\left(\frac{b}{a}\right)^{\left(\frac{1}{2p}-n-2\right)}\left(x^5(\tau)\right)^{\left(\frac{1}{2}-(n+2)p\right)};$$

$$q_{\frac{1}{2p}-2}=\frac{b^2}{ac\left(\frac{1}{2p}-1\right)};$$

$$q_{\frac{1}{2p}-3}=-\frac{b^4\left(\frac{1}{2p}-\frac{3}{2}\right)}{ac\left(\frac{1}{2p}-1\right)\left(\frac{1}{2p}-2\right)};$$

$$q_{\frac{1}{2p}-n-2}=-\frac{b^2\left(\frac{1}{2p}-n-\frac{1}{2}\right)q_{\frac{1}{2p}-n-1}+\left(\frac{1}{2p}-n\right)q_{\frac{1}{2p}-n}}{ac\left(\frac{1}{2p}-1-n\right)},$$

$$n=2,\ldots,\frac{1}{2p}-2;$$

$$\lambda=-\frac{q_0}{2}-q_1 \tag{C.73}$$

with

$$\begin{aligned} a &\equiv \pm\left(x_0^5\right)^p C_{32}^2; \\ b &\equiv A_2\left(x_0^5\right)^{1+p}; \\ c &= \pm\left(x_0^5\right)^{-p}\left(C_{12}^2+C_{22}^2-C_{02}^2\right). \end{aligned} \tag{C.74}$$

(1.4.2) $p=\frac{1}{2}$:

$$\begin{aligned} 0 &= \tau + A_1 \mp \left(x_0^5\right)^{(1+p)/2} \\ &\times\left[\widehat{\Theta}\left(\frac{ac}{b^2}\right)\widehat{\Theta}\left(\frac{1}{4}-\frac{ac}{b^2}\right)\frac{b}{\sqrt{ac}}\right. \\ &\times \ln\left|2\sqrt{\frac{c}{b}\left(\frac{ac}{b}\left(x^5(\tau)\right)^{2p}+\left(x^5(\tau)\right)^p+\frac{a}{b}\right)}+2\frac{c}{b}\left(x^5(\tau)\right)^p+1\right| \\ &+\delta\left(\frac{ac}{b^2}-\frac{1}{4}\right)2\ln\left|\frac{b}{2a}\left(x^5(\tau)\right)^p+1\right| \end{aligned}$$

$$+\widehat{\Theta}\left(\frac{ac}{b^2}-\frac{1}{4}\right)\frac{b}{\sqrt{ac}}\operatorname{arcsinh}\left(\frac{2\frac{c}{b}\left(x^5(\tau)\right)^p+1}{\sqrt{4\frac{ac}{b^2}-1}}\right)$$

$$-\widehat{\Theta}\left(-\frac{ac}{b^2}\right)\frac{b}{\sqrt{-ac}}\arcsin\left(\frac{2\frac{c}{b}\left(x^5(\tau)\right)^p+1}{\sqrt{-4\frac{ac}{b^2}-1}}\right)\Bigg] \tag{C.75}$$

with $a$, $b$, $c$ given by (C.70).

(1.4.3) $\frac{1}{2p}-1=-2,-3,\ldots \Longleftrightarrow p=-\frac{1}{2m}$, $m\in N$:

$$0=\tau+A_1\mp\left(x_0^5\right)^{(1+p)/2}$$

$$\times\Bigg\{-\sqrt{\frac{a}{b}\left[\left(x^5(\tau)\right)^{-2p}+\left(x^5(\tau)\right)^{-p}+\frac{c}{b}\right]}\widetilde{Q}(x^5(\tau);a,b,c,p)$$

$$-\kappa\Bigg[\widehat{\Theta}\left(\frac{ac}{b^2}-\frac{1}{4}\right)\operatorname{arcsinh}\left(\frac{2\frac{a}{b}\left(x^5(\tau)\right)^{-p}+1}{\sqrt{4\frac{ac}{b^2}-1}}\right)$$

$$+\delta\left(\frac{ac}{b^2}-\frac{1}{4}\right)\ln\left|\frac{2a}{b}\left(x^5(\tau)\right)^{-p}+1\right| \tag{C.76}$$

$$+\widehat{\Theta}\left(\frac{1}{4}-\frac{ac}{b^2}\right)\left(\begin{array}{c}\ln\left|\begin{array}{c}2\sqrt{\frac{a}{b}\left(\left(x^5(\tau)\right)^{-2p}+\left(x^5(\tau)\right)^{-p}+\frac{c}{b}\right)}\\ +2\frac{a}{b}\left(x^5(\tau)\right)^{-p}+1\end{array}\right|\\ =\operatorname{arctgh}\left(\frac{2+\frac{b}{a}\left(x^5(\tau)\right)^p}{2\sqrt{\frac{c}{b}\left(x^5(\tau)\right)^{2p}+\frac{b}{a}\left(x^5(\tau)\right)^p+1}}\right)\end{array}\right)\Bigg]\Bigg\},$$

where

$$\widetilde{Q}(x^5(\tau);a,b,c,p)$$

$$\equiv q_{-\frac{1}{2p}-1}\left(\frac{b}{a}\right)^{((1/2p)+1)}\left(x^5(\tau)\right)^{((1/2)+p)}$$

$$+q_{-\frac{1}{2p}-2}\left(\frac{b}{a}\right)^{((1/2p)+2)}\left(x^5(\tau)\right)^{((1/2)+2p)}$$

$$+\sum_{n=2}^{-\frac{1}{2p}-1} q_{-\frac{1}{2p}-n-1}\left(\frac{b}{a}\right)^{\left(\frac{1}{2p}+n+1\right)}\left(x^5(\tau)\right)^{((1/2)+(n+1)p)};$$

$$q_{-\frac{1}{2p}-1} = -2p;$$

$$q_{-\frac{1}{2p}-2} = \frac{2p(1+p)}{(1+2p)};$$

$$q_{-\frac{1}{2p}-n-1} = \frac{\left(-\frac{1}{2p}-n+\frac{1}{2}\right)q_{-\frac{1}{2p}-n}+\frac{ac}{b^2}\left(-\frac{1}{2p}-n-1\right)q_{-\frac{1}{2p}-n+1}}{\frac{1}{2p}+n},$$

$$n = 2,\ldots,-\frac{1}{2p}-1;$$

$$\kappa = -\frac{1}{2}q_0 - \frac{ac}{b^2}q_1, \tag{C.77}$$

and $a$, $b$, $c$ are given by (C.70).

(2) $p = 0$:

$$x^5(\tau) = \frac{(\tau+A_1)^2\left[A_2 x_0^5 \pm \left(C_{32}^2 + C_{12}^2 + C_{22}^2 - C_{02}^2\right)\right]}{4x_0^5}. \tag{C.78}$$

## C.6 Class (VI)

$$\tilde{\mathbf{q}}_{\mathrm{VI}} = (q,0,0,0,q-2)$$

One has

$$F_{\pm,\mathrm{VI}}(\zeta;q,A_2)$$
$$= \left\{\pm[C_{12}^2 + C_{22}^2 + C_{32}^2 \pm A_2(x_0^5)^{2-q}]\zeta^{2-q} \mp (x_0^5)^q C_{02}^2\zeta^{2-2q}\right\}^{-1/2}. \tag{C.79}$$

The solution writes

$$x^0(\tau) = C_{01} + C_{02}\int \mathrm{d}\tau\left(x^5(\tau)\right)^{-q}; \tag{C.80}$$

$$x^i(\tau) = C_{i1} + C_{i2}\tau, \quad i = 1,2,3, \tag{C.81}$$

with $x^5(\tau)$ given by:

(1) $q \neq 0$:

(1.1) $C_{12}^2 + C_{22}^2 + C_{32}^2 \pm A_2(x_0^5)^{2-q} \neq 0,\ C_{02} \neq 0$:

$$x^5(\tau) = \left\{ -\frac{(x_0^5)^q C_{02}^2}{C_{12}^2 + C_{22}^2 + C_{32}^2 \pm A_2(x_0^5)^{2-q}} \times \left[ -\frac{q^2 \left(C_{12}^2 + C_{22}^2 + C_{32}^2 \pm A_2(x_0^5)^{2-q}\right)^2}{4 (x_0^5)^2 C_{02}^2} (\tau + A_1)^2 - 1 \right] \right\}^{1/q} ; \tag{C.82}$$

(1.2) $C_{12}^2 + C_{22}^2 + C_{32}^2 \pm A_2(x_0^5)^{2-q} = 0,\ C_{02} \neq 0$:

$$x^5(\tau) = \left[ \pm q \sqrt{\mp (x_0^5)^q C_{02}^2}\, (x_0^5)^{(q-2)/2} (\tau + A_1) \right]^{(1/q)} ; \tag{C.83}$$

(1.3) $C_{12}^2 + C_{22}^2 + C_{32}^2 \pm A_2(x_0^5)^{2-q} \neq 0,\ C_{02} = 0$:

$$x^5(\tau) = \left[ \pm q \frac{\sqrt{\pm (C_{12}^2 + C_{22}^2 + C_{32}^2 \pm A_2(x_0^5)^{2-q})}}{2} (x_0^5)^{(q-2)/2} (\tau + A_1) \right]^{(2/q)} ; \tag{C.84}$$

(2) $q = 0$:

$$x^5(\tau) = \exp\left[ \pm \sqrt{\pm \left(C_{12}^2 + C_{22}^2 + C_{32}^2 \pm A_2(x_0^5)^{2-q} - (x_0^5)^q C_{02}^2\right)} (x_0^5)^{(q-2)/2} (\tau + A_1) \right] . \tag{C.85}$$

## C.7 Class (VII)

$$\widetilde{\mathbf{q}}_{\mathrm{VII}} = (q, -q, -q, -q, -q - 2) .$$

One has

$$F_{\pm,\mathrm{VII}}(\zeta; q, A_2) = \left\{ \pm (x_0^5)^{-q} [C_{12}^2 + C_{22}^2 + C_{32}^2] \zeta^{2+2q} \mp (x_0^5)^q C_{02}^2 \zeta^2 + A_2 (x_0^5)^{2+q} \right\}^{-1/2} . \tag{C.86}$$

The solution reads:

$$x^0(\tau) = C_{01} + C_{02} \int d\tau \, (x^5(\tau))^{-q} ; \tag{C.87}$$

$$x^i(\tau) = C_{i1} + C_{i2} \int d\tau \, (x^5(\tau))^q , \quad i = 1, 2, 3 \tag{C.88}$$

with $x^5(\tau)$ given by

(1) $q \neq 0$:

(1.1) $C_{12}^2 + C_{22}^2 + C_{32}^2 = 0$:

(1.1.1) $C_{02} \neq 0,\ A_2 \neq 0$:

$$2 + 2\sqrt{\mp \frac{A_2 (x_0^5)^{2+q}}{C_{02}^2} \left(\frac{x^5(\tau)}{x_0^5}\right)^q + 1 +} \\ \mp \left\{1 + \exp\left[\mp (\tau + A_1) \frac{q C_{02} \sqrt{\pm 1}}{x_0^5}\right]\right\} \frac{A_2 (x_0^5)^{2+q}}{C_{02}^2} \left(\frac{x^5(\tau)}{x_0^5}\right)^q = 0; \tag{C.89}$$

or

$$x^5(\tau) = \sqrt[q]{\mp \frac{C_{02}^2}{A_2 (x_0^5)^2} \left\{ tgh^2 \left[\mp (\tau + A_1) \frac{q C_{02} \sqrt{\pm 1}}{x_0^5} - 1\right]\right\}}; \tag{C.90}$$

(1.1.2) $C_{02} = 0,\ A_2 \neq 0$:

$$x^5(\tau) = \left[\mp \frac{q}{2} \sqrt{A_2}\, (\tau + A_1)\right]^{-2/q}; \tag{C.91}$$

(1.1.3) $C_{02} \neq 0,\ A_2 = 0$:

$$x^5(\tau) = \exp\left[\pm \sqrt{\mp 1} \frac{C_{02}}{x_0^5} (\tau + A_1)\right]; \tag{C.92}$$

(1.2) $A_2 = 0$:

(1.2.1) $C_{02} \neq 0,\ C_{12}^2 + C_{22}^2 + C_{32}^2 \neq 0$:

$$x^5(\tau) = \left[\mp q \sqrt{\pm (C_{12}^2 + C_{22}^2 + C_{32}^2)} (x_0^5)^{-1-q} (\tau + A_1)\right]^{-1/q}; \tag{C.93}$$

(1.3) $C_{02} = 0$:

(1.3.1) $A_2 \neq 0,\ C_{12}^2 + C_{22}^2 + C_{32}^2 \neq 0$:

$$x^5(\tau) = \left[\frac{A_2 (x_0^5)^{2+2q}}{\frac{q^2 A_2^2}{4} (x_0^5)^{2+2q} (\tau + A_1)^2 \mp (C_{12}^2 + C_{22}^2 + C_{32}^2)}\right]^{1/q}; \tag{C.94}$$

(1.3.2) $A_2 \neq 0,\ C_{12}^2 + C_{22}^2 + C_{32}^2 = 0$:

$$x^5(\tau) = \left[\mp \frac{q}{2} \sqrt{A_2}\, (\tau + A_1)\right]^{-2/q}; \tag{C.95}$$

(1.3.3) $A_2 = 0,\ C_{12}^2 + C_{22}^2 + C_{32}^2 = 0$:

$$x^5(\tau) = \left[\mp q\sqrt{\pm\left(C_{12}^2 + C_{22}^2 + C_{32}^2\right)(x_0^5)^{-1-q}}\,(\tau + A_1)\right]^{-1/q}; \tag{C.96}$$

(1.4) $C_{02},\ A_2,\ C_{12}^2 + C_{22}^2 + C_{32}^2 \neq 0$:

$$0 = \tau + A_1 \mp (x_0^5)^{\frac{2+q}{2}} \left\{ -\frac{1}{q\sqrt{a}}\widehat{\Theta}\left(\frac{ac}{b^2} - \frac{1}{4}\right) \right.$$

$$\times \left[ \begin{array}{c} \operatorname{arcsinh}\dfrac{2 + \frac{b}{a}\left(x^5(\tau)\right)^q}{\frac{b}{a}\left(x^5(\tau)\right)^q\sqrt{4\frac{ac}{b^2} - 1}} = \\ = \ln\left|2\dfrac{2 + \frac{b}{a}\left(x^5(\tau)\right)^q}{\frac{b}{a}\left(x^5(\tau)\right)^q\sqrt{4\frac{ac}{b^2} - 1}}\right| \\ = -\ln\left|\dfrac{\frac{b}{a}\left(x^5(\tau)\right)^q\sqrt{4\frac{ac}{b^2} - 1}}{2 + \frac{b}{a}\left(x^5(\tau)\right)^q + 2\sqrt{\frac{c}{a}\left(x^5(\tau)\right)^{2q} + \frac{b}{a}\left(x^5(\tau)\right)^q + 1}}\right| \end{array} \right]$$

$$+\frac{1}{q\sqrt{a}}\delta\left(\frac{ac}{b^2} - \frac{1}{4}\right)$$

$$\times \left\{ \begin{array}{c} \ln\left|\dfrac{\frac{b}{a}\left(x^5(\tau)\right)^q}{2 + \frac{b}{a}\left(x^5(\tau)\right)^q}\right| \\ \text{or} \\ -\widehat{\Theta}\left(\frac{b}{a}\left(x^5(\tau)\right)^q\right)\widehat{\Theta}\left(-2 - \frac{b}{a}\left(x^5(\tau)\right)^q\right) \\ \times 2\operatorname{arccotgh}\left(\frac{b}{a}\left(x^5(\tau)\right)^q + 1\right) \end{array} \right.$$

$$-\frac{1}{q\sqrt{a}}\widehat{\Theta}\left(\frac{ac}{b^2} - \frac{1}{4}\right)$$

$$\times \ln\left|\frac{2 + \frac{b}{a}\left(x^5(\tau)\right)^q + 2\sqrt{\frac{c}{a}\left(x^5(\tau)\right)^{2q} + \frac{b}{a}\left(x^5(\tau)\right)^q + 1}}{\frac{b}{a}\left(x^5(\tau)\right)^q}\right|,$$

or

$$\left\{\begin{array}{c}\widehat{\Theta}\left(1-\frac{2+\frac{b}{a}\left(x^5(\tau)\right)^q}{2\sqrt{\frac{c}{a}(x^5(\tau))^{2q}+\frac{b}{a}(x^5(\tau))^q+1}}\right)\\ \times\widehat{\Theta}\left(1+\frac{2+\frac{b}{a}\left(x^5(\tau)\right)^q}{2\sqrt{\frac{c}{a}(x^5(\tau))^{2q}+\frac{b}{a}(x^5(\tau))^q+1}}\right)\\ \times\left[arctgh\left(\frac{2+\frac{b}{a}\left(x^5(\tau)\right)^q}{2\sqrt{\frac{c}{a}(x^5(\tau))^{2q}+\frac{b}{a}(x^5(\tau))^q+1}}\right)+\frac{1}{2}\ln\left|\frac{4ac}{b^2}-1\right|\right]\\ =\widehat{\Theta}\left(1-\frac{2+\frac{b}{a}\left(x^5(\tau)\right)^q}{2\sqrt{\frac{c}{a}(x^5(\tau))^{2q}+\frac{b}{a}(x^5(\tau))^q+1}}\right)\\ \times\widehat{\Theta}\left(1+\frac{2+\frac{b}{a}\left(x^5(\tau)\right)^q}{2\sqrt{\frac{c}{a}(x^5(\tau))^{2q}+\frac{b}{a}(x^5(\tau))^q+1}}\right)\\ \times\left[\frac{1}{2}\ln\left|\frac{2+\frac{b}{a}\left(x^5(\tau)\right)^q+2\sqrt{\frac{c}{a}(x^5(\tau))^{2q}+\frac{b}{a}(x^5(\tau))^q+1}}{-\left(2+\frac{b}{a}(x^5(\tau))^q\right)+2\sqrt{\frac{c}{a}(x^5(\tau))^{2q}+\frac{b}{a}(x^5(\tau))^q+1}}\right|+\frac{1}{2}\ln\left|\frac{4ac}{b^2}-1\right|\right]\end{array}\right\}, \tag{C.97}$$

where

$$\begin{aligned} a &\equiv \mp\left(x_0^5\right)^q C_{02}^2;\\ b &\equiv A_2\left(x_0^5\right)^{2+q};\\ c &= \pm\left(x_0^5\right)^{-q}\left(C_{12}^2+C_{22}^2+C_{32}^2\right); \end{aligned} \tag{C.98}$$

(2) $q=0$:

$$x^5(\tau)$$

$$\exp\left[\pm\sqrt{\pm(x_0^5)^{-q}\left(C_{12}^2+C_{22}^2+C_{32}^2\right)+A_2(x_0^5)^{2+q}\mp(x_0^5)^q C_{02}^2}\,(x_0^5)^{(2+q)/2}(\tau+A_1)\right]. \tag{C.99}$$

## C.8 Class (VIII)

$$\widetilde{\mathbf{q}}_{\mathrm{VIII}}=(0,0,0,0,r)\,.$$

One has

$$F_{\pm,\mathrm{VIII}}(\zeta;A_2)=\left\{\pm\left[C_{12}^2+C_{22}^2+C_{32}^2-C_{02}^2\pm A_2(x_0^5)^{-r}\right]\right\}^{-1/2}\zeta^{r/2}. \tag{C.100}$$

The solution writes

$$x^\mu(\tau)=C_{\mu 1}+C_{\mu 2}\left(\tau+\chi_\mu\right)=\widetilde{C}_{\mu 1}+C_{\mu 2}\tau\ ,\ \ \mu=0,1,2,3, \tag{C.101}$$

where the $\chi_\mu$'s are integration constant, and $\widetilde{C}_{\mu 1} = C_{\mu 1} + C_{\mu 2}\chi_\mu$, $x^5(\tau)$ being given by

$$x^5(\tau) = \left[\pm(r+2)\sqrt{\pm\left[C_{12}^2+C_{22}^2+C_{32}^2-C_{02}^2\pm A_2(x_0^5)^{-r}\right]}\left(x_0^5\right)^{r/2}(\tau+A_1)\right]^{2/(r+2)}. \quad \text{(C.102)}$$

## C.9 Class (IX)

It is specified by the set

$$\widetilde{\mathbf{q}}_{\mathrm{IX}} = (0,0,n,0,n-2)\,. \quad \text{(C.103)}$$

One gets

$$F_{\pm,\mathrm{IX}}(\zeta;n,A_2) = \left\{\pm\left[C_{12}^2+C_{32}^2-C_{02}^2\pm A_2(x_0^5)^{2-n}\right]\zeta^{2-n} \pm (x_0^5)^n C_{22}^2\zeta^{2-2n}\right\}^{-1/2}. \quad \text{(C.104)}$$

The solution writes

$$x^\mu(\tau) = C_{\mu 1} + C_{\mu 2}\tau\ , \quad \mu = 0,1,3, \quad \text{(C.105)}$$

$$x^2(\tau) = C_{21} + C_{22}\int d\tau\, \left(x^5(\tau)\right)^{-n} \quad \text{(C.106)}$$

with $x^5(\tau)$ given by

(1) $n \neq 0$:

(1.1) $C_{12}^2 + C_{32}^2 - C_{02}^2 \pm A_2(x_0^5)^{2-n} \neq 0$, $C_{22} \neq 0$:

$$x^5(\tau) = \left\{\frac{(x_0^5)^n C_{22}^2}{C_{12}^2+C_{32}^2-C_{02}^2\pm A_2(x_0^5)^{2-n}}\left[\frac{n^2\left(C_{12}^2+C_{32}^2-C_{02}^2\pm A_2(x_0^5)^{2-n}\right)}{4(x_0^5)^2C_{22}^2}(\tau+A_1)^2-1\right]\right\}^{1/n}; \quad \text{(C.107)}$$

(1.2) $C_{12}^2 + C_{32}^2 - C_{02}^2 \pm A_2(x_0^5)^{2-n} = 0$, $C_{22} \neq 0$:

$$x^5(\tau) = \left[\pm nC_{22}\sqrt{\pm 1}(x_0^5)^{n-1}(\tau+A_1)\right]^{1/n}; \quad \text{(C.108)}$$

(1.3) $C_{12}^2 + C_{32}^2 - C_{02}^2 \pm A_2(x_0^5)^{2-n} \neq 0$, $C_{22} = 0$:

$$x^5(\tau) = \left[\pm\frac{n}{2}\sqrt{\pm\left(C_{12}^2 + C_{32}^2 - C_{02}^2 \pm A_2(x_0^5)^{2-n}\right)}(x_0^5)^{(n-2)/2}(\tau + A_1)\right]^{2/n}; \tag{C.109}$$

(2) $n = 0$:

$$x^5(\tau) = \exp\left[\pm\sqrt{\pm\left(C_{12}^2 + C_{32}^2 - C_{02}^2 \pm A_2(x_0^5)^{2-n} + (x_0^5)^n C_{22}^2\right)}(x_0^5)^{\frac{n-2}{2}}(\tau + A_1)\right]. \tag{C.110}$$

## C.10 Class (X)

$$\widetilde{\mathbf{q}}_{\mathrm{X}} = \begin{pmatrix} q_0 = q, q_1 = -\frac{pq+np+nq}{n+p+q}, q_2 = n, q_3 = p, \\ r = \frac{(n+p+q)(n+p+q-2)-(pq+np+nq)}{n+p+q} = (n+p+q-2) + q_1 \end{pmatrix} (n+p+q \neq 0).$$

One gets

$$F_{\pm,\mathrm{X}}(\zeta; n, p, q, A_2) = \zeta^{r/2}\left\{\mp\left[(x_0^5)^{q_0}C_{02}^2 - (x_0^5)^{q_1}C_{12}^2\zeta^{-q_1} - (x_0^5)^{q_2}C_{22}^2 - (x_0^5)^{q_3}C_{32}^2\right] + A_2(x_0^5)^{-r}\right\}^{-1/2}. \tag{C.111}$$

The solution reads

$$x^\mu(\tau) = C_{\mu 1} + C_{\mu 2}\int d\tau \left(x^5(\tau)\right)^{-q_\mu}, \quad \mu = 0,1,2,3 \text{ (ESC off)} \tag{C.112}$$

with $x^5(\tau)$ given by

$$\tau + A_1 = \pm(x_0^5)^{-r}\frac{2(n+p+q)}{2p(n+p+q)+nq+n^2+q^2}\int \frac{dy}{\sqrt{\sum_{K=0,1,2,3,5} c_K y^{\alpha_K(n,p,q)}}}, \tag{C.113}$$

where $y = \left(x^5(\tau)\right)^{(2p^2+2pq+2np+nq+n^2+q^2)/2(n+p+q)}$ and

$$\begin{aligned} c_0 &\equiv \mp(x_0^5)^q C_{02}^2, \\ c_i &\equiv \pm(x_0^5)^{q_i} C_{i2}^2, \quad i = 1,2,3, \\ c_5 &\equiv A_1(x_0^5)^r, \end{aligned} \tag{C.114}$$

$$
\begin{aligned}
\alpha_0(n,p,q) &= \frac{2p^2+2np-2nq-2q^2}{2p\,(n+p+q)+nq+n^2+q^2},\\
\alpha_1(n,p,q) &= \frac{2p^2++4pq+4np+2nq}{2p\,(n+p+q)+nq+n^2+q^2},\\
\alpha_2(n,p,q) &= \frac{2p^2+2pq-2nq-2n^2}{2p\,(n+p+q)+nq+n^2+q^2},\\
\alpha_3 &= 0,\\
\alpha_5(n,p,q) &= \frac{2p^2+2np+2pq}{2p\,(n+p+q)+nq+n^2+q^2}.
\end{aligned}
\tag{C.115}
$$

The Riemann integral in (C.113) is unknown, and therefore not even an implicit solution can be obtained for $x^5(\tau)$.

## C.11 Class (XI)

$$
\widetilde{\mathbf{q}}_{\mathrm{XI}} = \left(\begin{array}{c} q_0 = q, q_1 = -\dfrac{n(2q+n)}{2n+q}, q_2 = n, q_3 = n, \\ r = \dfrac{3n^2-4n+2nq-2q+q^2}{2n+q} \end{array}\right) \quad (2n+q \neq 0).
$$

The generating function is

$$
F_{\pm,\mathrm{XI}}(\zeta;n,q,A_2) = \zeta^{r/2}\left\{\mp\left[(x_0^5)^q C_{02}^2-(x_0^5)^{q_1}C_{12}^2\zeta^{-q_1}-(x_0^5)^n\left(C_{22}^2+C_{32}^2\right)\zeta^{-n}\right]+A_2(x_0^5)^{-r}\right\}^{-1/2}.
\tag{C.116}
$$

The solution reads:

$$
x^\mu(\tau) = C_{\mu 1}+C_{\mu 2}\int \mathrm{d}\tau\,\left(x^5(\tau)\right)^{-q_\mu}, \quad \mu = 0,1,2,3 \text{ (ESC off)} \tag{C.117}
$$

with $x^5(\tau)$ given by

$$
0 = \tau + A_1 \mp (x_0^5)^{-r}\frac{2(2n+q)}{5n^2+3nq+q^2}\int \frac{\mathrm{d}y}{\sqrt{\sum\limits_{K=0,1,2,3,5} c_K y^{\alpha_K(n,q)}}}, \tag{C.118}
$$

where $y = \left(x^5(\tau)\right)^{(5n^2+3nq+q^2)(2n+q)}$, and

$$
\begin{aligned}
c_0 &\equiv \mp(x_0^5)^q C_{02}^2,\\
c_i &\equiv \pm(x_0^5)^{q_i} C_{i2}^2,\ i = 1,2,3,\\
c_5 &\equiv A_1(x_0^5)^{-r},
\end{aligned}
\tag{C.119}
$$

$$\begin{aligned} \alpha_0(n,q) &= \frac{4n^2-2nq-2q^2}{5n^2+3nq+q^2}, \\ \alpha_1(n,q) &= \frac{6n^2+6nq}{5n^2+3nq+q^2}, \\ \alpha_2(n,q) &= \alpha_3(n,q)=0, \\ \alpha_5 &= \frac{4n^2+2nq}{5n^2+3nq+q^2}. \end{aligned} \tag{C.120}$$

The Riemann integral in (C.118) is unknown, and therefore not even an implicit solution can be obtained for $x^5(\tau)$.

## C.12 Class (XII)

It is specified by the coefficient set $(n+p+q \neq 0)$:

$$\widetilde{\mathbf{q}}_{\mathrm{XI}} = \left( \begin{array}{c} q_0 = q, q_1 = n, q_2 = n, q_3 = -\dfrac{n(2q+n)}{2n+q}, \\ r = \dfrac{p^2+pq-2p+np-2n+nq+n^2-2q+q^2}{n+p+q} \end{array} \right). \tag{C.121}$$

One gets

$$F_{\pm,\mathrm{XII}}(\zeta;n,p,q,A_2)$$
$$= \zeta^{r/2}\left\{\mp\left[(x_0^5)^q C_{02}^2 - (x_0^5)^n (C_{12}^2+C_{22}^2)\zeta^n - (x_0^5)^{q_3} C_{32}^2\right] + A_2(x_0^5)^{-r}\right\}^{-(1/2)}. \tag{C.122}$$

The solution reads:

$$x^\mu(\tau) = C_{\mu 1} + C_{\mu 2}\int \mathrm{d}\tau \left(x^5(\tau)\right)^{-q_\mu}, \quad \mu = 0,1,2,3 \text{ (ESC off)} \tag{C.123}$$

with $x^5(\tau)$ given by

$$0 = \tau + A_1 \mp (x_0^5)^{-r}\frac{2(n+p+q)}{p^2+pq+2np+2nq+2n^2+q^2}\int \frac{\mathrm{d}y}{\sqrt{\sum\limits_{K=0,1,2,3} c_K y^{\alpha_K(n,p,q)}}}, \tag{C.124}$$

where $y = \left(x^5(\tau)\right)^{(p^2+pq+2np+2nq+2n^2+q^2)/2(n+p+q)}$, and

$$\begin{aligned} c_0 &\equiv \mp(x_0^5)^q C_{02}^2, \\ c_i &\equiv \pm(x_0^5)^{q_i} C_{i2}^2, \quad i=1,2,3, \\ c_5 &\equiv A_1(x_0^5)^{-r}, \end{aligned} \tag{C.125}$$

$$\begin{aligned}
\alpha_0(n,q) &= \frac{-2pq+2np+2n^2-2q^2}{p^2+pq+2np+2nq+2n^2+q^2}, \\
\alpha_1(n,q) &= \alpha_2(n,q)=0, \qquad\qquad\qquad\qquad\qquad\qquad \text{(C.126)} \\
\alpha_3(n,q) &= \frac{2np+6nq+4n^2}{p^2+pq+2np+2nq+2n^2+q^2}, \\
\alpha_5 &= \frac{2np+2nq+2n^2}{p^2+pq+2np+2nq+2n^2+q^2}. \qquad\qquad \text{(C.127)}
\end{aligned}$$

The Riemann integral in (C.124) is unknown, and therefore not even an implicit solution can be obtained for $x^5(\tau)$.

# References


[1] B. Finzi: "Relatività Generale e Teorie Unitarie," in *Cinquant'anni di Relatività* (in Italian), ed. M. Pantaleo (Giunti, Firenze, Italy, 1955), pg. 194.

[2] E. Recami and R. Mignani: *Riv. Nuovo Cimento* **4**, n.2 (1974), and references therein.

[3] R. Penrose: *The Emperor's New Mind* (Oxford University Press, 1989).

[4] J. Ehlers, F.A.E. Pirani and A. Schild: in *Papers in Honour of J. L. Synge*, edited by L. O'Raifeartaigh (Clarendon Press, Oxford 1972).

[5] J.S. Anandan: in *Potentiality, Entanglement and Passion-at-a-distance – Quantum Mechanical Studies for Abner Shimony*, vol. 2, edited by R.S. Cohen, M. Horne and J. Stachel (Kluwer, Dordrecht, Holland 1997), p. 31.

[6] F. Cardone and R. Mignani: *Energy and Geometry – An Introduction to Deformed Special Relativity* (World Scientific Series in Contemporary Chemical Physics, vol. 22) (World Scientific, Singapore, 2004); and references therein.

[7] For a review of Finsler geometry, see e.g., Z. Shen: *Lectures on Finsler Geometry* (World Scientific, Singapore, 2001).

[8] For a review, see G.Yu. Bogoslovsky: *Fortsch. Phys.* **42**, 2 (1994).

[9] For a review of Lie-isotopic theories, see R.M. Santilli: *Found.Phys.* **27**, 625 (1997).

[10] H. Nielsen and I. Picek: *Nucl.Phys.* **B 211**, 269 (1983).

[11] See e.g., A.U. Klimyk and K. Schmudgen: *Quantum Groups and Their Representations* (Texts and Monographs in Physics) (Springer-Verlag, New York, 1997).

[12] See e.g., R. Miron and M. Anastasiei: *The Geometry of Lagrange Spaces: Theory and Applications* (Kluwer, 1994); R. Miron, D. Hrimiuc, H. Shimada and S.V. Sabau: *The Geometry of Hamilton and Lagrange Spaces* (Kluwer, 2002); and references therein.

[13] R. Miron, R.K. Tavakol, V. Bălan, and I. Roxburgh: *Publ. Math. Debrecen* **42**, 215 (1993).

[14] See e.g., G. Amelino-Camelia: *Int. J. Mod. Phys.* **D 11**, 1643 (2002), and refs. therein

[15] S. Coleman and S.L. Glashow: *Phys. Lett.* **B 405**, 249 (1997).

[16] T. Levi-Civita: *The Absolute Differential Calculus* (Blackie and Son, 1954), p.403.

[17] H.A. Lorentz: in *Lectures on Theoretical Physics*, Vol. III (Macmillan & Co., London,1931). p. 208, and refs. therein.

[18] H. Poincaré: *La Science et l'Hypothese* (Flammarion, Paris, 1902).

[19] T. Van Flandern: *MetaRes.Bull.* **12**, 33 (2003).

[20] See M.C. Combourieu and J.P. Vigier: *Phys. Lett.* **A 175**, 269 (1993), and references therein.

[21] M. Consoli and E. Costanzo: *Phys. Lett.* A **333**, 355; *Nuovo Cim.* **119 B**, 393 (2004).

[22] See *e.g* F. Selleri: in *Redshift and Gravitation in a Relativistic Universe*, ed. K. Rudnicki, (Apeiron Press, Montreal, 2001), p.63, and refs. therein.

[23] F.R. Tangherlini: *Suppl. Nuovo Cimento* **20**,1 (1961).

[24] See e.g., E. Di Grezia, S. Esposito and G. Salesi: *Mod. Phys. Lett.* A **21**, 349 (2006), and refs. therein.

[25] See e.g., V.A. Rubakov: *Phys-Usp* **44**, 871 (2001), and refs. therein.

[26] T.E. Hartman: *J. Appl. Phys.* **33**, 3427 (1962).

[27] J.R. Fletcher: *J. Phys.* **C 18**, L55 (1985).

[28] G. Barton and K. Scharnhorst: *J. Phys.* **A 26**, 2037 (1993), and refs. therein.

[29] See H.E. Puthoff: *Found. Phys.* **32**, 927 (2002), and refs. therein.

[30] I.T. Grummond and S.J. Hathrell: *Phys. Rev.* **D 2**, 345 (1980).

[31] A. Enders and G. Nimtz: *J. Phys. I (France)* **2**, 1693 (1992).

[32] A. Enders and G. Nimtz: *J. Phys. I (France)* **3**, 1089 (1993).

[33] A. Enders and G. Nimtz: *Phys. Rev.* **E 48**, 632 (1993).

[34] A. Ranfagni, D. Mugnai, P. Fabeni and G.P. Pazzi: *Appl. Phys. Lett.* **58**, 774 (1991).

[35] A. Ranfagni, P. Fabeni, G.P. Pazzi and D. Mugnai: *Phys. Rev.* **E 48**, 1453 (1993).

[36] S.H. Aronson, G.J. Bock, H.-Y. Chang and E. Fishbach: *Phys. Rev. Lett.* **48**, 1306 (1982); *Phys. Rev.* **D 28**, 495 (1983).

[37] N. Grossman *et al.*: *Phys. Rev. Lett.* **59**, 18 (1987).

[38] UA1 Collaboration: *Phys.Lett.* **B 226**, 410 (1989).

[39] C.O. Alley: in *Quantum Optics, Experimental Gravity, and Measurement Theory*, P. Meystre and M.O. Scully eds. (Plenum Press, 1983), p.363.

[40] L. Kostro and B. Lange: *Phys. Essays* **10** (1999), and references therein.

[41] F. Cardone, A. Marrani and R. Mignani: *Found. Phys.* **34**, 617 (2004).

[42] F. Cardone, A. Marrani and R. Mignani: *Found. Phys.* **34**, 1155 (2004).

[43] F. Cardone, A. Marrani and R. Mignani: *Found. Phys.* **34**, 1407 (2004).

[44] R. Miron, A. Jannussis and G. Zet: in *Proc. Conf. Applied Differential Geometry – Gen. Rel. and The Workshop on Global Analysis, Differential Geometry and Lie Algebra, 2001.*, Gr. Tsagas ed. (Geometry Balkan Press, 2004), p.101, and refs. therein.

[45] J.L. Anderson: *Principles of Relativity Physics* (Academic Press, New York, 1967), pg.150.

[46] See e.g., N. Steenrod: *The Topology of Fibre Bundles* (Princeton University Press, 1951).

[47] See e.g., J.D. Bjorken and S.D. Drell: *Relativistic Quantum Fields* (McGraw-Hill, N.Y., 1965) Sect. 11.1.

[48] R. Mignani and E. Recami: *Int. J. Theor. Phys.* **12**, 299 (1975).

[49] J.D. Bjorken: *Annals of Phys.* **24**, 174 (1963).

[50] D.I. Blokhintsev: *Phys. Lett.* **12**, 272 (1964); *Sov. Phys. Uspekhi* **9**, 405 (1966).

[51] L.B. Redei: *Phys. Rev.* **145**, 999 (1966).

[52] P.R. Phillips: *Phys. Rev.* **139**, B491 (1965).

[53] S. Coleman and S.L. Glashow: "Evading the GZK cosmic-ray cut-off", preprint\medskip\ HUTP-98/A075 Harvard University (hep-ph/9808446 ), 1998.

[54] R. Jackiw: "Chern-Simons violation of Lorentz and PCT symmetries in electrodynamics" (hep-ph/9811322),1998, and references therein.

[55] See e.g., *CPT and Lorentz Symmetry I, II, III*, V.A. Kostelecky ed. (World Scientific, Singapore, 1999, 2002 ,2004).

[56] See C.M. Will: *Theory and Experiment in Gravitational Physics* (Cambridge University Press, rev.ed. 1993), and references therein.

[57] H.P. Robertson: *Rev. Mod. Phys.* **21**, 378 (1949).

[58] R.M. Mansouri and R.U. Sexl: *Gen. Rel. Grav.* **8**, 497; 515; 809 (1977).

[59] S. Buchman *et al.*: in *Proc. 52nd Frequency Control Symp., IEEE* (Washington, DC, 1998) p.534.

[60] C. Lämmerzahl *et al.*: *Classical Quantum Gravity* **18**, 2499 (2001).

[61] See R.Y. Chiao and A.M. Steinberg: "Tunneling Times and Superluminality" , E.Wolf ed., *Progress in Optics* **37**, 346 (Elsevier Science, NY, 1997), and references therein.

[62] See G. Nimtz and W. Heitmann: *Progr. Quantum Electr.* **21**, 81 (1997), and references therein.

[63] F. Cardone, R. Mignani, W. Perconti and R. Scrimaglio: *Phys. Lett.* **A 326**, 1 (2004).

[64] F. Cardone, R. Mignani, W. Perconti, A. Petrucci and R. Scrimaglio: *Int. J. Modern Phys.* **B 20**, 85 (2006).

[65] F. Cardone, R. Mignani, W. Perconti, A. Petrucci and R. Scrimaglio: *Int. J. Modern Phys.* **B 20**, 1107 (2006).

[66] A. Petrucci: "La correlazione relativistica in Meccanica Quantistica" (in Italian), Graduate Thesis in Physics (R. Mignani and R. Scrimaglio supervisors), Dipartimento di Fisica "E.Amaldi", University "Roma Tre" (2006).

[67] A. Ranfagni, D. Mugnai and R. Ruggeri: *Phys. Rev.* E **69**, 027601 (2004).

[68] A. Ranfagni and D. Mugnai: *Phys. Lett.* **A 322**, 146 (2004).

[69] D. Mugnai, A. Ranfagni, E. Allaria, R. Meucci and C. Ranfagni: *Modern Phys. Lett.* B**19, 1** (2005).

[70] L. de Broglie: *La réinterpretation de la Mécanique Ondulatoire* (Gauthier-Villars, Paris, 1971), and references therein.

[71] D. Bohm: *Causality and Chance in Modern Physics (with a foreword by L. de Broglie)* (London, 1957), and references therein.

[72] J.S. Bell: *Speakable and Unspeakable in Quantum Mechanics* (Cambridge University Press, Cambridge, 1987); and references therein.

[73] L. de Broglie and J. Andrade y Silva: *Phys. Rev.* **172**, 1284 (1968).

[74] P.S. Wesson: *Gen. Rel. Grav.* **38**, 937 (2006).

[75] U. Bartocci, F. Cardone and R. Mignani: *Found. Phys. Lett.* **14**, 51 (2001).

[76] F. Cardone, R. Mignani and R. Scrimaglio: *Found. Phys.* **36**, 263 (2006).

[77] M.A. Heald: *Amer. J. Phys.* **52**, 522 (1984).

[78] U. Bartocci and M. Mamone Capria: *Found. Phys.* **21**, 787 (1991).

[79] A.K.T. Assis, W.A. Rodrigues Jr. and A. Mania: *Found. Phys.* **29**, 729 (1999).

[80] O.D. Jefimenko: *Amer. J. Phys.* **30**, 19 (1962).

[81] R. Sansbury: *Rev. Sci. Instrum.* **56**, 415 (1985).

[82] W.F. Edwards, C.S. Kenyon and D.K. Lemon: *Phys. Rev.* **D 14**, 922 (1976).

[83] S. Maglic and D.F Bartlett: *Rev. Sci. Instrum.* **61**, 2637 (1990).

[84] D.K. Lemon, W.F. Edwards and C.S. Kenyon: *Phys. Lett.* **A 162**, 105 (1992).

[85] J.H. Taylor: *Rev. Mod. Phys.* **66**, 711 (1994); and refs. therein.

[86] R.A. Hulse: *Rev. Mod. Phys.* **66**, 699 (1994); and refs. therein.

[87] T. Van Flandern and J-P. Vigier: *Found. Phys.* **32**, 1031 (2002).

[88] T. Van Flandern: *Phys. Lett.* **A 250**,1 (1998).

[89] W.D. Walker and J. Dual: in *Gravitational Waves – Proc. of the Second Edoardo Amaldi Conference,* E. Coccia, G. Pizzella and G. Veneziano eds. (World Scientific, Singapore, 1997).

[90] S.M. Kopeikin: *Astrophys. J.* **556,** L1 (2001).

[91] H. Asada: *Astrophys. J.* **574**, L69 (2002).

[92] C.M. Will: *Astrophys. J.* **590**, 683(2003).

[93] L. Crum and D.F. Gaitan: *Frontiers of Nonlinear Acoustics*, 12th Int. Symp. On Nonlinear Acoustics (Elsevier Applied Science, New York, 1990).

[94] C.E. Brennen: *Cavitation and bubble dynamics* (Oxford University Press, 1995).

[95] H. Metcalf: *Science* **279**, 1322 (1998).

[96] F. Cardone and R. Mignani: in *Proc. Int. Conf. On Cosmoparticle Physics "Cosmion 2001"* (Moscow, Russia, May 2001), *Grav. Cosmol. Suppl.* **8**, 216 (2002).

[97] F. Cardone and R. Mignani: *Int. J. Mod. Phys.* **B 17**, 307 (2003).

[98] F. Cardone, R. Mignani, W. Perconti, E. Pessa and G. Spera: *J. Radioanal. Nucl. Chem.* **265**, 151 (2005).

[99] L.I. Urutskoev, V.I. Liksonov and V.G. Tsinoev: *Appl. Phys. (Russia)* **4**, 83 (2000); *Ann. Fond. L.de Broglie.* **27**, 701 (2002).

[100] V.D. Kuznetsov, G.V. Myshinskii, V.I. Zhemennik and V.I. Arbuzov: "Test Experiments on Observation of Cold Transmutation of Elements", *Proc. 8-th Russ. Conf. on the Cold Transmutation of Nuclei of Chemical Elements* (Moscow, 2001), p.308.

[101] A.G. Volkovich, A.P. Govorun, A.A. Gulyaev, S.V. Zhukov, V.L. Kuznetsov, A.A. Rukhadze, A.V. Steblevskii and L.I. Urutskoev: *Bull. Lebedev Physics Inst.*, 2002 (8).

[102] For a review, see L.I. Urutskoev: *Ann. Fond. L. de Broglie* **29**, 1149 (2004).

[103] R.P. Taleyarkhan *et al.*: *Science* **295**, 1868 (2002); *Phys. Rev. E* **69**, 036109 (2004); *Phys. Rev. Lett.* **96**, 034301 (2006).

[104] D. Shapira and M. Saltmarsh: *Phys. Rev. Lett.* **89**, 104302 (2002).

[105] R.P. Taleyarkhan *et al.*: "Comments on the Shapira and Saltmarsh report", Oak Ridge National Laboratory Internal Correspondence Report, March 2, 2002.

[106] B.P. Barber, C.C. Wu, R. Löfstedt, P.H. Roberts and S.J. Putterman: *Phys. Rev. Lett.* **72**, 1380 (1994).

[107] W.C. Moss, D.B. Clark, J.W. White and D.A. Young: *Phys. Fluids* **6**, 2979 (1994).

[108] T. Horigouchi, H. Koura and J. Katakura, eds.: *Chart of Nuclides* 2000 (Jap. Nucl. Data Com., JNDC); Nucl. Data Evaluation Lab. Korea Atomic Energy Res. Inst. 2000–2002 BNL, USA version.

[109] F. Cardone and R. Mignani: *Int. J. Modern Phys. E* **15**, 911 (2006).

[110] See e.g., *Fusion Dynamics at the Extremes*, Yu. Ts. Oganessian and V.I. Zagrebaev eds. (World Scientific, Singapore, 2000), and references therein.

[111] Th. Kaluza: *Preuss. Akad. Wiss. Phys. Math.* **K1**, 966 (1921).

[112] O. Klein: *Zeitschr. Phys.* **37**, 875 (1926).

[113] G. Nordström: *Zeitschr. Phys.* **15,** 504 (1914).

[114] P. Jordan: *Zeitschr. Phys.* **157**, 112 (1959), and references quoted therein.

[115] Y.R. Thiry: *C.R. Acad. Sci. (Paris)* **226**, 216 (1948).

[116] For the early attempts to geometrical unified theories based on higher dimensions, see e.g., M.A. Tonnelat: *Les Théories Unitaires de l'Electromagnetisme et de la Gravitation* (Gautier-Villars, Paris, 1965), and references quoted therein.

[117] R.L. Ingraham: *Nuovo Cim.* **9**, 87 (1952).

[118] J. Podolanski: *Proc. Roy. Soc.* **201**, 234 (1950).

[119] See e.g., J. Polchinski: *String Theory, Vol. 1: An Introduction to the Bosonic String*; *Vol. 2: Superstring Theory and Beyond* (Cambridge University Press, Cambridge, UK, 1998); and references therein.

[120] See e.g., S. Weinberg: *The Quantum Theory of Fields, Vol. 3: Supersymmetry* (Cambridge University Press, Cambridge, UK, 1999), and references therein.

[121] See e.g., M.B. Green, J.H. Schwarz and E. Witten: *Superstring Theory, vol.1: Introduction*; *vol.2: Loop Amplitudes, Anomalies and Phenomenology* (Cambridge University Press, Cambridge, UK, 1988); and references therein.

[122] See e.g., T. Appelquist, A. Chodos and P.G.O. Freund (editors): *Modern Kaluza–Klein Theories* (Addison-Wesley, Menlo Park, California, 1987), and references therein.

[123] P.S. Wesson: *Space–Time–Matter – Modern Kaluza-Klein Theory* (World Scientific, Singapore, 1999), and references therein.

[124] See e.g., *Supergravity in Diverse Dimensions*, A. Salam and E. Sezgin eds. (North-Holland & World Scientific, 1989), and references quoted therein.

[125] J.M. Overduin and P.S. Wesson: *Phys. Rept.* **283**, 303 (1997), and references quoted therein.

[126] T. Fukui : *Gen. Rel. Grav.* **20**, 1037 (1988); *ibidem*, **24**, 389 (1992).

[127] N. Arkani-Hamed, S. Dimopoulos and G. Dvali: *Phys. Lett.* B **429,** 263 (1998).

[128] L. Randall and R. Sundrum: *Phys. Rev. Lett.* **83**, 3370, 4690 (1999).

[129] See e.g., K. Cheung, J. F Gunion and S. Mrenna (editors): *Particles, Strings and Cosmology (PASCOS 99)* (Proc. 7th Intern. Symposium, Lake Tahoe, California, 10–16 December 1999) (World Sci., Singapore, 2000).

[130] F. Cardone, M. Francaviglia, and R. Mignani: *Gen. Rel. Grav.* **30**, 1619 (1998); *ibidem*, **31**, 1049 (1999).

[131] F. Cardone, M. Francaviglia, and R. Mignani: *Found. Phys. Lett.* **12**, 281, 347 (1999).

[132] T.D. Lee: *Phys. Lett.* B **122,** 217 (1982); *J. Statis. Phys.* **46**, 843 (1987).

[133] P.A.M. Dirac: *Proc. R. Soc. (London)* **A333**, 403 (1973); *ibidem*, **A 338**, 439 (1974).

[134] See e.g., F. Hoyle and J.V. Narlikar: *Action at a Distance in Physics and Cosmology* (Freeman, NY, 1974).

[135] V. Canuto, P.J. Adams, S.-H. Hsieh and E. Tsiang: *Phys. Rev.* **D 16**, 1643 (1977).

[136] A. Marrani: "Aspetti matematici della Relatività Deformata in cinque dimensioni" (in Italian), Ph.D. Thesis in Physics (R. Mignani supervisor), Dipartimento di Fisica "E.Amaldi", University "Roma Tre" (2003).

[137] P.S. Wesson and J. Ponce de Leon: *Astron. Astrophys.* **294**, 1 (1995).

[138] P.S. Wesson: *Mod. Phys. Lett.* **A 10**, 15 (1995).

[139] D. Youm: *Phys. Rev. D* **62**, 084002 (2000).

[140] P.S. Wesson: *Gen. Rel. Grav.* **36**, 451 (2004).

[141] See Marc Hindry, M. Hindry and J.H. Silverman: *Diophantine Geometry* (Springer-Verlag, Heidelberg, 2000), and refs. therein.

[142] See e.g., G.L. Naber: *Topology, Geometry, and Gauge Fields: Interactions* (Springer-Verlag, Heidelberg, 2000), and refs. therein.

# Index

## Fundamental Theories of Physics

*Series Editor: Alwyn van der Merwe, University of Denver, USA*

1. M. Sachs: *General Relativity and Matter.* A Spinor Field Theory from Fermis to Light-Years. With a Foreword by C. Kilmister. 1982 ISBN 90-277-1381-2
2. G.H. Duffey: *A Development of Quantum Mechanics.* Based on Symmetry Considerations. 1985 ISBN 90-277-1587-4
3. S. Diner, D. Fargue, G. Lochak and F. Selleri (eds.): *The Wave-Particle Dualism.* A Tribute to Louis de Broglie on his 90th Birthday. 1984 ISBN 90-277-1664-1
4. E. Prugovečki: *Stochastic Quantum Mechanics and Quantum Spacetime.* A Consistent Unification of Relativity and Quantum Theory based on Stochastic Spaces. 1984; 2nd printing 1986 ISBN 90-277-1617-X
5. D. Hestenes and G. Sobczyk: *Clifford Algebra to Geometric Calculus.* A Unified Language for Mathematics and Physics. 1984 ISBN 90-277-1673-0; Pb (1987) 90-277-2561-6
6. P. Exner: *Open Quantum Systems and Feynman Integrals.* 1985 ISBN 90-277-1678-1
7. L. Mayants: *The Enigma of Probability and Physics.* 1984 ISBN 90-277-1674-9
8. E. Tocaci: *Relativistic Mechanics, Time and Inertia.* Translated from Romanian. Edited and with a Foreword by C.W. Kilmister. 1985 ISBN 90-277-1769-9
9. B. Bertotti, F. de Felice and A. Pascolini (eds.): *General Relativity and Gravitation.* Proceedings of the 10th International Conference (Padova, Italy, 1983). 1984 ISBN 90-277-1819-9
10. G. Tarozzi and A. van der Merwe (eds.): *Open Questions in Quantum Physics.* 1985 ISBN 90-277-1853-9
11. J.V. Narlikar and T. Padmanabhan: *Gravity, Gauge Theories and Quantum Cosmology.* 1986 ISBN 90-277-1948-9
12. G.S. Asanov: *Finsler Geometry, Relativity and Gauge Theories.* 1985 ISBN 90-277-1960-8
13. K. Namsrai: *Nonlocal Quantum Field Theory and Stochastic Quantum Mechanics.* 1986 ISBN 90-277-2001-0
14. C. Ray Smith and W.T. Grandy, Jr. (eds.): *Maximum-Entropy and Bayesian Methods in Inverse Problems.* Proceedings of the 1st and 2nd International Workshop (Laramie, Wyoming, USA). 1985 ISBN 90-277-2074-6
15. D. Hestenes: *New Foundations for Classical Mechanics.* 1986 ISBN 90-277-2090-8; Pb (1987) 90-277-2526-8
16. S.J. Prokhovnik: *Light in Einstein's Universe.* The Role of Energy in Cosmology and Relativity. 1985 ISBN 90-277-2093-2
17. Y.S. Kim and M.E. Noz: *Theory and Applications of the Poincaré Group.* 1986 ISBN 90-277-2141-6
18. M. Sachs: *Quantum Mechanics from General Relativity.* An Approximation for a Theory of Inertia. 1986 ISBN 90-277-2247-1
19. W.T. Grandy, Jr.: *Foundations of Statistical Mechanics.* Vol. I: *Equilibrium Theory.* 1987 ISBN 90-277-2489-X
20. H.-H von Borzeszkowski and H.-J. Treder: *The Meaning of Quantum Gravity.* 1988 ISBN 90-277-2518-7
21. C. Ray Smith and G.J. Erickson (eds.): *Maximum-Entropy and Bayesian Spectral Analysis and Estimation Problems.* Proceedings of the 3rd International Workshop (Laramie, Wyoming, USA, 1983). 1987 ISBN 90-277-2579-9
22. A.O. Barut and A. van der Merwe (eds.): *Selected Scientific Papers of Alfred Landé.* [*1888-1975*]. 1988 ISBN 90-277-2594-2

# Fundamental Theories of Physics

23. W.T. Grandy, Jr.: *Foundations of Statistical Mechanics.* Vol. II: *Nonequilibrium Phenomena.* 1988 ISBN 90-277-2649-3
24. E.I. Bitsakis and C.A. Nicolaides (eds.): *The Concept of Probability.* Proceedings of the Delphi Conference (Delphi, Greece, 1987). 1989 ISBN 90-277-2679-5
25. A. van der Merwe, F. Selleri and G. Tarozzi (eds.): *Microphysical Reality and Quantum Formalism, Vol. 1.* Proceedings of the International Conference (Urbino, Italy, 1985). 1988 ISBN 90-277-2683-3
26. A. van der Merwe, F. Selleri and G. Tarozzi (eds.): *Microphysical Reality and Quantum Formalism, Vol. 2.* Proceedings of the International Conference (Urbino, Italy, 1985). 1988 ISBN 90-277-2684-1
27. I.D. Novikov and V.P. Frolov: *Physics of Black Holes.* 1989 ISBN 90-277-2685-X
28. G. Tarozzi and A. van der Merwe (eds.): *The Nature of Quantum Paradoxes.* Italian Studies in the Foundations and Philosophy of Modern Physics. 1988 ISBN 90-277-2703-1
29. B.R. Iyer, N. Mukunda and C.V. Vishveshwara (eds.): *Gravitation, Gauge Theories and the Early Universe.* 1989 ISBN 90-277-2710-4
30. H. Mark and L. Wood (eds.): *Energy in Physics, War and Peace.* A Festschrift celebrating Edward Teller's 80th Birthday. 1988 ISBN 90-277-2775-9
31. G.J. Erickson and C.R. Smith (eds.): *Maximum-Entropy and Bayesian Methods in Science and Engineering.* Vol. I: *Foundations.* 1988 ISBN 90-277-2793-7
32. G.J. Erickson and C.R. Smith (eds.): *Maximum-Entropy and Bayesian Methods in Science and Engineering.* Vol. II: *Applications.* 1988 ISBN 90-277-2794-5
33. M.E. Noz and Y.S. Kim (eds.): *Special Relativity and Quantum Theory.* A Collection of Papers on the Poincaré Group. 1988 ISBN 90-277-2799-6
34. I.Yu. Kobzarev and Yu.I. Manin: *Elementary Particles. Mathematics, Physics and Philosophy.* 1989 ISBN 0-7923-0098-X
35. F. Selleri: *Quantum Paradoxes and Physical Reality.* 1990 ISBN 0-7923-0253-2
36. J. Skilling (ed.): *Maximum-Entropy and Bayesian Methods.* Proceedings of the 8th International Workshop (Cambridge, UK, 1988). 1989 ISBN 0-7923-0224-9
37. M. Kafatos (ed.): *Bell's Theorem, Quantum Theory and Conceptions of the Universe.* 1989 ISBN 0-7923-0496-9
38. Yu.A. Izyumov and V.N. Syromyatnikov: *Phase Transitions and Crystal Symmetry.* 1990 ISBN 0-7923-0542-6
39. P.F. Fougère (ed.): *Maximum-Entropy and Bayesian Methods.* Proceedings of the 9th International Workshop (Dartmouth, Massachusetts, USA, 1989). 1990 ISBN 0-7923-0928-6
40. L. de Broglie: *Heisenberg's Uncertainties and the Probabilistic Interpretation of Wave Mechanics.* With Critical Notes of the Author. 1990 ISBN 0-7923-0929-4
41. W.T. Grandy, Jr.: *Relativistic Quantum Mechanics of Leptons and Fields.* 1991 ISBN 0-7923-1049-7
42. Yu.L. Klimontovich: *Turbulent Motion and the Structure of Chaos.* A New Approach to the Statistical Theory of Open Systems. 1991 ISBN 0-7923-1114-0
43. W.T. Grandy, Jr. and L.H. Schick (eds.): *Maximum-Entropy and Bayesian Methods.* Proceedings of the 10th International Workshop (Laramie, Wyoming, USA, 1990). 1991 ISBN 0-7923-1140-X
44. P. Pták and S. Pulmannová: *Orthomodular Structures as Quantum Logics.* Intrinsic Properties, State Space and Probabilistic Topics. 1991 ISBN 0-7923-1207-4
45. D. Hestenes and A. Weingartshofer (eds.): *The Electron.* New Theory and Experiment. 1991 ISBN 0-7923-1356-9

## Fundamental Theories of Physics

46. P.P.J.M. Schram: *Kinetic Theory of Gases and Plasmas.* 1991 ISBN 0-7923-1392-5
47. A. Micali, R. Boudet and J. Helmstetter (eds.): *Clifford Algebras and their Applications in Mathematical Physics.* 1992 ISBN 0-7923-1623-1
48. E. Prugovečki: *Quantum Geometry.* A Framework for Quantum General Relativity. 1992 ISBN 0-7923-1640-1
49. M.H. Mac Gregor: *The Enigmatic Electron.* 1992 ISBN 0-7923-1982-6
50. C.R. Smith, G.J. Erickson and P.O. Neudorfer (eds.): *Maximum Entropy and Bayesian Methods.* Proceedings of the 11th International Workshop (Seattle, 1991). 1993 ISBN 0-7923-2031-X
51. D.J. Hoekzema: *The Quantum Labyrinth.* 1993 ISBN 0-7923-2066-2
52. Z. Oziewicz, B. Jancewicz and A. Borowiec (eds.): *Spinors, Twistors, Clifford Algebras and Quantum Deformations.* Proceedings of the Second Max Born Symposium (Wrocław, Poland, 1992). 1993 ISBN 0-7923-2251-7
53. A. Mohammad-Djafari and G. Demoment (eds.): *Maximum Entropy and Bayesian Methods.* Proceedings of the 12th International Workshop (Paris, France, 1992). 1993 ISBN 0-7923-2280-0
54. M. Riesz: *Clifford Numbers and Spinors* with Riesz' Private Lectures to E. Folke Bolinder and a Historical Review by Pertti Lounesto. E.F. Bolinder and P. Lounesto (eds.). 1993 ISBN 0-7923-2299-1
55. F. Brackx, R. Delanghe and H. Serras (eds.): *Clifford Algebras and their Applications in Mathematical Physics.* Proceedings of the Third Conference (Deinze, 1993) 1993 ISBN 0-7923-2347-5
56. J.R. Fanchi: *Parametrized Relativistic Quantum Theory.* 1993 ISBN 0-7923-2376-9
57. A. Peres: *Quantum Theory: Concepts and Methods.* 1993 ISBN 0-7923-2549-4
58. P.L. Antonelli, R.S. Ingarden and M. Matsumoto: *The Theory of Sprays and Finsler Spaces with Applications in Physics and Biology.* 1993 ISBN 0-7923-2577-X
59. R. Miron and M. Anastasiei: *The Geometry of Lagrange Spaces: Theory and Applications.* 1994 ISBN 0-7923-2591-5
60. G. Adomian: *Solving Frontier Problems of Physics: The Decomposition Method.* 1994 ISBN 0-7923-2644-X
61. B.S. Kerner and V.V. Osipov: *Autosolitons.* A New Approach to Problems of Self-Organization and Turbulence. 1994 ISBN 0-7923-2816-7
62. G.R. Heidbreder (ed.): *Maximum Entropy and Bayesian Methods.* Proceedings of the 13th International Workshop (Santa Barbara, USA, 1993) 1996 ISBN 0-7923-2851-5
63. J. Peřina, Z. Hradil and B. Jurčo: *Quantum Optics and Fundamentals of Physics.* 1994 ISBN 0-7923-3000-5
64. M. Evans and J.-P. Vigier: *The Enigmatic Photon.* Volume 1: The Field $\boldsymbol{B}^{(3)}$. 1994 ISBN 0-7923-3049-8
65. C.K. Raju: *Time: Towards a Constistent Theory.* 1994 ISBN 0-7923-3103-6
66. A.K.T. Assis: *Weber's Electrodynamics.* 1994 ISBN 0-7923-3137-0
67. Yu. L. Klimontovich: *Statistical Theory of Open Systems.* Volume 1: A Unified Approach to Kinetic Description of Processes in Active Systems. 1995 ISBN 0-7923-3199-0; Pb: ISBN 0-7923-3242-3
68. M. Evans and J.-P. Vigier: *The Enigmatic Photon.* Volume 2: Non-Abelian Electrodynamics. 1995 ISBN 0-7923-3288-1
69. G. Esposito: *Complex General Relativity.* 1995 ISBN 0-7923-3340-3

# Fundamental Theories of Physics

70. J. Skilling and S. Sibisi (eds.): *Maximum Entropy and Bayesian Methods.* Proceedings of the Fourteenth International Workshop on Maximum Entropy and Bayesian Methods. 1996 ISBN 0-7923-3452-3
71. C. Garola and A. Rossi (eds.): *The Foundations of Quantum Mechanics Historical Analysis and Open Questions.* 1995 ISBN 0-7923-3480-9
72. A. Peres: *Quantum Theory: Concepts and Methods.* 1995 (see for hardback edition, Vol. 57) ISBN Pb 0-7923-3632-1
73. M. Ferrero and A. van der Merwe (eds.): *Fundamental Problems in Quantum Physics.* 1995 ISBN 0-7923-3670-4
74. F.E. Schroeck, Jr.: *Quantum Mechanics on Phase Space.* 1996 ISBN 0-7923-3794-8
75. L. de la Peña and A.M. Cetto: *The Quantum Dice.* An Introduction to Stochastic Electrodynamics. 1996 ISBN 0-7923-3818-9
76. P.L. Antonelli and R. Miron (eds.): *Lagrange and Finsler Geometry.* Applications to Physics and Biology. 1996 ISBN 0-7923-3873-1
77. M.W. Evans, J.-P. Vigier, S. Roy and S. Jeffers: *The Enigmatic Photon.* Volume 3: Theory and Practice of the $\boldsymbol{B}^{(3)}$ Field. 1996 ISBN 0-7923-4044-2
78. W.G.V. Rosser: *Interpretation of Classical Electromagnetism.* 1996 ISBN 0-7923-4187-2
79. K.M. Hanson and R.N. Silver (eds.): *Maximum Entropy and Bayesian Methods.* 1996 ISBN 0-7923-4311-5
80. S. Jeffers, S. Roy, J.-P. Vigier and G. Hunter (eds.): *The Present Status of the Quantum Theory of Light.* Proceedings of a Symposium in Honour of Jean-Pierre Vigier. 1997 ISBN 0-7923-4337-9
81. M. Ferrero and A. van der Merwe (eds.): *New Developments on Fundamental Problems in Quantum Physics.* 1997 ISBN 0-7923-4374-3
82. R. Miron: *The Geometry of Higher-Order Lagrange Spaces.* Applications to Mechanics and Physics. 1997 ISBN 0-7923-4393-X
83. T. Hakioğlu and A.S. Shumovsky (eds.): *Quantum Optics and the Spectroscopy of Solids.* Concepts and Advances. 1997 ISBN 0-7923-4414-6
84. A. Sitenko and V. Tartakovskii: *Theory of Nucleus.* Nuclear Structure and Nuclear Interaction. 1997 ISBN 0-7923-4423-5
85. G. Esposito, A.Yu. Kamenshchik and G. Pollifrone: *Euclidean Quantum Gravity on Manifolds with Boundary.* 1997 ISBN 0-7923-4472-3
86. R.S. Ingarden, A. Kossakowski and M. Ohya: *Information Dynamics and Open Systems.* Classical and Quantum Approach. 1997 ISBN 0-7923-4473-1
87. K. Nakamura: *Quantum versus Chaos.* Questions Emerging from Mesoscopic Cosmos. 1997 ISBN 0-7923-4557-6
88. B.R. Iyer and C.V. Vishveshwara (eds.): *Geometry, Fields and Cosmology.* Techniques and Applications. 1997 ISBN 0-7923-4725-0
89. G.A. Martynov: *Classical Statistical Mechanics.* 1997 ISBN 0-7923-4774-9
90. M.W. Evans, J.-P. Vigier, S. Roy and G. Hunter (eds.): *The Enigmatic Photon.* Volume 4: New Directions. 1998 ISBN 0-7923-4826-5
91. M. Rédei: *Quantum Logic in Algebraic Approach.* 1998 ISBN 0-7923-4903-2
92. S. Roy: *Statistical Geometry and Applications to Microphysics and Cosmology.* 1998 ISBN 0-7923-4907-5
93. B.C. Eu: *Nonequilibrium Statistical Mechanics.* Ensembled Method. 1998 ISBN 0-7923-4980-6

# Fundamental Theories of Physics

94. V. Dietrich, K. Habetha and G. Jank (eds.): *Clifford Algebras and Their Application in Mathematical Physics.* Aachen 1996. 1998 ISBN 0-7923-5037-5
95. J.P. Blaizot, X. Campi and M. Ploszajczak (eds.): *Nuclear Matter in Different Phases and Transitions.* 1999 ISBN 0-7923-5660-8
96. V.P. Frolov and I.D. Novikov: *Black Hole Physics.* Basic Concepts and New Developments. 1998 ISBN 0-7923-5145-2; Pb 0-7923-5146
97. G. Hunter, S. Jeffers and J-P. Vigier (eds.): *Causality and Locality in Modern Physics.* 1998 ISBN 0-7923-5227-0
98. G.J. Erickson, J.T. Rychert and C.R. Smith (eds.): *Maximum Entropy and Bayesian Methods.* 1998 ISBN 0-7923-5047-2
99. D. Hestenes: *New Foundations for Classical Mechanics (Second Edition).* 1999 ISBN 0-7923-5302-1; Pb ISBN 0-7923-5514-8
100. B.R. Iyer and B. Bhawal (eds.): *Black Holes, Gravitational Radiation and the Universe.* Essays in Honor of C. V. Vishveshwara. 1999 ISBN 0-7923-5308-0
101. P.L. Antonelli and T.J. Zastawniak: *Fundamentals of Finslerian Diffusion with Applications.* 1998 ISBN 0-7923-5511-3
102. H. Atmanspacher, A. Amann and U. Müller-Herold: *On Quanta, Mind and Matter Hans Primas in Context.* 1999 ISBN 0-7923-5696-9
103. M.A. Trump and W.C. Schieve: *Classical Relativistic Many-Body Dynamics.* 1999 ISBN 0-7923-5737-X
104. A.I. Maimistov and A.M. Basharov: *Nonlinear Optical Waves.* 1999 ISBN 0-7923-5752-3
105. W. von der Linden, V. Dose, R. Fischer and R. Preuss (eds.): *Maximum Entropy and Bayesian Methods Garching, Germany 1998.* 1999 ISBN 0-7923-5766-3
106. M.W. Evans: *The Enigmatic Photon Volume 5: O(3) Electrodynamics.* 1999 ISBN 0-7923-5792-2
107. G.N. Afanasiev: *Topological Effects in Quantum Mecvhanics.* 1999 ISBN 0-7923-5800-7
108. V. Devanathan: *Angular Momentum Techniques in Quantum Mechanics.* 1999 ISBN 0-7923-5866-X
109. P.L. Antonelli (ed.): *Finslerian Geometries A Meeting of Minds.* 1999 ISBN 0-7923-6115-6
110. M.B. Mensky: *Quantum Measurements and Decoherence Models and Phenomenology.* 2000 ISBN 0-7923-6227-6
111. B. Coecke, D. Moore and A. Wilce (eds.): *Current Research in Operation Quantum Logic.* Algebras, Categories, Languages. 2000 ISBN 0-7923-6258-6
112. G. Jumarie: *Maximum Entropy, Information Without Probability and Complex Fractals.* Classical and Quantum Approach. 2000 ISBN 0-7923-6330-2
113. B. Fain: *Irreversibilities in Quantum Mechanics.* 2000 ISBN 0-7923-6581-X
114. T. Borne, G. Lochak and H. Stumpf: *Nonperturbative Quantum Field Theory and the Structure of Matter.* 2001 ISBN 0-7923-6803-7
115. J. Keller: *Theory of the Electron.* A Theory of Matter from START. 2001 ISBN 0-7923-6819-3
116. M. Rivas: *Kinematical Theory of Spinning Particles.* Classical and Quantum Mechanical Formalism of Elementary Particles. 2001 ISBN 0-7923-6824-X
117. A.A. Ungar: *Beyond the Einstein Addition Law and its Gyroscopic Thomas Precession.* The Theory of Gyrogroups and Gyrovector Spaces. 2001 ISBN 0-7923-6909-2
118. R. Miron, D. Hrimiuc, H. Shimada and S.V. Sabau: *The Geometry of Hamilton and Lagrange Spaces.* 2001 ISBN 0-7923-6926-2

# Fundamental Theories of Physics

119. M. Pavšič: *The Landscape of Theoretical Physics: A Global View*. From Point Particles to the Brane World and Beyond in Search of a Unifying Principle. 2001 ISBN 0-7923-7006-6
120. R.M. Santilli: *Foundations of Hadronic Chemistry*. With Applications to New Clean Energies and Fuels. 2001 ISBN 1-4020-0087-1
121. S. Fujita and S. Godoy: *Theory of High Temperature Superconductivity*. 2001 ISBN 1-4020-0149-5
122. R. Luzzi, A.R. Vasconcellos and J. Galvão Ramos: *Predictive Statitical Mechanics*. A Nonequilibrium Ensemble Formalism. 2002 ISBN 1-4020-0482-6
123. V.V. Kulish: *Hierarchical Methods*. Hierarchy and Hierarchical Asymptotic Methods in Electrodynamics, Volume 1. 2002 ISBN 1-4020-0757-4; Set: 1-4020-0758-2
124. B.C. Eu: *Generalized Thermodynamics*. Thermodynamics of Irreversible Processes and Generalized Hydrodynamics. 2002 ISBN 1-4020-0788-4
125. A. Mourachkine: *High-Temperature Superconductivity in Cuprates*. The Nonlinear Mechanism and Tunneling Measurements. 2002 ISBN 1-4020-0810-4
126. R.L. Amoroso, G. Hunter, M. Kafatos and J.-P. Vigier (eds.): *Gravitation and Cosmology: From the Hubble Radius to the Planck Scale*. Proceedings of a Symposium in Honour of the 80th Birthday of Jean-Pierre Vigier. 2002 ISBN 1-4020-0885-6
127. W.M. de Muynck: *Foundations of Quantum Mechanics, an Empiricist Approach*. 2002 ISBN 1-4020-0932-1
128. V.V. Kulish: *Hierarchical Methods*. Undulative Electrodynamical Systems, Volume 2. 2002 ISBN 1-4020-0968-2; Set: 1-4020-0758-2
129. M. Mugur-Schächter and A. van der Merwe (eds.): *Quantum Mechanics, Mathematics, Cognition and Action*. Proposals for a Formalized Epistemology. 2002 ISBN 1-4020-1120-2
130. P. Bandyopadhyay: *Geometry, Topology and Quantum Field Theory*. 2003 ISBN 1-4020-1414-7
131. V. Garzó and A. Santos: *Kinetic Theory of Gases in Shear Flows*. Nonlinear Transport. 2003 ISBN 1-4020-1436-8
132. R. Miron: *The Geometry of Higher-Order Hamilton Spaces*. Applications to Hamiltonian Mechanics. 2003 ISBN 1-4020-1574-7
133. S. Esposito, E. Majorana Jr., A. van der Merwe and E. Recami (eds.): *Ettore Majorana: Notes on Theoretical Physics*. 2003 ISBN 1-4020-1649-2
134. J. Hamhalter. *Quantum Measure Theory*. 2003 ISBN 1-4020-1714-6
135. G. Rizzi and M.L. Ruggiero: *Relativity in Rotating Frames*. Relativistic Physics in Rotating Reference Frames. 2004 ISBN 1-4020-1805-3
136. L. Kantorovich: *Quantum Theory of the Solid State: an Introduction*. 2004 ISBN 1-4020-1821-5
137. A. Ghatak and S. Lokanathan: *Quantum Mechanics: Theory and Applications*. 2004 ISBN 1-4020-1850-9
138. A. Khrennikov: *Information Dynamics in Cognitive, Psychological, Social, and Anomalous Phenomena*. 2004 ISBN 1-4020-1868-1
139. V. Faraoni: *Cosmology in Scalar-Tensor Gravity*. 2004 ISBN 1-4020-1988-2
140. P.P. Teodorescu and N.-A. P. Nicorovici: *Applications of the Theory of Groups in Mechanics and Physics*. 2004 ISBN 1-4020-2046-5
141. G. Munteanu: *Complex Spaces in Finsler, Lagrange and Hamilton Geometries*. 2004 ISBN 1-4020-2205-0

# Fundamental Theories of Physics

142. G.N. Afanasiev: *Vavilov-Cherenkov and Synchrotron Radiation*. Foundations and Applications. 2004 ISBN 1-4020-2410-X
143. L. Munteanu and S. Donescu: *Introduction to Soliton Theory: Applications to Mechanics*. 2004 ISBN 1-4020-2576-9
144. M.Yu. Khlopov and S.G. Rubin: *Cosmological Pattern of Microphysics in the Inflationary Universe*. 2004 ISBN 1-4020-2649-8
145. J. Vanderlinde: *Classical Electromagnetic Theory*. 2004 ISBN 1-4020-2699-4
146. V. Čápek and D.P. Sheehan: *Challenges to the Second Law of Thermodynamics*. Theory and Experiment. 2005 ISBN 1-4020-3015-0
147. B.G. Sidharth: *The Universe of Fluctuations*. The Architecture of Spacetime and the Universe. 2005 ISBN 1-4020-3785-6
148. R.W. Carroll: *Fluctuations, Information, Gravity and the Quantum Potential*. 2005 ISBN 1-4020-4003-2
149. B.G. Sidharth: *A Century of Ideas*. Personal Perspectives from a Selection of the Greatest Minds of the Twentieth Century. 2007. ISBN 1-4020-4359-7
150. S.-H. Dong: *Factorization Method in Quantum Mechanics*. 2007. ISBN 1-4020-5795-4
151. R.M. Santilli: *Isodual Theory of Antimatter with applications to Antigravity, Grand Unification and Cosmology*. 2006 ISBN 1-4020-4517-4
152. A. Plotnitsky: *Reading Bohr: Physics and Philosophy*. 2006 ISBN 1-4020-5253-7
153. V. Petkov: *Relativity and the Dimensionality of the World*. Planned 2007. ISBN to be announced
154. H.O. Cordes: *Precisely Predictable Dirac Observables*. 2006 ISBN 1-4020-5168-9
155. C.F. von Weizsäcker: *The Structure of Physics*. Edited, revised and enlarged by Thomas Görnitz and Holger Lyre. 2006 ISBN 1-4020-5234-0
156. S.V. Adamenko, F. Selleri and A. van der Merwe (eds.): *Controlled Nucleosynthesis*. Breakthroughs in Experiment and Theory. 2007 ISBN 978-1-4020-5873-8
157. F. Cardone and R. Mignani: *Deformed Spacetime*. Geometrizing Interactions in Four and Five Dimensions. 2007 ISBN 978-1-4020-6282-7